Advances in Berthing and Mooring of Ships and Offshore Structures

NATO ASI Series

Advanced Science Institutes Series

A Series presenting the results of activities sponsored by the NATO Science Committee, which aims at the dissemination of advanced scientific and technological knowledge, with a view to strengthening links between scientific communities.

The Series is published by an international board of publishers in conjunction with the NATO Scientific Affairs Division

A Life Sciences   Plenum Publishing Corporation
B Physics   London and New York

C Mathematical   Kluwer Academic Publishers
   and Physical Sciences   Dordrecht, Boston and London
D Behavioural and Social Sciences
E Applied Sciences

F Computer and Systems Sciences   Springer-Verlag
G Ecological Sciences   Berlin, Heidelberg, New York, London,
H Cell Biology   Paris and Tokyo

Series E: Applied Sciences - Vol. 146

Advances in Berthing and Mooring of Ships and Offshore Structures

edited by

Eivind Bratteland

Professor of Port Engineering,
Norwegian Institute of Technology,
Trondheim, Norway

Kluwer Academic Publishers

Dordrecht / Boston / London

Published in cooperation with NATO Scientific Affairs Division

Proceedings of the NATO Advanced Study Institute on
Advances in Berthing and Mooring of Ships and Offshore Structures
Trondheim, Norway
September 7–17, 1987

Library of Congress Cataloging in Publication Data
NATO Advanced Study Institute on "Advances in Berthing and Mooring of
  Ships and Offshore Structures" (1987 : Trondheim, Norway)
    Advances in berthing and mooring of ships and offshore structures
  / Eivind Bratteland [editor].
       p.   cm. -- (NATO ASI series. Series E, Applied sciences ; no.
  146)
    "Proceedings of the NATO Advanced Study Institute on 'Advances in
  Berthing and Mooring of Ships and Offshore Structures', Trondheim,
  Norway, September 7-17, 1987"--T.p. verso.
    "Published in cooperation with NATO Scientific Affairs Division."
    Includes index.
    ISBN 978-94-010-7129-1   ISBN 978-94-009-1407-0 (eBook)
    DOI 10.1007/978-94-009-1407-0

    1. Anchorage--Congresses. 2. Mooring of ships--Congresses.
  3. Offshore structures--Anchorage.   I. Bratteland, Eivind.
  II. North Atlantic Treaty Organization. Scientific Affairs
  Division. III. Title. IV. Series.
  VK545.N32 1988
  627'.22--dc19   88-15551
                                                      CIP

ISBN 978-94-010-7129-1

Published by Kluwer Academic Publishers,
P.O. Box 17, 3300 AA Dordrecht, The Netherlands.

Kluwer Academic Publishers incorporates the publishing programmes of
D. Reidel, Martinus Nijhoff, Dr W. Junk, and MTP Press.

Sold and distributed in the U.S.A. and Canada
by Kluwer Academic Publishers,
101 Philip Drive, Norwell, MA 02061, U.S.A.

In all other countries, sold and distributed
by Kluwer Academic Publishers Group,
P.O. Box 322, 3300 AH Dordrecht, The Netherlands.

CONTENTS

PREFACE

Two previous NATO Advanced Study Institutes (ASI) on berthing
and mooring of ships have been held; the first in Lisboa,
Portugal in 1965, and the second at Wallingford, England in
1973. These ASIs have contributed significantly to the understanding and development of fenders and mooring, as have works
by Oil Companies International Marine Forum (1978) and PIANC
(1984).
Developments in ship sizes and building of new specialized
terminals at very exposed locations have necessitated further
advances in the combined mooring and fendering technology.

Exploration and exploitation of the continental shelves have
also brought about new and challenging problems, developments
and solutions. Offshore activities and developments have influenced and improved knowledge about both ships and other
floating structures which are berthed and/or moored under
various environmental conditions.

The scope of this ASI was to present recent advances in berthing and mooring of ships and mooring of floating offshore
structures, focusing on models and tools available with a view
towards safety and reduction of frequencies and consequences
of accidents.

These proceedings contain invited lectures and other written
contributions basically as presented by the authors. The
articles have been judged and accepted on their scientific
quality, and language corrections may have been sacrificed in
order to allow dissemination of knowledge to prevail. It is
my hope that the editing done has not altered any content or
intention of the authors. The Institute dealt both with highly
theoretical models and approaches as well as the practical
application. Throughout the Institute lectures were followed
by discussions often including prepared written contributions.

This gave a unique opportunity to share with the other participants one's own experience and work, and to relate one's own
activities with corresponding activities elsewhere. Particularly the possibility of combining highly theoretical work with
practical experiences and demands was very engaging and promising for the developments to come.

Previous developments and research are often of vital importance to the understanding and ability to achieve further developments. This was underlined throughout the Institute.
Attention was given to the relation between fendering and
mooring and resulting damages and failures, furthermore environmental conditions of importance were discussed. An integrated
system approach in designing berthing, mooring and fendering

facilities and berth structures was emphasized and this
resulted in useful discussions on this topic. To achieve
optimum results, a combined effort of theoretical work,
physical model tests and field observations should be introduced.

The more theoretical part dealt with the environmental
conditions, the basic theories involved in calculating
fender forces related to ship impact and fender and berth
structures, and the forces related to the motion of ships
when moored at the berth. Station keeping systems and basic
principles in design were also included.

Practical application examples of the Institute involved
both field and laboratory measurements and analysis as well
as practical design work and equipment and prcedures suitable
for increasing safety in the berthing process and for the
ships when moored at the berth. Developments of fender systems
with lower recoilability are in progress and combined fender
and mooring action show promising results for future developments.

The support from NATO Scientific Affairs Division made this
ASI possible. I would also like to thank the Norwegian co-
sponsors who through their various contributions significantly
added to the successful organization and implementation of this
ASI. These are:

    The Norwegian Institute of Technology (NTH),
    The Foundation for Scientific and Industrial
    Research (SINTEF), Coast Directorate of Norway,
    The Royal Norwegian Council for Scientific and
    Industrial Research (NTNF), and
    Norwegian Hydrotechnical Laboratory (NHL)

I would like to thank my co-organizer prof. A.Tørum and the
programme advisers prof. F.Vasco Costa and prof. P. Bruun for
their keen interest and constructive suggestions related to
the organization of this ASI. The assistance of the Department
of Continuing Education at the Norwegian Institute of Technology is gratefully acknowledged, as are the efforts made by
the secretaries at this Department and at the Division of
Port and Ocean Engineering. Special thanks to Mrs. Britt Brun
Hansen who with enthusiasm guided us through the practical
arrangements.

                    Trondheim March 1988
                    Eivind Bratteland

HISTORICAL DEVELOPMENT OF BERTHING AND MOORING

F. Vasco Costa
Professor (ret.) Instituto Superior Técnico,
Lisbon, Portugal

INTRODUCTION

What is the use of history? Why bother about how, when and why past events occurred? Simply because by studying them we shall be in a better position to seek adequate solutions for the problems which will confront us in the future. This is especially true if conditions are changing and new solutions must be found.

There was a time when the berthing and mooring of ships posed no particular problems. As the ships were small they could navigate up rivers to locations not reached by sea waves. There they could be easily held quiet by relatively strong mooring cables, while small pieces of cargo were being loaded or unloaded.

But conditions suddenly changed in the late fifties. Not only did the draft of ships double in less than twenty years, but their mass increased from a maximum of 50 000 t to more than 500 000 t, and the heaviest sling load increased from 5 t to more than 50 t (see fig. 1). Just because the new ships require deeper water they must now be berthed in quite exposed locations, where they can be reached by very long and stable waves, that not only greatly affect berthing operations but, after a ship is already moored, can bring the ship to have oscillatory motions of large amplitude. To render the situation even more difficult, ships must now be kept quieter, because the much heavier cargo - up to 50t and even more - can more easily damage the ship than the former cargo loads of only 3.5 t. Besides, the pushing capacity of tug-boats and the strength of mooring ropes did not increase as fast as the mass of the ships, making the handling of large ships relatively more difficult. Last, but certainly not least, naval architects succeeded in designing ships with relatively thinner hulls compared to what they used to be in the small robust ships of the past.

The experience gained over long centuries in berthing and mooring small ships in well-protected harbours is of little help in solving the problems posed nowadays by very large ships with much larger areas exposed to the actions of winds and currents and that must be kept still while large and heavy pieces of cargo are being placed on board or unloaded from them.

The sudden increase in the size of ships, triggered by the Suez crisis, and the unitization and containerization of cargo provided for considerable reductions in transportation costs, but

1

E. Bratteland (ed.), Aavances in Berthing and Mooring of Ships and Offshore Structures, 1–5.
© *1988 by Kluwer Academic Publishers.*

created the problems posed by the need of berthing and mooring
of such very large ships in exposed locations. New solutions
must be found not by trying to extrapolate past experience but
by resort to the fundamental principle of Dynamics.

The pioneers. The study of the problems posed by the oscillatory motion of already moored ships in Table Bay, South Africa,
was entrusted to Dr.Basil W.Wilson. In his epoch-making papers, published just a few years before the Suez crisis, he
showed, just by resort to an analytical treatment, how long
waves of height so small as to pass unnoticed can bring the
ships to move in resonance with them. The rapid increase in
ship dimensions since the Suez crisis and the consequent problems this created, gave rise to further studies of the same
nature by Robert C. Russel and F.A. Kilner, regarding the behaviour of moored ships, and by Andre Pagés, F. Vasco Costa and
B.F. Saurin, in the study of berthing operations.
It must be pointed out that even before the Suez crisis the
problems posed by the berthing and mooring of ships was already arousing great interest and this was manifested in several papers submitted to the XVIII PIANC Congress (Rome, 1953)
by J.R.Ayers and R.C.Stokes.A.L.L.Baker, P.Callet, I:Descang
and L.de Kesel, A.Eggink, Foster and R.Lutz, F.Vasco Costa and
F.Visiol et al. As was to be expected the number of papers
presented at the two PIANC Congresses following the Suez
crisis was even larger. At the XX Congress (Baltimore, 1961)
papers were presented by. R.W.Abbet and A.Levinton, M.Bars and
M.Y.Khales, M.H.Deschenes and M.J. Dubois, N.N. Diunkorvski
and A.A.Kasparson, J.H.H.Gillespie et al, A.A.Gonzalez and
F.E.Agos, C.W.N.McGowan, L.Greco et al, E.Lackner and W.Hensen,
and F.Vasco Costa and J.F. Perestrelo. At the XXI Congress
(Stockholm, 1965) papers were presented by J.F. Agema et al,
P.O.Fagerholm, L.E. Van Houten, S.Kastner and K.Wendel, A.
Pagés and A. Girardin, C.F.Steward and F.Vasco Costa and J.C.
Leite.
In the interval between those two congresses a NATO-ASI was
held in Lisbon on the subject of the "Analytical Treatment of
Problems of Berthing and Mooring Ships" where lectures on the
mooring of ships were given by Basil W.Wilson, John T. O'Brien,
Francis Biesel, Robert C.H.Russel, Antonio G.Portela and Jean
Sommet; and on the berthing of ships by F.Vasco Costa, Jean
Sommet,Brendan F.Saurin, John T.O'Brien, André P.X.Pages,
Robert C.H.Russell and Albert Steenmeyer. Among the partici-
pants who more actively participated in the discussions the
names of S.Kastner, A.Torum, J.C.Lebreton, T.T.Lee, J.R.Rendle
J.Gervasio, A.F.Leite, P.Tryde, V.Bratianu and P.Giraudet are
to be mentioned. At this meeting the use of physical models, of
mathematic models, and the stochastic nature of the phenomena
under study were discussed at length.
Mentioned among the pioneers should also be O.Grim and Van
Loewven for studies of the added mass concept.

Institutions that promoted studies.   Among the institutions
that promoted   the study of the problems posed by the berthing and mooring of ships the most active was certainly PIANC-
Permanent International Association of Navigation Congresses.
Such subjects were not only discussed at the above-referred
Congresses but as well at the XXII Congress held in Paris in
1969 and also at the Congresses in 1973 (Ottawa) and 1977
(Leningrad). At those Congresses and at Committees especially
created for the purpose, important roles were played by A.F.
Dickson, J.M. Langeveld (Chairman of the International Com-
mission for Improving the Design of Fender Systems), J.Gervasio,
A.F.Leite (Chairman of the II Int. Oil Tanker Commission), J.V.
Brolsma, S. Nagai. A.Paape, E.H.Harlow and A.Furbother, among
several others.
The Hydraulics Research Station of Wallingford, the Delft
Hydraulics Laboratory, the Maritime Research Institute Nether-
lands (NSMB), the Waterways Experiment Station of the U.S.Army
Corps of Engineers at Vicksburg, the U.S.Naval Civil Enginee-
ring Laboratory, Port Hueneme, the W.M.Keck Laboratory,
Pasadena Ca., the Woods Hole Oceanographic Inst, Woods Hole,
Mass., promoted several studies that were presented in papers
submitted at various Coastal Engineering Conferences, Offshore
Technology Conferences and World Petroleum Congresses, or published in the PIANC Bulletin and the Proceedings of ASCE and
of the Institution of Civil Engineers, London. Among such
papers special reference is due to those presented by Brendan
F. Saurin at the 6th World Petroleum Congress held in Frank-
furt in 1963 and the one presented by S.Unoki and A.J.Watt at
the 10th Conference of Coastal Engineering in Tokyo in 1966,
and Dr.G.V.Oortmerssen thesis, presented in 1976, entitled
"The motions of a moored ship in waves".
In 1973 Dr.R.C.H.Russell organized a second NATO-ASI in Wallingford, England, at which papers were presented by Basil W.
Wilson, G.Goulston, Ib.A.Svendsen, Walter W.Massie, James M.
Keith, Ludwig H. Seidl, George H. Lean, J.D. Van den Bunt,
James F.Wilson, J.D. Randle, and F. Vasco Costa.
The oil companies, namely the British Petroleum, the Esso
Company and the Shell Oil Company have not only been active in
collaborating with the above mentioned institutions but have
also promoted studies and meetings on their own initiative for
dealing with problems in which they have particular interest,
such as those posed by the mooring of ships to a single point.
They created the Oil Companies International Forum and J.M.
Langeveld, A.F.Dickson,.J.D.Rendle and Roger Maari were particularly active in dealing with the single point mooring problem. The Forum also published in 1978 an easy to read book
"Guidelines and Recommendations for the Safe Mooring of Large
Ships at Piers and Sea Islands". Besides those already
mentioned from oil companies staffs, J.S.Balfour. J.C. Feben
and D.L. Martin of British Petroleum are deserving mentioning.
The British Standards Institution has been preparing a "Code
of Practice for the Design of Fendering and Mooring Systems
(BS 6349), the main parts of which have now been published.

Recent important contributions. Among the authors that published recent contributions the following are noteworthy of reading and of study by all those interested in the improvement of berthing and mooring operations;

- regarding the statistical approach - I.A.Svendsen, P. Tryde, G.Viggosson, D.H.Cooper, P.Tomlinsen and T. Andersen.
- regarding berthing manoeuvering and ship impacts - H. L.Fontijn, A. Vrigec, Ib.A.Svendsen, Shigeru Ueda, G. Goulston.
- regarding the behaviour of already moored ships - G. van Oertmerssen, J.Khanna and C.Birt, J.Moes and S.G. Holroyd, J.F.Wilson, L.H.Seidl and J.M. Keith.
- regarding a global approach to the referred problems Per Bruun, H.L.Fontijn, Paul Bastard, Eugene H.Harlow, A.Steenmeyer, Mario Grimaldi, Andreas Bohlin, Shiguro Kuroda, J.V.Brolsma, C.van der Burgt, W.H.R.Lawrence, S.Nagai, Nikerov et al and T.T.Lee.
- regarding the concept of added mass - Ball and Markham, Hayashi and Shirai, Ian Larsen, P.Tryde, H.L.Fontijn, A.Vrijer and Shigeru Ueda.
- regarding monomoorings - I.Fylling and Roger Maari.

We are lucky enough to have as lectures at this NATO-ASI some of the authors mentioned, and you will hear from them how new means available, namely monitoring equipment, transducers and the computer, can be used to improve berthing and mooring operation and to render them safer, even under adverse local conditions.

Forces applied by
tug-boats and
loads hoisted by
cranes

Displacement
of largest
ships

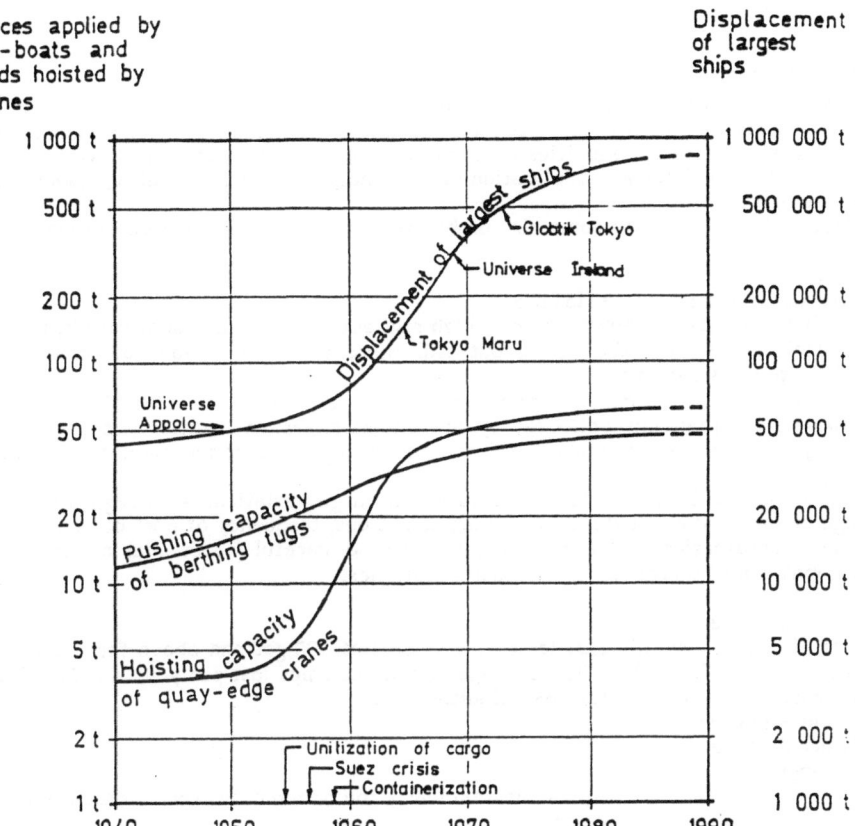

BERTHING AND MOORING ACCIDENT STATISTICS AND USE OF RISK ANALYSIS
IN BERTHING AND MOORING

BY TERJE ANDERSEN; PRINCIPAL ENGINEER; VERITAS OFFSHORE
TECHNOLOGY AND SERVICES A/S

1. INTRODUCTION
This paper is divided into three parts as follows:
• First some marine casualty statistics for the world wide fleet, the Norwegian fleet, as well as some data on casualties in entries and departures to ports are given.
• Further, some methods and application of risk analysis to shipping safety issues will be presented.
• At last an introduction to some risk problems associated with offshore mooring will be given.

2. MARINE CASUALTY STATISTICS
Navigating the seas has always been a high risk activity. The accident rate has gone down significantly over the years, but still there are high number of losses and casualties to ships each year.

Information and statistics on marine casualty types, causes and location for the worldwide fleet is mainly collected by the Underwriters. In addition more specific data may exist in national statistics for some of the fleets of the main countries of registry.

Ships are engaged in a multitude of trades with large variations with regard to size and age of ship, technical standard, sailing conditions, frequency of port calls, as well as crew qualifications. One should therefore be careful about making to wide conclusions from general loss and casualty statistics.

2.1 World wide fleet losses
Statistical data on world total losses for vessels of 500 grt and above for the years 1981-84 are given in Table 1. The average number of ships lost per year was 227. The most significant types of total loss casualties are:
- fire and explosions
- weather damage
- strandings
In tonnage terms fires and explosions are clearly the dominating type of casualty with more than 40% of the tonnage total losses.

6

E. Bratteland (ed.), Advances in Berthing and Mooring of Ships and Offshore Structures, 6–30.

Table 1: World total losses (500 gross tons and upwards) /1/

Type of casulaty	Total losses 1981-84		Total tonnage losses 1981-84	
	No of vessels	Percentage of total by type	1000 g.r.t	Percentage of total by type
- Weather damage	187	20.6	1075	18.8
- Foundering and abandonments	107	11.8	357	6.3
- Strandings	149	16.4	940	16.4
- Collisions	73	8.0	303	5.3
- Contact damage	32	3.5	126	2.2
- Fires and explosions	263	29.0	2380	41.6
- Missing	8	0.9	46	0.8
- Damage to machinery	22	2.4	238	4.2
- Other	67	7.4	254	4.5
Totals	908	100.0	5720	100.0

Average per year	227	1430115
World tonnage, average		412.3 mill g.r.t
Annual average loss ratio %		0.35

Source: Institute of London Underwriters; Casulalty Returns

The annual average loss ratio in tonnage terms are 0.35%, which is a very high loss ratio compared to fixed offshore structures or general landbased structures which have an annual loss ratio in the range 10^{-4} - 10^{-5}.

2.2 Norwegian fleet losses

In table 2 is presented some data on marine casualties for both total losses as well as partial losses for Norwegian registered ships in the years 1985 & 86. The data in Table 2 includes Norwegian vessels and drilling units of 25 gross tons and above. Table 3 gives information on registered Norwegian vessels during the same period. The loss rates as a percentage of the total Norwegian fleet is as follows:

	Annual loss percentages	
	No of vessels	Tonnages
Total losses:	0.28	0.04
Partial loss:	4.4	6.8

The lives lost due to marine casualties represent an annual rate of 0.27 per 1000 man employed.

Table 2:

Marine Casualties to Norwegian registered ships in 1985 and 1986[1] /2/

Type of Casualty	Total loss		Partial loss		Loss of life	
	Vessels (No)	Tonnage (Grt)	Vessels (No)	Tonnage (Grt)	Total loss casualties	Partial loss casualties
All types	25	6805	395	1 281 573	10	20
Grounding	5	708	215	448 846		8
Collision with vessel	1	295	70	264 818	-	10
Collision with drilling unit	-	-	2	1 651	-	-
Other collisions/Contact damage	2	27	36	167 163	-	1
Fire. explosion	6	299	42	256 842	-	1
Capsizing	4	4646	3	233	10	-
Leakage	7	531	6	959	-	-
Engine breakdown	-	-	7	108 752	-	-
Other damages	1	299	14	35 309	-	-
Unknown	-	-	-	-	-	-

1) Comprising Norwegian vessels and drilling units of 25 gross tons and over.

Source: Statistical Yearbook of Norway 1986 & 87.

Table 3:

Registered Norwegian Vessels /2/

End of year	Total		Merchant fleet		Others	
	Vessels	Tonnage (1000 Grt)	Vessels	Tonnage (1000 Grt)	Vessels	Tonnage (1000 Grt)
1985	4546	11 262	1454	10 809	3092	452
1986	4382	7 650	1335	7 197	3047	453

Loss rate:

Total losses:	0.28 %
Partial loss:	4.4 %
Loss of life:	30/110 000

In Table 4 the casualty data of Norwegian registered vessels in the years 1984 & 85 have been subdivided on casualty causes. The main causes of casualties seems to be faulty manoeuvering/navigation of vessels.

In Tables 5 and 6 are presented data on various circumstances related to place and conditions associated with grounding and collision casualties for Norwegian registered vessels above 25 g.r.t leading to partial loss. From the data in Table 5 and 6 it is interesting to see that most grounding and collision casualties seems to take place in clear weather with good visibility and calm and moderate wind conditions.

Table 4:
Serious casualties and total loss to Norwegian vessels 1984 & 85 by type and cause of casualty /3/

Casualty cause	Total		Grounding		Collision with vessel		Contact damage		Fire & explosion		Capsizing/ leakage		Engine breakdown		Other damages	
	Tot. loss	Ser. cas.	Tot. loss	Ser. cas.	Tot. loss	Ser. cas.	Tot. loss	Ser. cas.	Tot. loss	Ser. cas.	Tot. loss	Ser. cas.	Tot. loss	Ser. cas.	Tot. loss	Ser. cas.
All casualty causes	23	385	6	206	1	79	1	27	4	45	10	8		6	1	14
• Fire & explosion causes:																
– Spontaneous ignition		1								1						
– Faulty electrical installation		5								5						
– Welding/burning		7								7						
– Other causes of fire	2	26							2	26						
• Lack of stability:																
– Insufficient stability	1	1									1	1				
– Shifting of cargo	4	2									4	1				1
– Overloading																
• Technical failure:																
– Steering gear failure		8		8												
– Engine breakdown		21		8		1		5		2				5		
– Power blackout/ Automation failure		10		3		2		4		1						
• Navigation, watch e.g.:																
– Deficiency of navigation aids		2		2												
– Incorrect charts etc.		9		9												
– Manouevres/navigation of own vessel	6	158	4	120	1	25	1	12								1
– Insufficient watch	1	29	1	16		12										1
– Manouevres/navigation of other vessel		41		2		37		2								
• Various causes:																
– Hull damage	2	1									2	1				
– Human failure	3	30	1	26				1		1	2	1		1		
– Cause not established	1	6		2				1		1	1	2				
– Other causes	3	24		8				2	2	1		1			1	11
• Unknown		4		2		1						1				

Table 5:

Norwegian vessels partially lost by grounding or collision, by circumstance of casualty. 1984 & 85 /3/

Circumstance of casualty	Groundings and collisions, total	Groundings No.	Collision with vessel	Other collision/ Contact damage
Total	312	206	79	27
· Place of Casualty:				
Open sea	12	2	10	-
Coastal waters	127	95	29	3
Strait	86	64	17	5
Port	61	28	14	19
River, channel	26	17	9	-
· Light conditions:				
Light	142	81	48	13
Dark	148	111	26	11
Twilight	20	13	4	3
Unknown	·2	1	1	-
· Weather conditions:				
Fog	45	16	27	2
Drifting snow	22	14	5	3
Rain or haze	47	42	4	1
Clear weather	194	131	43	20
Unknown	4	3	-	1
· Visibility:				
< 1 nautical mile	68	33	31	4
1-1.9 nautical mile	21	19	2	-
2-4.9 nautical miles	27	22	3	2
> 5 nautical miles	191	129	43	19
Unknown	5	3	-	2
· Wind conditions:				
Calm/gentle breeze	157	98	48	11
Moderate/fresh breeze	114	72	29	13
Strong breeze/ Moderate gale	36	33	2	1
Full gale/hurricane	2	1	-	1
Unknown	3	2	-	1

Table 6:

Norwegian vessels partially lost by grounding or collision. Circumstance of casualty by percentage. 1984 & 85 /3/

Circumstance of casualty	Groundings and collisions, total (%)	Grounding No. (%)	Collision with vessel (%)	Other collision/ Contact damage (%)
Total	312(100)	206(66)	79(25)	27(9)
· Place of Casualty:				
Open sea	3.9	1.0	12.7	
Coastal waters	40.7	46.1	36.7	11.1
Strait	27.6	31.1	21.5	18.5
Port	19.6	13.6	17.7	70.4
River, channel	8.3	8.3	11.4	
· Light conditions:				
Light	45.5	39.3	60.8	48.2
Dark	47.4	53.9	32.9	40.7
Twilight	6.4	6.3	5.1	11.1
Unknown	0.6	0.5	1.3	
· Weather conditions:				
Fog	14.4	7.8	34.2	7.4
Drifting snow.	7.1	6.8	6.3	11.1
Rain or haze	15.1	20.4	5.1	3.7
Clear weather	62.2	63.6	54.4	74.1
Unknown	1.3	1.5		3.7
· Visibility:				
< 1 nautical mile	21.8	16.0	39.2	14.8
1-1.9 nautical mile	6.7	9.2	2.5	-
2-4.9 nautical miles	8.7	10.7	3.8	7.4
> 5 nautical miles	61.2	62.6	54.4	70.4
Unknown	1.6	1.5		7.4
· Wind conditions:				
Calm/gentle breeze	50.3	47.6	60.8	40.7
Moderate/fresh breeze	36.5	35.0	36.7	48.2
Strong breeze/				
Moderate gale	11.5	16.0	2.5	3.7
Full gale/hurricane	0.6	0.5		3.7
Unknown	1.0	1.0		3.7

2.3 Port of Hamburg Casualties /4/

Some data on collisions and groundings in the Elbe during the years 1970-81 are found in ref. /4/ and is summarized below.

Totally 135 collision and grounding casualties were reported with the following subdivision on types of casualties:

-	85 collisions between ships	63%
-	20 collisions with other objects/contact damages	15%
-	20 groundings/strandings	15%
-	7 secondary collisions and standings	5%
-	3 towing accidents	2%
	135 total	100%

The average annual number of port calls during the years 1975-80 was 16 880. If we assume the same traffic intensities during the other years for which the casualty data have been collected, the casualty rate per port call will be 7×10^{-4}, i.e. 7 casualties per 10 000 port calls.

A subdivision of the casualties according to causes has been carried out and the results are presented in Table 7.

Table 7: Causes of Collision and Grounding casualties in the Elbe in the years 1970-81

	Collisions between ships	Contact damages	Groundings
Failure of bridge instruments	2		
Failure of steering systems	3	4	3
Failure of propulsions		5	2
Misjudgement of situation	15		2
Erroneous navigation	10	3	3
Wrong speed range	7	2	
Violation of "Rules of the Sea"	6		
Human loss	3	1	3
Wrong communication	5		2
Low visibility - Fog	17	1	
Windage	1		3
Current drift	5	2	1
"Squat" effects	7	1	0
Failure of other ship	3		
Avoidance manoeuver	4	1	1
Acts of God	4		
Total	92	20	20

Some comments with regard to time and circumstances of casualties:

- The months of October, December and January have the highest casualty rates whereas the summer months of June. July and August have the lowest casualty rates.

- The most accident prone time of the day is between 8 pm and 4 am whereas the least accidents occur between noon and 4 pm.

- 17% more casualties occur during night or darkness.

- 46% of the casualties occurred during periods of reduced visibility whereas 54% occurred during periods of good visibility.

3. METHODS AND APPLICATION OF RISK ANALYSIS TO SHIPPING SAFETY
3.1 General Considerations

Several methods can be used to estimate the probability of marine accidents to tankers and the conditional probability that such accidents will result in cargo spills. These methods include:

- Statistical analysis of historical data for marine accidents and/or cargo spills;
- Analytical methods based on the geometry and kinematics of accident events as established for particular areas or routes, and verified or normalized by historical accident data; and
- Fault tree methods which relate the various possible event sequences leading to accidents and their probability occurrence.

The two first methods are the most useful for estimating the probability of marine accidents. Fault tree methods are, however, useful in spill probability assessments for terminal operations such as loading and unloading of hazardous materials.

Each of these methods generally require some subjective treatment, based on experience and engineering judgement, for applications to specific sites or routes or for particular types of cargo or vessel design.

The approach using historical casualty data requires two major assumptions:

- The basic causes of casualties have not changed over the time period for which accident data exists and therefore will not likely change in the future (except as discussed below). This premise implies that human error has been and will continue to be the major contributing factor in the occurrencce of marine casualties, and that the basic nature of human error has not and will not change over the period of interest. This assumption relates to the "base level" as defined for sailing procedures and currently established traffic measures. By introducing new sailing procedures and traffic restrictions the accident rate may be reduced.
- The basic processes responsible for marine casualties are primarily human factors and are, therefore, essentially independent of geographical region. The underlying causes which generate accidents in one harbour are the same as those which generate accidents in similar marine operations in other harbours and coastal areas anywhere in the world. This assumption is necessary since marine accidents are rare events, and the data base for any particular seaway or site is generally small.

The analytical methods developed to estimate the probability of marine casualties include the formulation of "collision models" based on assumptions concerning marine traffic and the disciplines governing the traffic movements in the particular area, as well as assumptions relating collision events to vessel size, speed, and port characteristics. Other similar techniques include the kinematics method developed by R.R. Solem /5/ and Macduff /6/ which can be formulated for both collisions and groundings. In some port/harbour situations, a casualty model can be developed to define the likelihood of ships moored at jetties being rammed, by considering the probability of mechanical breakdowns and the likely paths of the disabled vessel.

The probability of collisions and groundings are influenced by factors such as:

- fairway category
- traffic density
- navigational hazards as grounds and shouls
- fairway markings
- visibility
- environmental conditions as winds, waves and currents
- human factors
- reliability of ship machinery

Not all of the above factors are explicitly taken into account in this rather coarse risk model, but they should be considered in a more detailed shipping study.

3.2 Summary of studies

Several studies have attempted to estimate the probability of marine accidents for various ports. The results of some of them are briefly summarized in this chapter. Experience has shown that it is mainly impact damages due to collisions and groundings that are causing massive leaks and spills. In the results presented here we have therefore only included accidents with impact damages (i.e. collisions and groundings).

When comparing the results from the various studies and ports, one should bear in mind the following:

- the definition of "accident" varies among the references,
- incident reporting and data handling vary for the various ports and areas considered,
- the harbour areas vary in terms of fairway characteristics, time spent per movement, environmental constraints etc.
- the size of the ships included in the data base may vary,
- the sailing procedures vary for the areas considered.

In a recent study carried out by A.S VERITEC /7/. the probability for marine accidents (collisions and groundings) in four different harbour areas were quoted.

The results are presented in Table 8.

Table 8: The probability of marine accidents in selected harbours.

Harbour area	P(accident per movement) x 10^4		Total P(impact accident per port call) x 10^4
	Collision	Grounding	
Porsgrunn /8/	1.1	0.7	3.5
British /9/	3.0	2.9	11.8
Rotterdam /10/	1.3	0.4	3.4
Tokyo /11/	2.8	0.7	7.0
Hamburg /4/	5.8	1.1	7.0

The annual traffic volume in the harbour areas considered area as follows:

Harbour:	Movements per year:
Porsgrunn	11 400
British harbours (average for 10 harbours)	18 700
Rotterdam	70 000
Tokyo	836 000
Hamburg	17 000

As can be seen from the figures there seems to be very little correlation between the number of movements per year and the probability of collision per movement.

In another study carried out by Arthur D. Little, Inc. /12/, the probability of marine accidents in a number of other ports or from other references were summarized. The results are presented in Table 9.

Table 9: Summary of estimated impact accident probabilities.

Location or reference	In port P(Collision per port call)x10^4	Harbour transits		Total P(impact accident per port call)x10^4
		P(collisions per movement)x10^4	P(groundings per movement)x10^4	
DnV-Porsgrunn /8/		1.1	0.7	3.6
Thames (1965-76) /13/	0.1	0.46	0.36	1.7
North Sea (1963-74) /14/	-	0.5	0.17	1.3
& /15/	-	0.67	0.4	2.1
J.J. Henry (1971-72) /16/	0.66	4.6	3.9	17.7
IMCO (1968-79) /17/ >10000dwt	0.1	0.71	0.91	3.4
OIW (1969-72) /18/ >7000dwt	0.74	6.2	4.8	22.7
ITOPF /19/ > 10000 dwt	-	4.7	8.3	26

A brief presentation of the data basis and the results for some of the studies referenced in Table 9 is given here.

3.2.1 Porsgrunn Harbour Area /8/

A risk analysis study of the hazardous cargo movements to and from the petrochemical plants at Rafnes and Herøya was carried out by Det norske Veritas in 1977-78.

A total of 5700 arrivals per year of various types of ships are expected in the harbour area per year. Nearly 1100 arrivals per year will be of ships carrying hazardous cargoes, of which the more important are ammonia, chlorine, LPG, fuel oil, sulphuric acid and vinyl chloride.

The calculations are based on accident statistics available for the Porsgrunn harbour area 1960-77, for other ports in Europe, and for the world-wide operation of liquified gas carriers.

3.2.2 Marine Casualty Rates in Thames /13/

As part of an assessment of the risk associated with refinery operation on Canvey Island, an analysis was carried out to develop marine casualty rates for ship traffic in the Thames /13/. Total ship traffic, distributed by size, was obtained for the 12-year period 1965-1976. During this period, a total of 121 marine casualties were recorded, distributed into 21 groundings, 27 collisions with other ships, 52 ship-to-jetty collisions, 3 ship-to-ship at jetty collisions, 6 fires, 7 collisions with objects, and 5 adrift. Over this 12-year period, the total casualty rate was determined to be 4.1×10^{-4} casualties per port call, or 2.1×10^{-4} casualties per movement. By type of casualty, the rates would be as follows:

Groundings: 3.6×10^{-5} grounding/movement
Ship collisions: 4.6×10^{-5} collisions/movement
Ship-to-ship at jetty: 1.0×10^{-5} collisions/port call
Rammings of jetty: 1.8×10^{-4} rammings/port call
Other accidents: 3.1×10^{-5} casualties/movement

Of the total 121 casualties, the London Port Autority classed only one as a "major" casualty, 29 as "moderate" casualties, and the remaining as "minor" casualties, with little or no damage. A significant factor in this relatively low level of damage incurred in these marine casualties is likely the transiting velocity in the Thames which has been controlled to eight knots within some of the regions of the fairway.

3.2.3 Dutch Coast/Dover Straits Casualty Rate Studies /14/ & /15/

Many studies have been conducted involving marine casualties in the coastal appraoches and harbour areas along the Dutch coast and in the Dover Straits of the southern portion of the North Sea.

A study developing risk levels for LNG importation to the Netherlands considered marine casualties over a 12-year period (1963-1974) and concluded that the casualty proability for all tonnage classes and vessel types were 5×10^{-5} collisions per transit and 1.7×10^{-5} groundings per transit. Another study concludes that the collision probability for port approaches in the southern North Sea is 1 in 15.000 or 6.7×10^{-5} collisions per transit, and the grounding probability is 4×10^{-5} groundings per transit.

A study of ship collisions in the English Channel, Dover Strait, and southern North Sea analyzed a worldwide data base of 20 years (1957 through 1976) and resulted in a total of 3506 collisions, with an average number of ships in service equal to 27.800. This result suggests a yearly collisions probability of 0.6% per year, consistent with the IMCO study discussed later, which resulted in a collisions probability of 0.43% per year.

3.2.4 J.J. Henry Co., Inc. /16/

In this study, 1587 tanker casualties that occurred worldwide during the years 1971 and 1972 were analyzed by type and location. Of these casualties, 376 (24%) resulted in an oil spill. Considered in this study were all tankers over 100 gross tons (approximately 170 DWT); 6783 were in service during this period. The sources of these data included Lloyd's Weekly Casualty Reports, Lloyd's Register Quarterly, and U.S. Coast Guard Casualty Reports.

Of the 1587 casualties, 28% were collisions, 22% were groundings, 14% were structural failures, 13% were rammings, 12% were breakdowns, and the remaining 12% were due to fires, explosions, or capsizings. Of the 376 spills, 26% were collisions, 17% were either fires or explosions, and the remaining 7% were due to breakdowns or capsizings.

In terms of location, the 1587 tanker casualties were distributed with 22% in coastal waters, 10% in entrances, 34% in harbours, 11% at piers, and the remaining 25% at sea or unknown. For the 34% in harbours, 11% at piers, and the remaining 25% at sea or unknown. For the 376 spills, the distribution was 29% coastal waters, 12% entrances, 25% harbours, 15% piers, and '19% at sea or unknown. For both casualties and spills, the data were broken down further into distributions for the various location categories by type of casualty. Thus, the 447 collisions were distributed into 28 at piers (6%), 225 in harbours (50%), 57 in entrances (13%), 110 in coastal areas (25%), 12 at sea (3%) with the remaining 15 unknown.

Since exposure of these tankers, in terms of transits along particular waterways or time spent in the location categories, was not studied in this survey, casualty rates cannot be determined in any direct manner. However, an estimate has been developed, using reasonable assumptions about tanker operations.

3.2.5 Oceanographic Institute of Washington (OIW) /18/

In this study, the casualty rates in 7 U.S. ports (New York, Delaware Bay, Chesapeake Bay, Gulf Coast, Los Angeles, San Fransisco, and Puget Sound) were analyzed for U.S. flag, self-propelled tankers for the 4-year period of 1969-1972. A strong correlation was found between tanker casualties and port calls. For all 7 areas, the data totaled 185 casualties out of 41.908 tanker port calls (i.e., round trips), leading to an average casualty rate of 4.4×10^{-3} casualty per port call or 2.2×10^{-3} casualty/transit (in or out of the port). The data included vessels with gross weights in excess of 5000 tons (approximately 7000 DWT) or lengths greater than 500 feet. A similar analysis for the casualty frequency of all self-propelled tankers in these 7 ports for the 4-year period, without the size or flag constraints, also led to a similar strong correlation with port calls, with a casualty rate of 3.6×10^{-3} casualties per transit.

In this Oceanographic Institute report, it is also mentioned that a survey of worldwide tanker casualties indicates that 31% of casualties in coastal regions lead to oil spills, as do 28% of casualties in entrances and 21% of casualties in harbours. A weighted average of these indicates that approximately 24% of all tanker casualties result in oil spillage.

For distribution of the casualties by type, this report refers to the data given in the J.H. Henry report /16/. On the basis that the J.J. Henry data are applicable in this regard, the corresponding collision, grounding, and collisions-at-pier probabilities would be:

Collisions rate = $0.28 (2.2 \times 10^{-3}) = 6.2 \times 10^{-4}$ collisions/transit
Grounding rate = $0.22 (2.2 \times 10^{-3}) = 4.8 \times 10^{-4}$ grounding/transit
Collision-at-pier rate = $0.12 (6.2 \times 10^{-4}) = 7.4 \times 10^{-5}$ collision/port call

3.2.6 IMCO Tanker Casualty Report /17/

In 1978, IMCO published a report containing data on tanker casualties compiled by the United Kingdom Tanker Safety Group. These statistics, based on information in Lloyd's List, includes serious casualties to selfpropelled, ocean-going, oil and chemical tankers and combination carriers in oil trading, above 10.000 DWT. Also included were liquefied gas carriers of greater than 10.000 cubic meter capacity. "Serious" casualties were defined as any of the following:

o A total loss,
o A breakdown necessitating towage or shore assistance,
o A marine casualty (fire/explosions, collision, grounding, heavy weather damage, ice damage, hull defect) resulting in either loss of life, pollution of any quantity, or significant damage to the hull, engines, or accommodations.

Since the 1978 report, which included statistics for the 1968-1977 period, IMCO has updated this survey yearly. The most recently published analysis, covering the 12-year period 1968-1979, indicates a total of 933 casualties over 39,245 tanker-years, giving a casulty rate of 0.0238 per tanker-year. Of these casualties, 279 caused pollution spills. Thus, the spill rate was 0.0071 per tanker-year, or alternatively the conditional spill probability was 30%.

A distribution of the casualties into type was made for the 1968-1977 data. During this 10-year period, a total of 753 serious conditions were reported, and were distributed as follows: 176 groundings (23%), 134 collisions (18%), 150 breakdowns (20%), 201 fire/explosion (27%), and 92 other (12%).

The exposure of those tanker vessels, in terms of port calls or time in restricted waters, was not indicated in the IMCO survey. However, an estimate has been developed, using reasonable assumptions about the operations of this tanker fleet.

3.2.7 International Tanker Owners' Pollution Federation ITOPF /19/

This study analyzed oil pollution incidents in Great Britain and on a worldwide basis using data for the 5-year period 1968-1972. The study was restricted to tanker sizes above 10,000 DWT. The overall rate for collisions and groundings (combined) in port areas, based on 727 such incidents and an average of 3.079 tankers in the commercial fleet, was determined to be 4.7×10^{-2} casualties per tanker-year. Assuming, as before, that tankers above 10.000 DWT will average 10.7 round trips per vessel-year, and that each round trip involves 4 harbour transits, the casualty rate becomes 1.1×10^{-3} casualties per transit.

Based on the IMCO data discussed earlier, we can assume that 57% of these casualties are groundings and 43% are collisions, leading to the following rates:

Grounding casualties = 8.3×10^{-4} per tranist
Collision casualties = 4.7×10^{-4} per transit

3.2.8 Edmondsson & Blything /20/

The mean incident rates per voyage were calculated and the results are as follows:

Carrier < 5000 m^3 (Total 77-81): 8.4×10^{-4}
Carrier > 5000 m^3 (Total 71-81): 2.9×10^{-3}

The stranding and collision rates per year were given in the same publication:

Carrier size	Average incident rate/yr & carrier	
	Strandings	Collisions
< 5000 m^3	7 x 10^{-3}	2.6 x 10^{-2}
> 5000 m^3	27 x 10^{-3}	3.8 x 10^{-2}

If the incident rates per year were transformed into incident rates per port call the difference between small and large tankers would be even more pronounced, since the smaller vessels would on average make more port calls per year than the larger vessels.

4. ANALYTICAL MODELS FOR COLLISIONS AND GROUNDINGS

The probabilities of collisions and of groundings in restricted waterways can also be estimated through the use of analytical models. A number of such models exists of which 3 are briefly described here: These models are: R.R. Solem /5/, Macduff /6/and Krappinger and Sharma /21 & 22/.

4.1 R.R. Solem /5/

In this model the probability of a ship accident (grounding, collision) is interpreted as a product of two probabilities. One is the probability of encountering an obstruction in the course, which in the absence of a manoeuvre would lead to an accident. The second is the probability of a mismanoeuvre in the face of an impending accident.

The encounter probability is calculated for various collision scenarios. For grounding scenarios, necessary course alternations are interpreted as corresponding to encounter events.

Available accident statistics makes it possible, in conjunction with calculated encounter rates, to obtain an estimate of the mismanoeuvre probability. A model for this probability as a function of parameters associated with the manoeuvre is suggested.

The probability models described, may be of use in predicting ship accident rates in waters for which no accident records are available.

The resulting models for calculating the probability of collision and grounding are as follows:

Type of incident	P(incident per movement)
Collision with meeting/overtaking traffic	$P(C) \times P(E) \times 3 \times 10^{-8} \times L x n_1 x (V_1+V_2)(b_1+b_2)/(V_1 x V_2 x B)$
Collision with crossing traffic	$P(C) \times P(E) \times 6 \times 10^{-8} x n_2 x((l_1+b_2)/V_1 + (l_2+b_1)/V_2)$
Grounding on channel bank	$P(MC) \times P(E) \times 0.3 \times L/B$
Grounding on rocks, shoals	$P(C) \times P(E) \times (b_1+d)/B$

Collision with
stationary vessel
situated near
channel bank

$$P\left(\begin{array}{c}\text{grounding on the channel}\\\text{bank per movement}\end{array}\right) \times l_1/L$$

where P(C) = probability of unsuccessful avoidance manoeuvre by vessel'screw when on collision/grounding course = 2×10^{-4}

P(MC) = probability of mismanoeuvre by vessel's crew per course alternation = 10^{-5}

P(E) = probability of unsuccessful "external" control when on collision/grounding course
= 1 in areas with no regulations other than the general "rules of the road"
= 0.3 - 0.5 in areas with additional regulations including
 - compulsory pilotage
 - restrictions on navigation if reduced visibility
 - compulsory guidance from traffic control centre
= 0.05 - 0.15 in areas with traffic separation systems as well as all above mentioned regulations (only applicable to collision with other moving vessels)

n_1 = number of movements per year (applicable to meeting/overtaking traffic)

n_2 = number of movements per year (applicable to crossing traffic)

L = length of fairway considered (m)

B = breadth of fairway considered (m)

V_1 = speed of vessel under consideration (knots)

V_2 = speed of other vessels in the area (knots)

l_1 = length of vessel under consideration (m)

l_2 = length of other vessels in the area (m)

b_1 = breadth of vessel under consideration (m)

b_2 = breadth of other vessels in the area (m)

d = breadth of rocks, shoals in the fairway (m)

4.2 Macduff kinematic method /6/

The MacDuff model, /6/ is based on analog to molecular collision theory and computes a Mean Free Path as:

$$F = \frac{VD^2 \ 1850}{L \ v \ \sin \frac{\vartheta}{2}}$$

where

D = average distance between ships, miles (density measure)
L = length of ship, metres
ϑ = angle between track of single ship and stream of ships
V = single ship speed, knots
v = relative speed between ship and stream, knots
X = actual length of single ship path, miles

For crossing situations and for head-on meeting situations, v = 2V. The "Geometric Probability of Collision" =

$$P_g = \frac{X}{F} = \frac{x \, L \, \sin \frac{\vartheta}{2}}{925 \, D^2}$$

The Real Probability of Collision = $P_g \times P_{causation}$
After the institution of traffic lanes in the British Channel in June 1967, five-year statistics yielded values of P_C as follows:

Head-on Collision: .000315
Crossing Collision: . 000095

This P_C is similar to the P(C) probability used in the Solem Model. Based on worldwide accident data, accident data specific to several American and European ports, and upon an analysis of the MacDuff collisions, the value is estimated to be 1.0 x 10^-4 for average trafficated harbours for both types of cases.

4.3 Different games model, Krappinger & Sharma /21 & 22/
A method of estimating collision rates using differential games modelling was developed by Krappinger and Sharma. In this method, based on the determination of an optimum path of maneuvrability for a particular ship transit, the probability of collision with another ship is deduced from the statistical patterns of the traffic environment.
To follow this development, for estimation of collision involving the target ship as the struck ship, let

$$V_R = \sqrt{V_A{}^2 + V_B{}^2 - 2V_A V_B \cos \vartheta_r}$$

where
V_R = relative speed of ship A with respect to ship B
V_A = speed of ship A
V_B = speed of ship B (target ship)
$\vartheta_r = \vartheta_B - \vartheta_A$ = angle of movement of ship A with respect to ship B

Following Sharma's development, the collision probability λ_c is given by

$$\lambda_c = \rho \int_{-M}^{M} \int_{0}^{2\pi} \int_{0}^{v} V_r \cdot f(v, \vartheta) \, P_f \, dv \, d\vartheta \, dM$$

where
ρ = traffic density, (ships/unit area)
M = a specified value of range of miss-distance defining collision encounter events
$f(V, \vartheta)$ = frequency function (or bivariate probability density function) of speed and angle of movement in traffic. It is assumed that this function is independent in V and and hence can be expressed simply as the product of these two uniform density functions.
P_f = probability of failure which is basic cause for a particular encounter (e.g., engine trouble, faulty navigation, etc.)
V = the upper limit of the transit speed of the target ship

It is shown by Sharma that the collision probability is approximated in a conservative fashion by the relation:

$$\lambda_c = 3 p p_f \, M \, \bar{v}$$

The p_f is similar to the P_c in the MacDuff model and the $P(c)$ in the Solem model. A reasonable value for p_f is 10^{-4}.

4.4 Fog Collision Risk Index

The probabilities of groundings and collisions as calculated by the Solem model can be further adjusted for the particular visibility conditions in the fairway at question. The effect of visibility on the ship traffic collision risk has been discussed by R.G. Lewison /23/.

The visibility was divided in the following ranges:

range 1 = clear (better than 4 km)
range 2 = mist or fog (200 m - 4 km)
range 3 = thick or dense (less than 200 m visibility)

By using the results from /23/ and /5/ we can deduce the following visibility adjustment factors (VAF) to the collision probabilities calculated by the Solem model.

Groundings: VAF = $0.6 \, V_1 + 3 \, V_2 + 60 \, V_3$
Collisions: VAF = $0.3 \, V_1 + 3 \, V_2 + 100 \, V_3$

where V_i is the fraction of the time the visibility is in range i.

5. MOORING OF OFFSHORE STRUCTURES
5.1 Anchor chain failures

5.1.1 Introduction. The main parameters that influence the performance and reliability of the mooring chain system are the following:
- Selection of chain material
- The soundness of the chain material especially with respect to toughness
- Location and size and depth of defects
- Increased stresses in chain links when passing over cablelifter/fairlead
- Cablelifter/fairlead design and material selection
- Chain accessories material selection and design
- Corrosion and corrosion protection
- Winch and control system
- Wear and workhardening of chain and accessories as well as cablelifter and fairleaders.
- Service inspection arrangement and frequency
- Chain handling

Most of the above parameters are well controllable during the mooring system design or the chain production process. The corrosion and wear processes are at present not fully controlled. However, the degradation process by these factors are slow and the structural soundness of the chain is controllable by inspection.

Chain handling must be considered a fairly uncontrolled operation and should be restricted as much as possible.

5.1.2 Anchor line fracture frequencies. The experienced chain anchor line fracture frequency is as indicated in fig. 1 /24/.

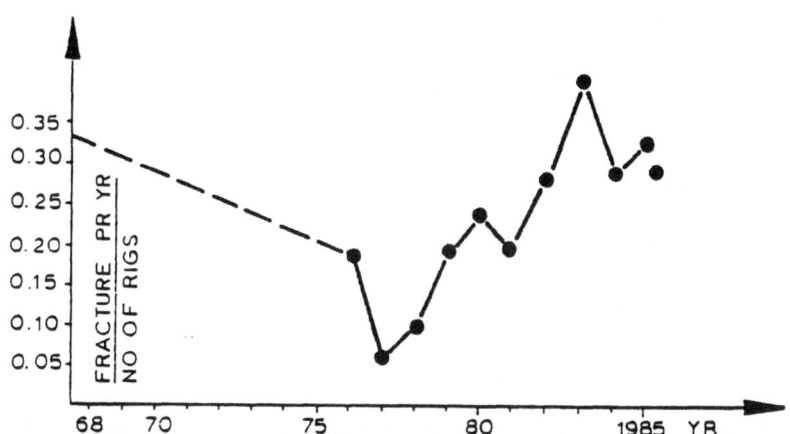

FIGURE 1. Fractures divide by no. of semisubs /24/.

Previous work in Det norske Veritas/Veritec indicates the following:
- The probability of line failure increases if hauling of the rig is performed, i.e. the proposed slackening/hauling procedure may increase the probability of line failure.
- The frequency of fractures of K4 quality chains per chain year for rigs classed with Det norske Veritas was approximately 0.02 in 1984 and 1985 /24/, /25/.
- For 1986 - 1987 the frequency has been reduced.
- Approximately 40% of the reported fractures (K3 and K4 quality chains) were with chains of less than 2 years age /24/.
- The chain fractures occur at almost all tension levels.
- The frequency of multiple line failure has been rather high, in the range 10^{-2} - 10^{-3} per rig year.

24

5.1.3 Location of fractures. Results from /24/ giving position of fracture, their number and the quality of the chain is reproduced in Fig. 2.

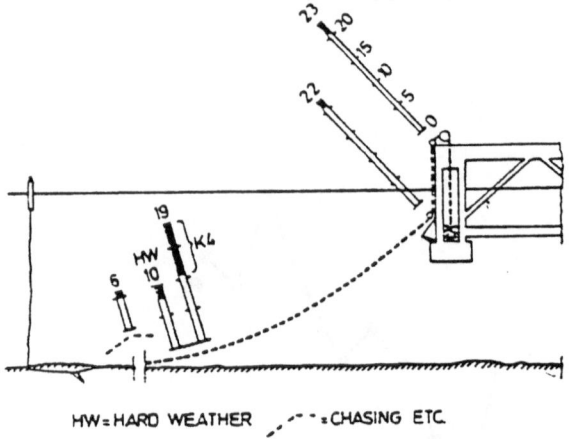

HW = HARD WEATHER   --- = CHASING ETC.

FIGURE 2. Location etc. of 80 chain fractures /24/.

5.1.4 Anchor system handling. The probability of line failure is dependent on anchor handling. In ref. /26/ this is discussed in more detail and the conclusion indicates that the probability of chain failure during handling is proportional to the number of chain links which pass the fairlead.
Taraldsen /24/ show that 45 of 80 investigated chain fractures occured at or near fairlead or windlass/winch system.

5.1.5 Chain quality. Investigation of chain fractures has shown that most chain failures are brittle fractures.
Fracture test at Veritec of a poor quality 3" K3 quality chain has shown: The failure probability of 3500' of this chain, loaded at proof load level, is approximately 40%. The failure probability is approximately 100% at minimum breaking load (MBL).
This test should not be interpreted as typical behaviour of K3 chain. In service experience has shown that some rigs never get line failures while for other rigs the line failure rate is very high. This indicates a significant spread of chain quality, the test referenced above clearly shows that the breaking strength of a low quality K3 chain is significantly lower than MBL.
Similar test for K4 chain has not been performed. However, inservice experience again indicates a significant scatter of chain quality. The conclusion derived from the inservice failure investigation, the Veritec test, and discussion with manufacturers, are:
- The main reason for the high mooring line failure rate is the scatter in quality (a few failures may be due to fatigue or overloading). Failures on chains manufactured later than 1983/84 have not occured. Most probably due to increased quality in the manufacturing processes.

- Substandard quality has been the reason for the large majority (99%) of the experienced anchor line failures. Substandard quality is here defined as chain having low toughness (i.e. CTOD values less than 0.3), and/or defects. A satisfactory manufacturing process is absolutely essential in order to obtain a reliable chain. No amount of NDE, inspection and testing can substitute for an inferior manufacturing process.
- A chain is never stronger than the weakest link and a good QA procedure is required. The stress distribution in the links will result in the largest stresses at the outer surface of the link, i.e. at a location were inspection is simple and reliable. The failure probability for an example case is discussed in /27/, and is found to be approximately 10^{-4} for crack depths less or equal to 3 mm.

The conclusion of this is that a QA system which is capable of detecting outer surface cracks of 3 mm depths and more, and ensuring a CTOD value of more than 0.3 is essential.

5.1.6 Fairlead Design. Many of the reported fractures are caused by cracks developed when the chain passes the fairlead.

Optimal fairlead design will reduce the additional stress level in the chain and thus the probability of developing such cracks. Optimizing of fairlead design may among others be to use more pockets in the fairlead and to avoid the use of kentner shackles. A fairlead designed for chain only may have smaller tolerances and smaller fairlead induced stresses than a fairlead designed both for chain and shackles.

5.2 Mooring System Failure Probability

5.2.1 Mathematical Formulation. The catenary mooring system for a mobile offshore structure can be modelled as a redundant load-sharing system. After simplifications a system can be modelled as shown below.

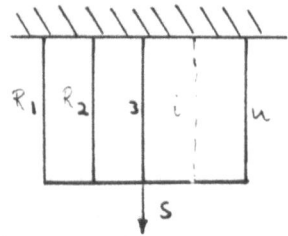

Element i. fails if the strength Ri is less than the load Si. Hence, the event of failure Ei is defined as:

$$Ei = \left\{ Ri - Si < 0 \right\}$$

The system probability of failure P_s is then

$$P_s = P\left[E_1 \cap E_2 \cap \dots \cap E_n\right]$$

where $\cap$ denotes intersection.

If we assume mutually independent failure events,

$$P_s = \prod_{i=1}^{n} P(E_i)$$

Further assumptions are as follows:
i)   uncorrelated line strengths
ii)   identical lines
iii) mutually exclusive failure sequences
iv)   The load is deterministic.
v)   After failure of any line, the remaining lines will evenly share the load.

The failure probability of a line system consisting of n-1 remaining lines is then given as:

$$P_s \;=\; (n-1)! \int_0^{\frac{r}{n-1}} \dots \int_{y_2}^{\frac{r}{2}} f(y_1) \dots f(y_{n-2}) \left[F(r) - F(y_1)\right] dy_1 \dots dy_{n-2} \qquad (1)$$

where $F(r)$ is the conditional failure probability of a chain given that all other have failed.

$F(\frac{r}{n})$ is the failure probability of any one chain in the n element intact system.

5.2.2 Thruster Failures. In addition, we included the system failure probability in case of thruster blackout, P^T. This probability can also be obtained from equation (1) by substituting

$$\frac{r + F_T}{n-1}$$

for the load per line,

$$\frac{r}{n-1}$$

in the remaining n-1 line system. Here F_T is the force provided by the thrusters which in case of thruster blackout represent an equivalent load to be carried evenly by the remaining n-1 lines.

5.2.3 Single Line Strength Distributions.   Two different mooring line strength probability distributions have been applied (both of the Weibull type) to see how this affects the results:

I. Fitted Weibull distribution according to test data for anchor chain test specimens. The distribution is as follows:

$$F(r) = 1 - \exp - (\frac{r}{320.6}) 11.3$$

A proof load of 300 metric tons has been assumed to represent a safety factor of 1.

II. Modified fitted Weibull distribution of mooring wire test data. This distribution is as follows:

$$F(r) = \begin{cases} 0 & \text{if Safety Factor} > 2 \\ 0.01 & \text{if Safety Factor } 2 \leq SF \leq 1 \\ 1 & \text{if Safety Factor} < 1 \end{cases}$$

The reason for distribution II is: The mooring wire test data showed small variations in breaking strength. This gave a fitted Weibull distribution with a cut-off at approximately 400 tons, i.e. below 397 tons the breaking probability was = 0 and above 397 tons = 1. To approach our calculations in a conservative manner, we maintain that in the safety factor range of 1 to 2 (below 397 tons) there is a fixed probability of failure of 0.01.

For all calculations we have assumed a conditional probability of thruster blackout of $P_T = 0.01$.

5.2.4 Mooring Failure Probability Results The calculation results are presented in Figures 3 and 4 , strength distribution I and II respectively. The mooring failure probability curve in Fig. 4 has been smoothed out to see the tendencies more clearly.

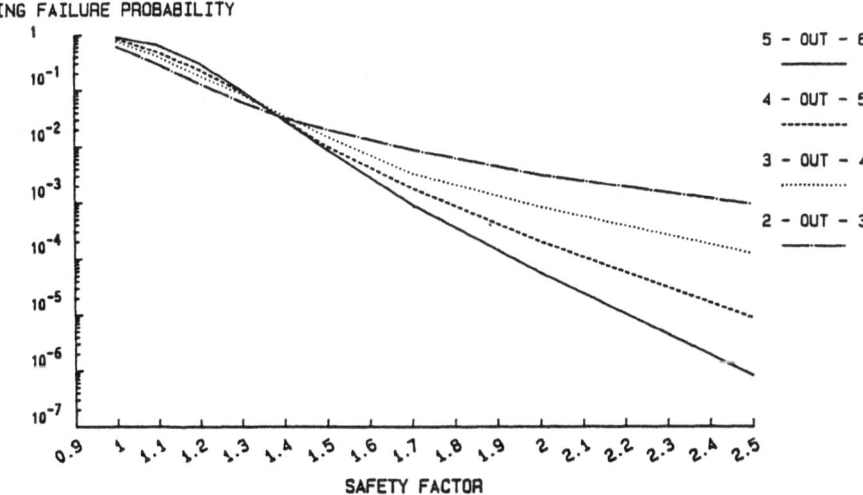

MOORING FAILURE PROBABILITY VERSUS INITAL SAFETY FACTOR IN REMAINING SYSTEM AFTER ONE LINE IS BROKEN (STRENGTH DISTRIBUTION AS FOR CHAINS).

Fig. 3

28

MOORING FAILURE PROBABILITY VERSUS INITAL SAFETY
FACTOR IN REMAINING SYSTEM AFTER ONE LINE IS
BROKEN (STRENGTH DISTRIBUTION AS FOR WIRE ROPE)

MOORING FAILURE PROBABILITY

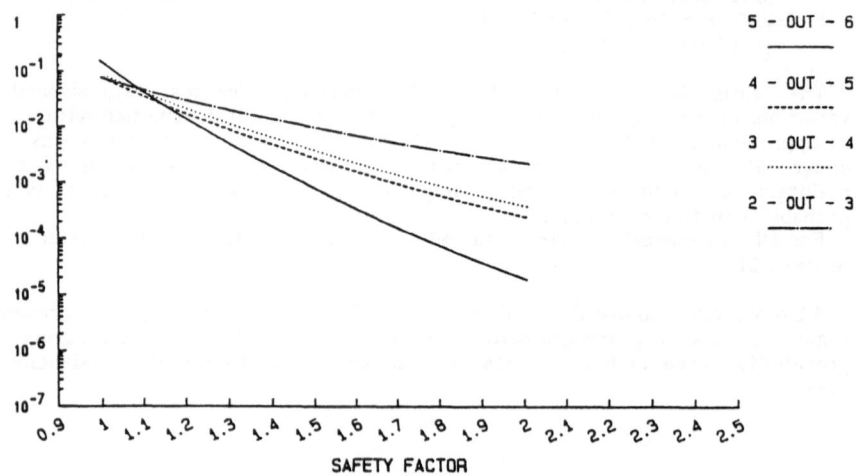

SAFETY FACTOR

NOTE: These curves are smoothed by a power function (y = ax^b)

Fig. 4

Some conclusions with regard to the results of the calculations:
- The mooring system failure probability is considerably reduced with increasing safety factor in particular for systems with several parallel load-sharing elements.
- For systems with low overall safety factors, the mooring system failure probability is increased with increasing number of lines, whereas for high safety factors, the system failure probability is reduced with the increasing number of lines.
- For the same load distribution and number of lines, a wire system is in general more reliable than a chain system with the same overall safety factors.

REFERENCES

/1/: Institute of London Underwriters; Casualty Returns.
/2/: Statistical Yearbook of Norway 1986 & 87.
/3/: Norges Offisielle Statistikk B614
     Sjøulykkesstatistikk 1984 & 1985
     Marine Casualties.
/4/: HANSA-Schiffahrt-Schiffbau-Hafen
     120. Jahrgang - 1983 Nr. 4 page 303
     Analyse der Kollisionen und Strandungen auf dem Elbrevier.
/5/: Solem, R.R.:
     "Probability Model of Groundins and Collisions"
     ASSOPO '80, Trondheim, June 1980
/6/: MacDuff, T.:
     "The Probability of Vessel Collisions"
     Ocean Industry, September, 1974
/7/: VERITEC Report No 84-3113
     "Gas Terminal at Kårstø
     Risk Analysis of the Transportation of Gas and Petrochemicals with Ship"
     Statoil, April, 1984
/8/: Det norske Veritas
     "Risikoanalyse av Skipstrafikk i Grenlandsområdet"
     DnV Report No 78-154 and 79-0670
/9/: National Ports Council:
     "Analysis of Marine Accidents in Ports and Harbours 1976"
     ISBM 0-91058-93-9, NPC,
     England, September 1976
/10/: Netherland Maritime Institute:
      "Statistical Analysis of LNG-Carrier Accident Risk in the Approaches and
      Entrances to Eurport/New Rotterdam Waterway"
      Report No R-05 NMI, Nederland, August 1975
/11/: Y. Fujii:
      "The Degree of Damage Caused by Collision"
      The Journal of Navigation, No 1, 1978
/12/: Arthur D. Little, International
      "Risk Analysis of Ships Serving Expanded Mongstad Refinery"
      Statoil, May 1983
/13/: Study conducted by Arthur D. Little, Inc. on Marine Operations in Thames
/14/: Ligthart, V.H.M.
      "Maritime Risk Analysis for the Importation of LNG into the Netherlands"
      Supplement, 3rd International Symposium on MTS, Liverpool 1978
/15/: Wepster, A:
      "Developments in Marine Traffic Operations and Research an Introduction"
      Supplement, 3rd International Symposium on Marine Traffic Systems,
      Liverpool, 1978
/16/: J.J. Henry Co. Inc.:
      "An Analysis of Oil Outflows Due to Tanker Accidents, 1971-1972"
      Prepared for U.S. Coast Guard AD-780-315, November, 1973
/17/: "Tanker Casualties Report"
      IMCO No 78-16 E, London 1978

/18/: "Offshore Petroleum Transfer Systems for Washington State.
A Feasibility Study"
Oceanographic Institute of Washington Report
No. OIW-OCW-7401, December 16, 1974

/19/: J.W. Smith:
"Occurrence, Cause and Avoidance of the Spilling of Oil by Tankers"
Prevention & Control of Oil Spills, 1973

/20/: Edmondson J.N. & Blything, K.W.
"Incident Rates on Liquid Gas Carriers with Special Reference to Fire &
Explosion"
4th International Symposium on Loss Prevention and Safety Promotion in the
Hazardous Industries
Harrogate, England, 12-16 September, 1983

/21/: Milok, T. and S.D. Sharma:
"Maritime Collisions Avoidance as a Differential Game"
Insitute für Schiffbau der Universität Hamburg,
Bericht No 329 August, 1976

/22/: Sharma, S.D.:
"On Ship Manoeuverability and Collision Avoidance"
Safety at Sea, Paper No 16 (pp 189-200), 1977

/23/: Lewison, G.R.G.
The Estimation of Collision Risk for Marine Traffic in UK waters.
Journal of Navigation, p. 317-28, September 1980.

/24/:Taraldsen:
"Anchor Chain Fractures", OTC 5059
OTC, Houston, Texas, 1985

/25/:Carlsen, C.A.; Knutsen, B.:
"Analysis Methods and In-Service Experience for Semi-submersible Platforms"
Presented at 3. Congresso Brasileiro de Petrola, Rio de Janeiro,
5-10th Oct., 1986

//26/:Dahle, E.A.:
"Probability of Anchor System Failure during Shifting Operations"
Det norske Veritas Report 86-1146, Sept. 1986

/27/:Palm, S.:
"Offshore Chain Mooring Systems, Sub Project 3, Reliability Analysis"
VT Report 87-3245, Høvik 29.04.87

MARINE TERMINAL TECHNOLOGY. WINCH BERTHING AND MOORING. RECENT
DEVELOPMENTS

PER   BRUUN
DR.TECH.SC., CHAIRMAN (RET.), NORWEGIAN INSTITUTE OF TECHNOLOGY,
TRONDHEIM, 7034, NORWAY.

ABSTRACT
This paper reviews present knowledge about BERTHING/MOORING/
FENDERING/DEBERTHING, including forces, procedures and new
developments in research and technology. Particular attention
is paid to the use of winches installed on piers or quays for
berthing and mooring. A practical illustration of its content
by an AUDIO-VIDEO movie was given at the ASI.

INTRODUCTION
The character of sea transport may be devided in bulk, container bulk and general cargo with bulk concentrating on oil, gas,
ore, coal and grain. It is a common misunderstanding that developing countries are mainly the exporters of bulks and im-
porters of manufactured products. The USA, Canada, Australia
and Sweden all export large bulk quantities (ore and grain)
while typical developing countries like India, Pakistan and
China have very little bulk export.
The tremendous increase in the carrying capacity of bulk vessels has brought large areas of the world, previously inaccess-
ible due to their distance from developed area, into the production system.
Due to the trend of the large-scale development of ports, one
may expect that future superports will be arranged within a
large frame relying on deep natural or artificial channels con-
necting the open ocean with the port. The port area may be
very large in water as well as land area and it may consist of
a number of single facilities including bulk terminals of
various kinds of oil, gas, ore, grain, container, general cargo
and special terminals for fruits, frozen products, etc. While
navigational channels, harbor basins, and possibly breakwaters
and/or training walls, will probably be government-owned (either national, state, province, county or municipal) the single
terminals will probably be owned by private corporations or
individual firms. An exception may be fishing ports and small
craft harbors, which for reasons of economics, are usually pub-
licly owned. In this way, modern ports are becoming represent-
atives for the abilities and accomplishment of free enterprise
rather than examples of the general tendencies towards socia-
lization.
Advantages of bulk handling include: minimum labor requirements,
no packing or packaging, one bill of lading, usually one port
of call, and the ability to haul large volumes over long distances. Disadvantages of bulk handling include: not all ports

31

E. Bratteland (ed.), Advances in Berthing and Mooring of Ships and Offshore Structures, 31–61.
© 1988 by Kluwer Academic Publishers.

can service bulk vessels, limited variety of cargo, requirements for large storage facilities, many bulk cargoes difficult to unload, and the need for extensive clean-up of equipment for different type of cargoes using the same unloading
equipment.
Bulk cargo operations depend upon the existence of adjacent
industrial plants such as, flour and lumber mills, refineries,
iron and steel mills which require water front access to export their products. The distance from the water front to
such plants depends upon the handling techniques involved.
Liquid (pipe) commodities may be located several miles from
the water front. Bulk cargo handled by conveyors, may be located several hundred yards from the water front.
Bulk terminals generally require deeper channels and larger
turning basins than other types of terminals due to the draft
of the vessels.
Liquid terminals utilize special wharf, pier, and other mooring depending on the water depth, bottom and soil conditions,
and the needed rate of loading/unloading. They are often located on the deeper side of other commercial docking areas.

DEEP   WATER   PORTS
The growing demand for deepwater berths in more exposed sea
areas has subsequently prompted the installation of still
larger breakwaters located in 20 to 30 meters depth. The total
height of such breakwater may therefore be about 40 m, and
with a crown width of 10 m and slopes of 1:1.5 the cross sectional areas would be as much as 3,000 m^2, or about 5,000 ts of
(rock) fill per meter. At a cost of $ 50,000 to $ 150,000 per
m length, the cost for one kilometer of structure, only, may
be from $ 50 to $ 150 million.
It is obvious that it may sometimes be difficult to justify
the cost of such breakwaters and that it therefore becomes necessary to find less expensive alternative(s) or to reduce their
length to a minimum. This raises the question as to whether, in
specific cases, it will be possible to avoid the use of break-
water(s) altogether, in favor of an unprotected berth, and thus
to construct only a pier - unprotected.
A breakwater is placed to protect the ships against wave action above an acceptable limit. Converted into practical port
engineering, this means that the movements of a vessel at
berth must not exceed certain functional limits. "Functional"
in this respect is the ability to operate the loading/unload-
ing gear without being hampered by excessive movements of the
vessel, ultimately jeopardizing safety at berth. With respect
to handling, for example, under ideal conditions it should be
possible to unload up to about 30 containers per hour. This
limit is seldom reached. Most container operations handle
some 20 containers per hour, while others may sometimes come
down to just a few per hour. This happens in particular where
waves are very long and tend to set up seiche motions in harbor basins or roadsteads. Ports on the Pacific coast are particularly vulnerable in this respect and, in some cases, container operations have to be stopped entirely. The average
output thereby drops to a few containers per hour against nor-

mally 25 to 28 - delaying the vessel, sometimes for several
days. This may cost from $ 15,000 to $ 30,000 a day! Seen
from an operational point of view, protection of a vessel at
berth should be interpreted as a hindrance to or decreasing of
movements of the vessel. Such movements, however, can be stopped or minimized directly by means of moorings without the
assistance of protective works.
Another way of solving the problem is to consider relative movements only and eventually design an unloading/loading system,
where the relative movement of the vessel and the handling or
lifting equipment is equal to zero. While such a procedure has
been followed in numerous cases in industry, it has not yet
been used in port technology, undoubtedly because it introduces
some additional risks.
Tying the vessel down to "zero" or to "elastic movements" in
fenders and moorings only, seems to be a more acceptable idea.
The question, however, is of how much movement can be allowed
without introducing adverse effects on the efficiency of operation and on safety at berth.
The movements which we are concerned with are: surge (parallel to quay), sway (perpendicular to quay), yaw (turning around
the vertical center line of vessel) and pitch (turning around
the horizontal center line perpendicular to the vessel).
The various movements are not equally important for different
kinds of vessels. Surge is relatively less important for tankers. Bruun (4,5,9) discusses the socalled "allowable movements" for various kinds of vessels but at the same time points
out that any movement of a vessel at berth is undesirable, because it is always adverse to operations as well as to safety
at berth. Innumerable practical cases have proven how easily
a mooring system can fail, most often due to resonance effects
of the outside forces and the forces in the mooring ropes.
It is logical to assume that a proper mooring and fendering
system will be able to replace a breakwater - or at least part
of it.
According to Bruun (4) this was actually done at HADERA,
ISRAEL, as mentioned in detail later. This open pier intalla-
tion for coal bulkers up to 120,000 DWT has now functioned satisfactorily since 1981.
The HADERA terminal was first planned with a 600 m long shore-
parallel breakwater connected to land by a 2.3 km trestle. As
the breakwater had to be located in 24 m depth with wave action
up to $H_s = 6$ m, often arriving at angles with the shore of ± 15
degrees, it became very expensive. After long discussions and
testing it was decided to build an unprotected facility with a
400 m pier at 22-24 m depth and a 1,700 m trestle carrying con-
veyor belts. Design details are described in a later paragraph
entitled "Replacing Breakwater by Adequate Mooring and Fendring"
Only little tug assistance is needed. Similar piers or terminals with no breakwater protection are planned i.e. at Glatved,
Denmark and at Hälsingborg, Sweden. A general purpose unprotected terminal at NOME, ALASKA was constructed in 1986. At
HADERA, tension-mooring was provided by ships winches, but tests
were run for higher tensioned pier-based winches which can be
put in, when it is considered desirable or necessary.

The following is a brief discussion on rational design of a
mooring/fendering system, which provides considerably more
safety than conventional systems. At the same time it has several operational advantages, particularly for terminals in
exposed locations. The success of an unprotected terminal depends upon its exposures including orientation, and certainly
upon its mooring/fendering system. Berth availability should
at least be 80 to 90 % averagely, and not less than 70 % at any
particular month of the year, unless the terminal is located
in arctic seas with ice covers for part of the year.

PORT TERMINALS IN GENERAL
A terminal is a loading or unloading station for handling of
many types of goods including oil products, ores, coal, grains,
containerized or palletized goods, or passengers. A terminal
may include one or more berths equipped with loading/unloading
equipment like cranes, ramps, elevators, hose-batteries, slewing bridges, etc. It may be located inside or outside a normal port area, depending upon space availability and depth requirements. The oldest terminals were the ferry terminals
which were usually rail-connected and placed inside a port
area. The development of transportation systems for oil products made it - for reasons of safety - necessary to establish tanker terminals either in remote areas of existing ports
or outside existing ports. This immediately brought up the
question of site selection for facilities handling inflammable
and explosive cargoes. The situation further aggrevated with
the introduction of LNG and LPG-vessels, which increased the
demands to operational safety. Safety requirements in the
principal areas are recognized universally, but so far no international standards have been adopted. Attempts are now being made to do so through committees working on national and
international levels. With respect to dry bulks it has been
common practice through decades to store coal, coke, ores, fer-
tilizers, grains, etc. inside an existing port - and to accept
whatever nuisance which resulted. But gradually - as experience became available - concerns regarding pollution grew, and
at present many rules regulating industrial development, smo-
king or smelling commodities, have been established on local
and national levels. At the same time the absolute size of
the loads increased, particularly influencing depth requirements. This established another very important boundary condition for new, more effective and environmentally acceptable
facilities.
The ultimate result was the introduction of the socalled "deep
water terminals", which are referred to as bulk, mainly oil
and gas products,but during recent years also to coal and ore
handling facilities. Whenever it was possible their berths
were placed in protected waters which caused a depth problem.
Depth and exposure, however, often go together, apart from at
those places, which nature fitted with deep fiords or bays as
they are found in Norway and on the North Pacific Coast. In
most cases the desire of increased depths to accomodate vessels
of larger draft had to be fulfilled by the acceptance of higher
exposures to environmental forces, particularly forces by waves

and currents. A good example is the Texas Deep Water Project
(9), where the dredged navigational channel won the contest.
The demands to structural design as well as to higher operational efficiency without reducing considerations to safety,
as a result of the increased exposure, became mandatory. This
did not mean that conventional principles of design and operation were cancelled. Rather attempts were made to adjust and
expand them to fulfill more rigid demands. In this respect
conservation modes have for long been in the lead.
Unfortunately, it has been a tendency to "extrapolation" and
less attempts on the introduction of nonconventional and new
improved principles. As it happened in numerous other fields
this often proved to be an inadequate and dangerous practice resulting in major mistakes and often in even fatalities. The
experiences mentioned in the section on moorings on the use of
conventional mooring and fendering facilities to tie up bulk
carriers of conventional type at conventional terminals are
examples among many others proving that extrapolations are not
always possible but could be very dangerous if safety shall be
preserved. A "pier was a pier" regardless of where it was placed. Increase in gravitational loads without proper considerations to the increase of dynamic loads including increased
variances, however, disregarded a very important factor which
is man's ability to predict the load in a system strongly influenced by the environment, in a much larger scale than before.
Additional forces and man's reactions to these forces became
much more crucial than before. As an example early tankers
had 12 to 14 m draft. The drafts may now be between 18 and 22
meters. Berths, therefore, have to be built in water depths
of 24 to 28 meters. Most likely currents are stronger there
and they do not necessarily run to the same direction at the
surface and at the bottom. Wave exposure increases genera-
ting problems related to larger vessels with little bottom clearance when exposed to wave and often concentrated current
actions which may cause severe yawing of the vessel.
Sedimentary problems also arose in a different scale, when wide
and deep channels and basins had to be dredged. Problems related to bottom and foundation-stability were often encountered due to locations, when sedimentary layers had not been
consolidated to the same extent as in the nearshore areas. Facilities therefore, had to be designed based on more stringent
criteria than before. In addition it often proved difficult
to establish practical design criteria with a desirable reliability. Examples on this are numerous in the offshore area,
where shortcomings in knowlegde of environmental forces have
caused a great number of accidents and fatalities.
Construction, like design, had to cope with particular problems
calling for the use of much larger and sturdier structures and
mooring and fendering equipment. Steel as well as concrete
technology faced new problems, when weldings of thick walls
failed and concrete cracked in large units due to the heavy increase of mass, thereby of a variety of internal stresses. New
technologies like pre-and post-stressing were developed as well
as new type tougher and denser concrete,based on higher quality

and better controlled materials. The cement industry had to improve its outputs to meet new challenges of production, unifor-
mity, curing, aging, deformation and chemical resistance of concrete and man had to adjust himself to designing, building, controlling, steering, maneuvering and handling problems in scales
which necessitated more meticulous planning and performance
than before (9).

BASIC ASPECTS OF LAYOUT AND DESIGN OF TERMINALS
Types of Terminals   Terminals are designed to serve a variety
of purposes. The most important types are:
>   Terminals for Oil Products, which are transported in
>   tankers, LNG and LPG vessels
>   Terminals for Coal, Ores, Cement, Grains and Fertilizers
>   Terminals for Unit Transports such as Containers, and
>   Terminals for passengers including ferry terminals
The size range for these are:
>   Tankers' 20,000 to 350,000 DWT with many vessels of
>   130,000 to 220,000 DWT
>   LNG and LPG: 20,000 to 130,000 DWT
>   Coal and Ore Bulkers: 40,000 to 220,000 DWT, recently
>   (1987)   365,000 DWT ore-bulker,
>   TUBARARO to ROTTERDAM
>   Unit Transport vessels: 15,000 to 40,000 DWT with some
>   smaller and a few larger. In addition RO/RO and ferries
>   of a variety of sizes from a few hundreds to 20,000 DWT.
Considering the large dimensions of some of these vessels it
should be realized that the forces which have to be encountered
in design and construction are also large. The particular design first of all depends on the usage of the terminal. Oil
and gas products are loaded (unloaded) through pipe lines. Coal
and ore bulks are handled by cranes, continuous unloaders and
conveyors. For unit transports special heavy duty cranes are
used. All aspects of terminal design including forces on terminal structures are dealt with thoroughly in Bruun(9), which also covers berthing,mooring and fendering all aspects included.

Berth Requirements   There are many berth requirements. The
vessel should be safe at the berth under most(but not nececca-
rily all) conditions of waves and currents (see later Fig. 12).
"Safe" means that the vessel does not break its mooring and
start drifting damaging itself, other vessels, or pier structures (9,27). Next the "berth availability" must be reasonable high. The "availability" may be considered in two different ways:
>   (a) Navigational availability which describes how often
>   the vessel is able to call the port or terminal, safely.
>   This subject is dealt with in relation to environmental
>   factors and channel geometries in refs. 3,4 and 5.
>   (b) Operational availibility which describes how often
>   the vessel is able to operate, i.e. load or unload, at
>   the berth or terminal (4,5,6,9,19,24).

Ships Movement to the Berth-Berthing Procedures On its way to
the berth the vessel usually moves through a navigation channel

or a fairlane. Dredged channels are always relatively narrow
(9). As the vessel during its passage through the channel may
be subject to cross-winds as well as cross-currents it will
need assistance by tug boats. Tugs are also widely used for
the final berthing procedure. The number of tugs needed depends upon the size of the vessel. Table 1 shows the "moderate
number of tugs needed for berthing" (9,15).

Table 1. The moderate Number of Tugboats for Berthing
         (Kikutany, 15).

Size of ship (10^4 DWT)	6	6-12	12-17	17-22
Number of tugs	2	3	4	5-6

These figures are in the best agreement with universal experience which does not mean that the number of tugs cannot be
reduced under special circumstances. This is sometimes a MUST
when tugs are difficult to obtain due to remote location of the
terminal. The availability of tugs (or the lack of tugs) is
the main reason for the development of systems using quay-
based winches for hauling as well as for mooring. The winch
method has proven to be, not only practial, but economical.
The mooring winches on the quay or pier provied additional
safety to the entire operation as well as to safety at berth.
Tug boats, however, are still rather weak instruments for maneuvering. They are only able to pull or push about 1 - 1,25 %
of their horsepower. That means only 40 ts for a 4,000 hp tug.
And, it is a fact that current and wind forces for even moderate velocities exert pressures and forces on a large vessel
which are much larger than normally available tug forces,e.g.100
ts! Thrusters which are sometimes installed in the bow and
possibly also in the stern of larger vessels, particularly
those in relatively short runs, are excellent for maneuvering,
but generally too weak to hold a vessel up against current or
winds of any magnitude. A solution to the problem is the use
of quay-based hauling-winches, which may also be placed on
dolphins and operated automatically by the vessel or by harbor
personnel in charge of the berthing operation.

Ship's Movement at the Berth Bruun (5,9) reviews ships movements at berth classifying these movements and their relative
danger in relation to the type of vessel (Table 2).

Table 2. Ranges for Allowable Movements for Various Class
         Vessels (4,5,9)

Type of Vessel	SURGE m	SWAY m	HEAVE m	ROLL degrees	YAW degrees
		► All movements are plus/minus ◄			
Tankers[1]	2.3	1	0.5	4	3
Ore Carriers (cranes with clam shells)	1.5	0.5	0.5	4	2
Grain Carriers (elevator or suction)	0.5	0.5	0.5	1	1
Container, LO/LO (normal locks)	0.5	0.3	0.3	3	2
Container, RO/RO[2] (side)	0.2	0.2	0.1	0-1	0-1
Container, RO/RO (bow or stern)	0,1	0	0.1	0	0
General Cargo[3]	1	0.5	0.5	3	2
LNG/LPG	Values for tankers are sometimes accepted. Others reduce movements to 1/3 for reasons of safety.				

1. larger for SBM and SPM systems
2. depending on design of ramps
3. depending upon hoisting equipment and cargo.

Requirements to minimum movements are highest for the more
"delicate vessels" which are the RO/RO's and in many instances
also the LNG/LPG's. Basic requirement to any mooring/fendering
system is that it provides safety and efficiency of operations
at berth (5,6,9). Mathematical modelling is explained in refs.
5,9,10,11,20 and 25. A rational approach based on probabilities
was developed based on tests in EUROPORT (1).
Fig. 1 shows a 1 in 3,000 probability against displacement diagram (1). It indicates the considerable larger unit energies
for smaller vessels. Balfour (1) emphasizes that the results
from the PORT OF ROTTERDAM field tests do not necessarily apply
to other more exposed ports. But the trend however, is probably
general. It is also suggested that the energy value required
for a jetty or pier situated in a harbor basin can be read directly off Fig. 2. which shows the total energy values versus
vessel displacement.

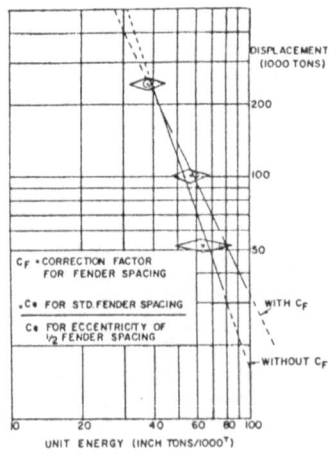

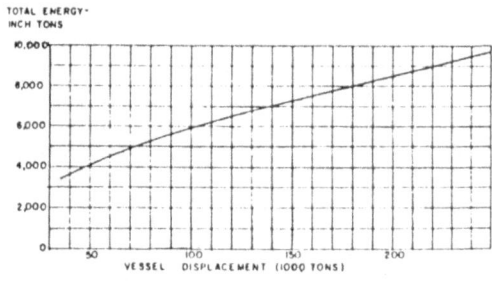

Fig. 1 1:3,000 Probability against Displacement (Jetties 1 and 2 (1,5,6,9)

Fig. 2 Design Energies for Enclosed Harbor (1,5,9).

The dynamic behavior of ships is dealt with by many authors like Cuthbert (10), Fontijn (11,12), van Oortmerssen (20) and Seidl (25). A resumé-article is found in the February, 1984-issue of THE DOCK AND HARBOUR AUTHORITY. The movements:surge, sway, heave, roll, pitch and yaw are not equally important for different types of vessels. Surge is relatively less important for tankers when loading or unloading through a platform-mounted battery of hoses, than it is for a container vessel where containers must be hooked onto a hoist. The same is true for sway. For bulkers(dry) the situation depends upon the character of the bulk (ore or grain). For ore carriers hatch size is a determining factor, because clamshells have to pass freely through the hatch. Classification-companies have detailed records of hatch sizes. Heave is not important as long as movements up and down are relatively slow. For container vessels working on the LO/LO-principle surge, sway and roll are the most pertinent parameters, while heave is not too important, as long as movements are relatively slow. Twist locks in the hoist will locate apertures in the container frame, even if they are moving up or down slowly, but in a vertical plane. This is particularly true, if magnetic locks are used. Roll is important wherever heavy loads have to be raised or lowered, and where during such movements their cranage has a chance of bumping against the hatch, if the vessel moves excessively. Table 2 (6,9) is an attempt to give reasonable ranges of movements for bulk carriers when unloading.Wave periods were up to 120 sec (long-period). The data of table 2 were based on discussions with captains, supervisors of port operations and port engineers. They are subject to variances in accordance with local circumstances. For containers, some claim that the moevements of table 2 are on the high side, while others accept movements, which are even higher assuming that the containers are on deck and that very experienced crane

operators are available. For containers below deck the angle
between the guides and the vertical should never exceed about
3 degrees. For a 40 ts container this gives a horizontal force
of about 2 ts. With a 30 ft (9m) hoisting distance to deck
this corresponds to 0.6 m max movement horizontally or a little
more than listed in table 2. Skilled personnel is essential to
perform such operations safely. Accidents have happened by
which a container became jammed in a hatch due to the movements
of the vessel, and too slow release of hoisting cables, resulting in the crane turning over. The same has happended with
clam shells. Criteria for ship movements referring to working
conditions suggested by the NORDIC RESEARCH COMMITTEE ON SHIP
MOVEMENTS, 1984-1987, are given in table 3. These figures are
not too different from those in refs. 5,6,9 and table 2. For
safe mooring under non-operation conditions the corresponding
figures including some max velocities for movements, are given
in table 4. Table 3's figures for container vessels refer to
LO/LO only. For RO/RO no figures were given in the NORDIC
REPORT, apart from "ferries", where figures refer to a Swedish
Baltic Island connection with relatively small car ferries only.
For the larger RO/RO ferries in Denmark the figures were given
by the DANISH STATE RAILWAYS, table 5a for rail and table 5b
for car ferries.

Table 3 Criteria for ship movements(loading/unloading operations).

Type of vessel	surge (m)	sway (m)	heave (m)	yaw (deg)	pitch (deg)	roll (deg)
Fishing vessels[1]						
(L_{oa}=25-60 m)						
LO-LO	1.0-1.5	1.0-1.5	0.4-0.6	3-5	4	3-5
Eleyator crane	0.15	0.15				1.5
Suction pump	2.0-3.0					
Freighters, Coasters[1]						
(L_{oa}= 60-130 m)						
Crane on the vessel	1.0-2.0	1.2-1.5	0.6-1.0	1-3	1-2	2-3
Crane on the quay	1.0-2.0	1.2-1.5	0.8-1.2	2-4	1-2	3-5
Ferries[2]						
(L_{oa}= 100-150 m)		0.8	1.0	1.0	1.0	2.0
Container Vessels[1]						
(L_{oa} = 100-200 m)						
90-100% efficiency	0.6-1.0	0.6-0.8	0.6-0.9	0.5	1.5	3.0
50% efficiency	2.0	2.0	1.2	1.5	2.5	6.0

1.2. see next page.

The movements are max peak-peak
1. frequency of these movements should be less than one week
   per year (2 % of the time)
2. frequency for these movements should be less than three
   hours per year (0.3 % of time).

Table 4 Criteria for vessel movements for safe mooring conditions at berth. The movements are peak-peak values.
For the berth to be acceptable, the frequency of these
movements should be less than 3 h/year.

Type of vessel	surge (m)	sway (m)	heave (m)	yaw (deg)	pitch (deg)	roll (deg)
Fishing vessel (L_{oa} = 25-60 m) Movement	1.2-1.5	1.0-2.0	0.6-1.0	6	4	8
Freighters/coasters (L_{oa} = 60-120 m) Movement	1.0-2.0	1.5-2.0	1.0-1.5	3-5	2-3	6
Velocity Size of vessel						
about 1000 DWT	0.6 m/s	0.6 m/s		2.0 deg/s	2.0 deg/s	
about 2000 DWT	0.4 m/s	0.4 m/s		1.5 deg/s	1.5 deg/s	
about 5000 DWT	0.3 m/s	0.3 m/s		1.0 deg/s	1.0 deg/s	

Table 5 a.   Movement Criteria for Rail Ferry Wedge Berth (DSB,
Danish State Railways)

	Surge (m)	Sway (m)	Heave (m)	Yaw (deg)	Pitch (deg)	Roll (deg)
Rail Ferry Wedge Berth	0.1	0.1	0.4	-	1	1

Table 5 b.   Movement Criteria for Car Ferry Berths at Corner
Berths and for Link Spans (DSB, Danish State Railways).

	Surge (m)	Sway (m)	Heave (m)	Yaw (deg)	Pitch (deg)	Roll (deg)
Corner Berths	0.3	0.3	0.8	1	1	2
Link Span	0.4	0.6	0.8	3	2	4

RELATION TO BREAKWATER PROTECTION
A breakwater is supposed to reduce wave action thereby prohibiting excessive movements. Mooring and fendering systems may
be designed "stiff" (rigid) of "soft" (flexible) referring to
their ability to keep the movements of the vessel down and to
absorb energy. In this respect the low extensibility of wire
ropes is an advantage when movements shall be lessened. To
keep movements down below "acceptable limits" a combination of
(some) breakwater protection and a stiff mooring of conventional type may have to be used. Tension Mooring, however, is
always better, as explained in the following sections.

MOORING
The most important publication on mooring is probably the report by OIL COMPANIES INTERNATIONAL FORUM (19) which gives a
comprehensive review of all factors pertaining to "good" mooring, including forces, layouts, equipment, cables and ropes.
It also discusses mooring management. Ship's mooring systems
are normally designed to hold a vessel safely at a berth against loadings resulting from occurring wind and current conditions. No appreciable allowance is made in the design for
dealing with dynamic loadings arising from wave motion.
Additional restraint at berths subject to surge and swell conditions is usually obtained by augmenting the vessel's mooring
capability and providing the necessary additional elasticity
to absorb shock loads. This is done by means of heavy nylon
springs or hawsers. These may be positioned on a jetty or pier
andshackled to the ends of a vessel's mooring wires, usually
back springs are fitted to dolphins or shore bollards as a
part of the shore mooring system. The size and type of the
hawsers used should be suitable for providing the necessary elasticity in the mooring system designed and arranged, as indicated in Fig. 3. Due to their great length and consequent
higher elasticity and poor orientation, head and stern lines
are normally not very effective in restraining a vessel at
berth. If adequate mooring facilities are available for good
breast and spring lines, a vessel can be moored most efficient-
ly, virtually within ist own length.

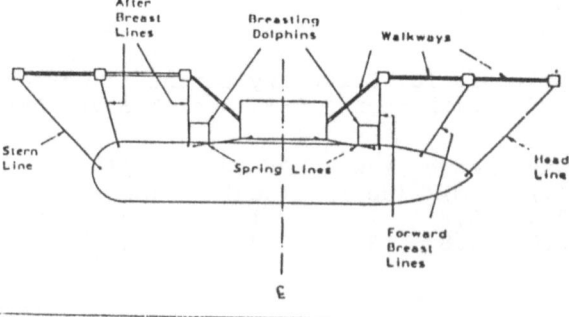

Fig. 3 TYPICAL MOORING ARRANGEMENT (9,19).

REDUCTION OF SHIPS MOVEMENTS AT ITS BERTH
There are many different ways of decreasing ships movements at
berth. A few are listed below. They include:

a. Keep propellors running on spring lines (but this may cause
   bottom scour).
b. Use side propellors (only few vessels have).
c. Use balance-tanks to reduce roll.
d. Keep ship turned a few degrees away from quay to reduce roll.
e. Turn ship's cranes perpendicular to vessel to stabilize
   against roll.
f. Use non-or low-recoiling fenders.
g. Use tension mooring by ships own - or better - by quay-based
   winches. This is the most effective method. It is described in the following paragraph.

Pretension of Mooring Lines Pretension of mooring lines is a
means of pressing the ship against the fenders so as to utilize
fender friction, and fender and line energy absorption, to reduce ship motions and thereby lower forces in mooring lines. A
reduction in mooring forces allows a reduction in the number of
mooring lines.
Pretension is a modern term for the common practice of tightening some of the ship's moorings at the time of mooring the
ship against a fixed berth. It is common practice to pull the
ship against a berth, if adverse conditions are expected. "The
captain tightens his ropes".
The magnitude of pretensioning required in the mooring lines is
not precisely fixed. It is known from hydraulic and mathematical model testing that the pretensioning of the breast lines is
most beneficial because it presses the ship against the fenders.
Oil Companies report (1977,19) claims that from a practical
stand point, a small amount of pretension on the order of ten
tons is expected to be satisfactory. Later research and practical experiences have clarified that tensions may have to be
considerably larger than 10 ts to obtain the necessary safety
against movements of larger vessels. Model and other experiments have proven that tensioned moorings will not only be able
to decrease movements considerably. An added advantage is that
tensioned moorings cause lower forces in mooring cables during
peak loading conditions. This is mainly due to lower variances
in forces. Examples are given in refs. 5 and 9.
One generally adopted tension method is to employ wires of
40/45 mm diameter permanently attached to the shore mooring
points or to buoys or anchor cables. These wires are handed
aboard the vessel by suitable messengers, stoppered off and
made fast to conventional bitts. Due to the difficulties in
satisfactorily using chain stoppers when securing large wires
to bitts, it is usually preferable for the shore to provide
carpenter's stoppers for use during the mooring operations
where a number of shore wires are to be secured.

Shore Positioned Winches and Wires. Powered winches on shore
(berth, pier, quay) from which wires are hove out to the vessel
by messenger after it has been secured in the berth and made
fast to bitts on board enables a better control to be exercised

on the shore moorings. In this way they can be used for greater effect. In order to deal with emergency situations it is desirable that shore lines be attached to some form of slip hook or Senhouse slip as an alternative to securing the bitts, so that, if necessary, the wires can be quickly released. Most ships do not have such hooks as a standard part of their equipment.

It should be recognized that it would be highly unlikely that the loads assumed by the shorebased lines would be equal, unless the lines were pretensioned to the same extent. This pretensioning would not be likely, unless the wires were on shorebased winches.

A strong expression of the importance of pretension of mooring lines is found in Wood (29) who says: "Pretensioning is a modern term for the common practice of tightening some of the ship's mooring lines at the time of mooring a ship against a fixed berth. It is a common practice to pull the ship against a berth, if adverse conditions are expected". The most recent development in the mooring of ships is probably the UK-MLM system (Mooring Load Monitor). It is described in a small note in DREDGING AND PORT CONSTRUCTION, June, 1987. The MLM-system can be used on docks providing anchorage for vessels of 40,000 to 250,000 DWT or more. When a tanker of this type is moored, it is tied to the dock by a number of cables. The hooks these cables are attached to on the dockside can swivel to be in line with the cable so that any pull in the rope can be directly transmitted to the hook. The hooks pivot up and down on large bearing shafts. The STRAINSTALL CO. has replaced these with load sensitive shafts, instrumented by strain gauges, that give out an electrical signal directly related to the pull of the cable. With the MLM-system in place, the pull in each cable, as the winches start pulling the ship tight to the dock side, can be displayed in a local side control room and also on an equivalent unit on board the ship. The cable loads are displayed on a screen and any changes as the ship is tied up, or when mooring is complete, are shown by the rise and fall of a coloured bar. Each bar represents one hook or cable. The person in charge of mooring the ship can watch the increase in load in each cable and make sure that the loads are spread evenly, or to a specific mooring pattern. This is especially important in the case of wind and tide changes. Using the screen display, the ship's personnel can adjust the mooring pattern to take account of the changing conditions.

FENDERING

Fender problems are dealt with in refs. 8,9,21. The following was abstracted from ref. 8.

"Fendering is the final step in the process of bringing a vessel safely to the berth. It is preceded by berthing. Mooring is the final step in berthing, next fendering. Mooring and fendering have been separated to such an extent that committees were established on mooring, not including fendering and vica versa. "Guidelines on Moorings" by Oil Companies (19) is an example of the former and the PIANC 1984 Fender Report is an example of the latter. For a scientific practitioner or

for a practical scientist in port engineering this does
not make much sense. One does not need to be either a
mathematician or an engineer versed in applied physics
to realize that mooring/fendering are so closely related
that they must be treated together engineering-wise and
by experiments".
The reason for this misconception lies in the fact that mooring
was (is) handled by the ship, fendering by the pier (port), so
that the two were never planned together. Furthermore fendering
was often interpreted solely as a measure against impacts by
the vessel without the realization that fenders spend most of
their time protecting vessel and pier, while the vessel is
moored. The close relationship between mooring and fendering
is well demonstrated by physical as well as mathematical models,
e.g. Fontijn (11) and van Oortmerssen (20). It is easily ob-
servable in practice. Bruun (5,6,9) discusses tension mooring
and non-recoiling fenders to improve safety at berth. Other
fender studies are presented by Padron and Han (21) and discussed by all three authors in refs. 7,12,13. - None of the
above mentioned reports distinguished clearly between IMPACT
and COVERING FENDERS. The former is paid primary attention by
PIANC although experience has shown that the COVERING FENDER
could become a severe source of accidents with the vessel at
berth, if it contributes to oscillations of the vessel.
The PIANC Committee (23) made inquiries about the type and performance of fender systems at 200 terminals around the world.
Most fenders today are made from rubber and come in a variety
of geometries and sizes. Section 3.3 (p.49) of the PIANC 1984
"Fender Report" dicusses design criteria including friction
coefficients between fender and hull (average 0.39), berthing
energies and reaction forces which depend upon the type of
fender and therefore is highest for the rigid systems like
batter rfc. piles, often causing severe damage to vessel and/
or piles.
Flexible systems like steel dolphins give the lowest reaction
forces, but they are limited upward by structural characteristics. The hydraulic system differs from other systems because it absorbs the energy in a stroke. Consequently, the
structural elements do not need to be designed for more than
one known or for "set forces" which will be smaller than for
other types of fenders. This fact is often overlooked.
Chapter 4 (p.66) is an "Inventory of Fender Systems", comprising no less than 15 different types of fenders. The gravity,
buoyancy, spring, torsion, flexible pipe and hydraulic fenders
all enjoy the grade "as required for the project". The buckling
type fenders which may be cylindrical or cone-shaped have the
best energy-absorbing diagram of all rubber fenders. Their
weakness lies in their varying degrees of recoilability and low
resistence against shear forces that always occur under an
approach angle, and are particularly detrimental if the vessel
during berthing keeps moving along the berth. Many rubber
fenders have sheared off. Use of small teflon sheets prohibit
that, but they also eliminate the advantage of friction between vessel and quay when vessel is "at rest". Hydraulic
piston fenders have vast reserves of resistance against shears.

The best fenders for very large vessels is suggested to be the hydraulic fender mentioned in Section 4.3.15. "Since almost any reaction/reflection relationship can be arranged, and since it is possible to vary the relationship, even during the berthing impact, this type of fender has potential for development for use at locations exposed to appreciable wave action". "Wave action" may be interpreted better as oscillating forces by winds, short as well as long period waves, and currents. The most recent development in hydraulic fendering is the Japanese hydraulic fender operating on water flowing through a hydraulic nozzle.

Flexible (steel) pile dolphins are highly energy-absorbing. They may be combined with rubber block fenders, but hydraulic pistons are preferable. Rigid dolphins like rfc. piles are equally dangerous to themselves, the fenders they carry and to the vessel. They may puncture its hull! Chapter 6 deals with "Approach Velocity and Energy Absorption Recording Devices - Principles and Experience in Use". This gives a description of available instrumentation, designed to control vessel movements in the final berthing procedure, including electro-mechanical, sonar and radar techniques.

Chapter 7 deals with "The Estimation of the Allowable Pressure on a Ship's Hull upon Impact with a Fender". Chapter 8 "Geo-technical Aspects of Flexible Breasting Dolphin Design" which has become an important subject as rigidity is undesirable due to the accompanying fragility of the system. The article by Padron and Han (21) reporting the results of the USDC study is an appraisal of the condition of fender system currently in use in the United States. Suggestions are made for improving existing technology. The worst problems are "high berthing energy", "fender systems wear" and "deterioation by marine organisms".

Bruun, Haldeman and Harlow (7,12,13) discuss the report by Padron and Han which is accepted as a thorough review of existing US designs and their performance. Pardon and Han in their closure (22) explain that most fenders are placed to serve relatively small vessels, not much change in present systems may be expected, but timber systems will, of course, be less important in the future, and be replaced in the US by other systems. There was unanimous agreement among those involved in the discussion that the best fender system is one which is inexpensive and absorbs energy with a low "coefficient of restitution" (low recoiling effect). In this respect, rubber fenders deviate much in performance. Figure 4a shows principles of performances for various types of fenders. Fig. 4b shows characteristics of a U-fender (Trellex).

Accidents are not only related to type of fender, but also to the environment and physical conditions and to the geometry of basins or ships. Currents not running parallel to the pier-face always cause problems, as winds do. But the worst problem could be a (too) restricted basin geometry making berthing difficult, mainly due to the lack of proper backholds for tug operations. Tugs find it hard to operate in narrow basins and quay-based winches usually do not exist. Consequently, vessels with large areas of their hull exposed to currents (loaded bulkers, submarines etc.) or vessels with large windage areas (unloaded tankers, aircraft carriers) may start to behave erratically.

A lack of proper berthing aid controls, by sonar or radar, as mentioned later, makes berthing management more difficult. Today there are only few terminals for handling of large vessels which have berthing aid systems.

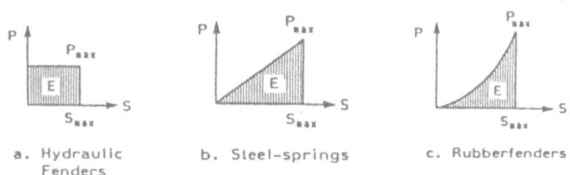

a. Hydraulic   b. Steel-springs   c. Rubberfenders
   Fenders

Fig. 4 a DEFORMATION DIAGRAM FOR FENDERS (5,8,9).

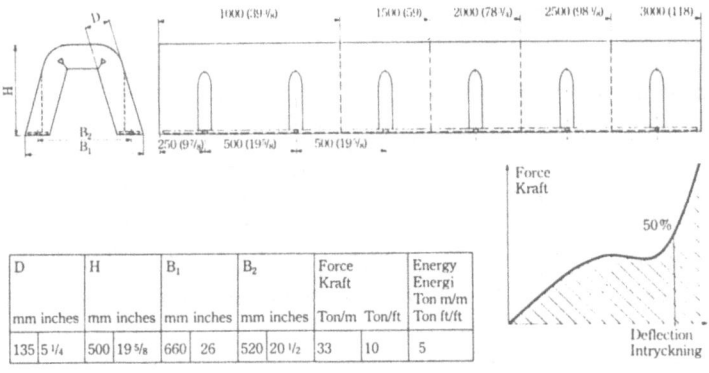

D		H		B₁		B₂		Force Kraft		Energy Energi Ton m/m Ton ft/ft
mm	inches	mm	inches	mm	inches	mm	inches	Ton/m	Ton/ft	
135	5 ¼	500	19 ⅝	660	26	520	20 ½	33	10	5

Impacts, still, are not the only reason for damage to fenders. Many accidents have occurred, while the vessel was at the berth. Reports from these accidents describe how fenders contributed to or reinforced oscillating movements of the vessel by "buffeting action".
It is apparent from reports and experience that most of the presently available fenders are inadequate, but in varying degrees. PIANC (23),perhaps more indirectly then directly, advocates systems of large ranges or "as required for the project" - but this focuses on certain systems. The hydraulic systems are given a high grade for their ability to make adjustments "even during impacts". The non-or low-recoiling systems are the safest, but there are not many on the market. Hydraulic systems are at present apparently the only thoroughbred non-recoiling sy-

stem. Certain buckling fenders including "the cones" which reflect about 50 % energy (Nikerov, 18) are better than those reflecting 80 - 90 %. The importance of energy dissipation is emphasized by many authors (9,14,28). Rubber manufactures are developing fenders of lower recoilability, but none are available at this time (1987). "Cones" and "Phi's" seem to be the best. The weakness with all rubber fenders is still their relatively low resistance against shear forces. This may in some cases be mitigated by shear beams (best) or by smooth front pads, which let the vessel "run". This, however, is not a good practice with respect to SAFETY AT THE BERTH.

The mooring safety system described in references 2and 3 is designed with due consideration to the fact that mooring and fendering must be considered as one entity. This system must not contribute to oscillations due to swells, wind gustiness or shifting currents. Monitoring devices provide all the information that any master or port manager requires when considering the safety of the vessel at berth as well as the stability of the pier. Expanded to its fullest, the system will control all mooring and fender forces, activating winch controls, alarm systems and even cargo pumps and disengagement systems for loading arms.

Where environmental forces are less severe, one can relax on requirements to anti-oscillation measures. For large vessels the answer is obvious: tension, non-recoil (or minimum recoil), high shear fenders and berthing aid systems, sonar or radar, to stop problems developing at an early stage.

Today we have an excellent opportunity to look into the performance of various mooring/fendering systems by using mathematical models (11,20,24,25), subtituting in many cases for the more laborious and time-consuming physical (hydraulic) models. Many laboratories are in existence to provide the qualitative or quantitative answers. Both naval architectural and coastal engineering laboratories are able to combine their efforts to provide economical answers.

PRACTICAL ASPECTS OF A MOORING/FENDERING SYSTEM BASED ON TENSIONS.
In practic one may face meteorological conditions which make it difficult or impossible to keep a vessel at berth. Sometimes it may be necessary to stop all handling operations, because strong winds influence the operations of hose batteries, clamshells, slewing bridges, grain elevators, etc. exessively. Forces by the vessel on the pier may become exessively large, e.g. due to strong winds pushing the vessel towards or away from the pier. For off-winds this problem may be solved by properly designed moorings, but wind gustiness could set up oscillations in forces on the vessel which then have to be absorbed by the mooring and fendering system, i.e. by constant tension winches and by the use of non-recoiling fenders. Wave action could also cause large forces and excessive movements in surge, sway, heave, roll and yaw.

A properly designed mooring system using heavy tension winches, e.g. 50 or 75 tons and proper fenders may be able to decrease the adverse effects, but forces may still become excessive or uneconomical for the normal pier structure.

Mooring Patterns and Guidelines. Proceedings by Oil Companies
International Marine Forum (19) gives valuable information on
the mooring of large vessels. Examples are shown in Bruun (9).
According to reference 19:
"Mooring lines should be arranged as symmetrical as possible
about the midship point of the vessel. A symmetrical arrangement is more likely to insure a good distribution than an unsym-
metrical aggangement.
Breast lines should be oriented as perpendicular as possible
to the longitudinal centerline of the vessel and as far aft and
forward as possible".
Tension mooring concept is shown in Figure 5. While practice
has shown that vessel-mounted mooring winches are sometimes not
well-maintained, winches on the quay can be maintained adequately by the port's technical administration.

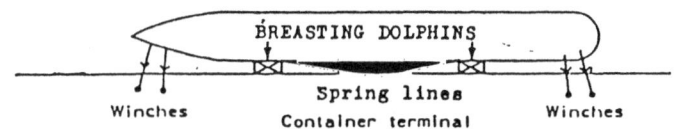

Fig. 5. THE PRINCIPLE OF TENSION MOORING

Such winches should probably be of the constant-tension traction type which can be set at a certain force to be adjusted
to the actual condition of vessel size and occurring forces.
Some ports have already installed winches on their quay, and
they are often used at ferry berths and at Ro/Ro berths.At some
ports, i.e. Antifér in France (Le Havre), capstans are installed on the back side of the breasting dolphins, which are
then used to tighten up the mooring cables. The distance between the capstans and the vessel, however, is usually so small
that the breasting lines become too steep. Mooving the mooring dolphins a distance (30-40 m) away from the vessel makes
a better utilization of the winch-powers, but the breasting
ropes may have to be elevated on a pile or tower to avoid problems with traffic on the pier. With respect to the character
of the fendering system it is obvious that fenders must be of
the non- or low-recoiling type (2,3,9). Nikerov et al (18)
found by extensive testing that the cone-type is the best of
available rubber fenders (1981). It has a long stroke, is
therefore rather "soft" with lower reaction forces than other
more compact geometries.
The necessity of having highly energy absorbing fender system
is pointed out in several references: 2,3,4,5,8,9,23,24. With
high tension forces goes the requirement of a low friction coefficient vertically, while the fenders must have a relatively
high friction coefficient for horizontal movements. This is
e.g. possible by the Kleber rubber roller pads (5), mentioned
later. Bruun 5,9) discusses structural aspects of a mooring/
fendering system based on tension including friction by the
fendering. Berth availability can be improved by combining
tension mooring with no (or low) recoiling fenders.

But mooring winches may also function as hauling winches during
berthing. All Danish and most Norwegian RO/RO ferry berths
apply tension mooring, some of them by quay-based (ramp-)
winches. The Norwegian SLAGEN TANGEN oil terminal mentioned
later (Fig 10) has also installed a winch on the pier. STIG-
SNAES in Denmark has two winches used for berthing/mooring/
deberthing (Fig. 6).

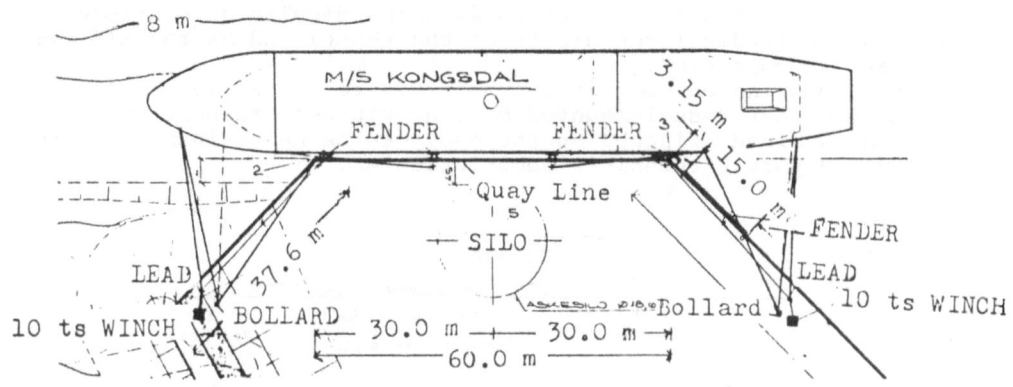

Fig. 6 THE CEMENT/FLYASH WINCH-BERTH AT STIGSNAES, DENMARK.

REPLACING BREAKWATERS BY ADEQUATE MOORING AND FENDERING
The problem with "designers" has often been that they firmly
believed that a breakwater was able to protect against wave
action without realizing that the same breakwater could increase wave action by its presence due to reflections of waves
from its walls (9). Another misunderstanding has been that a
single shore-parallel breakwater sometimes was believed to
offer adequate protection without understanding that wave
diffraction still could cause problems. Such breakwater, instead of offering better mooring conditions, causes surge movements to become sway motions which are even worse than surge
for safety at berth and operational efficiency in loading/un-
loading.
Breakwaters are built to protect against wave action. In some
cases they also act as walls against sediment transports. Some
breakwaters are connected to land and may be provided with quay
walls on the inside. Others are located in the open ocean.
They serve no other purpose than as protection against wave
action and are usually built in relatively deep water, with no
loading or unloading equipment installed.
Assuming that for economic reasons an expensive breakwater cannot be considered (like HADERA, ISRAEL) - Fig. 7a, and a layout like Fig. 7b is selected, a mooring and fendering system
must be designed which will be able to tie the vessel down, so
that excessive movements (such as surge) can be avoided.
A large coal terminal up to about 150,000 DWT, employing the
principle shown ing Fig.7b has been built in Israel. This
terminal was first planned as shown in Fig. 7a with a 600 m

long breakwater parallel to shore connected to land by a 2,3
km long trestle. As the breakwater had to be located in 24 m
depth in wave action up to H_s = 6 m, often arriving at angles
to the shore of ± 15 degrees, it became very expensive. After
long discussions and testing it was decided to build an unprotected facility similar to Fig. 7b with a 400 m long unloading
pier at 22/26 m depth and a 1,700 m trestle. Analyses of functional stability showed that the pier would be "available"
for 60,000 to 120,000 DWT vessels with max wave height 2,5 m
for berthing and handling using tug boats, for no less than 95 %
of the year. The excessive wave heights occurred principally
during winter months. The mooring system is a highly energy
absorbing system of ropes and mammoth fenders with max forces
per rope of about 100 ts using 15 - 25 ts pretension winches.
Fender forces are up to about 300 ts with pile breasting dolphins of ab. 350 tm.

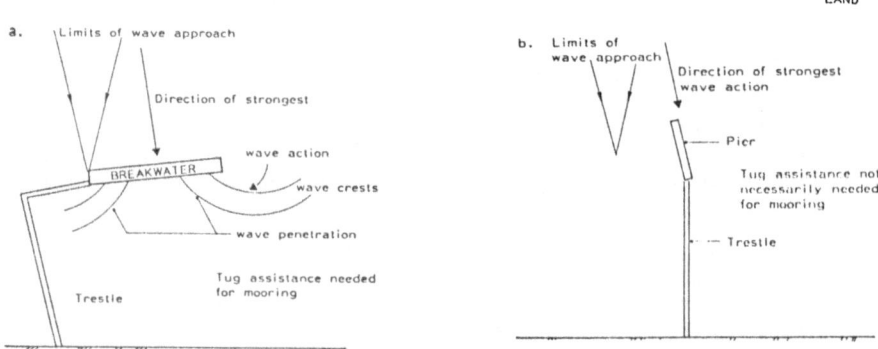

Fig. 7 a Shore parallel Breakwater   Fig. 7b Pier connected
connected to Land by Trestle (4,9)   to Land by Trestle (4,9).

The pier is of a steel pile design, with 20 m reinforced concrete spans and the trestle is a steel pile structure with 30.5m
reinforced concrete spans. The pier is easy to approach and
easy to leave and tugs are not necessarily needed, or not needed
at all at some departures.
Similar piers are built elsewhere and may be designed as any
combination of a breakwater and a mooring pier. Due to the
fact that most deep water installations will still be located
relatively close to shore, where refraction has turned the waves
to some extent more perpendicular to shore this filtering of
direction of wave approach makes pier installations easier to
design and operate. There is no dobut that similar unprotected
piers will be built in a variety of designs in the future.
Large savings will result when this concept is selected. Another alternative, of course, is to build the pier with a slig-
th curvature and next mooring the vessel where it is best for
heading the waves (Bruun,9,chapter 4). The HADERA PIER has now

been in operation since 1982 and so far operations have been
smooth. Berth availability has been 96 % and only minor problems have been experienced. Strong tension by ship's own winches has been applied. Although the terminal was designed to
accomodate quay-based winches, they have not been installed
to-date. Without question, the combination of tension mooring-
best by quay-based winches - and fenders of low recoilability
will be able to provide the optimum solution to safety.

RECENT EXAMPLES ON THE USE OF TENSION MOORING
Some recent examples are described in Bruun (5,9,24). One
large "model" was installed at the STIGSNAES, DENMARK, terminal,
Fig. 6. It is described by Roll (9, chapter 4) as follows:
"Quay winches and fenders of lower recoilability are installed
serving 5,000 to 10,000 DWT bulkers mainly transporting fly-
ash and cement. Fig. 6 gives particulars of the facility located in THE GREAT BELT, where wave height may be between 1 and
2 m and period 3 to 6 sec. Currents may run up to 5 knots in
the approach.Density currents as well as ice occur during the
winter".
The experience with the established system is that it has increased operational efficiency as well as safety at berth. A
research project at the Techn. Univ. of Denmark, unpublished
report, 1984, Jensen and Tryde, demonstrated the importance
of non-recoiling fenders to decrease sway and yaw. Tests on
hydraulic fenders have been undertaken in Japan and are planned
in Norway.

Model Experiments and Calculation of Mooring Forces in a
Combined System.
The following is in citation from refs. 8 and 9:
      "To keep the vessel from surging due to seiches, alterna-
      ting currents and wind gustiness may require considerable
      forces. Such forces may be determined by models, mathe-
      matical or physical. An easy first approximation to for-
      ces caused by seiches may be obtained by placing the
      vessel on a slope extending from the crest to the trough
      of a seiche or other long-period wave. As such waves are
      very low the slope may be of the order of say 1 in 3,000
      if it is not amplified by resonance. Consider, for example,
      a seiche of 60 seconds on 20 m depth. The wave is about
      8,000 m long and the crest to trough distance is about
      4,000 m. Harmonics are 2,000 1,000 and 500 m. A 250,000
      DWT bulker fully loaded weighs about 330,000 ts. With a
      slope of 1 in 2,000, the quay-parallel surge force would
      be 165 ts. This force may be absorbed by a fender fric-
      tion force assuming a friction coefficient fender/hull
      of 0.7 = 165/0.7 = 240 ts or 120 ts for each breasting
      unit. Two 75 ts tension winches per unit is 150 ts. If
      an off-wind blows simultanously exerting 100 ts pressure
      on the hull away from the berth an additional two 75 ts
      winches are needed or a total of six 75ts winches..Ships
      own spring lines will of course be helpful too. - The
      earlier (conventional) practice was to use ship's own
      winches and to install constant tension winches on the

ship. As experience has shown that although many
vessels have installed tension winches these winches
are often not properly maintained, it is now considered
more practical to install winches on the quay or on dolphins and let them operate by the harbor captain. Winches
available include rope winches of constant tension type
(i.e Pusnaes and Norwinch, Norway). A 50 ts winch has a
"locked" brake power of about 90 - 100 ts. A 75 ts winch
has about 160 - 170 ts. In the above mentioned case two
50 ts and two 75 ts winches will be able to handle the
problem, if no winds are present, but it is probably
better to put in six 50 ts to provide a safety margin
for high winds of gustiness factors exceeding 30 %. With
respect to winds parallel to the quay, forces are relatively small. Wind forces above 25m/sec may be considered
extreme (25m/sec = 90 km/hour = 55 - 60 knots, which
is close to the limiting hurricane velocity of 70 knots.
It is not likely, however, that the vessel will stay at
the berth under such conditions. Decreasing velocities
to 45 knots (20 m/sec) lowers the forces to about 2/3.
This is a more practical criteria for design. Operational
limits for unloading may be 20 - 25 knots only for crane
operations. This is another limit, but the vessel may
stay at the quay even under high winds prohibiting operations, which, of course, is an advantage. - The most
dangerous wind, however, is not the perpendicular beam
wind, but a wind blowing at a 70 degree angle with the
vessel from the bow side, where the vessel usually is
highest in unloaded condition. In that case forces at
the bow would exceed forces on the stern, but the total
holding power of four 75 ts winches is 4 x 160 ts =640ts,
while forces on the bow breasting lines could be about
250 ts and 120 ts on the stern breasting lines. In this
connection it should be remembered that ship's own mooring lines are not "abandoned" and that they will be able
to provide say another 100 ts to assist in keeping the
vessel in place during such extreme situation, where
seiches may occur. - It will be practical to use additional winches, if recoiling fenders are used and also to
use wire ropes. Highly recoiling fenders should, however, be avoided any time! A hard pressure between the
vessel and the fender in compression may, however, have
certain disadvantages for the vertical (heave) movements
and for roll and pitch. In a port with only short period waves heaving will hardly exist, apart from at smaller
boats, but heave for long period waves could occur. Other
movements include tidal oscillations and the movements
associated with loading and unloading. Rolling and pitching will also occur for long period waves. These movements are damped by tension moorings. - Fenders usually
have a front plate or pad to distribute forces on the hull.
These front plates may be covered by synthetic materials
or rubber, hardwood or the plates are provided by rubber
rollers as described below. Some companies are selling
TEFLON sheets for covers to make friction low and enable
sliding during impacts to avoid shearing off of the fenders.

Such seets may, however, easily be torn off. And sliding during a non-operation condition is not desirable.
Tension mooring is based on a high lateral resistance
against surge but a low(er) vertical resistance. Hardwood like Azobé, has a 0.3 lateral friction factor only,
and 0.2 factor for vertical movements. The best systems
in this respect are those, which have rubber-rollers
with horizontal axes providing a 0.7 friction coefficient
laterally and a 0.1 or less coefficient vertically (Fig.
8, Kleber, France). - It is of course a requirement that
fender forces shall not exceed design values for ships
hull. That is usually of the order of 20 to 25 ts per
square meter of perhaps 30 ts for new vessels."

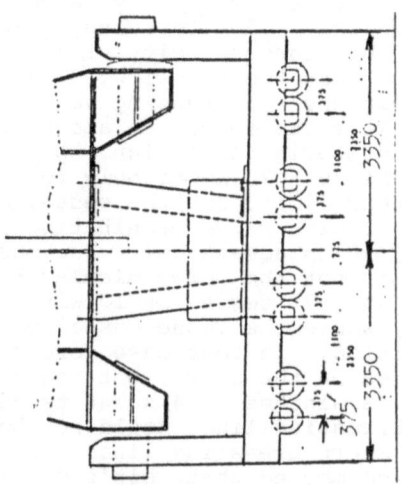

Fig. 8   300 mm Rubber Roller Pad
          by KLEBER   6 4,5,6,9).

Bruun (5,9) gives a number of examples on the design of a
tension mooring system using various types of models and design
procedures for a variety of locations, including HADERA,
ISRAEL, GLATVED, DENMARK, TANJONG BERHALA, MALAYSIA etc. All
pertinent forces by winds, currents, waves of short or long
period type are given. Tension mooring requires proper moo-
ring/hauling winches. Some of the best are the Norwegian
PUSNAES winches, hydraulic, mechanical, tension-traction
winches of normal capacities from 30 up to 75 ts in hauling
and about twice as much in brake. Redering may be arranged at
the desirable levels which may be designed for local environmental conditions.
Fig. 9 is an example of an extremely heavy fendering used at
ferry berths for rail ferries COPENHAGEN, DENMARK, to
HALSINGBORG, SWEDEN.
Fig. 10 shows the layout of the tanker berth at SLAGENTANGEN,
NORWAY, where a 50 ts mooring winch was installed on the quay
to hold the ship in strong quay-parallel currents. Further details on the SLAGENTANGEN terminal are given in Fig. 12.

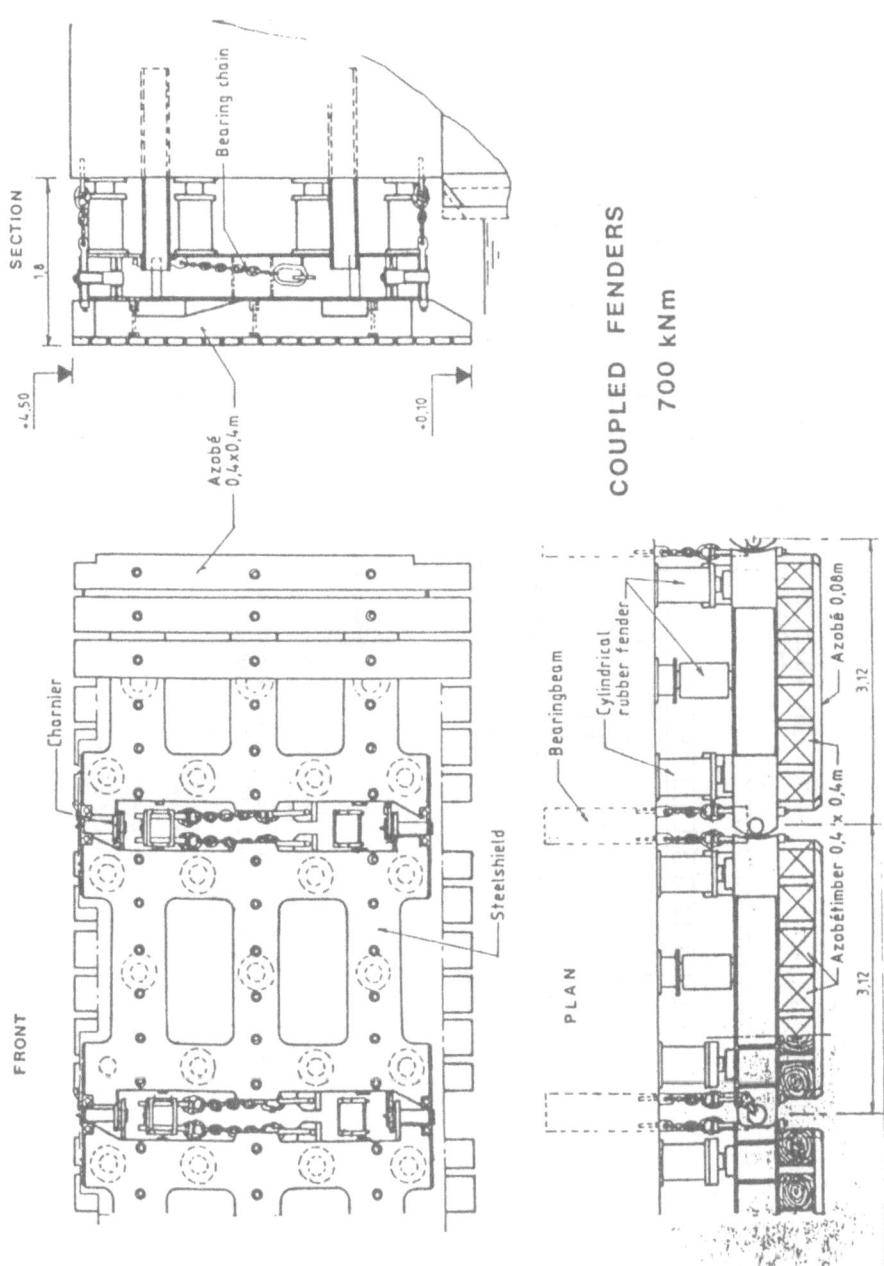

Fig. 9. Heavy Fendering used at Rail-ferry Berths in Copenhagen
and Halsingborg.
Danich and Swedish State Railways.

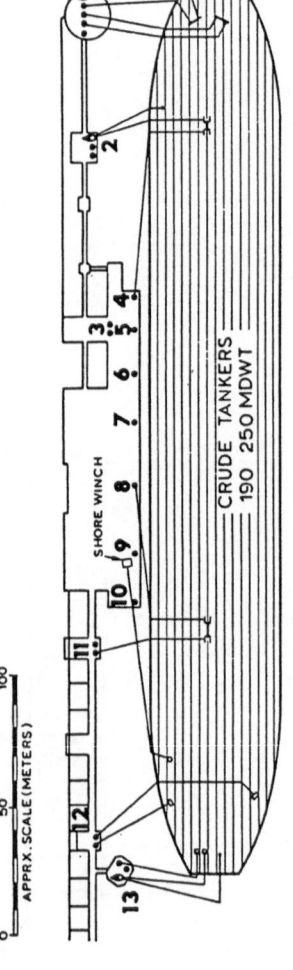

SLAGEN MARINE TERMINAL MOORING SKETCH

APPRX. SCALE (METERS)

CRUDE TANKERS 190 250 MDWT

SHORE WINCH

LEGEND:
- QUICK RELEASE HOOK
- QUICK RELEASE PULLEY
- SHORE BASED MOORING WINCH

Mooring Philosophy

Hawser loads calculated on ER&E computer program taking into consideration vessel's condition — light/loaded — angles — wind — current velocities and direction — fairlead — friction etc.

Mooring lines assumed to be 1 5/8" diam. steel wire 90 to 100 metric tons braking strength — fitted with 10" circumf. synthetic tail, 110 m/ton braking strength.

Mooring winches: 25 m/ton tractive power
55 m/ton brake power

Brake mooring winch assumed to release at 55 m/ton.

Hawser not to be stressed beyond 55 % of breaking strength i.e. 50 – 55 m/ton

Hawser starts to yield at 50 – 60 % of ultimate.

Efficient line tending assumed at all times, by ship personnel.

Self tensioning not allowed.

Mixing of wire/synthetic lines in the same direction not allowed.

Synthetic tails on steel wires, irrespective of condition, are not accepted if older than 12 months. Tails can be made available if advance notice given by owners/vessel.

Fig. 10   LAYOUT OF THE SLAGEN TANGEN OIL TERMINAL, NORWAY

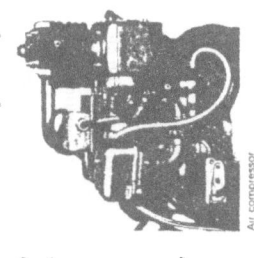

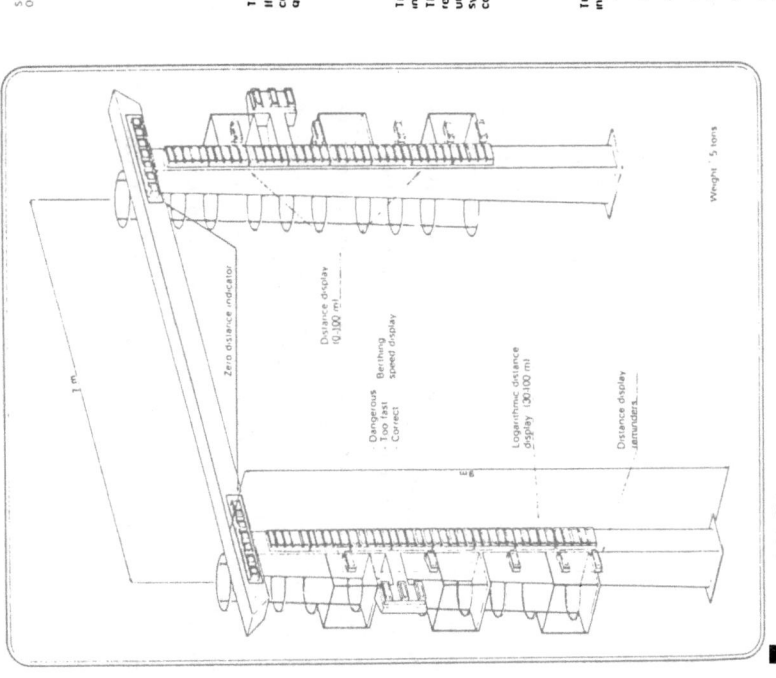

SHIP'S ANGLE
OF APPROACH

The angle of a ship's longitudinal axis, relative to the quay, is also signalled during berthing. If this angle becomes excessive, a warning orange lamp situated at the base of the vertical light column will commence to flash. The sensitivity of this alarm varies with the ship's distance from the quay.

RADAR SYSTEM

The radar unit has been widely used in the aerospace industry.

The SYMINEX system provides maximum reliability with respect to speed and distance information without the underwater interference problems associated with sonar systems. Each unit is individually mounted in a pressurised container to meet the most stringent safety regulations.

The assembly of this equipment is modular and a complete installation comprises

- Two Radar Transmitter/Receiver units operating in the GHz frequency band
- Control console containing processing electronic equipment
- Pressurizing system for radar unit containers

- Accuracy   1% of distance $\pm$ 0.05m
- Measurement range   distance 0 to 2000 m
     speed $\pm$ 50 cm/s
- Beam aperture (3 dB)   vertically $\pm$ 7.5°
     horizontally $\pm$ 15°
- Operating temperature 20°C to +70°C
- Units can be adapted to non European standards i.e. 60 Hz

Single view of
radar unit

Height   1.00 m
Width   0.50 m
Length   0.80 m
Weight   90 kgs

Air compressor

Zero distance indicator

Distance display
(0-100 m)

Dangerous
Too fast
Correct

Berthing
speed display

Logarithmic distance
display (30-100 m)

Distance display
(remainder)

Weight 5 tons

1 m

6 m

SIGNALLING GANTRY

The Display System for the signalling gantry functions at two different light intensities adjusted automatically for day or night conditions

Fig. 11 EXAMPLE OF A BERTHING AID SYSTEM (5,9)

Mooring

Normally vessel is moored port side to pier.

In normal weather condition 5 to 7 tugs representing approx. 12000. Hp will be used during berthing operation.

Two mooring launches, 100/45 Hp are utilized to pull the lines ashore.

12 mooring lines required from vessel. Ship's aft spring line will be supplemented by one 1 7/8" wire, braking strength 113 m/ton on one-layer.

Mooring sketch to be followed closely.

Safety Precautions

Relative position of ship/shore manifold to obtain max range for our loading arms.

Before arms are connected — correct setting of winch brakes to be checked by shore and ship personnel as follows:

A. Mechanical brake by Torque Wrench — thereafter checking every second hours.

B. Hydraulic brake — setting of brakes and checking procedure to be agreed upon in each case.

One safety/oil pollution avoidance/mooring/turnaround inspector will be placed onboard while vessel is moored alongside berth.

Two tugboats will be stand by.

VHF's will be given to: Safety Inspectors, OM&S controlroom and Chief Officer.

Weather forecast available on request from our harbour office.

Approach velocity is continously measured by Radar Doppler on pier and communicated to pilot/master during berthing.

Communication between: pier — pilot — tugs and mooring boats maintained by walkie talkie (VHF) on channel 14.

Max approach speed normal to pier as shown in table on the back of this folder.

Only 1 max 2 layers to be allowed on working drum when moored if split-drum.

If standard drum, max number of layers to be discussed in each case.

Weather tolerance for normal operation of the pier:

Wind velocity 20 knots average 10 min. periods.

Current velocity 0,6 knots — outgoing.

If reaching the above tolerance:

Stop discharge and drain all arms.

Close all tank lids and ullage plugs.

Tugs to be ordered to the ship's side or in towing hawsers.

Summon Harbour Master and mooring crews.

Master to confirm that ship's engine is ready for use.

If reaching wind velocity 25 knots or current velocity 0,7 knot — arms to be disconnected and vessel ordered to be stand by.

The Harbour Master together with the Ship's Captain and Harbour Pilot will decide whether to take the ship out.

Fig. 12 MOORING AND SAFETY PRECAUTIONS RULES FOR SLAGEN TANGEN OIL TERMINAL, NORWAY

BERTHING AID SYSTEMS

A lack of proper BERTHING AID CONTROLS, by sonar or radar
makes berthing-management even more difficult and risky. The
fact, however, is that there are only a few terminals for handling large vessel which have berthing aid systems installed on
the pier or quay.
Fig. 11 shows a modern berthing aid system including a dis-
play-gantry and a radar control of high resolution. For details see Bruun 8 and 9).

THE TERMINAL MOORING MANAGEMENT

Terminal mooring management consists of pre-planning for the
range of ships which may call the terminal, establishing a
basis for the ships to maintaining a safe mooring, and then
monitoring the ship's mooring system to assure maintenance of
a safe at-berth mooring system. Each terminal is limited
either by structural design or by site conditions to a max
sized vessel it is able to accomodate. Thus there is a range
of vessel sizes that can call at the terminal and be safely
berthed at its facilities. These ships can have a wide range
of mooring equipment onboard. For example, there can be ships
with all synthetic moorings, or mixed moorings (synthetics and
wire) or all wire (with or without synthetic tails). Also the
brake holding capacity of the winches can vary from 30 - 90%
of the original minimum breaking load of the line. In addition, winch and fairlead locations on the ship can vary significantly for the same size vessel. And, lastly, ship crews
may differ in expertise and in philosophies concerning maintenance of the mooring and the maintenance and replacement of
critical items of mooring equipment.
To cope with ships having varying types of mooring equipment
and state of maintenance, the terminal can employ several techniques to reduce the possibility of ship breakout followed
by risks of wrecking and pollution. Administrative under-
takings include:
Developing operating guidelines and mooring limits.
Obtaining ship's mooring equipment information prior to arrival
and revising guidelines as it may become necessary following
the development.
Inspecting ship's mooring equipment to determine if a reduction
in operating limits is required because of poor maintenance,
improperly trained crews (often in developing countries), etc.
For further information on mooring management the reader should
consult refs. 9 and 19.
Fig. 12 gives MOORING AND SAFETY PRECAUTIONS for the SLAGEN-
TANGEN oil terminal in Norway. These rules are strictly enforced by the terminal administration. The experience has
been very satisfactory.

60

REFERENCES

1. Balfour, J.S.et al., "Fendering Requirements - Design Fender Impact Criteria", PORTS 80, ASCE (American Society of Civil Engineers), 1980.
2. Brummenaes,I, "The IRVING Marine Fender, Non-Recoiling", Ship Equipment, Haugesund, Norway, 1978. See also HAZARDOUS BULLETIN, June, 1978.
3. Bruun, P., "An Alternative to Conventional Fendering", Schiff und Hafen, Kommando brücke, 32. Jahrgang, Heft 1,1980.
4.Bruun, P., "Breakwaters versus Mooring", THE DOCK AND HARBOUR AUTHORITY, Vol. XLII, Nor. 730, Sept.,1981,pp126-129.
5. Bruun, P., "Mooring and Fendering. Rational Principles in Design", 8th Internat. Harbour Conference, Antwerp, 1983. Koninklijke Vlaamse Ingenieursvereningen, Technologisk Institue, Antwerp, Belgium.
6. Bruun, P., "Mooring and Fendering. Rational Prinsiples in Design", THE DOCK AND HARBOUR AUTHORITY, Fefr., 1984,pp.200 - 212.
7. Bruun, P., "Discussion of Paper by D.V.Padron and E.H.Y. Han, 1983", Proc. ASCE, Journ. Waterway, Port and Coastal Engineering, Vol. 110, No. 4, 1984, pp. 520 - 524.
8. Bruun, P., "Fendering - 82", THE DOCK AND HARBOUR AUTHORITY, March, 1985.
9. Bruun, P., "Port Engineering IV", The GULF PUBLISHING CO, Houston, TX, 2 volumes, 1,700 pages.
10. Cuthbert, D.R. and Seidl, L.H., "Mathematical Analyses of Ship Mooring Systems. Comparison with Hydraulic Scale Model Investigations", Proc. 4th Conf. on PORT AND OCEAN ENGINEERING UNDER ARTIC CONDITIONS, Memorial University of New Foundland, St. Johns, 1978.
11. Fontijn, H.L. "The Berthing Ship Problem" Report No.78-2, Communication on Hydraulics, Delft University of Technology, Delft, Netherlands, 1978. See also Proc. ASCE, Journ. Waterway, Port and Ocean Engineering Division, Vol. 106, No.2, 1984, pp. 239 - 250.
12. Haldeman, B., "Discussion of Paper by D.V.Padron and E.H.Y.Han, 1983, Proc. ASCE, Journ. Waterway, Port and Coastal Engineering Division, Vol. 110, No. 4, 1984, pp. 524 - 526.
13. Harlow, E.H., "Discussion" of Paper by D.V. Padron and E.H.Y.Han, 1983, Proc. ASCE, Journ. Waterway, port and Ocean Engineering Division, Vol. 110, No.4, 1984, pp. 526-527.
14. Khanna, J. and Ottesen Hansen, N.E., "Moored Ships exposed to combined Waves and Currents", Proc. POAC 77, St. John's Memorial University, New Foundland, Canada, 1977, pp. 253 - 263.
15. Kikutany, H., et al, "Some Requirements for the Design of Sea-Berths from the Viewpoint of Ship-Handling", PIANC, S II, Vol. 3, 1977.
16. Mulcahy, M.W., "Mooring Instrumentation for Large Marine Terminals", Proc. POAC -77, St. Johns Memorial University, New Foundland, Canada, 1977.
17. Nagai, S. et al, Impact Forces, Mooring Tankers and Motions of Supertankers at offshore Terminals subjected to Wave Action", PIANC, S II, Vol. 4, 1977.

18. Nikerov, P.S. et al, "Improving the Methods of determining the Loads applied by Berthing Ships, the Effects of Wave Disturbance, and Examination of Flexible Fenders", PIANC, SII, Vol. 1, 1981.

19. Oil Companies Internat. Marine Forum, "Guidelines and Recommendations for the Safe Mooring of Large Ships at Piers and Sea Islands", Witherby and Co., London, 1978, 88 pages.

20. Oortmerssen, van G., "The Motions of a moored Ship in Waves", NSMB Publ. 510, Wageningen, Hol. 79.

21. Padron, D.V.and Han, H.Y., "Fender System Problems in US Ports", Proc. ASCE, Journ. Waterway, Port and Coastal Engineering Division, Vol. 109, No.3, 1983, pp. 296 - 309.

22. Padron, D.V. and Han, E.H.Y., "Closure" of Discussion, Proc. ASCE, Journ. Waterway, Port and Coastal Engineering Division, Vol. 109, No.3, pp. 296-309, Closure, pp. 527 - 529, Vol. 110, No.4. 1984.

23. PIANC, "Report of the Internat. Commission for Improving the Design of Fender Systems", Suppl. to Bulletin, No. 45, Brussels, Belgium, 1984, 158 pages.

24. Bruun, P., "Can We Save on Breakwaters by Providing Better Mooring-Fendering?", PIANC, Bull. 53,85.

25. Seidl. L.H. and Lee, I., "Correlation between theoretical and experimental Values of Motions and Mooring Forces of Ships Moored at a Sea Berth", 3rd. Internat. Ocean Developm. Conf, Tokyo, 1973.

26. Pardon, D.V. and Han, E.H.Y., "Fender System Problems in US Ports", US Dept. of Commerce, Maritime Administration and American Association of Port Authorities, Proc. ASCE, Journ. Waterway, Port and Coastal Engineering Division, Vol. 109, No. 3, 1983, pp. 296 - 309.

27. Vasco Costa, F., "The Berthing Ship", Foxlow Publications, London, 1964, 120 pages.

28. Vasco Costa, F., "Fenders and Energy Dissipators", THE DOCK AND HARBOUR AUTHORITY, September, 1978.

Wood,F.S. etal, "Design of an Offshore Berth for Mooring large Ships in strong Currents", Proc. 25th PIANC Congress, Edinburgh, S II - 1, 1981.

ON THE PREDICTION OF FENDER FORCES AT BERTHING STRUCTURES
Part I: CALCULATION METHOD(S)

H.L. Fontijn,
Senior Scientific Officer, Department of Civil Engineering, Delft
University of Technology, The Netherlands.

1. INTRODUCTION

1.1. General description of berthing-ship phenomenon

During the last decades ships have grown larger and larger. As a
consequence berthing facilities have to be newly constructed or adapted to
the larger units. Up to now, reliable, theoretically founded design
criteria are hardly available. The lack of good design criteria is the
prime reason for making researches into the possibilities of an
experimental and/or theoretical determination of berthing forces.

Generally a berthing facility consists of one or more elastic elements
(fenders) attached to a rigid structure (finger pier, caisson-type jetty,
quay-wall, etc.). The fenders absorb the berthing forces and form a
protection for ship and berthing structure. As the maximum permissible
berthing force against the side of e.g. a mammoth tanker is distinctly
lower than what is acceptable for the berthing structure, the ship is
therefore the prevailing factor for fender design. Ref. [1] gives a review
of various types of open berthing structures; besides it presents a
classification of the countless systems of fenders with special regard to
their properties and applicabilities. For an inventory of fender systems it
further is referred to ref. [2].

The phenomena occurring during the berthing manoeuvre of a ship are
complicated and the fender loads are influenced by a lot of parameters: the
configuration of the berthing site, the geometry and the rigidity of the
(hull of the) ship, the mechanical properties of the fender(s), the speed
of approach, the forces exerted by tugs, wind, current and waves, the mode

E. Bratteland (ed.), Advances in Berthing and Mooring of Ships and Offshore Structures, 62–94.
© 1988 by Kluwer Academic Publishers.

of motion (in general translation in the horizontal plane combined with rotation), the keel clearance.

As far as the lay-out of the berthing site is concerned, two situations can be distinguished:

a - a situation of water with relatively large horizontal dimensions, which implies an open jetty-type berthing facility not interfering with flow and pressure fields around the ship; and

b - a situation with a closed berth, i.e. a berthing structure with a solid front; now the berth does interfere with the flow and pressure fields around the ship, so that the hydrodynamic phenomena are more complicated than in the situation mentioned sub a.

The laterally moving ship pushes ahead of itself a positive pressure field, more or less noticeable as a raised water level. In case of a closed berth this pressure field is reflected by the solid front of the structure, further raising the water level between ship and wall. The rise in water level becomes greater the nearer the ship gets to the berth. When the ship slows down and stops on the berth, the underkeel flow, which keeps going for a time, sucks water out of the gap between berth and ship (i.e. the quay clearance) thus drawing down the water level there. In case of a berthing structure with a solid front there thus appears to be two opposing effects: 1) as the ship closes on the berth, the reflected pressure wave increasingly cushions the impact by raising the water level in the quay clearance, 2) as the ship slows down on the fender, the inertia of the underkeel flow draws down the water level in the quay clearance and 'sucks' the ship harder onto the berth. In advance it is not simply clear which effect will dominate: it requires careful analytical and experimental research to establish the net effect on ship motions and fender loads.

The behaviour of a berthing ship and the resulting fender loads can be determined either by means of experiments with scale models or by way of an analytical treatment of the phenomenon. Of course a combination of both methods is possible as well.

On the one hand model testing has a few drawbacks. Model tests are expensive and time consuming. The test set-up is complicated; it is essential that the elastic properties of the fenders are simulated very carefully and, sometimes, sophisticated facilities are needed to simulate the relevant environmental conditions. For these reasons test programs are usually restricted to those final design configurations and selected conditions which are assumed to be the most critical. Besides, the insight gained from model tests into the fundamentals of the problem remains limited: only the resulting output is measured without yielding much knowledge of the mechanism which causes the output. On the other hand a general analytical treatment of the problem is rather complicated.

1.2. Review of previous studies

When designing a berthing structure generally an approach is used in which it is assumed that the energy to be absorbed by the fender(s) equals the kinetic energy of the ship. Usually the mode of motion of the ship then consists of a translation - with or without forward speed - combined with a rotation. To include the effect of the entrained water a certain constant added mass(-moment of inertia) is introduced (see e.g. refs. [1 through 29] for the open jetty-type berthing facility and refs. [1 through 4, 6, 8, 12, 15, 16, 23, 27, 29 through 34] for the closed berthing structure). This is also the case in refs. [35, 36] where, in addition, an account is given of research on the slowing down of a ship in approaching laterally a closed quay-wall. Refs. [37, 2] present a review of the most common expressions for the added mass. In this context, guidelines for fender system design are given in ref. [2].

This approach, in fact, involves the use of Newton's second law

$$\frac{d}{dt}(m\dot{x}) = f(t) \quad , \tag{1.1^a}$$

describing the motion(s) $x(t)$ of a freely floating ship with mass(-moment of inertia) m in response to some external force or moment $f(t)$; t represents the time co-ordinate. Since m may be regarded as a constant the equations of motion become:

$$m\ddot{x} = f(t) \quad . \tag{1.1^b}$$

In the following the concept 'force' has to be understood in a generalized sense meaning force or moment. In general the external force $f(t)$ in $(1.1^{a,b})$ is composed of:
- forces, e.g. due to waves, varying arbitrarily in time,
- hydrodynamic and hydrostatic restoring forces, which are a function of the motions of the ship,

- (restoring) forces due to the fender and/or mooring system, which are a
  function of the instantaneous position of the ship.

In the classical theory of ship motions it is common practice to formulate
the equations (of motion) as follows:

$$(m+a)\ddot{x} + b\dot{x} + cx = f(t) \quad , \tag{1.2}$$

where a is the added mass(-moment of inertia), b the hydrodynamic
coefficient of the damping force and c the hydrostatic restoring
coefficient; the coefficients a and b represent the hydrodynamic effects.
(1.2) has the form of a linear differential equation of the second order
with constant coefficients; due to its linear character it can only reflect
linearized hydrodynamic phenomena.

Applying the assumption of linearity, it is obvious that a ship, under the
action of a harmonically oscillating force at one specific frequency, comes
at a harmonic motion with the same frequency as that of the excitation. The
distribution of the hydrodynamic stress on the wet ship hull then also
presents a harmonic behaviour with the same frequency. Experimentally and
theoretically it can be shown that harmonic ship motions lead to frequency-
dependent coefficients a and b, the so-called hydrodynamic coefficients;
the coefficient c is considered to represent the hydrostatic restoring
effects and is defined as being independent of the frequency (see e.g.
refs. [38 through 66]). This frequency dependence only emerges, when a free
water surface is present, also if viscous effects are neglected; in absence
of a free water surface the hydrodynamic coefficients are constants.
Therefore, it is stated that the occurrence of frequency-dependent
hydrodynamic coefficients completely can be ascribed to the existence of a
boundary in the form of a free water surface.

Introduction of hydrodynamic coefficients with a frequency-dependent
behaviour generates a formulation in the time domain, which differs
fundamentally from (1.2): instead of forces acting only instantaneously in
time, now also a 'memory effect' appears on the scene, i.e. each occurrence
becomes, in fact, dependent on all preceding occurrences. Actually the
'memory effect' reflects the dissipative property of the free water surface
(wave radiation). Then (1.2) takes the form:

$$\{m + a(\omega)\}\ddot{x} + b(\omega)\dot{x} + cx = f(t) \quad , \tag{1.3}$$

where ω represents the circular frequency. (1.3) states that a harmonically oscillating excitation with $f(t) = \hat{f} \exp(i\omega t)$ has as its response a harmonic motion $x(t) = \hat{x} \exp(i\omega t)$; the circumflex means 'amplitude of', $i = \sqrt{-1}$. Now (1.3) is not any longer a real equation of motion in the sense that it relates the variables of the instantaneous motion to the instantaneous values of the exciting forces. On the contrary, (1.3) merely represents a set of algebraic equations fixing the amplitudes and phases of the (six) oscillations of the ship under the action of an exciting oscillating force at one specific frequency; in other words, this set of equations is valid only if the right-hand members all vary sinusoidally at a single frequency and if the 'constant' coefficients a and b on the left have the values appropriate to that frequency. Therefore (1.3) can only be used as a representation in the frequency domain of a steady oscillating motion, since the hydrodynamic coefficients a and b depend on the frequency of the motion itself. Substitution of $f(t) = \hat{f}(\omega)\exp(i\omega t)$ and $x(t) = \hat{x}(\omega)\exp(i\omega t)$ into (1.3) yields an expression which has to be considered as a description of the ship-fluid system in the frequency domain:

$$[-\omega^2\{m + a(\omega)\} + i\omega\, b(\omega) + c]\hat{x}(\omega) = \hat{f}(\omega) \quad . \tag{1.4}$$

This can be rewritten as

$$R(\omega)\ \hat{x}(\omega) = \hat{f}(\omega) \tag{1.5}$$

with

$$R(\omega) = -\omega^2\{m + a(\omega)\} + i\omega\, b(\omega) + c \quad ; \tag{1.6}$$

$R(\omega)$ relates the harmonically oscillating excitation with its response. The analytical (and experimental) work dealing with the berthing of ships, as mentioned in refs. [1 through 36], in principle is based on (1.3). In these the coefficient a is supposed to be independent of the frequency, while the coefficient b is neglected. Since berthing manoeuvres mainly take place in the horizontal plane, the hydrostatic restoring coefficient c is left out of consideration. (1.3) then reduces to a simplified form of (1.2):

$$(m+a)\ddot{x} = f(t) \quad . \tag{1.7}$$

(1.7) can be regarded as a differential equation, representing a set of equations of motion, which is only adequate to describe the motion of a body in a fluid without a free water surface. However, when a free water surface is present, (1.7) will yield incorrect results due to the occurring 'memory effect'. In (1.7) the hydrodynamic influences are reflected only by the added mass(-moment of inertia), which is assumed to be constant during the motion of the ship; the effect of the hydrodynamic damping has not been taken into account. Besides, the choice of a value for a is a problem, the more so as it appears from literature (see e.g. refs. [38 through 66]) that the hydrodynamic coefficients are very much dependent on the frequency, especially in shallow water and in the vicinity of a closed wall: generally it holds true that the concept of constant hydrodynamic coefficients is not justifiable. Consequently, to determine the fender forces as a result of the berthing of a ship, a time-domain description of the behaviour of the moving ship is needed, which is able to make allowance for the frequency dependence of the fluid reaction forces, i.e. a method has to be used in which the hydrodynamic coefficients are taken into account as functions of the frequency.

1.3. The present research

1.3.1. Objective of investiqation. The present investigation aims at the formulation of a mathematical model which is sufficiently accurate both to describe the behaviour of a ship berthing to an open jetty-type facility or a closed structure (either fitted with fenders) and to determine the response of the fenders themselves in a theoretical way; all essential features are to be maintained and quantitative results of sufficient accuracy are to be produced for most practical applications.
To achieve this end, a system approach is followed, which has the restriction that the combination of ship and fluid is supposed to be linear. In addition to the fender loads other (external) forces upon the ship, such as forces exerted by wind, waves, current, tugs and mooring lines can be incorporated in the model as well.

1.3.2. Ship-fluid system and linearity concept. When applying a system approach to the problem under consideration, for obvious reasons the berthing ship as a whole, i.e. the combination of ship, fluid and fender structure, has not to be taken for 'the system'. By isolating the freely floating ship in still water, ship and fluid combined can be conceived of

as the system to be considered. The fender loads then are thought to belong
to the category of external forces.

On account of several investigations (see e.g. refs. [43, 46, 50, 54,
58]) it can be stated that the ship-fluid system is linear. In addition to
the references mentioned, a good survey on this point as well as a
(comprehensive) description of character and behaviour of the linear ship-
fluid system is given in ref. [67]. All (experimental) data indicate that
this basic linearity assumption is a well-working approximation for small
to moderate displacements of real ship forms. Therefore it is hypothesized
that the assumption of linearity of the ship-fluid system holds absolutely.
With regard to fluid idealization the facts point in two directions. While
it is practically sure that the restriction to a homogeneous,
incompressible fluid, free from surface tension, is not a serious
limitation, the viscosity may lead to complications, notwithstanding the
fact that viscous terms are basically linear. On the one hand, in dealing
with (ship) motions it is of great advantage and in some cases (e.g. a ship
in waves) necessary to consider the water as inviscid; it is a logic
consequence of the validity of the linearization, for, if the viscosity
would have a great influence, then the linearity would be impaired as well.
On the other hand, due to interaction between the viscosity and the (non-
linear) convective terms flow separation and consequent eddy formation may
occur, which phenomena are distinctly perceptible, especially with larger
ship motions. It makes itself primarily felt in additional damping and in a
change in the hydrodynamic coefficients which couple the motions mutually.
As long as the ship motions (i.e. displacements or velocities, or both)
remain small, viscous effects can be taken into account without violating
the basic linearity concept of the ship-fluid system.
Beside linearity, the further requirements to be made upon this system
approach are time independence and stability.

1.3.3. Approach to be followed. The berthing-ship problem is concerned
with fixing those quantities as function of time, which are essential for
the motion of the ship and, especially, the interaction between ship and
fender. In order to be able to make allowance for the time-dependent ship-
water interaction, the frequency dependence of the hydrodynamic
coefficients completely has to be taken into account. By means of a
Fourier-transform technique a formally correct representation of the ship-
fluid interaction in the time domain can be drawn up, which is equivalent

to its formulation in the frequency domain. This representation in the time
domain holds good for external forces arbitrarily varying in time (in the
sense of transient disturbances of restricted duration). The condition
attached is that the ship-fluid system behaves linearly. The influence of
the water, and particularly of its free surface ('memory effect'), can only
be represented correctly by using the full information embedded in the
frequency dependence of the hydrodynamic coefficients. In literature –
though for non-horizontal motions – this is confirmed theoretically and
experimentally (see e.g. ref. [67]). Particularly the keel clearance and
the vicinity of a closed wall do highly affect the sensitivity of the ship-
fluid interaction to frequencies.

Considering the above allegations now two approaches can be followed:

I – The description of the linear ship-fluid system in the time domain can
be determined as the inverse Fourier transform of (1.4). On certain
conditions with respect to the transformed functions, this yields an
equation of motion for the variable x(t) in the form of an integro-
differential equation, viz.:

$$m\ddot{x} + \int_{-\infty}^{t} \ddot{x}(\tau)\, A(t-\tau)d\tau + \int_{-\infty}^{t} \dot{x}(\tau)\, B(t-\tau)d\tau + cx = f(t) \quad , \qquad (1.8)$$

where τ represents an integration variable (time). This expression
includes convolution integrals containing the so-called retardation
functions $A(t)$ and $B(t)$, which are the inverse Fourier transforms of
$a(\omega)$ and $b(\omega)$, respectively. The convolution products thus arise from
the frequency dependence of the hydrodynamic coefficients and,
therefore, represent the memory effect as generated by the free water
surface (see further refs. [68, 67, 58]).

II– Starting from (1.5),(1.6) the inverse Fourier transform takes – on
certain conditions – the form

$$f(t) = \int_{-\omega}^{t} x(\tau)\, r(t-\tau)d\tau \quad , \qquad (1.9)$$

where $r(t)$, with $r(t) \equiv 0$ for $t < 0$, is the inverse Fourier transform
of $R(\omega)$. In (1.9) the response of the ship to arbitrary motions is

fully characterized by the function r(t). This compact formulation
supposes a generalized-function concept: r(t) consists, among other
things, of contributions from delta or Dirac functions and their
derivatives. According to the specific notation (1.9) the system of the
ship-fluid interaction is regarded as a black box, relating the
excitation (input signal) and the response (output signal) of the
system without reflecting the physical processes behind it. In this,
r(t) has to be conceived of as the impulse response function of the
system, i.e. the response to a unit pulse, on the understanding that
response and excitation represent force and motion, respectively. The
requirement that $r(t) \equiv 0$ for $t < 0$, ensues from the fact that the
linear ship-fluid system - like each physical system - is causal.
With regard to the interaction between berthing ship and fender(s) it
is obvious - contrary to the above - to interchange response and
excitation: now the forces f(t), exerted somewhere upon the ship, are
conceived of as input signals (excitation), whereas the ship motion
x(t) (displacement and rotation or derived quantities) is considered to
be the output signal (response). Then - provided the ship-fluid system
is linear - input signal and output signal are connected by means of a
convolution integral over the entire time history of the forcing
function(s) according to

$$x(t) = \int_{-\infty}^{t} f(\tau)\, k(t-\tau)\, d\tau \quad , \tag{1.10}$$

where k(t) represents the impulse response function, i.e. the response
to a unit pulse (Dirac function at t=0). Naturally (1.10) has a similar
form as (1.9); k(t), with $k(t) \equiv 0$ for $t < 0$, is the inverse Fourier
transform of $1/R(\omega)$. The linear ship-fluid system is fully
characterized if k(t) is known, i.e. the response x(t) to an arbitrary
forcing function f(t) can be found in terms of k(t). The external
forces, e.g. fender loads, may be linear or non-linear and can be
incorporated in the forcing function. According to (1.10) the ship-
fluid interaction again is regarded as a black box (see fig. 1.1).
The approach outlined above is denominated as 'impulse response
function'-technique.

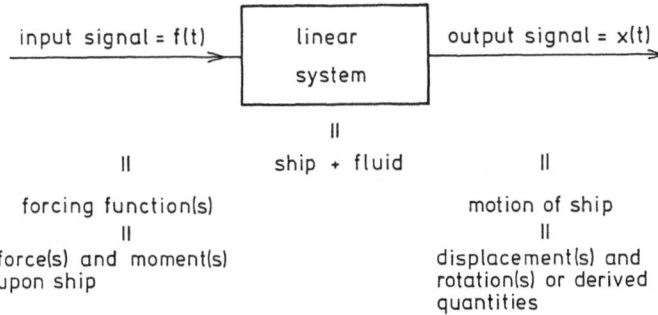

FIGURE 1.1 - Schematic representation of ship-fluid system ('black box').

For approach I as well as approach II the respective descriptions of
the linear ship-fluid system in time and frequency domain are related by
means of Fourier-transform techniques. A good review of the ship-fluid
system in the time domain and the frequency domain is given in ref. [67];
further it can be referred to refs. [69, 68].

When the hydrodynamic coefficients in the frequency domain once are
known, both approach I and approach II is appropriate to be applied to
time-dependent problems: the situation of a ship being initially at rest as
well as the situation of a ship with a uniform motion can be considered.
For, both situations are to be conceived of as initial states of
equilibrium, from which - according to the supposed linearity - small
disturbances are occurring. The hydrodynamic influence of a given, initial
velocity finds merely expression in $a(\omega)$, $b(\omega)$ and c, and consequently is
only reflected by the retardation functions (approach I) and the impulse
response function (approach II).

Practical applications of approach I are presented in refs. [70, 58],
which are concerned with ship motions on water with relatively large,
horizontal dimensions. Ref. [70] deals with ship berthing, viz. a centric
impact to a jetty fitted with a linear, undamped fender. In a more
universal way the motions of a moored ship in waves are described in ref.
[58].

Due to its black-box formulation the 'impulse response function'-
technique (approach II) is less appropriate to analyse the response (i.e.
the motion) of the ship than approach I: for, making use of approach II it

is difficult to discriminate between the respective contributions of
inertia and damping effects. If it is only a question of the response
itself and not of its analysis, then the 'impulse response function'-
technique is an appropriate approach, offering the possibility to
incorporate in an efficient way all kind of factors which are of importance
for the ship-fluid system. In ship berthing the main point is with the
resulting course of the ship motion and its related history of the load on
the berthing facility. That is the reason why in the present investigation
a choice is made for applying the 'impulse response function'-technique
with the forcing function(s) as input signal and the ship motion as output
signal.

   1.3.4. Simplification of problem. For the specific case of a ship
berthing to a fender structure the following assumptions and
simplifications are made.
The open berth is of the jetty-type; the closed berth consists of a
straight, impermeable, vertical wall of infinite length. Both berthing
facilities are fitted with one single fender without mass of its own, or at
most with a mass which is small with respect to that of the ship. The
fender has a horizontal line of action situated in the plane of the water
surface at rest; for the closed berth the line of action is perpendicular
to the front of the (quay-)wall.
The characteristics of the fender are assumed to be undamped and (non-)
linear. The frictional force between the hull of the ship and the fender is
neglected.
Only berthing operations on sheltered locations (e.g. harbours) are
considered, i.e. the influences of waves, current and wind are not taken
into account.
As berthing manoeuvres and the ship-fender interactions take place mainly
in the horizontal plane, only the surge, sway and yaw motions of the ship
do play a part; so, in this context it is assumed implicitly that dynamic
effects due to eventual vertical ship motions (heaving and pitching) and
rolling - which in a way do occur in reality - are of minor importance and
do not influence the motion in the horizontal plane.
The vessel is considered as a rigid, prismatic body with a rectangular
cross-section and a symmetrical distribution of mass. This schematization
is justified by the fact that many sea-going vessels and the most inland

ships have a more or less box-like shape, being slightly streamlined at bow and stern.

The ship's forward speed is supposed to be zero or negligibly small, which ensues from the fact that during a berthing operation the forward speed is indeed small or zero.

Further, in case of a closed berthing facility it is assumed that the ship maintains a lateral motion with its longitudinal axis of symmetry parallel to the face of the berth. It implies a centric impact in which only the sway motion plays a part, and no rotation. This assumption arises from the geometrical situation of berthing manoeuvres at closed structures in general and the fact that the influence of the sway motion on fender loads predominates the effect of surging and yawing.

Diffraction phenomena and flow around bow and stern are not considered. Special attention is paid to the case when shallowness of the water is of dominant importance, for, berthing facilities are often located in shallow water. The bottom is horizontal and impermeable. In case of the open jetty-type berth the fluid domain is supposed to be relatively large in the horizontal directions; the same applies to the fluid domain in front of the (quay-)wall.

Besides it is assumed that the fluid is incompressible.

A very important assumption with respect to the ship is that displacements and rotations or derived quantities remain small; for, this makes it possible to regard the problem of the ship motions as linear, so that ship and fluid combined can be conceived of as being a linear system. Further it is supposed that this ship-fluid system is time-dependent and stable.

With the supposed linearity of the ship-fluid system and under the simplifications mentioned above the problem of a ship berthing to a fender structure now has been reduced greatly. What remains is the formulation of a mathematical model based on the linearity of the ship-fluid system, which is able to describe the force(s) exerted by a (schematized) ship with a horizontal motion (swaying and yawing) upon some berthing facility at calm, shallow water with relatively large, horizontal dimensions.

1.4. Outline of successive Sections

In Section 2 the 'impulse response function'-technique is dealt with in a general mathematical formulation and its features are discussed. The approach is applied to systems with three degrees of freedom: only ship

motions in the horizontal plane are considered. The linear ship-fluid
system is described in the frequency domain as well as in the time domain.
Both the stability of the system and the causality condition is considered.
Then expressions are derived for the respective impulse response functions.

Part II deals with the application of the 'impulse response function'-
technique to the berthing of ships.
Section 3 is concerned with ship berthing to open fender structures. The
'impulse response function'-technique is applied to the horizontal modes of
motion of a schematized ship on (shallow) water with relatively large
horizontal dimensions. At first the hydrodynamic coefficients are
determined, theoretically as well as experimentally, from which the
corresponding impulse response functions are calculated. Then a
mathematical approach is presented to simulate the berthing of the
(schematized) ship to a jetty and to determine the relevant quantities. For
certain situations the results of theoretical and experimental
investigations are compared and discussed.
In Section 4 a similar procedure is followed for the case of a
(schematized) ship berthing to a closed fender structure. Further, for this
situation, a direct time-domain approach is presented in which non-
linearities (in the hydrodynamics) can be taken into account.
Section 5 concludes with a summary and conclusions.

The work to be presented mainly is based on the results of research
published as refs. [71, 54, 72, 73, 74, 66, 75, 76].

2. THE 'IMPULSE RESPONSE FUNCTION'-TECHNIQUE

2.1. Introduction

In this section a general formulation of the horizontal ship motion is
presented in the frequency domain as well as in the time domain. Use is
made of the 'impulse response function'-technique.
The two important assumptions made are that the ship behaves as a rigid
body and that the motions of the ship remain small. In order to facilitate
the formulation three further restrictions are made: firstly, the ship's
form is transversely symmetric with respect to its vertical centre plane,
longitudinal symmetry is not assumed; secondly, at rest the ship is

floating upright in stable equilibrium; and thirdly, the ship has a
constant (mean) velocity with two components, viz. a forward speed and a
transverse speed, parallel and perpendicular to the above plane of
symmetry, respectively.

2.2. Co-ordinate systems

Analogous to ref. [46] the following co-ordinate systems are
introduced:

$OX_1X_2X_3$ = space-fixed right-handed system of Cartesian co-ordinates with
origin O; OX_1X_2 coincides with the water surface at rest; the
vertical OX_3-axis is positive upwards; the forward speed V_1 and
the transverse speed V_2 of the ship is parallel to the positive
OX_1-axis and the positive OX_2-axis, respectively.

$ox_1x_2x_3$ = right-handed Cartesian co-ordinate system parallel with $OX_1X_2X_3$,
but translating with the (constant) ship's speeds V_1,V_2; at rest
the origin o coincides with the ship's centre of gravity G; the
longitudinal ox_1-axis is positive in forward direction, the ox_2-
axis is positive to port-side, the ox_3-axis is positive upwards.

Gxyz = moving right-handed Cartesian co-ordinate system with origin G
and fixed with respect to the ship; Gxz coincides with the
longitudinal plane of symmetry of the ship; the Gy-axis is
positive to port-side, the Gz-axis is positive upwards.

Assuming G to be situated in the plane of the water line, the relations
between $OX_1X_2X_3$ and $ox_1x_2x_3$ are:

$$X_1 = x_1 + V_1t \quad , \quad X_2 = x_2 + V_2t \quad , \quad X_3 = x_3 \quad .$$

On account of its definition $ox_1x_2x_3$ is an inertial system. In
principle, within the linearity concept small disturbances are considered
from an initial state of motion of the ship. Relating to berthing this
implies small ship motions with respect to the translating $ox_1x_2x_3$-co-
ordinate system, which acts as state of equilibrium.
The relevant motions of the ship now can be represented by the motion
variable $x_j(t)$, where j = 1,2,6; x_1 and x_2 stand for the respective
translations in the X_1-direction (surge motion, positive forwards) and the
X_2-direction (sway motion, positive to port-side); x_6 denotes the rotation
around the OX_3-axis (yaw motion, positive with bow moving to port-side).

In consequence of the ship's rotation formally virtual forces (due to Coriolis and centrifugal effects) as well as an inertial contribution (due to the angular acceleration) are introduced. These influences can be neglected, since - within the context of the linear approach - they become small of the second order. If these effects nevertheless should be taken into account, they are to be classed in the forcing function (i.e. input signal) of the linear ship-fluid system.

2.3. Ship-fluid system in frequency domain

Due to the linearity of the ship-fluid system, (1.3) can be extended for the general case of coupled ship motions; i.e. the mass, damping and restoring forces resulting from the distinct directions of motion may be superimposed to counterbalance the exciting force in the relevant direction (see refs. [68, 67]):

$$\sum_{j}^{1,2,6} [(m_{jk} + a_{jk}(\omega))\ddot{x}_j + b_{jk}(\omega)\dot{x}_j + c_{jk}x_j] = f_k(t) \quad , \quad k = 1,2,6, \qquad (2.1)$$

where m_{jk} = inertia matrix (i.e. generalized horizontal mass) of ship (m_{jk}=0 for j≠k),

$a_{jk}(\omega)$ = hydrodynamic coefficient of mass term in k-equation as result of motion in j-direction,

$b_{jk}(\omega)$ = hydrodynamic coefficient of damping force in k-equation as result of motion in j-direction,

c_{jk} = hydrostatic restoring coefficient in k-equation as result of (static) displacement in j-direction at zero speed,

$f_k(t)$ = external exciting harmonic force upon ship in k-direction; the double subscript j,k relates the force in the k-direction to the motion in the j-direction.

Expression (2.1) is a description of the linear ship-fluid system in the frequency domain. Accordingly it is not a set of real differential equations in the time, and it does not represent a set of equations of motion in the sense that instantaneous quantities of the motion are related to instantaneous values of the external force. (2.1) exclusively holds good for steady harmonic oscillations at a specific frequency and their corresponding 'constants' in the left-hand member (see also refs. [69, 68, 67]).

The exciting harmonic force $f_k(t)$ has the form

$$f_k(t) = \hat{f}_k(\omega) \, e^{i\omega t} \quad . \tag{2.2}$$

Then the motion variable $x_j(t)$ has to be written as:

$$x_j(t) = \hat{x}_j(\omega) \, e^{i\omega t} \quad . \tag{2.3}$$

Substitution of (2.2) and (2.3) into (2.1) yields

$$\sum_{j}^{1,2,6} R_{jk}(\omega) \, \hat{x}_j(\omega) = \hat{f}_k(\omega) \quad , \tag{2.4}$$

with

$$R_{jk}(\omega) = -\omega^2 \{m_{jk} + a_{jk}(\omega)\} + i\omega \, b_{jk}(\omega) + c_{jk} \quad . \tag{2.5}$$

Now consider the motion in one direction, say the i-direction; then $\hat{x}_j(\omega)=0$ for $i{\neq}j$, and $R_{ik}(\omega)$ can be shown to be the harmonic transfer function for the k-direction in response to a harmonic (motion) excitation in the i-direction.

As a result of the condition $\hat{x}_j(\omega)=0$ for $i{\neq}j$, it can be stated that with respect to (1.3) - in a formal sense - excitation and response have been interchanged: now $x_i(t)$ is to be considered as the excitation and $f_k(t)$ as its response. This view corresponds with the common practice in forced harmonic oscillation experiments. From (2.5) it then follows for $a_{ik}(\omega)$ and $b_{ik}(\omega)$:

$$\left.\begin{aligned} a_{ik}(\omega) &= \frac{1}{\omega^2} \, \{c_{ik} - \mathrm{Re}[R_{ik}(\omega)]\} - m_{ik} \, , \\[2mm] b_{ik}(\omega) &= \frac{1}{\omega} \, \mathrm{Im}[R_{ik}(\omega)] \quad , \end{aligned}\right\} \quad \text{provided } \hat{x}_j(\omega)=0 \text{ for } i{\neq}j, \tag{2.6}$$

with

$$\left.\begin{aligned} \mathrm{Re}[R_{ik}(\omega)] &= \frac{|\hat{f}_k(\omega)|}{|\hat{x}_i(\omega)|} \, \cos(\theta_{ik}) \quad , \\[6mm] \mathrm{Im}[R_{ik}(\omega)] &= - \, \frac{|\hat{f}_k(\omega)|}{|\hat{x}_i(\omega)|} \, \sin(\theta_{ik}) \quad , \end{aligned}\right\} \tag{2.7}$$

where $\theta_{ik}(\omega)$ = phase shift between harmonic motion and (force) response; the symbolic notation Re[...] and Im[...] means 'real part of' and 'imaginary part of', respectively.

Due to linearity it holds good for a ship with $V_1, V_2 = 0$ that (ref. [46])

$$a_{jk}(\omega) = a_{kj}(\omega) \quad , \; b_{jk}(\omega) = b_{kj}(\omega) \quad , \; c_{jk} = c_{kj} \quad . \tag{2.8}$$

In the event of a ship with a given speed, in principle also in the horizontal plane hydrodynamic effects of the form $c_{jk}x_j$ can occur, so that in this case $c_{jk} \neq 0$ for $j,k = 1,2,6$.

The respective hydrodynamic coefficients in fact represent in-phase (with x_i and $\ddot{x}_i$) and out-of-phase (with $\dot{x}_i$) components of the hydrodynamic force. For the hydrodynamic damping coefficient $b_{ik}(\omega)$ this simply is obvious. The in-phase component includes two contributions: the added mass and the restoring force coefficient. In order to avoid ambiguity with respect to their determination, - by definition - c_{ik} now is conceived of as being frequency independent: c_{ik} is determined in the first place by the geometry of the hull of the ship, and further it may vary with the ship's speed. So, the indication of 'hydrostatic coefficient' for c_{jk} is not altogether right, since at a given ship's speed in general hydrostatic as well as hydrodynamic effects play a part. Nevertheless, in the generalized expression (2.1) c_{jk} in principle is maintained as a frequency-independent quantity. Accordingly c_{jk} is characteristic of the ship itself, so that

$$c_{jk} \equiv 0 \quad \text{for} \quad j,k = 1,2,6 \quad . \tag{2.9}$$

$a_{jk}(\omega)$ and $b_{jk}(\omega)$ are in the first place characteristics of the flow around the ship and therefore can be influenced by external conditions, such as the position of the bottom, the presence of a quay-wall, etc.
Now the hydrodynamic coefficients $a_{jk}(\omega)$ and $b_{jk}(\omega)$ formally can be determined by means of a harmonic analysis of the linear ship-fluid system making use of (2.6), (2.7) and (2.9).

2.4. Ship-fluid system in time domain

2.4.1. General description. The combination of ship plus fluid can be conceived of as a linear, time-invariant, stable, physical system: the

forces exerted somewhere upon the ship are regarded as input signals,
whereas the motion of the ship (displacement and rotation or derived
quantities) is considered to be the output signal. In consequence of the
'memory effect' associated with the influence(s) of the free surface (and
of the vorticity, respectively), it is necessary to represent the transient
ship motion - arising from a set of forces - in terms of a convolution
integral over the entire time history of the forcing functions. Thus the
three components of the horizontal motion have to be considered to be of
the general form (see refs. [77, 79, 78]):

$$u_j(t) = \sum_i^{1,2,6} \int_{-\infty}^{t} f_i(\tau) \, k_{ij}(t-\tau)d\tau = \sum_i^{1,2,6} \int_0^{\infty} k_{ij}(\tau)f_i(t-\tau)d\tau, \quad j=1,2,6, \quad (2.10)$$

where $u_j(t)$ = response of system in j-direction to set of input signals
$\{f_i(t)\}$,

$k_{ij}(t)$ = response for j-direction to unit pulse (i.e. Dirac function
at t=0) in i-direction = impulse response function (i.r.f.).

The time independence of the system implies that the system parameters
do not depend on t, i.e. the input-output relation does not change in time.
The 'black-box' approach according to (2.10) defines the characteristic
features of the system by means of the relation between input signal and
output signal.

Since a physical system is causal, it must hold good that (principle
of causality):

$$k_{ij}(t) \equiv 0 \quad \text{for} \quad t < 0 \quad , \tag{2.11}$$

i.e. the future behaviour of $f_i(\tau)$ for $\tau > t$ does not affect $u_j(t)$.

The requirement of stability implies that the difference between the
responses of the system to distinct excitations, for $t \to \infty$ always converges
to a finite value. In addition to the stability of the system, it is
naturally required that $|f_i(t)|$ remains bounded; then $u_j(t)$ is bounded too.
The i.r.f. $k_{ij}(t)$ is a real function of t which depends on the geometry of
the ship as well as on the boundaries of the fluid domain and its physical
properties. The matrix $\{k_{ij}(t)\}$ represents the 'memory effect' due to the
presence of the free surface and fully characterizes the response of the
ship to an arbitrary excitation. Apart from convergence of the

(convolution) integrals, the only assumption required in this is that the
ship-fluid system behaves linearly. The input signals need not be linear.

2.4.2. Stability. The ship-fluid system behaves stable, if at least the
following condition is met:

$$\lim_{t \to \infty} k_{ij}(t) = \text{constant} \quad . \tag{2.12}$$

A horizontal ship motion in the j-direction has a zero restoring force (see
(2.9)). This implies that the ship - in response to a unit pulse in the i-
direction at t=0 - does not return to its original position. Dependent on
the existence of damping in the system for the steady equilibrium
situation (t → ∞), two cases can be distinguished.
- Zero damping from the fluid for t → ∞: the velocity of the ship in the j-
  direction approaches asymptotically to a constant value in conformity
  with a certain equilibrium situation; the displacement in the j-direction
  does not remain bounded for t → ∞.
- Non-zero damping from the fluid for t → ∞: the velocity of the ship in
  the j-direction approaches to 0 for t → ∞; the corresponding displacement
  then approaches asymptotically to a certain constant ≠ 0.
Therefore the ship-fluid system anyhow behaves stably, if for all modes of
motion the velocity is considered to be the output signal, i.e.

$$u_j(t) = \dot{x}_j(t) \quad . \tag{2.13}$$

For an elaborate definition and a further explanation of the concept
'stability' in case of the linear ship-fluid system it is referred to ref.
[73].

2.4.3. Frequency response versus impulse response. The Fourier transform
of (2.10) combined with (2.13) yields a description of the linear ship-
fluid system in the frequency domain of the form:

$$\mathcal{F}\{\dot{x}_j(t)\} = i\omega \, \mathcal{F}\{x_j(t)\} = \sum_{i}^{1,2,6} \mathcal{F}\{f_i(t)\} \, \mathcal{F}\{k_{ij}(t)\} \quad , \tag{2.14}$$

where $\mathcal{F}\{f(t)\} = \int_{-\infty}^{\infty} f(\tau) \, e^{-i\omega\tau} d\tau$ = Fourier transform of f(t),

$\mathcal{F}\{k_{ij}(t)\} = K_{ij}(\omega)$ = harmonic transfer function for j-direction in response to (harmonic force) excitation in i-direction = frequency response function (f.r.f.).

As long as response and excitation can be considered as transient quantities, (2.14) represents per Fourier component the equations for arbitrary ship motions as a result of an arbitrary external force. (2.14) is only then a meaningful expression, if $f_i(t)$ is bounded in time and the ship-fluid system is stable, i.e. $k_{ij}(t \to \infty)$ = constant $\neq 0$, c.q. = 0 and $\dot{x}_j(t \to \infty)$ = constant $\neq 0$, c.q. = 0. These very properties of $f_i(t)$, $k_{ij}(t)$ and $\dot{x}_j(t)$ are an absolute requirement for the existence of their corresponding Fourier transforms in (2.14).

The description of the ship-fluid system in the frequency domain by (2.14) is equivalent to that by (2.4). As a result of the mapping per Fourier component by (2.14), in (2.14) and (2.4) $\mathcal{F}\{f_i(t)\}$ and $\hat{f}_k(\omega)$ for k = i as well as $\mathcal{F}\{x_j(t)\}$ and $\hat{x}_j(\omega)$ are identical. Now by putting

$$\left. \begin{array}{l} U_j(\omega) = \mathcal{F}\{\dot{x}_j(t)\} = i\omega\, \mathcal{F}\{x_j(t)\} = i\omega\, \hat{x}_j(\omega) \quad , \\[4mm] \\[4mm] F_k(\omega) = \mathcal{F}\{f_k(t)\} = \hat{f}_k(\omega) \quad , \end{array} \right\} \tag{2.15}$$

the following set of equations is generated:

$$\left. \begin{array}{ll} U_j(\omega) & = {}^{1,2,6}\!\!\!\sum_i F_i(\omega)\, K_{ij}(\omega) \quad , \\[6mm] F_k(\omega) & = {}^{1,2,6}\!\!\!\sum_j R^*_{jk}(\omega)\, U_j(\omega) \quad , \\[6mm] F_k(\omega) & = 0 \quad \text{for } k \neq i \quad , \end{array} \right\} \quad k = 1,2,6 \quad , \tag{2.16}$$

where on account of (2.5) and (2.9)

$$R^*_{jk}(\omega) = i\omega\{m_{jk}\, \delta_{jk} + a_{jk}(\omega)\} + b_{jk}(\omega) \quad , \quad j,k = 1,2,6 \quad , \tag{2.17}$$

with δ_{jk} = Kronecker delta: δ_{jk} = 1 for j = k, δ_{jk} = 0 for j ≠ k; the superscript * indicates that the quantity concerned is a reduced version of its original.

As indicated in ref. [74] in principle $K_{ij}(\omega)$ is to be solved from (2.16) and expressed in $R^*_{jk}(\omega)$, so that $K_{ij}(\omega)$ becomes a function of $a_{jk}(\omega)$ and $b_{jk}(\omega)$. The i.r.f. $k_{ij}(t)$ then can be determined as the inverse Fourier transform of $K_{ij}(\omega)$. If this inverse Fourier transform exists in a general sense, i.e. in terms of the generalized function theory, the f.r.f. and the i.r.f. are related by a Fourier transform (refs. [80, 81]):

$$K_{ij}(\omega) = \int_{-\infty}^{\infty} k_{ij}(\tau)e^{-i\omega\tau}d\tau = \int_{0}^{\infty} k_{ij}(\tau)e^{-i\omega\tau}d\tau \quad , \qquad (2.18^a)$$

or,

$$k_{ij}(t) = \frac{1}{2\pi} \int_{-\infty}^{\infty} K_{ij}(\omega)e^{i\omega t}d\omega \quad . \qquad (2.18^b)$$

Since $k_{ij}(t)$ is a real and causal time function it can be derived:

$$k_{ij}(t) = \frac{2}{\pi} \int_{0}^{\infty} \text{Re}[K_{ij}(\omega)] \cos(\omega t)d\omega =$$

$$= -\frac{2}{\pi} \int_{0}^{\infty} \text{Im}[K_{ij}(\omega)]\sin(\omega t)d\omega \qquad \text{for} \quad t > 0 \quad ,$$

$$k_{ij}(0) = \frac{1}{\pi} \int_{0}^{\infty} \text{Re}[K_{ij}(\omega)]d\omega = \frac{1}{2} k_{ij}(0^+) \quad ,$$

$$k_{ij}(t) \equiv 0 \qquad \text{for } t < 0 \quad .$$

$$\left. \right\} \qquad (2.18^c)$$

In determining the integrals in (2.18^c) any singularities occurring in the functions $K_{ij}(\omega)$ may play a part. The steady state as attained for $t \to \infty$ in case of transient motions corresponds in the frequency domain with $\omega = 0$. For ship motions without restoring force (i.e. c_{ij} = 0) $K_{ij}(\omega)$ may show a singular behaviour dependent on the existence of damping in the ship-fluid system for $\omega \to 0$. With zero damping from the water for $\omega \to 0$, i.e. $b_{ij}(\omega \to 0) = 0$, $K_{ij}(\omega)$ appears to behave singularly for $\omega = 0$; the system is stable with $\dot{x}_j(t \to \infty) = $ constant $\neq 0$; this case corresponds with $k_{ij}(\infty)$ = constant $\neq 0$, i.e.

$$\int_0^\infty |\,k_{ij}(t)\,|dt$$

does not exist. With non-zero damping from the water for $\omega \to 0$, i.e. $b_{ij}(\omega \to 0) \neq 0$, $K_{ij}(\omega)$ is a regular function; the system is asymptotically stable with $\dot{x}_j(t \to \infty) = 0$; this case corresponds with $k_{ij}(\infty) = 0$, i.e.

$$\int_0^\infty |\,k_{ij}(t)\,|dt$$

does exist.

2.5. Determination of impulse response function

Besides by means of an approach as presented in ref. [73] using the generalized function theory (see refs. [81, 82]), expressions for the f.r.f. and the i.r.f. can also be determined by applying certain aspects of the theory of Laplace transforms related to the Fourier integral of a causal function (ref. [74]).
Let the Laplace transform associated with the generalized f.r.f. $K_{ij}(\omega)$ be denoted by $H_{ij}(s)$, which function on account of ref. [77] can be understood as the general transfer function for the j-direction in response to a (force) excitation in the i-direction:

$$H_{ij}(s) = \mathcal{L}\{k_{ij}(t)\} = \int_0^\infty k_{ij}(\tau)e^{-s\tau}d\tau \tag{2.19}$$

where $s = \lambda + i\omega$ = complex variable with $\text{Re}[s] = \lambda$, $\text{Im}[s] = \omega$,

$$\mathcal{L}\{f(t)\} = \int_0^\infty f(t)e^{-s\tau}d\tau = \text{unilateral Laplace transform of } f(t) \text{ with region of existence } (\text{Re}[s] > \text{Re}[s_1]),$$

s_1 = certain complex number.

If $\mathcal{L}\{f(t)\}$ does exist and $f(t)$ has a limit for $t \to \infty$ then it holds (ref. [83]):

$$\lim_{s \downarrow 0} s\, \mathcal{L}\{f(t)\} = \lim_{t \to \infty} f(t) \quad .$$

Making use of this lemma one obtains:

$$\lim_{t \to \infty} k_{ij}(t) = K_{ij}(\infty) = \lim_{s \downarrow 0} s\, \mathcal{L}\{k_{ij}(t)\} = \lim_{s \downarrow 0} s\, H_{ij}(s) \quad . \tag{2.20}$$

Now the Fourier transforms in (2.16) can be replaced by Laplace transforms. Hereby it has to be borne in mind that only then a set of meaningful expressions is generated, if the respective Laplace transforms of $\dot{x}_j(t)$, $f_i(t)$ and $k_{ij}(t)$ do exist and at least whether the Laplace transform of $f_i(t)$ or that of $k_{ij}(t)$ in the right-hand member of the Laplace transformed convolution integral converges absolutely. Then in (2.16) ω replacing by s, at first $U_j(s)$ is eliminated for the respective cases i=k=1, i=k=2 and i=k=6. Making use of the fact that – due to the extant symmetry of usual ship forms – there is only a coupling between swaying and yawing, it follows for $H_{ij}(s)$:

$$H_{11}(s) = \frac{1}{R_{11}^*(s)} \quad ,$$

$$\left. \begin{array}{l} H_{1i}(s) = H_{1j}(s) = 0 \quad , \\[2ex] H_{i1}(s) = H_{j1}(s) = 0 \quad , \end{array} \right\} \text{ for } i=2,\ j=6 \quad ,$$

$$\left. \begin{array}{l} H_{ii}(s) = \dfrac{1}{R_{ii}^*(s) - \dfrac{R_{ij}^*(s)R_{ji}^*(s)}{R_{jj}^*(s)}} \quad , \\[5ex] H_{ij}(s) = \dfrac{1}{R_{ji}^*(s) - \dfrac{R_{ii}^*(s)R_{jj}^*(s)}{R_{ij}^*(s)}} \quad , \end{array} \right\} \text{ for } i=2,\ j=6 \text{ and } i=6,\ j=2, \tag{2.21}$$

where $R_{ij}^*(s)$ has the form:

$$R_{ij}^{*}(s) = s\{m_{ij}\delta_{ij} + a_{ij}(s)\} + b_{ij}(s) \quad ,$$

$$\left.\begin{array}{l} \\ \\ \\ R_{12}^{*}(s) = R_{21}^{*}(s) = R_{16}^{*}(s) = R_{61}^{*}(s) = 0 \quad , \end{array}\right\} \quad i,j=1,2,6. \qquad (2.22)$$

Since (2.4) combined with (2.9) describes the linear ship-fluid system in the 'real frequency' or ω-domain, formally $H_{ij}(s)$ - being a function of the 'complex frequency' s - is only known for $s = i\omega$. Therefore it is obvious to work with its related (generalized) Fourier transform of $k_{ij}(t)$, viz. $K_{ij}(\omega)$.

With respect to the relation between $H_{ij}(s)$ and $K_{ij}(\omega)$ generally three cases can be considered. The region of existence of $H_{ij}(s)$ was denoted by Re[s] > Re[s₁]. If the region of convergence of $H_{ij}(s)$ contains the $i\omega$-axis in its interior, i.e. if Re[s₁] < 0 (first case), then

$$K_{ij}(\omega) = H_{ij}(s)\big|_{s=i\omega} \quad . \qquad (2.23)$$

This case applies generally to modes of motion with a restoring force, implies that $k_{ij}(\infty) = 0$ and is left further out of consideration.

If the $i\omega$-axis is outside the region of convergence of $H_{ij}(s)$, i.e. if Re[s₁] > 0 (second case), then $K_{ij}(\omega)$ does not exist: the function $k_{ij}(t)$ has no Fourier transform in terms of the generalized function theory and the linear ship-fluid system behaves unstably.

The last case is Re[s₁] = 0; the function $H_{ij}(s)$ is analytic for Re[s] > 0, but at least one of the singular points lies on the $i\omega$-axis. This case applies to modes of motion without restoring force, implies that $k_{ij}(\infty)$ = constant $\neq$ 0 and is connected with singularities (poles) in $H_{ij}(s)$ for s=0, i.e. λ=0, $i\omega$=0. From a physical point of view a pole for s=iω=0 means that, in case of a translation (rotation) in the horizontal plane with constant (rotational) velocity, i.e. ω=0, no force (moment) is required (see ref. [73]). In this context, now, a function $H_{ij}(s)$ is considered with n simple poles, $i\omega_1$, $i\omega_2$, ..., $i\omega_n$ and no other singularities in the half plane Re[s] $\geq$ 0. This function then can be written as (ref. [80])

$$H_{ij}(s) = G_{ij}(s) + \sum_{m=1}^{n} \alpha_{ij}^{(m)}(s - i\omega_m)^{-1} \quad ,$$

in which $G_{ij}(s)$ = a function free from singularities for Re[s] ≥ 0,

$$\alpha_{ij}^{(m)} = \lim_{s \to i\omega_m} (s - i\omega_m) H_{ij}(s) \quad .$$

From (2.23) it follows that the Fourier transform corresponding to $G_{ij}(s)$ is given by $G_{ij}(i\omega)$; therefore (ref. [80])

$$K_{ij}(\omega) = H_{ij}(s)\big|_{s=i\omega} + \pi \sum_{m=1}^{n} \alpha_{ij}^{(m)} \delta(\omega - \omega_m) \quad ,$$

where $\delta(\omega)$ = delta function or Dirac function.

With $i\omega_m = 0$ and n=1 this leads directly to an expression for the generalized f.r.f., viz.

$$K_{ij}(\omega) = H_{ij}(i\omega) + \pi \alpha_{ij} \delta(\omega) \quad , \tag{2.24}$$

where

$$H_{ij}(i\omega) = H_{ij}(s)\big|_{s=i\omega} \quad , \tag{2.25^a}$$

$$\alpha_{ij} = \lim_{s \downarrow 0} s\, H_{ij}(s) \quad . \tag{2.25^b}$$

On account of (2.20) it also holds good that

$$\alpha_{ij} = k_{ij}(\infty) \quad . \tag{2.25^c}$$

In this context $H_{ij}(i\omega)$ is to be conceived of as the non-generalized f.r.f. From (2.21), (2.22) and (2.25^a) it can be derived for $H_{ij}(i\omega)$:

$$H_{11}(i\omega) = \frac{b_{11}(\omega)}{\{m_{11}+a_{11}(\omega)\}^2 \omega^2 + b_{11}^2(\omega)} - i\, \frac{\omega\{m_{11}+a_{11}(\omega)\}}{\{m_{11}+a_{11}(\omega)\}^2 \omega^2 + b_{11}^2(\omega)} \quad , \Big]$$

$$H_{1i}(i\omega) = H_{1j}(i\omega) = 0 \quad , \;\Big]$$

$$\qquad\qquad\qquad\qquad\qquad \Big\} \quad \text{for } i=2,\ j=6$$

$$H_{11}(i\omega) = H_{j1}(i\omega) = 0 \quad , \;\Big]$$

$$H_{ii}(i\omega) = \frac{p_{ii}(\omega)b_{jj}(\omega)+\omega q_{ii}(\omega)\{m_{jj}+a_{jj}(\omega)\}}{p_{ii}^2(\omega) + q_{ii}^2(\omega)} \quad +$$

$$- i \frac{q_{ii}(\omega)b_{jj}(\omega)-\omega p_{ii}(\omega)\{m_{jj}+a_{jj}(\omega)\}}{p_{ii}^2(\omega) + q_{ii}^2(\omega)} \quad ,$$

for i=2,j=6
and i=6,j=2,

$$H_{ij}(i\omega) = \frac{p_{ij}(\omega)b_{ij}(\omega)+\omega q_{ij}(\omega)a_{ij}(\omega)}{p_{ij}^2(\omega) + q_{ij}^2(\omega)} \quad +$$

$$- i \frac{q_{ij}(\omega)b_{ij}(\omega)-\omega p_{ij}(\omega)a_{ij}(\omega)}{p_{ij}^2(\omega) + q_{ij}^2(\omega)} \quad ,$$

(2.26)

with

$$p_{ii}(\omega) = b_{ii}(\omega)b_{jj}(\omega) - b_{ij}(\omega)b_{ji}(\omega) +$$

$$- \omega^2 [\{m_{ii}+ a_{ii}(\omega)\}\{m_{jj}+a_{jj}(\omega)\} - a_{ij}(\omega)a_{ji}(\omega)],$$

$$q_{ii}(\omega) = \omega[b_{ii}(\omega)\{m_{jj}+ a_{jj}(\omega)\} + b_{jj}(\omega)\{m_{ii}+ a_{ii}(\omega)\} +$$

$$- \{a_{ji}(\omega)b_{ij}(\omega)+ a_{ij}(\omega)b_{ji}(\omega)\}] \quad ,$$

for (2.27)
i=2,j=6;

$$p_{ii}(\omega) = p_{jj}(\omega) \quad ,$$

$$q_{ii}(\omega) = q_{jj}(\omega) \quad ,$$

$$p_{ij}(\omega) = p_{ji}(\omega) = - p_{ii}(\omega) \quad ,$$

$$q_{ij}(\omega) = q_{ji}(\omega) = - q_{ii}(\omega) \quad ,$$

further $m_{11} = m_{22}$.

Any possible singularities in $H_{ij}(i\omega)$ for $\omega \neq 0$ due to particular combinations of $a_{jk}(\omega)$ and $b_{jk}(\omega)$ are beforehand excluded. If they nevertheless should be present, they probably lie in the left half-plane; the presence of hydrodynamic damping points that way. From a physical point of view $a_{jk}(\omega)$ and $b_{jk}(\omega)$ must be even functions of ω. In a more general sense this can be derived mathematically from the fact that $Re[K_{ij}(\omega)]$ is an even function of ω and $Im[K_{ij}(\omega)]$ an odd function. It is an obvious supposition, affirmed by refs. [46, 54, 58, 66] that $a_{jk}(0) =$ finite $\neq 0$ and $b_{jk}(0) = 0$. For (very) small values of ω $a_{jk}(\omega)$ and $b_{jk}(\omega)$ then can be represented by

$$\left. \begin{array}{l} a_{jk}(\omega) = a_{jk}(0) + a_{jk}^{(2)}\omega^2 + O(\omega^4) \ , \\[2em] b_{jk}(\omega) = \qquad\qquad b_{jk}^{(2)}\omega^2 + O(\omega^4) \ , \end{array} \right\} \quad \text{for } \omega \to 0 \quad , \qquad (2.28)$$

respectively,

where $a_{jk}^{(n)}$ = coefficient of term with order n in power series development
        for $a_{jk}(\omega)$,

    $b_{jk}^{(n)}$ = coefficient of term with order n in power series development
        for $b_{jk}(\omega)$.

On account of (2.28) - bearing in mind that $k_{ij}(t)$ and $b_{ij}(\omega)$ are real functions - it must hold good that

$$\lim_{s \downarrow 0} \frac{b_{ij}(s)}{s} = 0 \quad .$$

Appying this result, combination of (2.21), (2.22) and (2.25^{b,c}) yields for α_{ij}:

$$\alpha_{11} = k_{11}(\infty) = \frac{1}{m_{11}+a_{11}(0)} \quad ,$$

$$\left. \alpha_{11} = k_{1i}(\infty) = \alpha_{1j} = k_{1j} \ \infty) = 0 \quad , \right\} \quad \text{for i=2, j=6,}$$

$$\alpha_{i1} = k_{i1}(\infty) = \alpha_{j1} = k_{j1}(\infty) = 0 \quad , \Bigg] \qquad (2.29)$$

$$\alpha_{ii} = k_{ii}(\infty) = \frac{m_{jj} + a_{jj}(0)}{\{m_{ii}+a_{ii}(0)\}\{m_{jj}+a_{jj}(0)\}-a_{ij}(0)a_{ji}(0)} \quad ,$$

for $i=2, j=6$ and $i=6, j=2$.

$$\alpha_{ij} = k_{ij}(\infty) = \frac{-a_{ij}(0)}{\{m_{ii}+a_{ii}(0)\}\{m_{jj}+a_{jj}(0)\}-a_{ij}(0)a_{ji}(0)} \quad ,$$

The fact that the respective expressions for $\alpha_{ij} = k_{ij}(\infty)$ are independent of b_{ij} is caused by the parabolic behaviour of $b_{ij}(\omega)$ near by the point $\omega=0$. Since it holds good that $k_{ij}(\infty) = $ constant $\neq 0$, the linear ship-fluid system indeed behaves stably in the case under consideration (see ref. [73]).

The real and imaginary part of the generalized f.r.f. $K_{ij}(\omega)$ presented in (2.24) reads as

$$Re[K_{ij}(\omega)] = Re[H_{ij}(i\omega)] + \pi\alpha_{ij}\delta(\omega) \quad ,$$

$$Im[K_{ij}(\omega)] = Im[H_{ij}(i\omega)] \quad ,$$

(2.30[a,b])

respectively. From (2.26),(2.27) together with (2.28) it is to be derived that

$$\lim_{\omega\to0} Re[H_{ij}(i\omega)] = \text{finite} \quad \text{and} \quad \lim_{\omega\to0} Im[H_{ij}(i\omega)] = \infty \quad ;$$

since for $\omega\to\infty$ it can be shown that $a_{jk}(\infty)$ remains finite and $b_{jk}(\infty)$ asymptotically tends to zero (see e.g. refs. [85, 84, 86, 54, 60, 73]), it holds at the same time

$$\lim_{\omega\to\infty} Re[H_{ij}(i\omega)] = 0 \quad \text{and} \quad \lim_{\omega\to\infty} Im[H_{ij}(i\omega)] = 0 \quad ;$$

therefore

$$\int\limits_0^\infty \text{Re}[K_{ij}(\omega)]d\omega$$

does converge absolutely and

$$\int\limits_0^\infty \text{Im}[K_{ij}(\omega)]d\omega$$

does not. Then, on account of the lemma of Riemann-Lebesgue (refs. [87, 83]) it holds good that

$$\lim_{t\to\infty} \int\limits_0^\infty \text{Re}[H_{ij}(i\omega)] \cos(\omega t)d\omega = 0$$

so that (2.18^C), using (2.30^a), (2.25^C) and

$$\int\limits_0^\infty \delta(\omega) \cos(\omega t)d\omega = \frac{1}{2}$$

indeed leads to

$$\lim_{t\to\infty} k_{ij}(t) = k_{ij}(\infty) = \text{constant} \neq 0 \quad.$$

Due to the behaviour of $\text{Im}[H_{ij}(i\omega)]$ for $\omega \to 0$, the determination of $k_{ij}(t)$ using (2.30^b) requires an asymptotic expansion for small values of ω. In evaluating the i.r.f., therefore in (2.18^C) the expression containing $\text{Im}[K_{ij}(\omega)]$ is left further out of consideration. Naturally, in the limit for $t \to \infty$ (2.18^C) and (2.30^b) combined must yield $k_{ij}(\infty)=\text{constant}\neq 0$.

The i.r.f. $k_{ij}(t)$ now can be evaluated by means of (2.18^C) with $\text{Re}[K_{ij}(\omega)]$ given by (2.30^a). $\text{Re}[H_{ij}(i\omega)]$ and $\alpha_{ij} = k_{ij}(\infty)$ occurring in the expression for $\text{Re}[K_{ij}(\omega)]$ are to be determined according to (2.26), (2.27) and (2.29), respectively. A necessary condition then is that $a_{ij}(\omega)$ and $b_{ij}(\omega)$ are known functions of ω.

2.5.1. Special cases. A special case arises when the hydrodynamic coefficients are independent of V_1 and V_2; this situation occurs for instance, with (very) small values of V_1 and V_2. For this case it can be shown that the sequence of the respective subscripts i,j - representing directions - may be mutually interchanged, which eventually induces simplified expressions for the i.r.f.

A further simplification is achieved if at the same time the modes of motion of the ship are supposed to be uncoupled. Then it can be derived from the above for i=1,2,6:

- description of the linear ship-fluid system in the frequency domain:

$$\{m_{ii} + a_{ii}(\omega)\}\ddot{x}_i + b_{ii}(\omega)\dot{x}_i = f_i(t) \quad ; \tag{2.31}$$

- description of the linear ship-fluid system in the time domain:

$$\dot{x}_i(t) = \int_{-\infty}^{t} f_i(\tau)k_{ii}(t-\tau)d\tau = \int_{0}^{\infty} k_{ii}(\tau)f_i(t-\tau)d\tau \quad ; \tag{2.32}$$

- impulse response function:

$$k_{ii}(t) = \frac{2}{\pi} \int_{0}^{\infty} Re[K_{ii}(\omega)] \cos(\omega t)d\omega =$$

$$= -\frac{2}{\pi} \int_{0}^{\infty} Im[K_{ii}(\omega)] \sin(\omega t)d\omega \quad \text{for } t > 0,$$

$$k_{ii}(0) = \frac{1}{\pi} \int_{0}^{\infty} Re[K_{ii}(\omega)]d\omega = \frac{1}{2} k_{ii}(0^+) \quad ,$$

$$k_{ii}(t) \equiv 0 \quad \text{for } t < 0 \quad ; \tag{2.33}$$

- generalized frequency response function:

$$K_{ii}(\omega) = H_{ii}(i\omega) + \pi \alpha_{ii} \delta(\omega) \quad , \tag{2.34}$$

with

$$Re[K_{ii}(\omega)] = Re[H_{ii}(i\omega)] + \pi \alpha_{ii} \delta(\omega) \quad , \left.\vphantom{\begin{array}{c}1\\1\\1\\1\end{array}}\right\}$$

$$Im[K_{ii}(\omega)] = Im[H_{ii}(i\omega)] \quad ; \left.\vphantom{\begin{array}{c}1\\1\end{array}}\right\} \qquad (2.35^{a,b})$$

- non-generalized frequency response function:

$$H_{ii}(i\omega) = \frac{b_{ii}(\omega)}{\{m_{ii}+a_{ii}(\omega)\}^2 \omega^2 + b_{ii}^2(\omega)} - i \frac{\omega\{m_{ii}+a_{ii}(\omega)\}}{\{m_{ii}+a_{ii}(\omega)\}^2 \omega^2 + b_{ii}^2(\omega)} \quad ; \qquad (2.36)$$

- $\alpha_{ii} = k_{ii}(\infty)$:

$$\alpha_{ii} = k_{ii}(\infty) = \frac{1}{m_{ii}+a_{ii}(0)} \quad . \qquad (2.37)$$

For the case under consideration, with negligible influence of V_1, V_2 on the hydrodynamic coefficients, the uncoupling of the ship motions can be materialized by schematizing the ship to a rigid, prismatic body with a rectangular cross-section and a symmetrical distribution of mass. Besides, in case of shallow water the uncoupling of the motions requires a horizontal bottom. When a closed wall is present, the ship motions are only uncoupled if one of the horizontal body axes of the (schematized) ship is parallel to the wall.

2.6. Causality

Like each physical system also the linear ship-fluid system is a causal system, which finds expression both in the frequency domain and in the time domain.

The respective hydrodynamic coefficients of the mass term and the damping force are mutually dependent, since they are to be derived from one and the same physical quantity, viz. the distribution of the hydrodynamic stress on the wet ship hull. Considering, for instance, a ship-fluid system with a total number of n hydrodynamic coefficients, this implies that n/2 relations exist to be satisfied by $a_{ij}(\omega)$ and $b_{ij}(\omega)$. The mutual dependence in the frequency domain mentioned above is equivalent to the causal behaviour of the ship-fluid system in the time domain. This very attribute

gives rise to the so-called 'memory effect' materialized by the wave
radiation at the free water surface: only waves already generated (i.e.
'the past') do influence the interaction between the moving ship and the
surrounding water, the waves to be generated (i.e. 'the future') do not. It
can be established that the memory effect in the frequency domain is
expressed by the frequency dependence of the hydrodynamic coefficients, and
that this frequency dependence in its turn is due to the wave radiation at
the free water surface.

Regarding the general description of the linear ship-fluid system in
the time domain (2.10) and (2.13) combined, it is observed that the memory
effect is represented by means of the convolution integrals and especially
by the i.r.f. $k_{ij}(t)$. Of great importance for the relation between the
respective system descriptions in the frequency domain and the time domain
is the property that the system behaves causally, which finds expression in
the real, causal time function $k_{ij}(t)$, viz. $k_{ij}(t) \equiv 0$ for $t < 0$. In fact,
causality conditions for the real and imaginary part of $K_{ij}(\omega)$ should be
introduced as from the first expression in (2.16), since this expression in
itself passes over the information embedded in (2.11). These very
conditions should be taken along throughout all further derivations until
(2.18^C), where they allow the inverse Fourier transform applied. (2.18^C)
then directly expresses the causality conditions and is equivalent to the
explicit relations between the real and imaginary part of the (generalized)
f.r.f. of a causal system referred to as the so-called Hilbert transforms
(refs. [80, 81]), in their modified form also indicated as the Kramers-
Kronig relations (refs. [67, 46]); see just as well refs. [88, 89]. As a
matter of fact, it generally applies that for a causal function explicit
relations exist between the real and imaginary part of its Fourier
transform. Such formulae are obtainable whenever the system response obeys
a linear law and there is a clear causality relation between input and
output.

Considering (2.4) together with (2.5) and (2.9) it is stated that the
real and imaginary part of $R_{ik}(\omega)$ have to satisfy the Hilbert transforms
too. The causal time function associated with $R(\omega)$ is now indicated as
retardation function and the description of the linear ship-fluid system in
the time domain has the form of an integro-differential equation for the
motion variable $x(t)$ (see (1.8) and further refs. [68, 67, 58]). This

illustrates that the eventual form of the causality relations is dependent on the way of describing the linear ship-fluid system in the time domain. Starting from $R_{ik}(\omega)$ as Fourier transform, the Hilbert transforms lead to explicit relations between the respective hydrodynamic coefficients $a_{ik}(\omega)$ and $b_{ik}(\omega)$ (see refs. [67, 46]); due to its black-box approach the i.r.f.- technique does not.

2.7. Concluding remarks

From the foregoing it is obvious that the main interest concerns the i.r.f., since the determination of transient ship motions requires knowledge of the behaviour of this very quantity. In this context the following can be remarked.

The respective descriptions of the linear ship-fluid system in the time domain and the frequency domain are completely equivalent. Both can be used in order to define the response to transient disturbances; there is no specific advantage attached to either of them. If the linear ship-fluid system has been formulated mathematically the i.r.f. and/or the f.r.f. can be evaluated, but if this is not possible they can also be determined in an experimental way.

The f.r.f. can be determined experimentally using a harmonically varying input signal. This actually amounts to direct determination of the hydrodynamic coefficients as functions of the frequency. Since the i.r.f. and the f.r.f. are related by means of a Fourier transform, a mere determination of the f.r.f. is sufficient. Direct determination of an i.r.f., however, would be far more efficient: a few experiments, using a pulse or/and an arbitrary function of time, are sufficient. On the other hand many tests have to be carried out in order to find the f.r.f. over a sufficiently long interval of the frequency. Compared with transient pulse tests, experiments to determine f.r.f. are much easier, since the pulse technique presents more specific problems and demands a higher degree of accuracy of the measuring equipment. For these reasons the choice in favour of a determination of the f.r.f. is obvious.

ON THE PREDICTION OF FENDER FORCES AT BERTHING STRUCTURES
Part II: SHIP BERTHING RELATED TO FENDER STRUCTURE

H.L. Fontijn,
Senior Scientific Officer, Department of Civil Engineering, Delft
University of Technology, The Netherlands.

3. SHIP BERTHING TO OPEN FENDER STRUCTURE

3.1. Introduction

This section deals with the application of the impulse response
function-technique to the situation of a ship berthing to an open jetty-
type fender structure.

The assumptions and simplifications made for this specific case have
already been presented in Section 1.3.4; it is obvious that, due to the
schematization of the ship, any coupling between the sway and yaw mode of
motion does not exist.

Since the i.r.f.-technique as applied requires knowledge of the
hydrodynamic coefficients, these quantities are determined theoretically as
well as experimentally for the respective cases of pure swaying and yawing
at zero forward and transverse speed (i.e. $V_1, V_2 = 0$). Hereby it is assumed
implicitly that in berthing the transverse velocity of the ship is so small
that it does not affect the hydrodynamic coefficients and a restoring force
is not generated. With regard to the theoretical determination of the
hydrodynamic coefficients, the fluid motion is supposed to be two-
dimensional (strip-theory); further it is assumed that the fluid is
inviscid and moves irrotationally.

The theoretical results for the hydrodynamic coefficients as well as for
the simulated berthing operations were verified by model tests.

A detailed account of theory and experiments is given in refs [54, 73]; see
also ref. [74].

E. Bratteland (ed.), Advances in Berthing and Mooring of Ships and Offshore Structures, 95–166.
© 1988 by Kluwer Academic Publishers.

3.2. Hydrodynamic coefficients and i.r.f.

The ship motions take place with respect to the $ox_1 x_2 x_3$-co-ordinate system, which is now space-fixed. At rest the ox_3-axis coincides with the ship's longitudinal plane of symmetry. The keel clearance of the ship is constant and the side-walls maintain a vertical position. The assumption of a two-dimensional fluid motion implies that merely motions in planes perpendicular to the ship's longitudinal plane of symmetry are considered and that the calculations relate to the unit length, i.e. a strip-theory approach is applied.

The water depth at rest (i.e. the mean water level) is represented by h, the draught, the beam and the length of the ship by D, B and L, respectively. The respective fluid velocities in the x_2- and x_3-direction are denoted by v and w. The fluid domain can be divided into three regions; the subscripts a, b and c are used to indicate that the dependent variables concerned must be related to these respective regions (see fig. 3.1).

On account of the above the mathematical approach may be formulated in terms of a velocity potential $\Phi = \Phi(x_2, x_3, t)$. The horizontal and vertical velocity component of a fluid particle with co-ordinates x_2, x_3 at time t is

$$v = \frac{\partial \Phi}{\partial x_2} \quad , \quad w = \frac{\partial \Phi}{\partial x_3} \quad , \tag{3.1}$$

respectively. The velocity potential must satisfy the Laplace equation

$$\nabla^2 \Phi = \frac{\partial^2 \Phi}{\partial x_2{}^2} + \frac{\partial^2 \Phi}{\partial x_3{}^2} = 0 \tag{3.2}$$

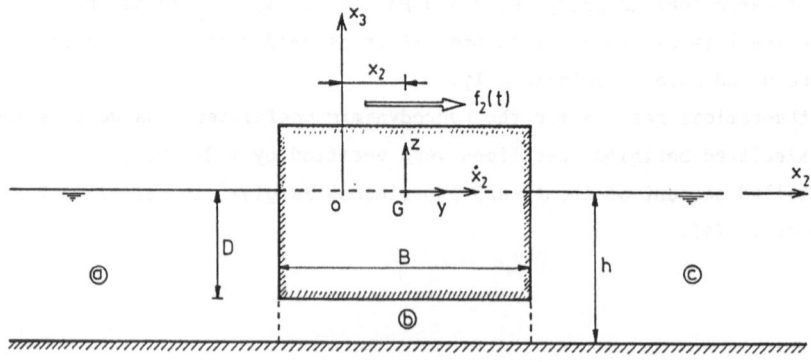

FIGURE 3.1 - Definition sketch for sway motion on horizontally unrestricted water.

in the field of flow, subject to relevant boundary conditions on all boundary surfaces and at infinity.

On each side of the ship a velocity potential exists. As only small ship motions are considered, these respective velocity potentials are antimetric. Coupling of the fields of flow on both sides of the ship is done by applying the law of conservation of momentum to the mass of water in the keel clearance. As a consequence, it is sufficient to determine the velocity potential only on one side of the ship.

Now define a fluid region R coinciding with region c, in which the Laplace equation is to be solved:

$$\left. \begin{array}{l} x_2 > \frac{1}{2}B \quad , \quad -h \leq x_3 \leq 0 \quad , \\[2ex] x_2 = \frac{1}{2}B \quad , \quad -D \leq x_3 \leq 0 \quad , \quad -h \leq x_3 < -D \quad . \end{array} \right\} R \qquad (3.3)$$

Ignoring surface tension the free-surface boundary condition reads – in linearized form – as:

$$\frac{\partial^2 \Phi}{\partial t^2} + g \frac{\partial \Phi}{\partial x_3} = 0 \quad \text{on} \quad x_2 \geq \frac{1}{2}B \quad , \quad x_3 = 0 \quad , \qquad (3.4)$$

where g = acceleration due to gravity.

In region b, the keel clearance, it is assumed that B >> h-D. Pressure gradients or accelerations in the x_2-direction then being large as compared with the corresponding quantities in the x_3-direction, this leads to $w_b = 0$, a hydrostatic pressure and a uniform velocity distribution over the height. The boundary condition on the plane $x_2 = \frac{1}{2}B$ now can be written as:

$$\frac{\partial \Phi}{\partial x_2} = \{U(x_3+D) - U(x_3)\}\dot{x}_2 + \{U(x_3+h) - U(x_3+D)\}v_b \quad \text{on} \quad x_2 = \frac{1}{2}B \quad , \qquad (3.5)$$

where U(x) = unit step or Heaviside function, defined as

$$U(x) = \begin{cases} 0 \text{ for } x < 0 \quad , \\[1ex] \frac{1}{2} \text{ for } x = 0 \quad , \\[1ex] 1 \text{ for } x > 0 \quad , \end{cases}$$

v_b = fluid velocity in x_2-direction in region b.

The assumption of an impermeable bottom leads to the boundary condition:

$$\frac{\partial \Phi}{\partial x_3} = 0 \quad \text{on} \quad x_2 \geq \frac{1}{2}B \quad , \quad x_3 = -h \quad . \qquad (3.6)$$

The boundary condition at infinity states that

$$\Phi(x_2,x_3,t)\big|_{x_2 \to \infty} \longrightarrow \text{ outgoing dispersive wave} \quad , \tag{3.7}$$

or, at infinity only simple-harmonic waves propagating in positive x_2-direction are permissible.

As a supplementary condition it is supposed that in region R the function $\Phi(x_2,x_3,t)$ together with its respective first partial derivatives remain finite:

$$\Phi(x_2,x_3,t) \quad , \quad \Phi^1(x_2,x_3,t) \text{ being finite in R} \quad ; \tag{3.8}$$

the superscript 1 means first partial derivative.

The hydrodynamic problem as formulated above now is solved for the specific case that a simple-harmonic motion is imposed on the ship, viz.

$$x_2(t) = -i\hat{x}_2 e^{i\omega t} \quad , \tag{3.9}$$

where $\hat{x}_2$ = amplitude of ship motion in sway ($\hat{x}_2>0$, limited and real). Since the ship motion is harmonic in time, the fluid velocity in the underkeel clearance has to be of the form:

$$v_b(t) = \hat{v}_b e^{i(\omega t-\theta)} \quad , \tag{3.10}$$

where $\hat{v}_b$ = amplitude of underkeel fluid velocity ($\hat{v}_b>0$, limited and real),
θ = phase shift between underkeel fluid velocity and sway velocity of ship (θ=constant and real).

Now the velocity potential $\Phi(x_2,x_3,t)$ is a simple-harmonic function of time, which has to satisfy the Laplace equation (3.2) plus the set of non-homogeneous boundary conditions (3.4), (3.5), (3.6) and (3.7) and the supplementary condition (3.8), with $x_2(t)$ and $v_b(t)$ prescribed according to (3.9) and (3.10), respectively. The solution of (3.4), (3.5), (3.6) and (3.7) specifies a mixed boundary-value problem for the Laplacian (3.2). From ref. [54] the general solution for $\Phi(x_2,x_3,t)$ is seen to be given by:

$$\Phi(x_2,x_3,t) = i \frac{\omega}{m_0}(A_0+B_0 e^{-i\theta})\cosh\{m_0(x_3+h)\}e^{i(\omega t-m_0 x_2+\frac{1}{2}m_0 B)} +$$

$$- \sum_{n=1}^{\infty} \frac{\omega}{m_n} (A_n + B_n e^{-i\theta}) e^{-m_n x_2 + \frac{1}{2} m_n B} \cos\{m_n (x_3 + h)\} e^{i\omega t} \quad ; \qquad (3.11)$$

the expressions for m_0, m_n, A_0, A_n, B_0 and B_n are given in Appendix A.

By means of the linearized equation of Bernoulli for unsteady flow

$$\frac{\partial \Phi}{\partial t} + \frac{p}{\rho} + g x_3 = 0 \quad ,$$

where p = fluid pressure,

ρ = specific mass density of fluid,

it can be derived for the total horizontal hydrodynamic force on the ship per unit length, $f_{2,s}(t)$ - making use of the antimetry of the respective fields of flow in the regions a and c:

$$f_{2,s}(t) = -2\rho \frac{\omega^2}{m_0^2} (A_0 + B_0 e^{-i\theta})\{\sinh(m_0 h) - \sinh\{m_0 (h-D)\}\} e^{i\omega t} +$$

$$-2\rho \sum_{n=1}^{\infty} i \frac{\omega^2}{m_n^2} (A_n + B_n e^{-i\theta})\{\sin(m_n h) - \sin\{m_n (h-D)\}\} e^{i\omega t} \quad . \qquad (3.12)$$

Similarly the total horizontal hydrodynamic force on the mass of water in the keel clearance per unit length, $f_{2,kc}(t)$, becomes:

$$f_{2,kc}(t) = -2\rho \frac{\omega^2}{m_0^2} (A_0 + B_0 e^{-i\theta}) \sinh\{m_0 (h-D)\} e^{i\omega t} +$$

$$-2\rho \sum_{n=1}^{\infty} i \frac{\omega^2}{m_n^2} (A_n + B_n e^{-i\theta}) \sin\{m_n (h-D)\} e^{i\omega t} \quad . \qquad (3.13)$$

The subscripts s and kc indicate that the quantity concerned relates to the 'ship' and the 'keel clearance', respectively.

The so far unknown (constant) quantities $\hat{v}_b$ and θ are determined by coupling the fields of flow on both sides of the ship. To that end the law of conservation of momentum is applied to the mass of water in the underkeel clearance:

$$f_{2,kc}(t) - 2\gamma B(v_b - \frac{1}{2}\dot{x}_2) = \rho B(h-D) \frac{dv_b}{dt} \qquad (3.14)$$

where Y = proportionality coefficient for shear stress in case of laminar flow.

For a predominantly oscillating flow in the keel clearance with a relatively thin laminar boundary layer a Stokes' type friction formula can be used, implying (see ref. [90])

$$Y = \rho \sqrt{\upsilon\omega} \quad , \qquad (3.15)$$

where υ = (coefficient of) kinematic viscosity of fluid.

Combining (3.9), (3.10), (3.13), (3.14) and (3.15) expressions can be derived for $\hat{v}_b$ and θ (see Appendix A).

The simple-harmonic sway motion (3.9) imposed on the ship requires an external exciting force $f_2(t)$. The 'equation of motion' in the x_2-direction for the oscillating schematized ship then is

$$m_{22}\ddot{x}_2 = Lf_{2,s}(t) + Y BL(v_b - \dot{x}_2) + f_2(t) \quad , \qquad (3.16)$$

with m_{22} = ρLBD. According to (2.31) the description of the ship-fluid system in the frequency domain for the sway motion reads as

$$\{m_{22} + a_{22}(\omega)\}\ddot{x}_2 + b_{22}(\omega)\dot{x}_2 = f_2(t) \quad . \qquad (3.17)$$

Now (3.9), (3.10), (3.12), (3.15), (3.16) and (3.17) together yield expressions for $a_{22}(\omega)$ (added mass for swaying motion) and $b_{22}(\omega)$ (sway damping force coefficient). Due to the strip-theory approach $a_{66}(\omega)$ (added mass-moment of inertia for yawing motion) and $b_{66}(\omega)$ (yaw damping moment coefficient) are determined from $a_{22}(\omega)$ and $b_{22}(\omega)$. Since the influence of the underkeel friction can be shown to be of minor importance (see ref. [54]), it is further left out of consideration. Expressions for the hydrodynamic coefficients are presented in Appendix A.

The hydrodynamic coefficients were determined not only in an analytical way, but also experimentally as functions of ω with h as parameter. The main particulars of the schematized ship (model) used were as follows: L = 2.438 m, B = 0.375 m, D = 0.150 m, block coefficient = 1.00, m_{11}, m_{22} = 137.24 kg, m_{66} = 50.99 kg m^2. Two water depths were involved, viz. h = 0.200 m, 0.175 m. Further ρ = 1000 kg m^{-3},

$g = 9.81 \text{ m s}^{-2}$. Figs. 3.2[a] and 3.2[b] show theoretical results for the hydrodynamic sway coefficients. By way of example, for one water depth, in figs. 3.3 and 3.4 a_{ii}, b_{ii} (i=2,6) are presented along that part of the frequency range, for which also experimental results are available. In these figures results derived analytically by means of a long-wave approximation for the motion of the water – being basically one-dimensional – are also inserted. Since the theoretical results for the hydrodynamic coefficients were determined using strip theory (i.e. a two-dimensional approach), three-dimensional effects such as the circulation around 'bow' and 'stern' could not be taken into account. This is the main cause of the discrepancy between the theoretical and the experimental results in the lower frequency range. For higher frequencies the strip-theory approach is sufficiently accurate – also on shallow water – to determine the hydrodynamic coefficients. As a consequence, for the low frequencies a_{ii}, b_{ii} (i=2,6) as calculated two-dimensionally have been adapted to the (three-dimensional) experimental values; for the higher frequencies they are maintained (see further refs. [54, 73]).

The respective i.r.f. for the sway motion and the yaw motion are calculated from the hydrodynamic coefficients using (2.33), (2.35[a]), (2.36) and (2.37). In figs. 3.5 and 3.6, as typical examples, results are presented as functions of time with h as parameter; the frequency range applied was $0 \leq \omega \leq 80 \text{ s}^{-1}$. These figures show three and two curves, respectively: the dot-and-dash line represents $k_{ii}(t)$ (i=2,6) as calculated from hydrodynamic coefficients determined theoretically (i.e. two-dimensionally) along the whole frequency range; the full line represents $k_{ii}(t)$ as calculated from hydrodynamic coefficients that, in case of low frequencies, were adapted to experimental (i.e. three-dimensional) values and further, if necessary, were extrapolated linearly and that, in case of higher frequencies were determined theoretically (i.e. two-dimensionally); the broken line represents $k_{22}(t)$ as calculated from an expression derived analytically using a long-wave approximation. From figs. 3.5 and 3.6 it can be seen that the i.r.f. approaches rather quickly to a constant value as t increases; this means that in the associated convolution integrals much emphasis is laid on the very near past of the time history of the forcing functions (see further ref. [73]).

102

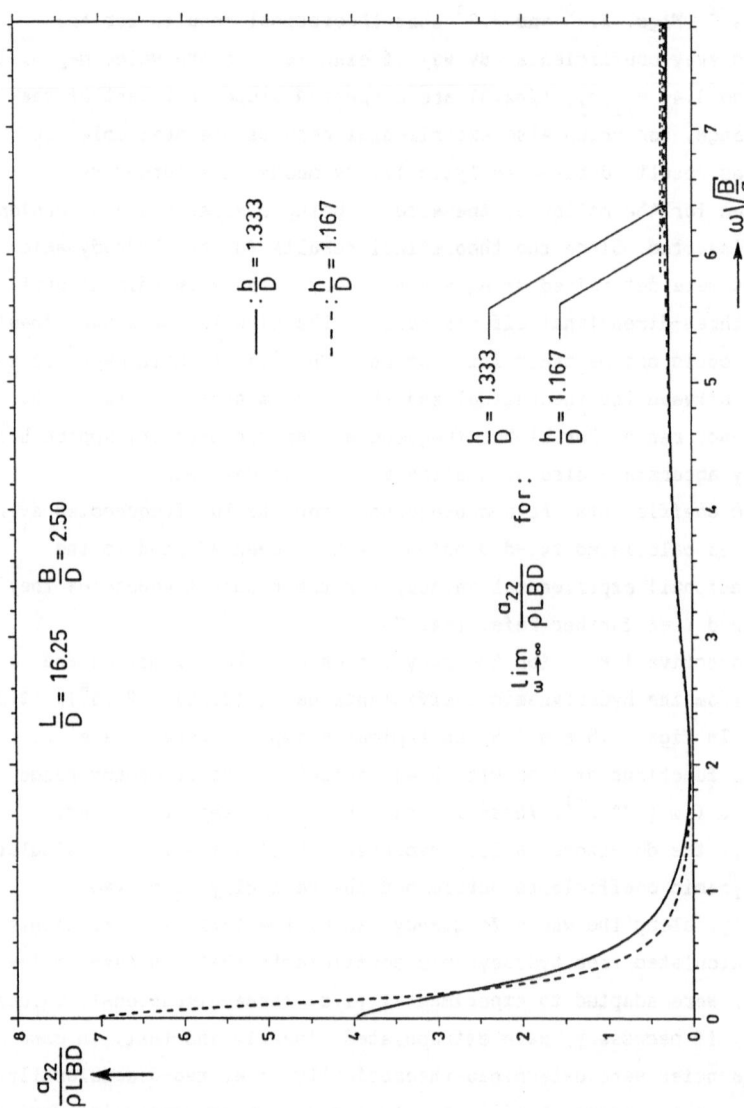

FIGURE 3.2[a] – Added mass for swaying motion on horizontally unrestricted water (non-zero keel clearance).

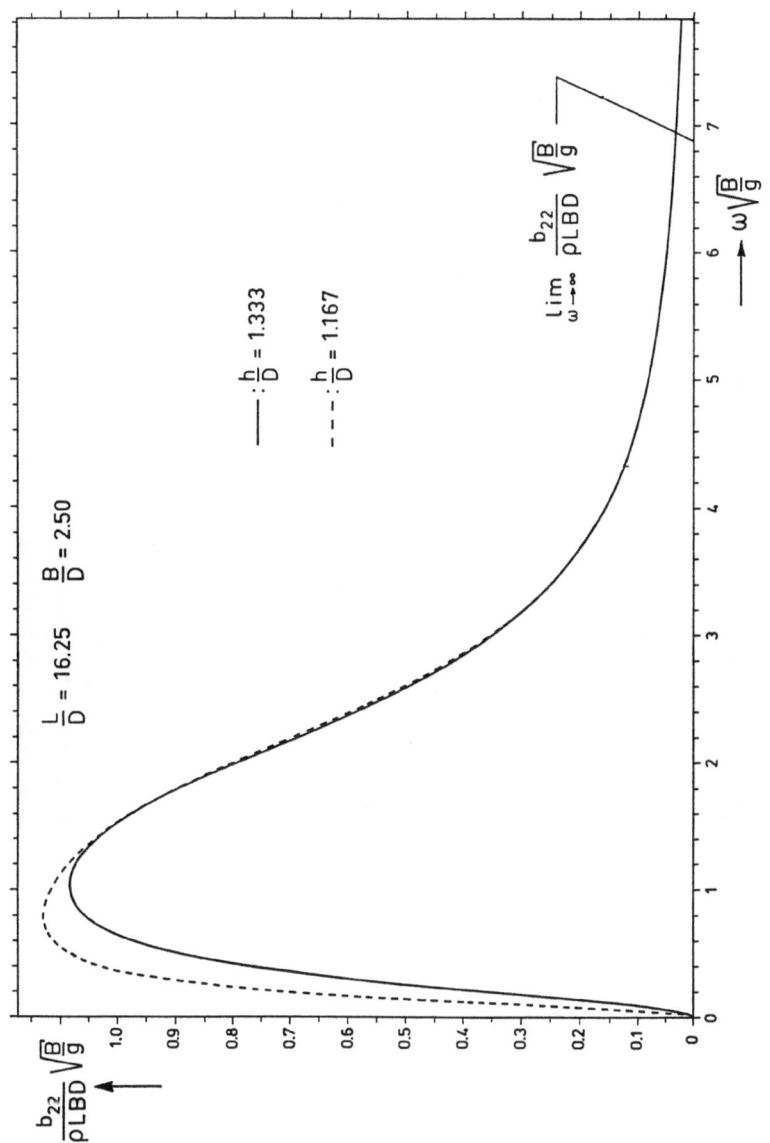

103

FIGURE 3.2^b - Sway damping force coefficient on horizontally unrestricted water (non-zero keel clearance).

104

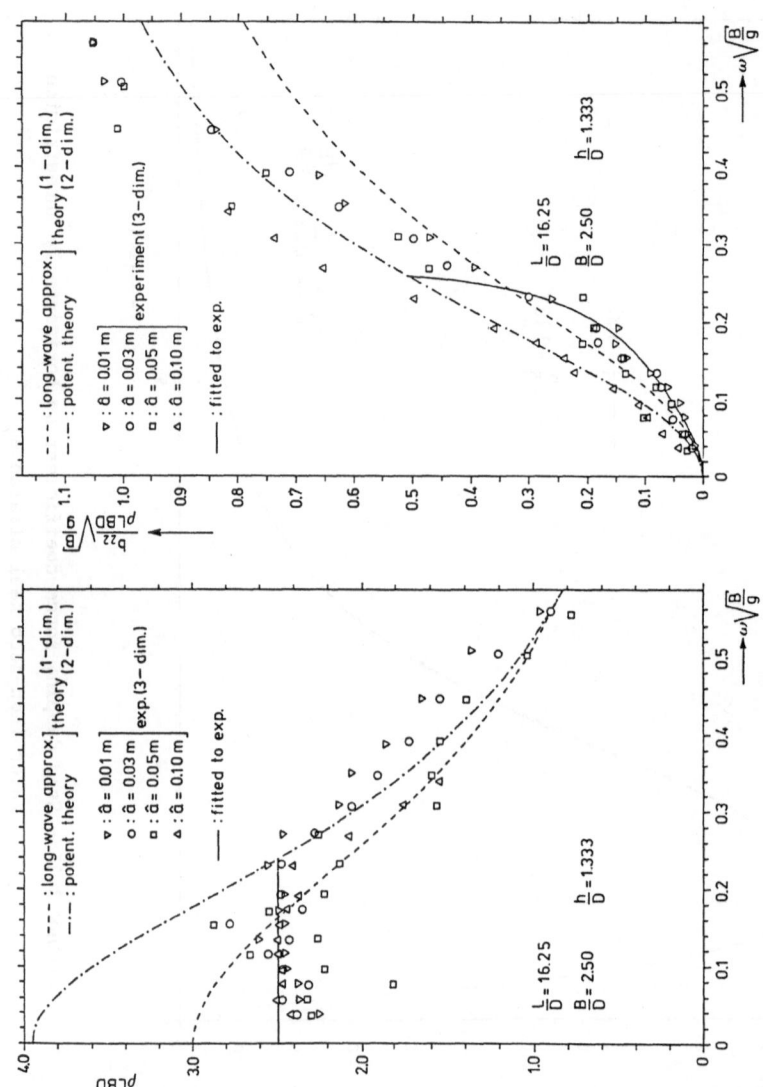

FIGURE 3.3 - Added mass for swaying motion and sway damping force
coefficient on horizontally unrestricted water (h/D=1.333).

105

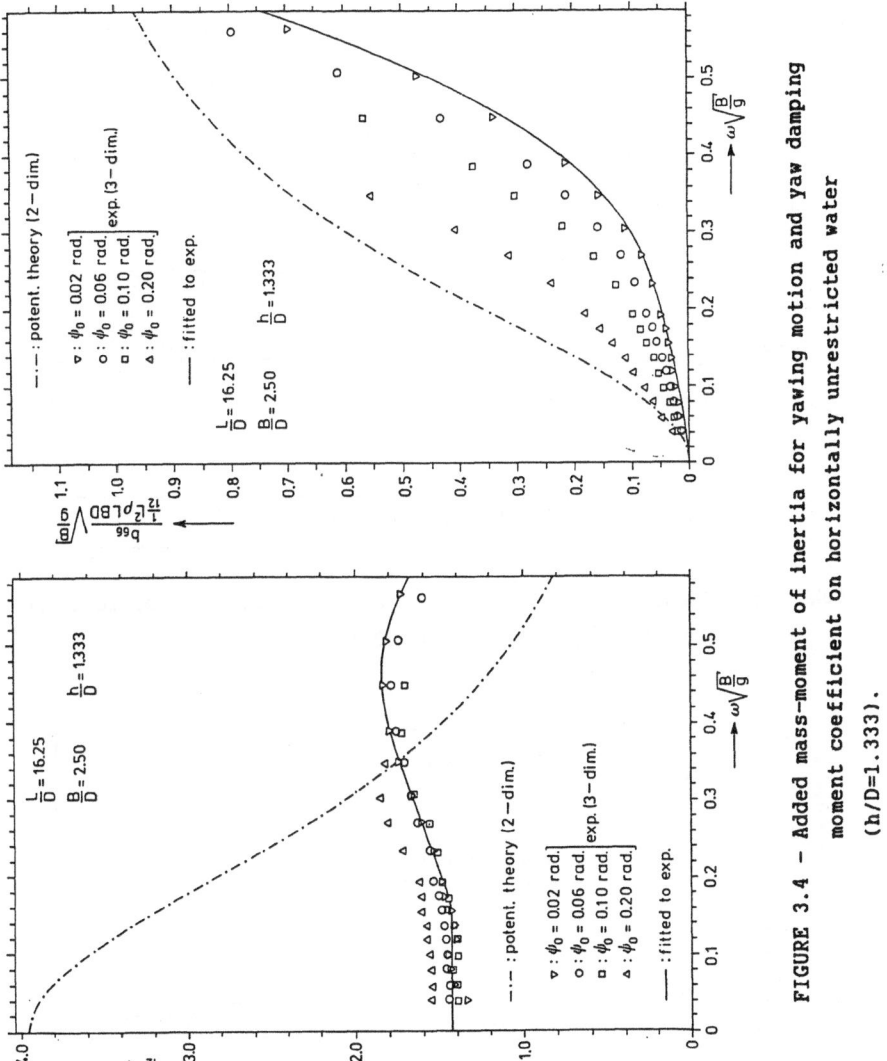

FIGURE 3.4 – Added mass-moment of inertia for yawing motion and yaw damping moment coefficient on horizontally unrestricted water (h/D=1.333).

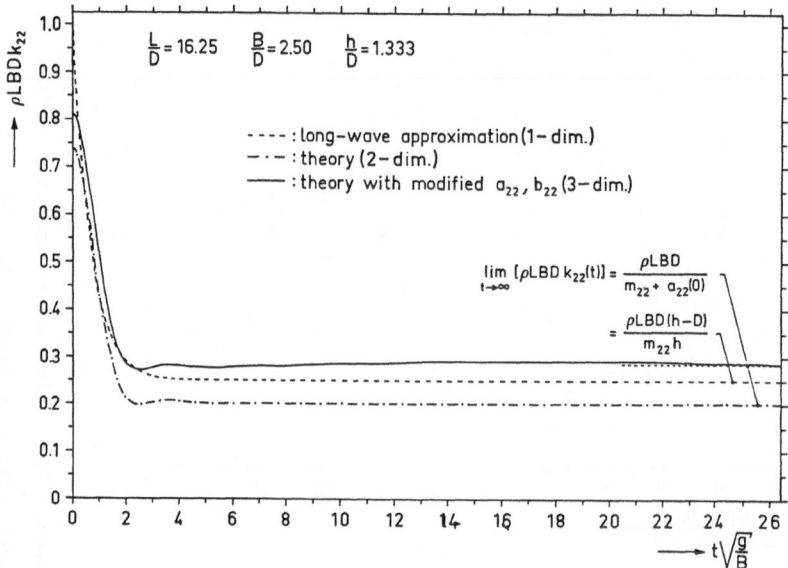

FIGURE 3.5 - Impulse response function for sway motion on horizontally
unrestricted water (h/D=1.333).

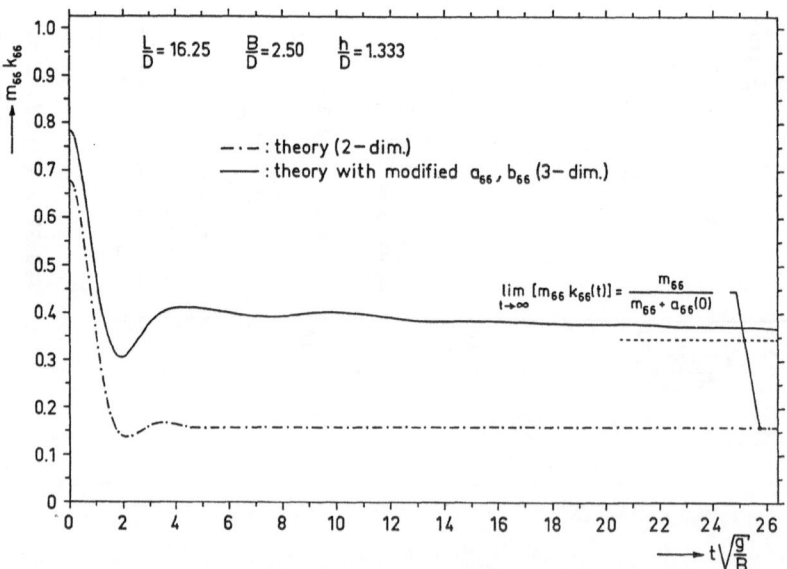

FIGURE 3.6 - Impulse response function for yaw motion on horizontally
unrestricted water (h/D=1.333).

3.3. Outline of mathematical approach

Consider the schematized ship berthing to an open jetty fitted with one single fender, represented by an undamped spring with a horizontal line of action situated in the plane of the water surface at rest (see fig. 3.7). The mass of the fender is small with respect to that of the ship. This implies that the effect of the initial impact may be neglected, i.e. the given state of motion does not change. If the fender mass is not negligibly small, then the fender gets a sudden acceleration at the

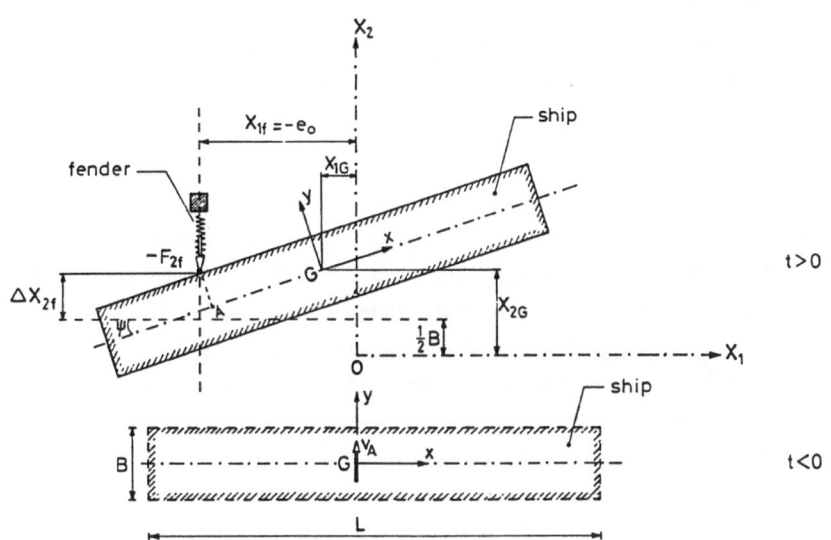

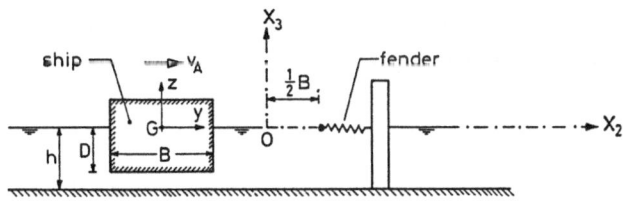

FIGURE 3.7 - Plan and cross-section of open berthing lay-out.

first moment of contact. Due to a redistribution of the momentum of the approaching ship, in that case a new initial value for the joint velocity of ship and fender arises, though without any change of the initial position of the ship. Further the frictional force between the hull of the ship and the fender is neglected.

Initially, i.e. before the first contact between ship and fender, the ship moves laterally towards the berth with a zero forward speed (i.e. $V_1 = 0$), a constant speed of approach $V_2 = v_A$ and without rotation. The first contact between ship and fender takes place at point of time $t = 0$. Then the line of action of the fender is perpendicular to the longitudinal axis of symmetry of the ship; its initial distance to the ship's centre of gravity G is denoted by e_0. At $t = 0$ the space-fixed $OX_1 X_2 X_3$-co-ordinate system is assumed to coincide with the translating $ox_1 x_2 x_3$-co-ordinate system; the ship fixed $Gxyz$-co-ordinate system then also coincides with $OX_1 X_2 X_3$. When $e_0 \neq 0$, at $t = 0$ the ship starts rotating, so that for $t > 0$ the motion of the ship consists of a translation and a rotation around the OX_3-axis. The co-ordinates of G are indicated by $X_{1G}(t)$ and $X_{2G}(t)$; the angle of the ship's longitudinal axis of symmetry around the OX_1-axis is $X_6(t) = \Psi(t)$. The co-ordinates of the point of the fender are $X_{1f}(t) = -e_0$ for all t, and $X_{2f}(t)$ with $X_{2f}(t) = \frac{1}{2}B$ for $t \leq 0$. The added subscripts G and f indicate that the quantity concerned must be related to the ship's centre of gravity G and the fender, respectively. The deflexion of the fender is denoted by $\Delta X_{2f}(t) = X_{2f}(t) - \frac{1}{2}B$ and can be expressed as:

$$\Delta X_{2f}(t) = X_{2G} - \frac{1}{2}B\{1 - \cos(\Psi)\} + \{X_{1f} - X_{1G} + \frac{1}{2}B \sin(\Psi)\}\tan(\Psi)$$

$$\Delta X_{2f}(t) \geq 0 \quad . \qquad (3.18)$$

The relation between $\Delta X_{2f}(t)$ and the corresponding reaction force in the fender $F_{2f}(t)$ can be represented as:

$$F_{2f}(t) = f(\Delta X_{2f}) \qquad \text{for} \quad t \geq 0 \quad . \qquad (3.19)$$

The resulting force and moment, as acting in and about G, then become:

$$f_2(t) = F_{2f} \cos(\Psi) \quad , \qquad \left.\rule{0pt}{18pt}\right]$$

$$f_6(t) = - \overline{AG} \; F_{2f}\cos(\Psi) \quad , \qquad \left.\begin{array}{c} \\ \\ \end{array}\right\} \quad t \geq 0 \quad , \qquad (3.20^{a,b})$$

respectively, where

$$\overline{AG} = \frac{X_{1G} - X_{1f}}{\cos(\Psi)} - \frac{1}{2}B \; \tan(\Psi)$$

with $X_{1f} = - e_0$, $\Psi < \frac{\pi}{2}$, $| \; \overline{AG} \; | \leq \frac{1}{2} L.$

The initial values of the berthing problem are given at $t = 0$ and read as:

$$\left.\begin{array}{l} X_{1G}(0) = 0 \quad , \quad X_{2G}(0) = 0 \quad , \\[2ex] \dot{X}_{1G}(0) = V_1 = 0 \quad , \quad \dot{X}_{2G}(0) = V_2 = v_A \quad , \\[2ex] X_{2f}(0) = \frac{1}{2}B \quad , \\[2ex] x_6(0) = \Psi(0) = 0 \quad , \\[2ex] f_1(0) = 0 \quad , \quad f_2(0) = 0 \quad , \quad f_6(0) = 0 \quad . \end{array}\right\} \qquad (3.21)$$

$\dot{X}_{1G}(0) = V_1 = 0$ and $X_{2G}(0) = V_2 = v_A$ describe the uniform translational motion of the $ox_1x_2x_3$-system.

According to (2.32) the (transient) velocities of the ship are described with respect to the $ox_1x_2x_3$-co-ordinate system, which translates with the respective constant speeds V_1 and V_2, and - in the case under consideration - for $t > 0$ also rotates with an angular velocity $\dot{\Psi}(t)$. Due to the rotation $ox_1x_2x_3$ cannot longer be considered as an inertial system, so that formally a correction has to be made in order to implicate its effect.

Within the context given above (2.32) can be rewritten as:

$$\dot{x}_i(t) = \int_0^t f_i(\tau) \; k_{ii}(t-\tau)d\tau \quad , \quad i = 1,2,6 \quad . \qquad (3.22)$$

110

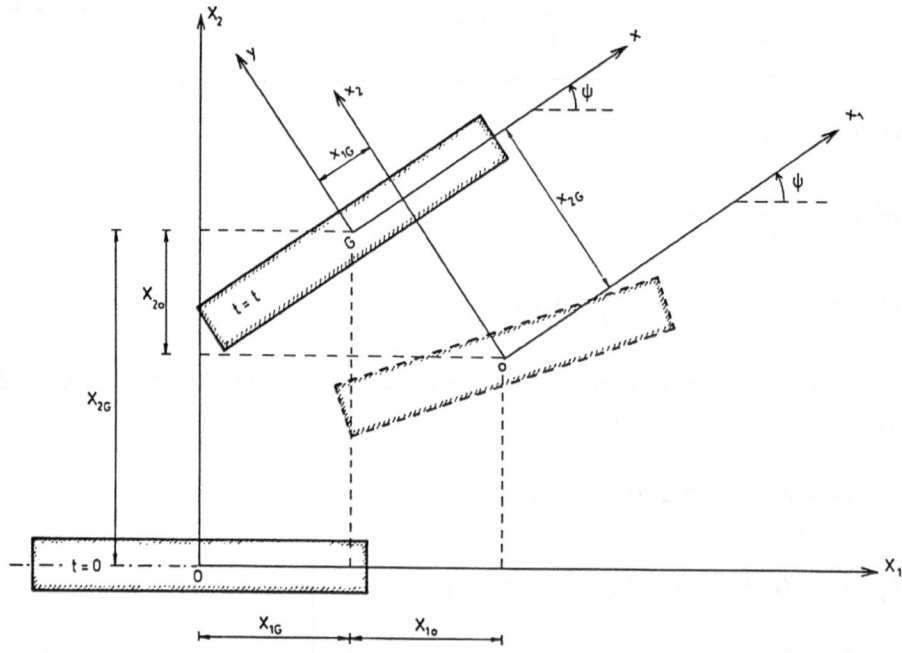

FIGURE 3.8 - Definition sketch of ship motion during berthing operation.

Now let $f_6(t)$ be known until the point of time t. Then $\dot{x}_6(t) = \dot{\psi}(t)$ can be evaluated by means of

$$\dot{\psi}(t) = \int_0^t f_6(\tau) \, k_{66}(t-\tau)d\tau \qquad (3.22^C)$$

in order to put the ship into its proper orientation given by $\psi(t)$. Under these conditions the absolute velocities of G are (see fig. 3.8):

$$\left. \begin{aligned} \dot{X}_{1G}(t) &= \dot{x}_1(t)\cos\{\psi(t)\} - \dot{x}_2(t)\sin\{\psi(t)\} + V_1 - X_{2o}(t)\,\dot{\psi}(t) \quad, \\[2ex] \dot{X}_{2G}(t) &= \dot{x}_1(t)\sin\{\psi(t)\} + \dot{x}_2(t)\cos\{\psi(t)\} + V_2 + X_{1o}(t)\,\dot{\psi}(t) \quad, \end{aligned} \right\} \qquad (3.23)$$

where $X_{1o}(t)$, $X_{2o}(t)$ = distance covered by G in X_1- and X_2-direction, respectively, with respect to uniformly travelling origin of $ox_1x_2x_3$;

the subscript o is added to indicate that the quantity concerned must be related to the origin of the $ox_1x_2x_3$-co-ordinate system; in (3.23) it holds good:

$$X_{1o}(0) = 0, \quad X_{2o}(0) = 0; \quad X_{1o}(t) \leq 0, \quad X_{2o}(t) \geq 0; \quad V_1 = 0, \quad V_2 = v_A. \qquad (3.24)$$

Supposing that $f_1(t)$ and $f_2(t)$ are known functions until point of time t, $\dot{x}_1(t)$ and $\dot{x}_2(t)$ are to be evaluated by

$$\dot{x}_1(t) = \int_0^t f_1(\tau)\, k_{11}(t-\tau)d\tau \quad , \qquad (3.22^a)$$

and

$$\dot{x}_2(t) = \int_0^t f_2(\tau)\, k_{22}(t-\tau)d\tau \quad , \qquad (3.22^b)$$

respectively. The first two terms in the right-hand members of (3.23) represent the relative velocities of the ship in the $ox_1x_2x_3$-system resolved in the absolute $OX_1X_2X_3$-system; the third and the fourth term are corrections due to the uniform translation of $ox_1x_2x_3$ and its rotational velocity, respectively. Introducing $x_{1G}(t)$ and $x_{2G}(t)$, defined as the co-ordinates of G within the (travelling) $ox_1x_2x_3$-system, it can be written:

$$\left. \begin{array}{l} X_{1o}(t) = x_{1G}(t)\, \cos\{\Psi(t)\} - x_{2G}(t)\, \sin\{\Psi(t)\} \quad , \\[2mm] X_{2o}(t) = x_{1G}(t)\, \sin\{\Psi(t)\} + x_{2G}(t)\, \cos\{\Psi(t)\} \quad . \end{array} \right\} \qquad (3.25^a)$$

Concerning the co-ordinates of the origin of $ox_1x_2x_3$ it can be stated that

$$V_1 t = X_{1G}(t) - X_{1o}(t) \quad \text{and} \quad V_2 t = X_{2G}(t) - X_{2o}(t) \quad ,$$

from which it follows:

$$\left. \begin{array}{l} X_{1o}(t) = X_{1G}(t) \quad , \\[2mm] X_{2o}(t) = X_{2G}(t) - v_A t \quad . \end{array} \right\} \qquad (3.26)$$

Due to the rotation of the $ox_1 x_2 x_3$-system formally additional forces have to be introduced into the description of the force balance. These additional forces can be determined by relating the accelerations in the $ox_1 x_2 x_3$-system to a new co-ordinate system $ox_1 x_2 x_3$, which merely translates. Then it holds good that

$$
\left.
\begin{aligned}
\ddot{x}_1(t) &= \ddot{\underline{x}}_1(t)\,\cos\{\Psi(t)\} + \ddot{\underline{x}}_2(t)\,\sin\{\Psi(t)\} + a_{1r}(t) \quad , \\[2em]
\ddot{x}_2(t) &= -\ddot{\underline{x}}_1(t)\,\sin\{\Psi(t)\} + \ddot{\underline{x}}_2(t)\,\cos\{\Psi(t)\} + a_{2r}(t) \quad ,
\end{aligned}
\right\}
\qquad (3.27)
$$

where $a_{1r}(t)$, $a_{2r}(t)$ = additional accelerations in $ox_1 x_2 x_3$ to be introduced due to rotation of $ox_1 x_2 x_3$ with respect to $o\underline{x}_1\underline{x}_2\underline{x}_3$; the subscript r indicates that the quantity concerned is due to the rotation. The first two terms in the right-hand members of (3.27) represent the instantaneous components of $\ddot{\underline{x}}_1$ and $\ddot{\underline{x}}_2$ in the respective x_1- and x_2- direction. Now $a_{1r}(t)$ and $a_{2r}(t)$ can be put into the form

$$
\left.
\begin{aligned}
a_{1r}(t) &= -2\dot{x}_2(t)\,\dot{\Psi}(t) + x_{1G}(t)\,\dot{\Psi}^2(t) - x_{2G}(t)\,\ddot{\Psi}(t) \quad , \\[2em]
a_{2r}(t) &= 2\dot{x}_1(t)\,\dot{\Psi}(t) + x_{2G}(t)\,\dot{\Psi}^2(t) + x_{1G}(t)\,\ddot{\Psi}(t) \quad ,
\end{aligned}
\right\}
\qquad (3.28)
$$

respectively, where the first term in each right-hand member represents the Coriolis-effect and the second term the centrifugal effect - both due to the angular velocity -, whereas the third term stands for the inertial contribution due to the angular acceleration. $x_{1G}(t)$ and $x_{2G}(t)$ can be expressed into $X_{1o}(t)$ and $X_{2o}(t)$ by

$$
\left.
\begin{aligned}
x_{1G}(t) &= X_{1o}(t)\,\cos\{\Psi(t)\} + X_{2o}(t)\,\sin\{\Psi(t)\} \quad , \\[2em]
x_{2G}(t) &= -X_{1o}(t)\,\sin\{\Psi(t)\} + X_{2o}(t)\,\cos\{\Psi(t)\} \quad ,
\end{aligned}
\right\}
\qquad (3.25^b)
$$

being equivalent to (3.25^a). The additional forces in the x_1- and x_2- direction, as a result of the rotation of $ox_1 x_2 x_3$, then consequently are

$$
f_{1r}(t) = m_{11}\,a_{1r}(t) \qquad\qquad f_{2r}(t) = m_{22}\,a_{2r}(t) \quad . \qquad (3.29)
$$

In other words, the respective additional force in the x_1- and the x_2-direction due to the rotation of the $ox_1 x_2 x_3$-system is the sum of the so-called virtual forces (consisting of the Coriolis-force and the centrifugal force) and the inertial force (resulting from the angular acceleration). If these additional forces are taken into account, they are to be classed in the respective (external) forcing functions $f_i(t)$ of the ship-fluid system.

Using the set of equations (3.18) through (3.29) as presented above, it is possible - in a formal way- to determine the fender force and to describe the ship trajectories during the berthing operation.

Within the linearity of the i.r.f.-technique (dealing with small disturbances with respect to a given uniform ship motion) the virtual forces as given in (3.28),(3.29) represent a second-order effect. Further the berthing situation under consideration leads to the supposition that $x_{1G}(t)$ and $x_{2G}(t)$ (and therefore $X_{1o}(t)$ and $X_{2o}(t)$, see (3.25^{a,b})) remain small, so that the inertial forces due to the angular acceleration also are to be considered as a second-order effect. Consequently the additional forces $f_{1r}(t)$ and $f_{2r}(t)$ may be neglected. With this simplification it can be stated that $f_i(t) = 0$ for all t, so that on account of (3.22^a) also $\dot{x}_1(t) = 0$. The contribution to $\dot{X}_{1G}(t)$ and $\dot{X}_{2G}(t)$, respectively, in (3.23) due to the rotational velocity of $ox_1 x_2 x_3$ just as well may be considered as a second-order effect, so that (3.23) eventually takes the form

$$\left.\begin{aligned}
\dot{X}_{1G}(t) &= -\dot{x}_2(t)\,\sin\{\psi(t)\} \quad, \\[2em]
\dot{X}_{2G}(t) &= v_A + \dot{x}_2(t)\,\cos\{\psi(t)\} \quad,
\end{aligned}\right\} \qquad (3.30^{a,b})$$

whereby it has to be borne in mind that according to (3.30^a) $\dot{X}_{1G}(t)$ likewise will remain small of the second order.

With these simplifications the mathematical formulation of the berthing problem now has been reduced to a set of equations consisting of (3.18), (3.19), (3.20^{a,b}), (3.21), (3.22^{b,c}) and (3.30^{a,b}). Actually this (simplified) approach amounts to a formulation related to an $ox_1 x_2 x_3$-coordinate system travelling along with the given initial velocities V_1, V_2 and without rotation.

Since in $(3.22^{b,c})$ the forcing functions $f_2(t)$ and $f_6(t)$, as acting
during the contact between ship and fender, are functions of the
displacements of the ship as well as of the deflexion of the fender,
(3.18), (3.19), $(3.20^{a,b})$, $(3.22^{b,c})$ and $(3.30^{a,b})$ combined form a closed-
loop system; $(3.22^{b,c})$ represents a set of two integro-differential
equations. Then, provided the relevant i.r.f. are known, it is possible to
determine fender loads and ship trajectories; naturally this can only be
done if the fender characteristics are given too.

Two kinds of fenders are considered: a linear fender represented by

$$F_{2f}(t) = \begin{cases} 0 & \text{for } \Delta X_{2f}(t) < 0 \quad, \\ \\ - c_0 \, \Delta X_{2f}(t) & \text{for } \Delta X_{2f}(t) \geq 0 \quad, \end{cases} \qquad (3.31^a)$$

and a non-lineair fender represented by

$$F_{2f}(t) = \begin{cases} 0 & \text{for } \Delta X_{2f}(t) < 0 \quad, \\ \\ - c_1 \, \Delta X_{2f}(t) & \text{for } 0 \leq \Delta X_{2f}(t) \leq d_{sc}, \\ \\ - c_1 \, \Delta X_{2f}(t) - c_2 \{\Delta X_{2f}(t) - d_{sc}\} & \text{for } \Delta X_{2f}(t) \geq d_{sc} \quad, \end{cases} \quad (3.31^b)$$

where c_0 = spring rate of linear fender,
c_1, c_2 = respective spring rates of the two linear springs
which combined form the non-linear fender,
d_{sc} = initial distance (i.e. at rest) between the two linear
spring elements of the non-linear fender.

The combined equations (3.18), (3.19), $(3.20^{a,b})$, $(3.22^{b,c})$ and
$(3.30^{a,b})$ with initial values (3.21) and the fender characteristics given
by $(3.31^{a,b})$ now have to be solved numerically. This is carried through
according to an iteration procedure. The numerical integrations are carried
out by means of the trapezoidal rule. The calculation is finished when the
ship loses contact with the fender. As a criterion for the convergence of
the computational scheme it can be derived for the time step Δt (see ref.
[73])

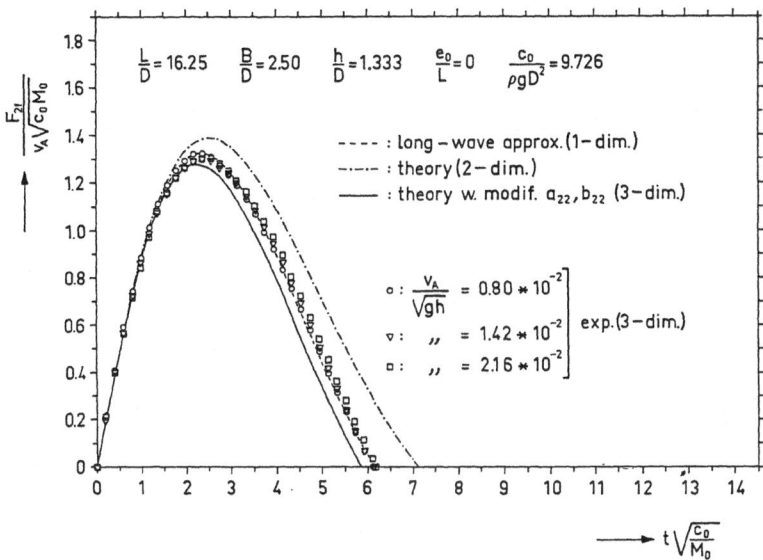

FIGURE 3.9 - Time history of fender force: open berth, linear fender, centric impact (h/D=1.333).

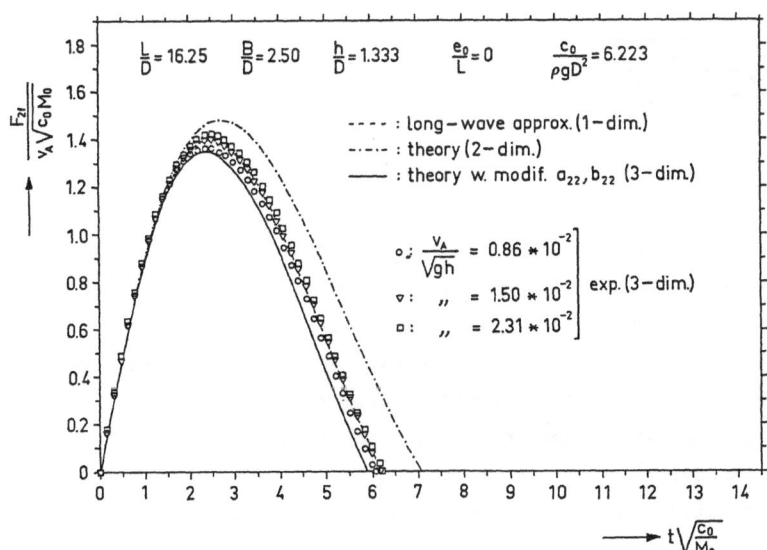

FIGURE 3.10 - Time history of fender force: open berth, linear fender, centric impact (h/D=1.333).

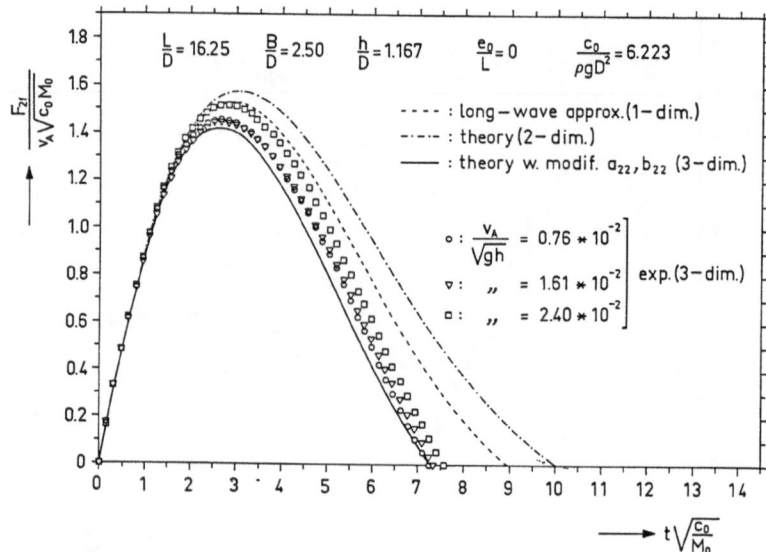

FIGURE 3.11 – Time history of fender force: open berth, linear fender,
centric impact (h/D=1.167).

$$\Delta t < 2\sqrt{\dfrac{1}{c_0\,k_{22}(0^+)}} \qquad ;$$

this actually only holds good for a centric impact to a linear fender.

3.4. Examples of berthing operations: experiment and theory

In order to examine the adequacy of the mathematical approach to the
simulation of berthing operations under conditions as described, an
extensive experimental program was carried out using the same schematized
ship model and the same water depths as stated previously.

In the experiments the following quantities were measured as functions
of time: the deflexion of the fender, the position of G, and the ship's
angle of rotation. In the numerical simulation of the berthing operations
the following quantities were calculated: $\dot{x}_6(t) = \dot{\psi}(t)$, $\dot{X}_{1G}(t)$ and $\dot{X}_{2G}(t)$,
$X_{1G}(t)$ and $X_{2G}(t)$, $x_6(t) = \psi(t)$, $\Delta X_{2f}(t)$ and $F_{2f}(t)$. The time

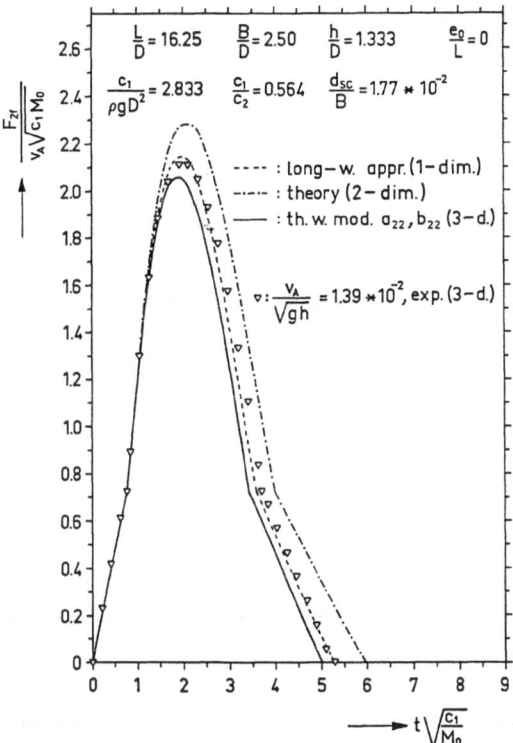

FIGURE 3.12 - Time history of fender force: open berth, non-linear fender, centric impact (h/D=1.333).

increment used was $\Delta t = 0.01$ s. $X_{1G}(t)$ further is left out of consideration, since this quantity turned out to be very small with respect to $X_{2G}(t)$, while its experimental value fell within the accuracy of the measurements. From a (rough) estimation of the rotational influences it appeared that these - when fully taken into account - indeed can be neglected as being second-order effects, at least for the berthing situation under consideration.

To describe the berthing of the (schematized) ship to a certain fender it is sufficient to consider only $F_{2f}(t)$, $X_{2G}(t)$ and $\Psi(t)$. The parameters that further play a part are the characteristics of the fender, h, e_0 and v_A. In addition to the experimental results that are plotted as centered symbols, the figures to be presented each show three curves representing the theoretical results: the dot-and-dash line represents the results as calculated by means of i.r.f., which have to be considered as two-

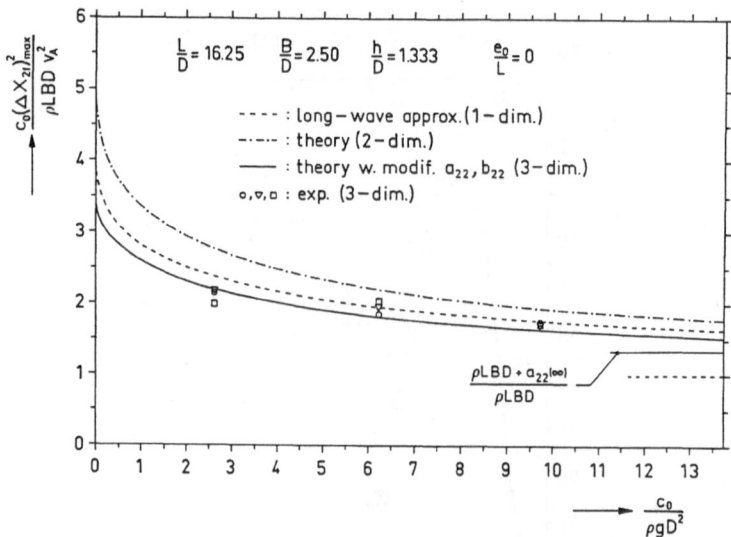

FIGURE 3.13 - Influence of fender elasticity on absorbed energy: open
berth, linear fender, centric impact (h/D=1.333).

dimensional; the full line represents the results as calculated by means of
i.r.f., which can be considered as three-dimensional; the broken line
represents the results as determined (analytically) from a long-wave
approximation for the motion of the water, which are basically one-
dimensional (see ref. [73]). Further in these figures the quantity M_0
occurs, defined as

$$M_0 = \rho LBD \, m_{66} \, (\rho LBD \, e_0^2 + m_{66})^{-1} \quad ,$$

which has to be interpreted as the reduced or effective mass of the ship
(model) for horizontal motion.

Examples of berthing operations in which $e_0 = 0$ ('centric impacts')
are given in figs. 3.9 through 3.13. Figs. 3.9 and 3.10 show calculated and
measured fender forces as functions of time in case of a linear fender with
certain spring rates; similar results are presented for a different water
depth (fig. 3.11) and for a non-linear fender (fig. 3.12). As expected, in
the case of the linear fender, the calculated fender forces are
proportional to v_A. The total amount of energy, E, absorbed by a fender

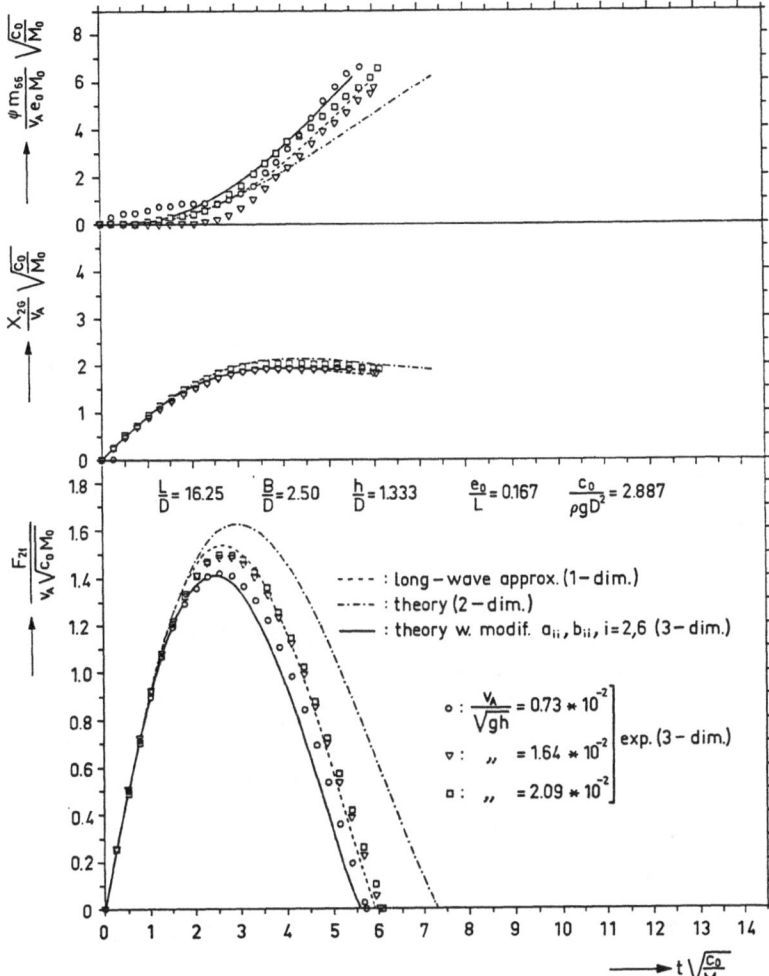

FIGURE 3.14 - Time histories of fender force, translation of G and angle of
rotation: open berth, linear fender, eccentric impact
(h/D=1.333).

with linear behaviour is given by

$$E = \int_0^{(\Delta X_{2f})_{max}} F_{2f}(t)\, d(\Delta X_{2f}) = \frac{1}{2}\, c_0\, (\Delta X_{2f})^2_{max} \quad ,$$

where $(\Delta X_{2f})_{max}$ = maximum deflexion of (linear) fender.

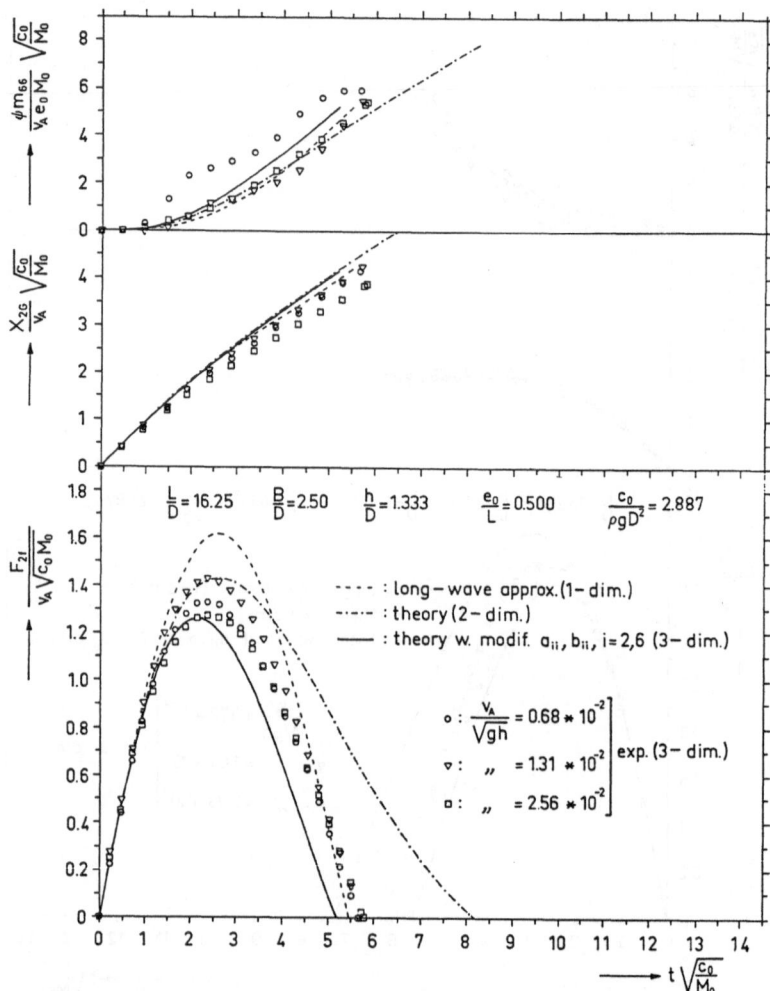

FIGURE 3.15 - Time histories of fender force, translation of G and angle of
rotation: open berth, linear fender, eccentric impact
(h/D=1.333).

This expression represents the influence of the fender stiffness in case of
a linear fender on the absorption of energy; see fig. 3.13, which shows
that a stiff fender absorbs less energy to stop the ship than a soft
fender. This effect is caused by the greater wave radiation in case of a
stiffer fender.

Examples of berthing operations in which $e_0 \neq 0$ ('eccentric impacts')
in case of a linear fender are given in figs. 3.14 through 3.16 for two

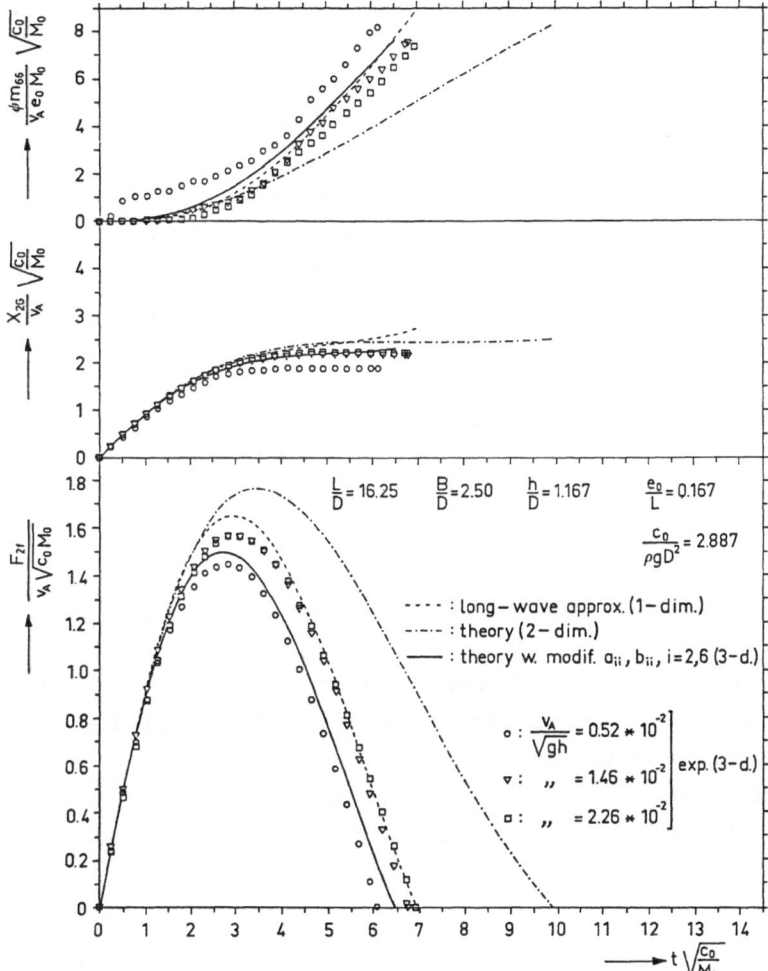

FIGURE 3.16 - Time histories of fender force, translation of G and angle of
    rotation: open berth, linear fender, eccentric impact
    (h/D=1.167).

different values of e_0 and h. The calculated fender forces, translations of
G, and angles of rotation can be considered to be proportional to v_A.

From the figures (see also ref. [73]) it can be seen that in case of
increasing fender stiffness, (the maximum value of) $F_{2f}(t)$ also increases
whereas the length of time of the contact between ship and fender decreases
and the point of time at which $F_{2f}(t)$ reaches its maximum occurs earlier.
The same trend applies to the non-linear fender in case of increasing value
of v_A. For a larger water depth, (the maximum value of) $F_{2f}(t)$ as well as

the length of time of the contact between ship and fender is smaller while
the point of time at which $F_{2f}(t)$ reaches its maximum occurs earlier; an
increasing value of e_0 shows the same trend. Results as determined from a
relatively simple long-wave approximation generally shows conservative
values for (the maximum value of) $F_{2f}(t)$.

In general it can be stated that the agreement between theory and
experiment is satisfactory. Notably by means of the theory using
hydrodynamic coefficients adapted to the three-dimensional situation the
relevant quantities are predicted well. In the mathematical approach
viscous effects have not been taken into account. Since in model tests
these effects are overestimated, it may be concluded from the good
agreement between calculated and measured results that the viscosity of the
fluid does not influence the fender forces significantly.

4. SHIP BERTHING TO CLOSED FENDER STRUCTURE
4.1. Introduction

In this section a mathematical model is presented that aims to
describe the behaviour of a ship berthing to a closed fender structure and
the prediction of the fender forces.
For the assumptions and simplifications made it is referred to Section
1.3.4. It should be borne in mind that some such idealized berthing
situation may lead to conservative fender forces.
In order to determine the fender loads as a result of ship berthing also in
this specific situation a time-domain description of the moving ship is
needed which makes allowance for the frequency dependence of the fluid
reaction forces. To this end, a set of equations is formulated describing
the transverse (i.e. sway) motion of the schematized ship in shallow water
at zero forward speed and parallel to a vertical wall. Non-linearities in
the displacement of the ship, friction effects and flow separation in the
underkeel region are taken into account. As during berthing manoeuvres low
frequencies (are supposed to) play a dominant part, in region c a long-wave
approximation for the motion of the water is applied. To solve the
governing equations, two separate procedures are followed. The first
approach, requiring a linearization, makes use of the i.r.f.-technique as
dealt with in Section 2 and already applied in Section 3. The hydrodynamic

coefficients for the swaying motion are determined and the corresponding
i.r.f. calculated. Just as in Section 3, it is hereby assumed implicitly
that in berthing the (transverse) velocity of the ship remains so small
that it does not affect the hydrodynamic coefficients and a restoring force
is not generated. Then berthing operations can be simulated. In the second
procedure, being a direct time-domain approach (T.D.A.), the influence of
the non-linearities can be evaluated. The equations are simplified to a
two-dimensional situation (strip theory) and solved directly in the time
domain.
The theoretical results were verified by model tests. A detailed account of
theory and experiments is given in ref. [66]; see also ref. [76].

4.2. Mathematical model

The ship motion is regarded with respect to the $ox_1x_2x_3$-co-ordinate
system, which - until further notice - is taken to be space fixed. In
accordance with Section 2.2, ox_1x_2 and Gxy are situated in the water
surface at rest; the ox_3- and Gz-axes are positive upwards. The coinciding
ox_2- and Gy-axes are at right angles to the vertical wall and positive
outwards. The origin o lies at $d_{sq}+\frac{1}{2}B$ in front of the wall, where d_{sq} is
the distance c.q. clearance between ship and wall when G coincides with o.
The sway motion variable of the ship again is represented by $x_2(t)$. The
fluid velocities in the x_1-, x_2- and x_3-direction are u, v and w,
respectively. ζ stands for the elevation of the water surface and η for the
height of a long wave, both with respect to the water depth at rest. The
fluid domain can be divided into four regions: the subscripts a, b, c and d
indicate that the dependent variables concerned must be related to these
respective regions. For a definition sketch see fig. 4.1.

For region a, representing the quay clearance, it is supposed that
$d_{sq}+x_2 \ll \frac{1}{2}L$ and $d_{sq}+x_2 \ll D$. The pressure gradients or accelerations
acting in the x_1- and x_3-direction can now be considered as large in
comparison with the corresponding quantities in the x_2-direction. This
implies a uniform velocity distribution along the x_2-direction. The law of
conservation of mass and the equations of motion in the x_1- and x_3-
direction then read as

$$\frac{\partial u_a}{\partial x_1} + \frac{\partial w_a}{\partial x_3} + \frac{\dot{x}_2}{d_{sq}+x_2} = 0 \quad , \tag{4.1}$$

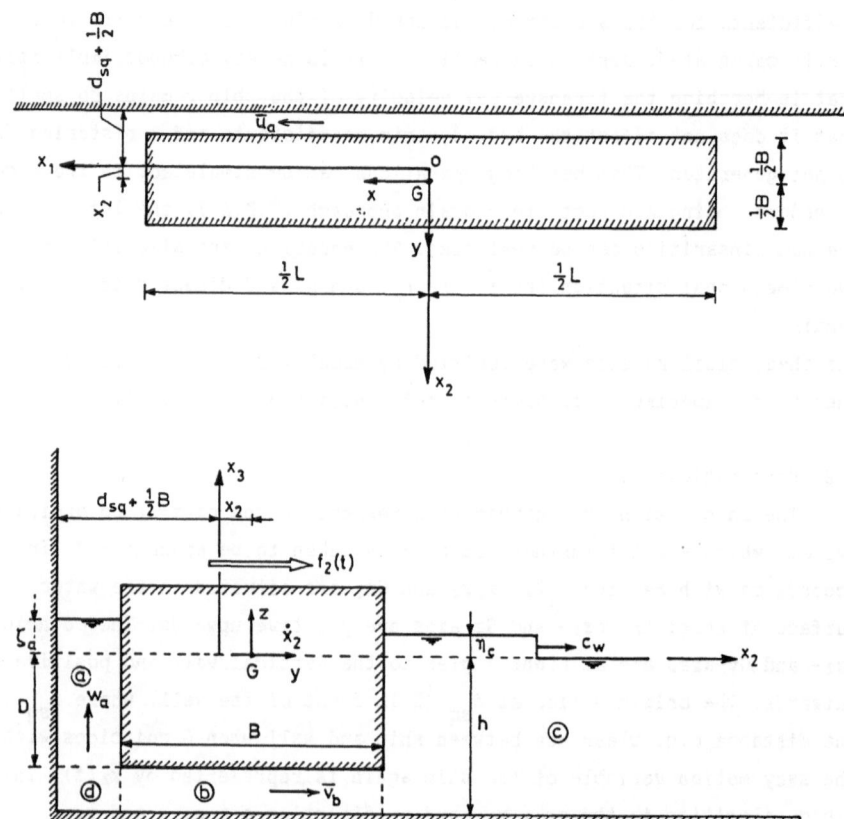

FIGURE 4.1 - Definition sketch for sway motion near a vertical wall.

$$\frac{\partial u_a}{\partial t} + \frac{1}{\rho} \frac{\partial p_a}{\partial x_1} = 0 \quad , \tag{4.2}$$

$$\frac{\partial w_a}{\partial t} + \frac{1}{\rho} \frac{\partial p_a}{\partial x_3} + g = 0 \quad , \tag{4.3}$$

respectively, where $u_a = u_a(x_1,x_3,t)$, $w_a = w_a(x_1,x_3,t)$ and $p_a = p_a(x_1,x_3,t)$. The boundary conditions are formulated as

$$w_a\Big|_{x_3=\zeta_a} = \frac{\partial \zeta_a}{\partial t} \quad , \quad p_a\Big|_{x_3=\zeta_a} = 0 \quad , \quad u_a\Big|_{x_1=0} = 0 \quad , \quad \zeta_a\Big|_{x_1=\pm\frac{1}{2}L} = 0 \quad , \qquad (4.4)$$

where $\zeta_a = \zeta_a(x_1,t)$. Only small values of u_a, w_a, ζ_a and their derivatives are considered.

Concerning the underkeel region b it is assumed $B \ll L$ and $h-D \ll B$. Pressure gradients or accelerations in the x_2-direction then being large as compared with the corresponding quantities in the x_1- and x_3-direction, this leads to $u_b = w_b = 0$, a hydrostatic pressure and a uniform velocity distribution over the height. The equation of motion in the x_2-direction now can be put into the form

$$B\frac{\partial \bar{v}_b}{\partial t} = -g\eta_c + \frac{1}{\rho} p_a(x_1,-D,t) - gD - R_b - H_b \quad , \qquad (4.5)$$

with

$$R_b = \alpha_1 Bgc_w^{-1}(\bar{v}_b - \tfrac{1}{2}\dot{x}_2), \quad \alpha_1 = 2\,\gamma c_w\{\rho g(h-D)\}^{-1} \text{ for laminar flow,}$$

$$R_b = \alpha_t B(h-D)^{-1}\{(\bar{v}_b - \dot{x}_2)^2 \text{sgn}(\bar{v}_b-\dot{x}_2) + \bar{v}_b^{\,2} \text{sgn}(\bar{v}_b)\} \text{ for turbulent flow,}$$

$$H_b = \tfrac{1}{2}\xi(\bar{v}_b-\dot{x}_2)^2 \text{sgn}(\bar{v}_b-\dot{x}_2), \quad \xi = \xi_e + \xi_o, \quad \xi_e = (\tfrac{1}{\mu} - 1)^2 \quad ,$$

where $\bar{v}_b = \bar{v}_b(x_1,t)$ and $\eta_c = \eta_c(x_1,t)$; a bar over a quantity means 'average value of'. R represents the influence of the friction and H/g the loss of energy head due to contraction and separation of flow. Further,

α = dimensionless friction coefficient,

c_w = velocity of propagation of long wave ($= \sqrt{gh}$),

ξ = general head loss coefficient,

ξ_e, ξ_o = respective head loss coefficients due to contraction at entrance and expansion at outlet,

$\mu\cdot$ = contraction coefficient;

the subscripts l and t indicate a laminar or turbulent flow regime. As the underkeel flow in principle can be laminar or turbulent, two expressions for R_b are introduced. Apart from the type of underkeel flow, the transient ship motion in berthing suggests the existence of a relatively thin boundary layer with large velocity gradients. With $\mu = 0.61$ (ref. [90]) and $\xi_o = 1$ (Borda–Carnot approach) $\xi = 1.44$.

Region d, forming a transition between the regions a and b, is supposed to be small with respect to these domains, which results in a hydrodynamic pressure not depending on x_2 and x_3. Application of the law of conservation of mass then yields

$$(d_{sq}+x_2)w_a(x_1,-D,t) + (h-D)\bar{v}_b(x_1,t) = 0 \quad . \tag{4.6}$$

By applying in the x_2-direction a one-dimensional long-wave approximation to the motion of the water in region c, which implies a hydrostatic pressure, $u_c = w_c = 0$ and $\eta_c \ll D$, it can be shown that

$$(h-D)\bar{v}_b + D\dot{x}_2 - c_w\eta_c = 0 \quad . \tag{4.7}$$

The equation of motion in the x_2-direction for the ship reads as

$$\rho LBD\ddot{x}_2 = 2 \int_0^{\frac{1}{2}L} \int_{-D}^{\zeta_a} p_a dx_3 dx_1 - \rho g \int_0^{\frac{1}{2}L} (\eta_c+D)^2 dx_1 + f_2(t) + R_{b,s} \quad , \tag{4.8}$$

with

$$R_{b,s} = \alpha_1 \frac{h-D}{h} c_w \rho B \int_0^{\frac{1}{2}L} (\bar{v}_b-\dot{x}_2)dx_1 \qquad \text{for laminar flow in region b} \quad ,$$

$$R_{b,s} = 2\alpha_t \rho B \int_0^{\frac{1}{2}L} (\bar{v}_b-\dot{x}_2)^2 \, \text{sgn}(\bar{v}_b-\dot{x}_2)dx_1 \quad \text{for turbulent flow in region b} \quad .$$

A further simplification is carried through by averaging the horizontal velocity u_a over the depth according to

$$(D+\zeta_a)\bar{u}_a = \int_{-D}^{\zeta_a} u_a dx_3 \quad , \tag{4.9}$$

where $\bar{u}_a = \bar{u}_a(x_1,t)$; in this operation use is made of Leibniz' rule.

Under the assumptions and simplifications made the set of eight governing equations (4.1),(4.2),(4.3),(4.5),(4.6),(4.7),(4.8) and (4.9) together with the boundary conditions (4.4), represents a general formulation in the time domain of the motion characteristics of ship and fluid.

4.3. Application of i.r.f.-technique

4.3.1. Hydrodynamic sway coefficients and i.r.f. Since application of the i.r.f.-technique requires a linear and time-invariant ship-fluid system, the above set of governing equations must be linearized. This implies that the displacement of the ship has to remain small with respect to a mean value (i.e. $x_2 \ll d_{sq}$) and that non-linear terms are neglected under the assumption of being small of the second-order.

According to (2.31) the behaviour of the ship-fluid system in the frequency domain in case of a pure sway mode of motion is described by

$$\{\rho LBD + a_{22}(\omega)\}\ddot{x}_2 + b_{22}(\omega)\dot{x}_2 = f_2(t) \quad . \tag{4.10}$$

After linearization of the governing equations the hydrodynamic sway coefficients can be determined by means of a harmonic analysis of the ship-fluid system. To that end, a simple-harmonic sway motion

$$x_2(t) = \hat{x}_2 \, e^{i\omega t} \quad , \quad \hat{x}_2 = \hat{x}_2(\omega) \quad , \tag{4.11^a}$$

is imposed on the ship, which requires an external exciting force in the x_2-direction of the form

$$f_2(t) = \hat{f}_2 \, e^{i\omega t} \quad , \quad \hat{f}_2 = \hat{f}_2(\omega) \quad . \tag{4.11^b}$$

Regarding the underkeel frictional effect, just as in Section 3.2 a Stokes' type friction formula is used for a predominantly oscillating flow with a relatively thin boundary layer, so that again

$$\gamma = \rho \, \sqrt{\nu\omega} \quad . \tag{3.15}$$

Now the hydrodynamic sway coefficients can be derived from the linearized

128

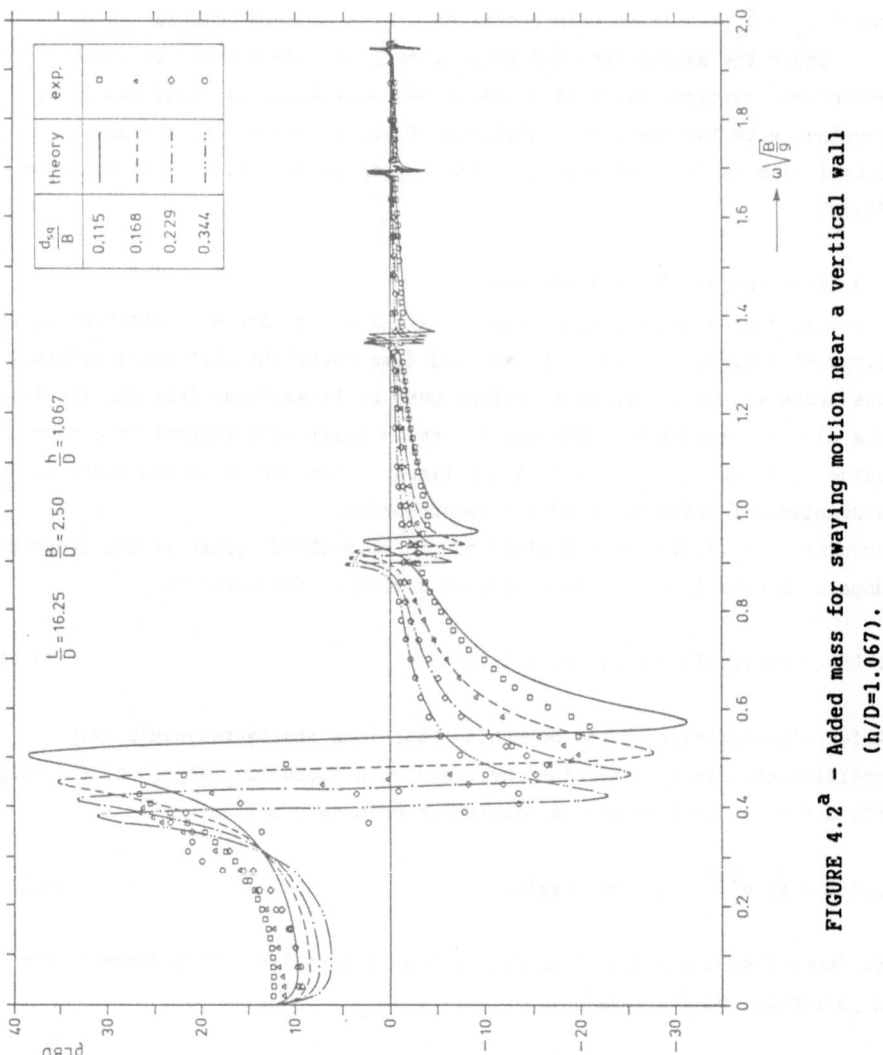

FIGURE 4.2^a - Added mass for swaying motion near a vertical wall (h/D=1.067).

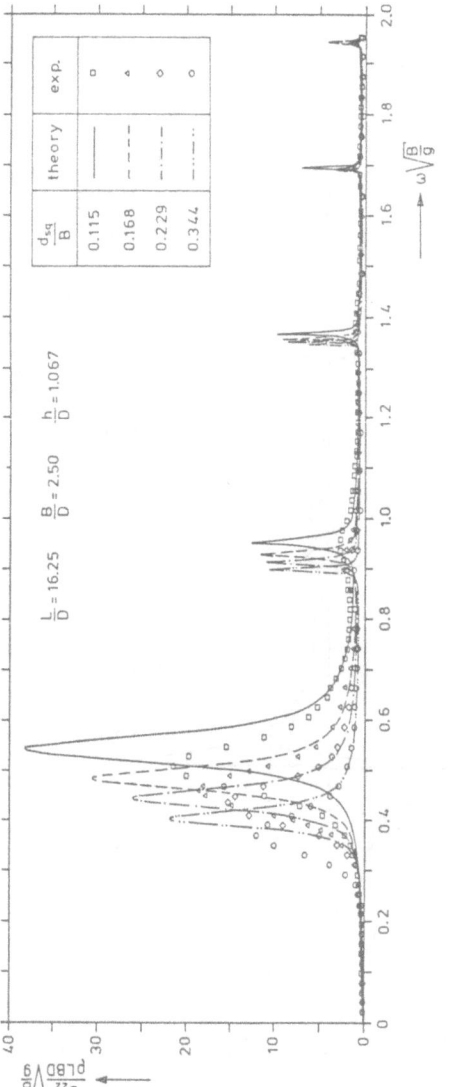

FIGURE 4.2^D – Sway damping force coefficient near a vertical wall (h/D=1.067).

130

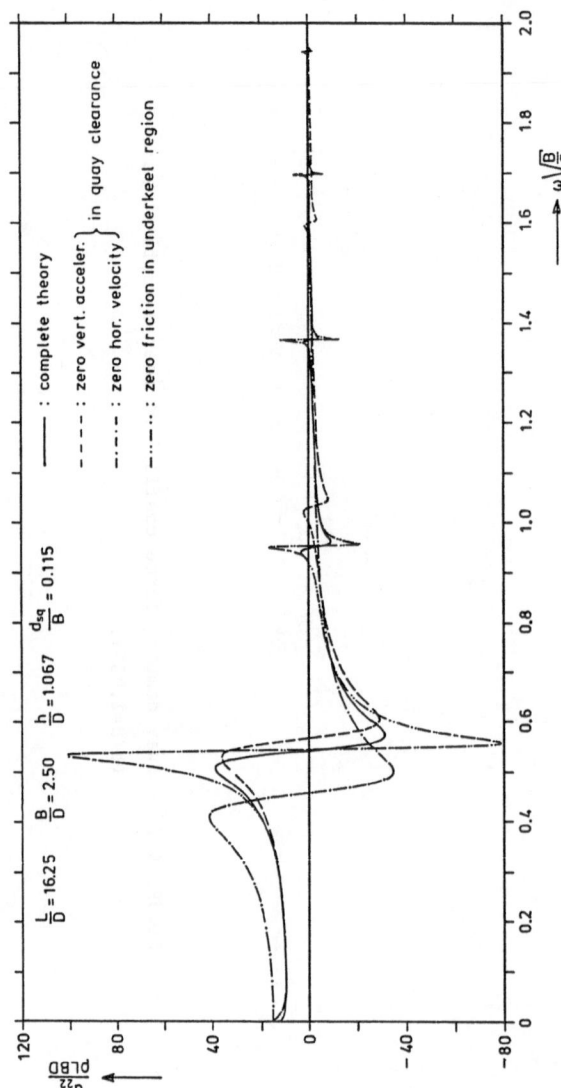

FIGURE 4.3^a – Added mass for swaying motion near a vertical wall (h/D=1.067, d$_{sq}$/B=0.115).

131

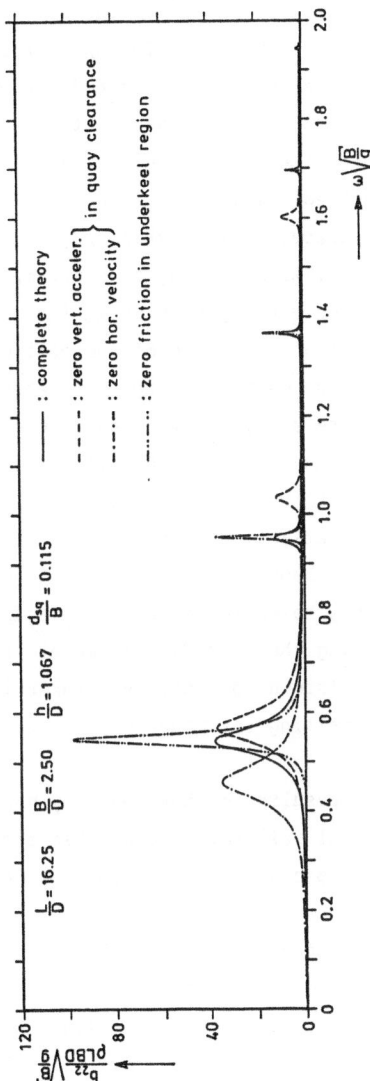

FIGURE 4.3^b - Sway damping force coefficient near a vertical wall
(h/D=1.067, d_{sq}/B=0.115).

governing equations combined with (4.10),(4.11^{a,b}) and (3.15). The
expressions for $a_{22}(\omega)$ and $b_{22}(\omega)$ are presented in Appendix B.

The hydrodynamic sway coefficients were determined not only
theoretically, but also in an experimental way as functions of ω with h and
d_{sq} as parameters. The same schematized ship model was used as in Section
3.2. Four water depths were involved, viz. h = 0.160 m, 0.175 m, 0.200 m,
0.250 m, and there were four distances between ship and quay-wall, viz.
d_{sq} = 0.043 m, 0.063 m, 0.086 m, 0.129 m. By way of example, for one water
depth, hydrodynamic sway coefficients are presented in figs. 4.2^a and 4.2^b.
The effects of the underkeel friction, and also of the horizontal velocity
and the vertical acceleration in the quay clearance, are shown in figs.
4.3^a and 4.3^b. As distinct from the situation with horizontally
unrestricted water, now the influence of the underkeel friction does appear
to be significant. The effect of the vertical acceleration in region a is
much less pronounced and the schematization of $\bar{u}_a$ seems to be too strong.
In general, the agreement between theory and experiment can be considered
as reasonable. The differences may be explained by the assumptions and
simplifications made in formulating the hydrodynamic model. For instance,
the circulation around 'bow' and 'stern' has not been taken into account.
Further, the concept used for modelling the underkeel friction might play a
part.

In comparison with corresponding results for horizontally unrestricted
water, the presence of the vertical wall has a remarkable effect on the
hydrodynamic sway coefficients, especially in the range of the lower
frequencies. Most interesting feature is the occurrence of (sharp) peaks
and negative values for the sway added mass. Observations during the tests
revealed that the peak values may be associated with the occurrence of
standing waves between ship and quay-wall with nodal lines perpendicular to
the quay-wall; this phenomenon was also found in ref. [58]. A physical
interpretation of negative (sway) added 'mass' is difficult. However, this
quantity is just the in-phase component of the fluid reactive force in the
frequency-domain description of the ship-fluid system (see Section 2.3).
Instead of combining this component with the inertia term (which is common
practice), it could also be considered as a displacement term and
denominated then as 'hydrodynamic spring coefficient'. Anyway it is obvious
that in the frequency range where the sway added mass is negative, the
water between quay-wall and ship acts like a spring. The hydrodynamic

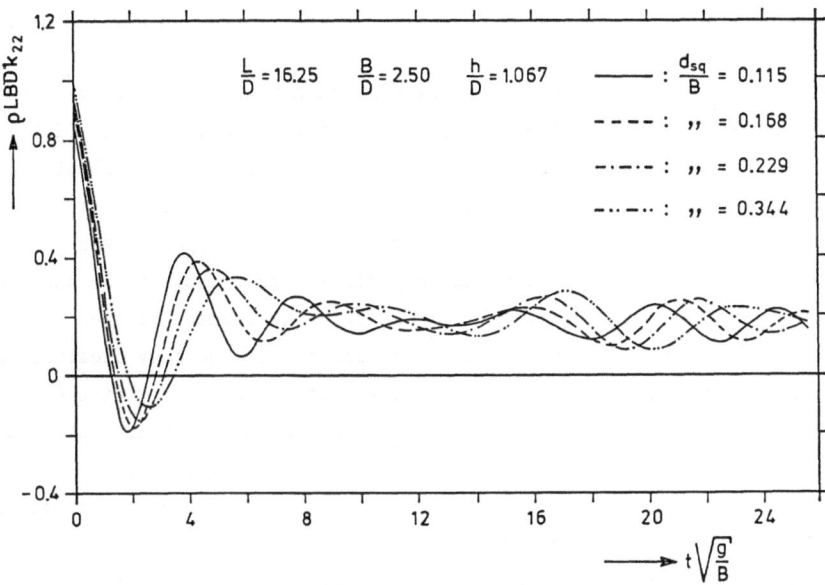

FIGURE 4.4 - Impulse response function for sway motion near a vertical wall
(h/D=1.067).

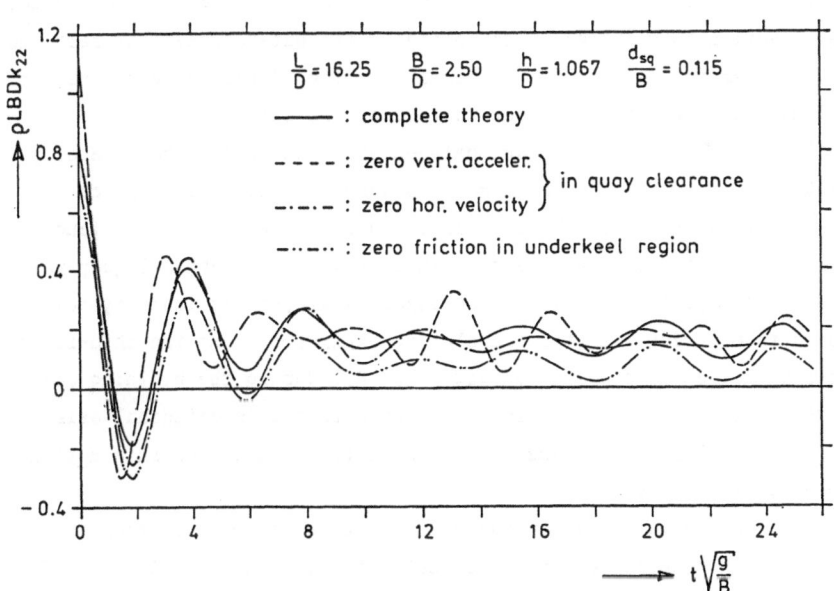

FIGURE 4.5 - Impulse response function for sway motion near a vertical wall
(h/D=1.067, d_{sq}/B=0.115).

damping near a vertical wall is considerably larger than on horizontally
unrestricted water.

As the hydrodynamic sway coefficients are now known, the corresponding
i.r.f. is determined directly from these quantities, using (2.33),(2.35[a]),
(2.36) and (2.37). In fig. 4.4, as a typical example, results calculated
from theoretical data for $a_{22}(\omega)$ and $b_{22}(\omega)$ are presented as function of
time with h and d_{sq} as parameters; the frequency range applied was
$0 \leq \omega < 100 \text{ s}^{-1}$. Fig. 4.5 shows that especially the effect of the underkeel
friction is significant, whereas the respective influences of the
horizontal velocity and the vertical acceleration in the quay clearance are
much less pronounced. From figs. 4.4 and 4.5 it can be seen that the i.r.f.
approaches (rather quickly) to a constant value as t increases; this means
that in the associated convolution integral, describing the linear ship-
fluid system in the time domain, the emphasis lies on the near past of the
time history of the forcing function.

4.3.2. Outline of mathematical approach. Consider the schematized ship
berthing to a closed structure; the geometrical situation and the
conditions are as described in Section 1.3.4. The berth is fitted with one
single fender, represented by an undamped, linear spring; its horizontal
line of action is situated in the plane of the water surface at rest and
perpendicular to the face of the berth (see fig. 4.6). Again the mass of
the fender is small with respect to that of the ship.
Shortly before the first contact between ship and fender the laterally
moving ship (i.e. $V_1 = 0$) has a certain constant speed of approach
$V_2 = - v_A$ towards the berth. The first contact between ship and fender
takes place at t = 0. Then the clearance between the vertical quay-wall and
the ship is d_{sq}. At t = 0 the space-fixed $OX_1X_2X_3$-co-ordinate system is
assumed to coincide with the translating $ox_1x_2x_3$-system; the ship-fixed
Gxyz-system then also coincides with $OX_1X_2X_3$. During the berthing operation
the ship maintains a translational motion with its longitudinal axis of
symmetry parallel to the face of the berth (i.e. uncoupled ship motions and
a 'centric impact').

Now a berthing situation has come about, which - apart from the
rotation - is comparable with that dealt with in Section 3.3. The deflexion
of the fender has the form $\Delta X_{2f}(t) = X_{2f}(t) + \frac{1}{2}B$ and can - see also (3.18) -
be expressed as:

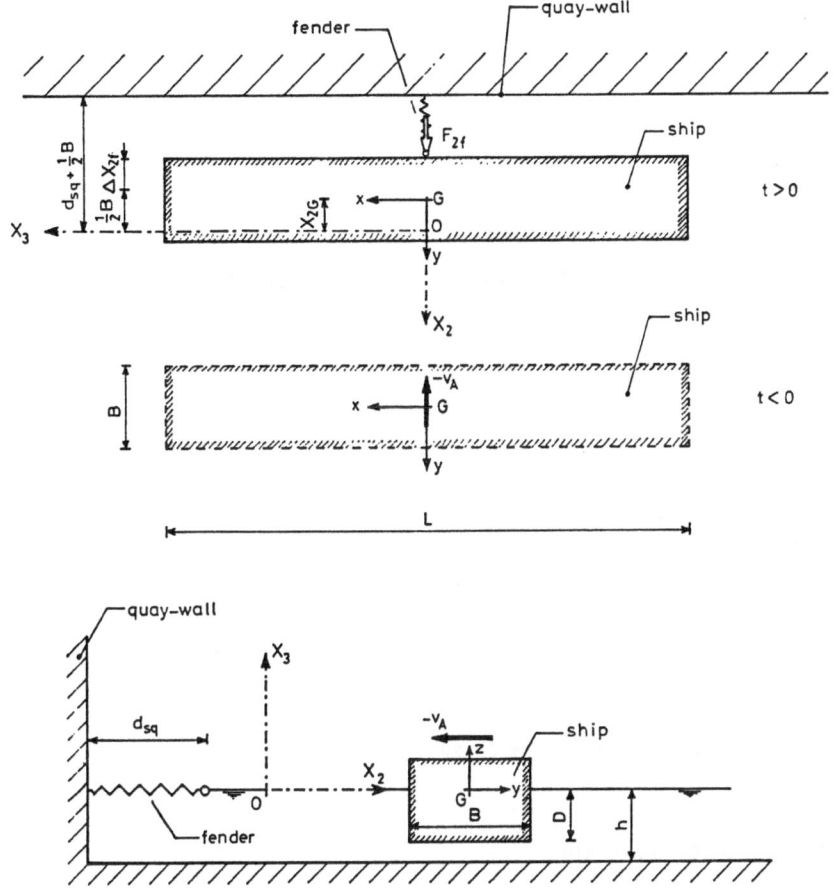

FIGURE 4.6 - Plan and cross-section of closed berthing lay-out.

$$\Delta X_{2f}(t) = X_{2G} \quad , \quad \Delta X_{2f}(t) \leq 0 \quad . \tag{4.12}$$

The relation between $\Delta X_{2f}(t)$ and the corresponding reaction force in the fender $F_{2f}(t)$ again is given by

$$F_{2f}(t) = f(\Delta X_{2f}) \quad \text{for } t \geq 0 \quad , \tag{3.19}$$

while the resulting force on the ship, as acting in G, becomes (see $(3.20^{a,b})$)

$$f_2(t) = F_{2f}(t) \quad , \quad t \geq 0 \quad . \tag{4.13}$$

The initial values of the berthing problem are given at $t = 0$ and read as:

$$\left.\begin{aligned} X_{2G}(0) &= 0 \quad , \quad \dot{X}_{2G}(0) = V_2 = -v_A \quad , \\ X_{2f}(0) &= -\tfrac{1}{2}B \quad , \qquad f_2(0) = 0 \quad . \end{aligned}\right\} \tag{4.14}$$

Since there is no rotational motion at all, it is obvious that the absolute velocity of the ship's centre of gravity - on account of $(3.30^{a,b})$ - becomes of the form

$$\dot{X}_{2G}(t) = - v_A + \dot{x}_2(t) \tag{4.15}$$

with $\dot{x}_2(t)$ given according to (3.22^b),

$$\dot{x}_2(t) = \int_0^t f_2(\tau) \, k_{22}(t-\tau)d\tau \quad . \tag{3.22^b}$$

The linear fender is represented by

$$F_{2f}(t) = \begin{cases} 0 & \text{for } \Delta X_{2f}(t) > 0 \quad , \\ - c_0 \, \Delta X_{2f}(t) & \text{for } \Delta X_{2f}(t) \leq 0 \quad . \end{cases} \tag{4.16}$$

Now it is possible to determine the fender force (and the ship trajectory), provided the relevant i.r.f. is known. The combined equations (4.12), (3.19),(4.13),(4.15) and (3.22^b) with initial values (4.14) and the fender characteristics given by (4.16), are solved numerically in a similar way as indicated in Section 3.3.

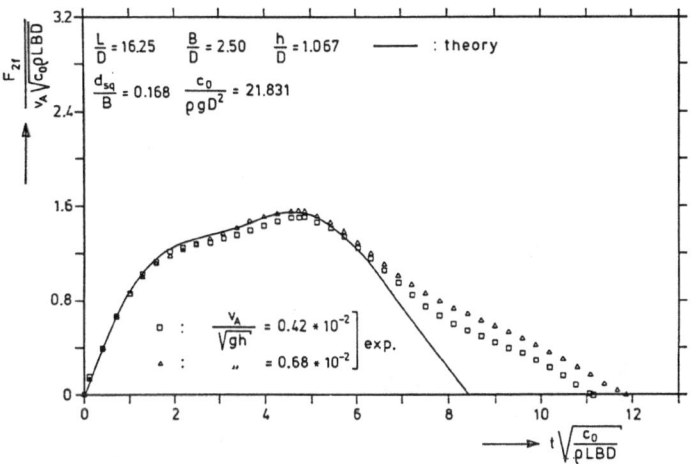

FIGURE 4.7 — Time history of fender force: closed berth (i.r.f.-technique, h/D=1.067).

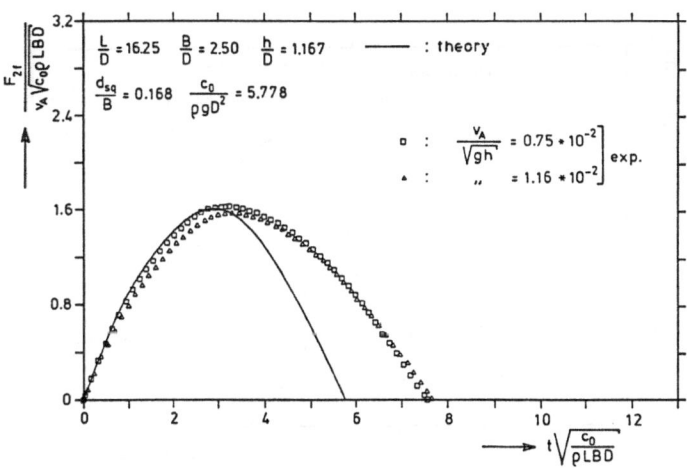

FIGURE 4.8 — Time history of fender force: closed berth (i.r.f.-technique, h/D=1.167).

138

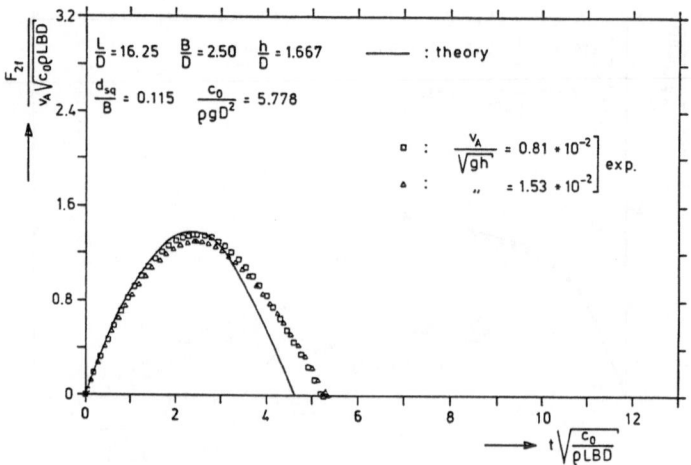

FIGURE 4.9 - Time history of fender force: closed berth (i.r.f.-technique, h/D=1.667).

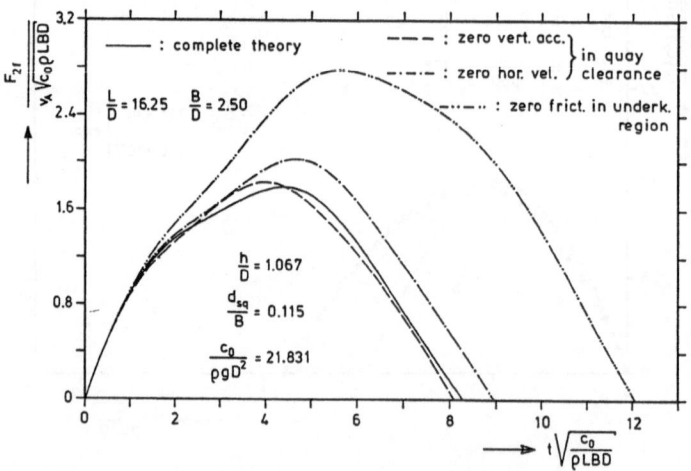

FIGURE 4.10 - Time history of fender force: closed berth (i.r.f.-technique, h/D=1.067).

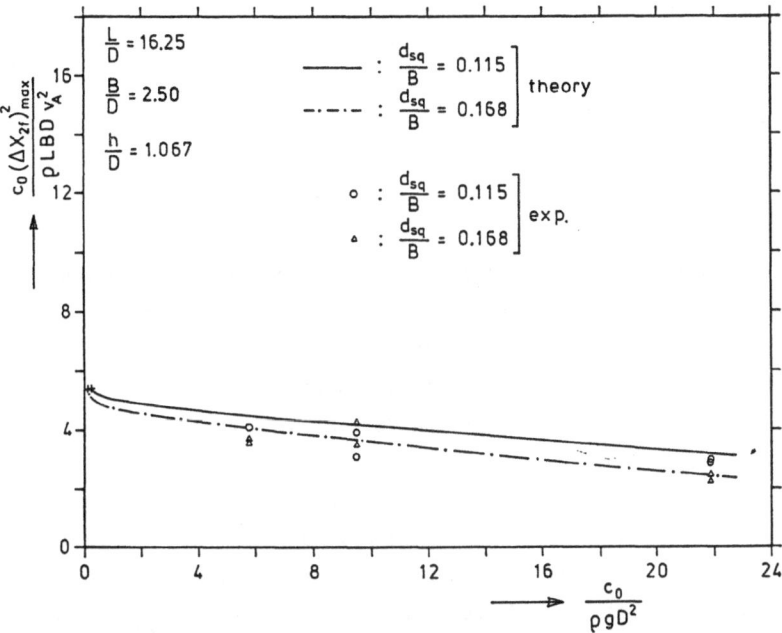

FIGURE 4.11 - Influence of fender elasticity on absorbed energy: closed
berth (h/D=1.067).

4.3.3. Examples of berthing operations: experiment and theory. Again an
extensive series of model tests was carried out to verify the adequacy of
the i.r.f.-technique for the simulation of berthing operations under
conditions as described. The same schematized ship model and water depths
were applied as stated previously. For d_{sq} only the two smallest values
were considered.

In the experiments the following quantities were measured as functions
of time: the deflexion of the fender, the position of G and the ship's
angle of rotation (merely for checking purposes), the vertical motion of
the water level in the quay clearance. In the numerical simulation $X_{2G}(t)$,
$\dot{X}_{2G}(t)$, $\Delta X_{2f}(t)$ and $F_{2f}(t)$ were calculated. The time increment used was
$\Delta t = 0.01$ s.
The berthing operation is characterized adequately by $F_{2f}(t)$. The
parameters that play a role are c_0, h, d_{sq} and v_A.

Examples of berthing operations are given in figs. 4.7, 4.8 and 4.9, showing fender forces as functions of time for various values of h, d_{sq} and c_0. As expected in the case of a linear fender, the calculated fender forces are proportional to v_A; the experimental results also point that way. Fig. 4.10 shows that the influence of the underkeel friction is significant, whereas the effects of $\bar{u}_a$ and the vertical acceleration in region a are much less pronounced. Fig. 4.11 presents the influence of the fender stiffness on the absorption of energy. Just as with horizontally unrestricted water, this figure shows - though to a less extent - that a soft fender absorbs more energy to stop the ship than a stiff fender. It has to be noted that the situation of an infinitely soft fender (i.e. $c_0 = 0$) cannot occur, since the deflexion of the fender must be smaller than d_{sq}.

From the figures it can be seen that in case of increasing fender stiffness, decreasing h or smaller values of d_{sq} (the maximum value of) $F_{2f}(t)$ increases. For larger c_0 or h, the duration of the contact between ship and fender decreases and the point of time at which $F_{2f}(t)$ reaches its peak value occurs earlier. These trends, except for d_{sq}, were also found for horizontally unrestricted water (see Section 3.4). The fender forces calculated by the i.r.f. determined from the theoretical data for $a_{22}(\omega)$, $b_{22}(\omega)$ are in satisfactory agreement with the berthing experiments: leading slope and peak value are predicted very well, the descending rear side shows a less convincing agreement. These observations apply for all parameters investigated.

4.4. Direct time–domain approach

4.4.1. Mathematical formulation. By means of a direct T.D.A., unlike the application of the i.r.f.-technique, in a general sense non-linearities (in the hydrodynamics) can be taken into account. For reasons of simplicity the horizontal velocity in the quay clearance - as being of minor importance (see above) - is left out of consideration, which implies a two–dimensional approach, or strip theory.

So, deleting u_a and $\bar{u}_a$ from the governing equations in Section 4.2 and supposing independence of x_1, a set of two ordinary differential equations in $\zeta_a(t)$ and $x_2(t)$ can be derived which are of the form

$$\ddot{\zeta}_a(t) = \{g_4(t)P_0(t) - g_3(t)Q_0(t)\}\{g_1(t)g_4(t) - g_2(t)g_3(t)\}^{-1} \quad , \qquad (4.17)$$

$$\ddot{x}_2(t) = \{-g_2(t)P_0(t) + g_1(t)Q_0(t)\}\{g_1(t)g_4(t) - g_2(t)g_3(t)\}^{-1} \quad , \quad (4.18)$$

where $P_0(t),Q_0(t)$ = function of $x_2,\dot{x}_2,\zeta_a,\dot{\zeta}_a$,

$\quad g_1(t),g_2(t),g_3(t),g_4(t)$ = function of x_2;

these time functions are presented in full detail in Appendix C. The underkeel friction effect is modelled using Blasius' formula adapted to the geometrical situation (see ref. [91]). The external forcing function upon the ship again is the linear fender force, now - on the analogy of (4.12), (3.19),(4.13) and (4.16) combined - to be represented by

$$f_2(t) = F_{2f}(t) = \begin{cases} 0 & \text{for } x_2(t) > 0 \quad , \\[2ex] -c_0 x_2(t) & \text{for } x_2(t) \leq 0 \quad . \end{cases} \quad (4.19)$$

The initial values of the problem are:

$$\zeta_a(0) = 0 \quad , \quad \dot{\zeta}_a(0) = 0 \quad , \quad x_2(0) = 0 \quad , \quad \dot{x}_2(0) = -v_A \quad . \quad (4.20)$$

The numerical solution of the set of differential equations (4.17) and (4.18) with $f_2(t)$ given by (4.19) and with initial conditions (4.20) is carried through by means of a fourth-order Runge-Kutta computational scheme (see ref. [92]). A time step of 0.01 s was found to be more than adequate to give reliable results. The calculation is finished when the ship loses contact with the fender. The linearized version of (4.17) and (4.18) has also been solved.

4.4.2. Presentation and discussion of results. For the numerical simulation the same berthing situations were selected as used in applying the i.r.f.-technique. In addition to the displacement and the velocity of the ship, the deflexion of the fender and the fender force, now also $\zeta_a(t)$ could be calculated. Further, the same parameters apply.

Examples of berthing operations are given in figs. 4.12, 4.13 and 4.14. In addition to the experimental results plotted as centred symbols, these figures each show three curves: the full line and the dashed line represent results as calculated from the direct T.D.A. taking into account non-linearities, for two different values of v_A; the dot-and-dash line represents the results as calculated from the linearized version of the direct T.D.A. By an appropriate choice of f_1 and f_t, for each combination of h and c_0, the underkeel friction was adapted such that the theoretical results in broad outline fit the experiments; for the sake of simplicity $f_1 = f_t$. In the linear version of the direct T.D.A. the same value of f_1 was applied as in the non-linear version. As expected, $F_{2f}(t)$ and $\zeta_a(t)$, as calculated from the non-linear version, are not proportional to v_A; naturally, the principle of proportionality holds good for the linearized version. Within the range of values applied for v_A, the test results do point in the direction of proportionality to this quantity. The behaviour of $F_{2f}(t)$ when varying c_0, h and d_{sq} is the same as found by applying the i.r.f.-technique. The (minimum) value of $\zeta_a(t)$ falls with increasing c_0 and decreasing d_{sq} and shows a rising trend in the case of larger h.

The friction formula used actually applies only to steady flow with fully developed boundary layers between parallel smooth walls. During berthing operations, however, the underkeel water motion has a transient character, which implies developing thin boundary layers with large velocity gradients not influenced by the height of the keel clearance. As the boundary layer controls the shear stress, this explains why the values of f_1, f_t as chosen for curve fitting differ substantially from their standard value 1 and grow larger with increasing h. It must be recognized that calibrating the theory with experimental results in terms of the selection of the friction factors covers any errors arising from the two-dimensional approach adopted and other theoretical (and experimental) imperfections in addition to its effect on the underkeel friction. Anyhow, the theoretical results are in satisfactory agreement with the experiments: leading slope and peak value of the fender force are represented reasonably well, the descendant rear side shows a less convincing agreement; in judging the theoretical results for ζ_a it has to be considered that the influence of the strip theory on the vertical motion of the water level in the quay clearance is relatively strong. In comparison with the non-linear version of the direct T.D.A. the linearized version shows consistently conservative values for $F_{2f}(t)$ and $\zeta_a(t)$, but the effect of the non-linearities is small. The underkeel friction effect can be shown to be significant, whereas the influences of the vertical acceleration in region a and the contraction and separation of flow under the ship are much less pronounced (see figs. 4.15 and 4.16). These observations apply for all values of h, c_0, d_{sq} and v_A investigated.

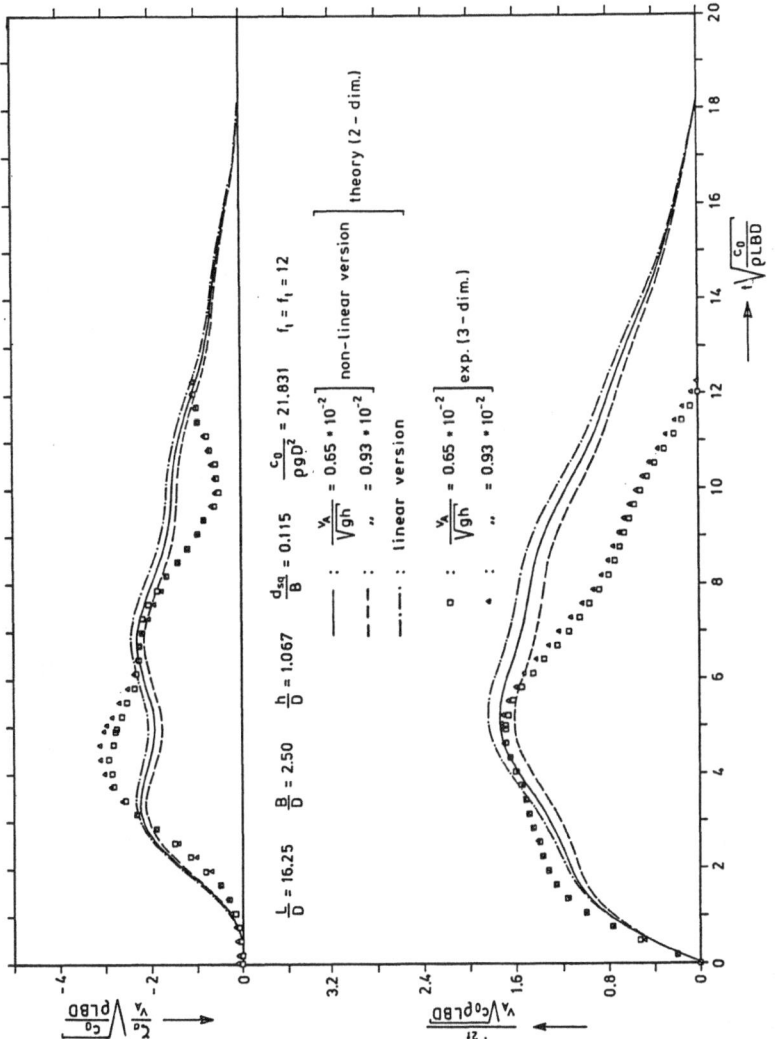

FIGURE 4.12 - Time histories of fender force and water-surface elevation in quay clearance: closed berth (direct T.D.A., h/D=1.067).

144

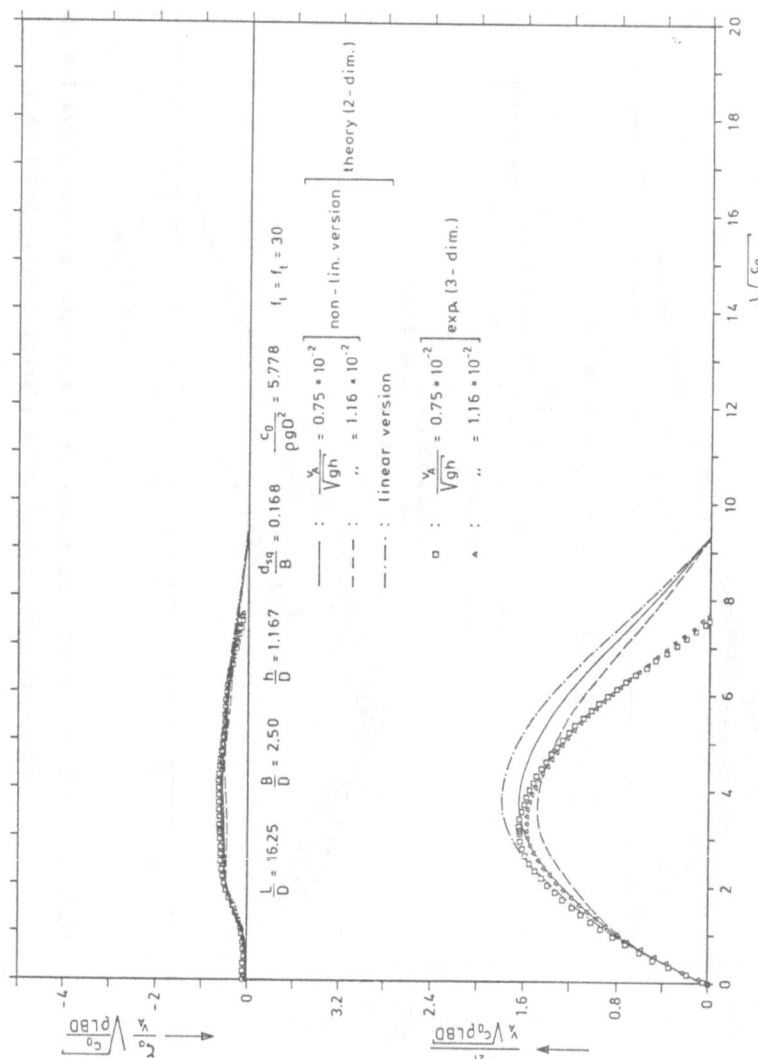

FIGURE 4.13 — Time histories of fender force and water-surface elevation in quay clearance: closed berth (direct T.D.A., h/D=1.167).

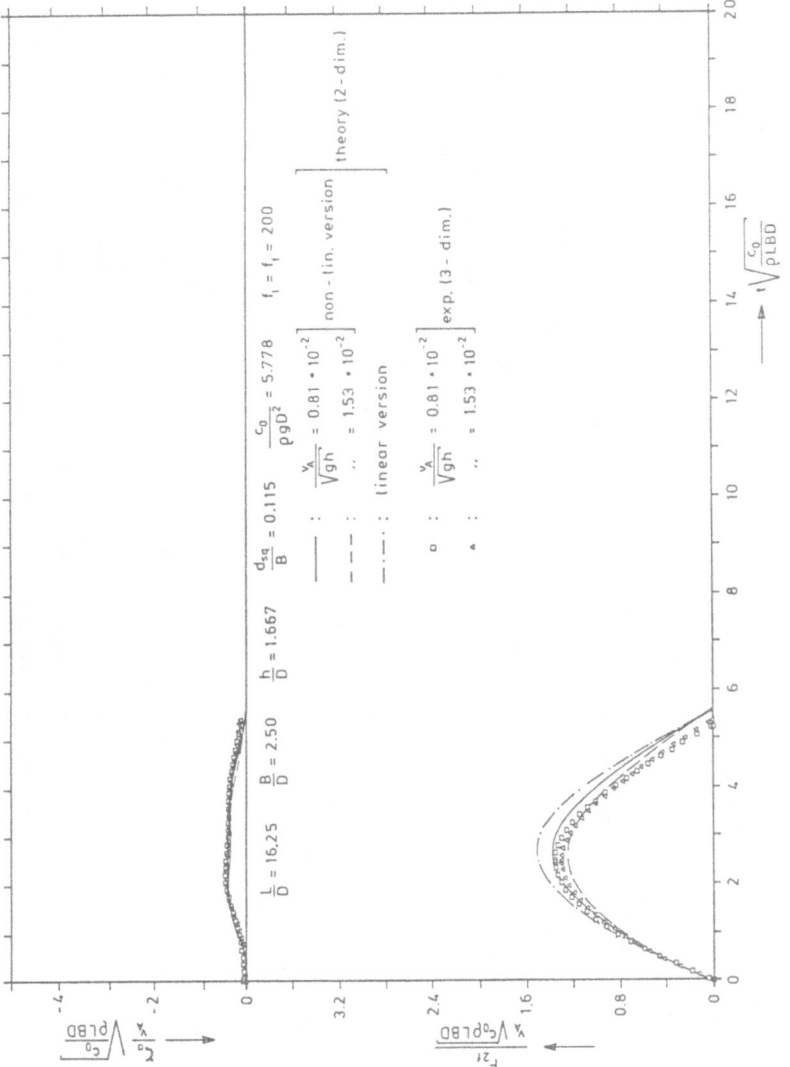

FIGURE 4.14 – Time histories of fender force and water-surface elevation in quay clearance: closed berth (direct T.D.A., h/D=1.667).

146

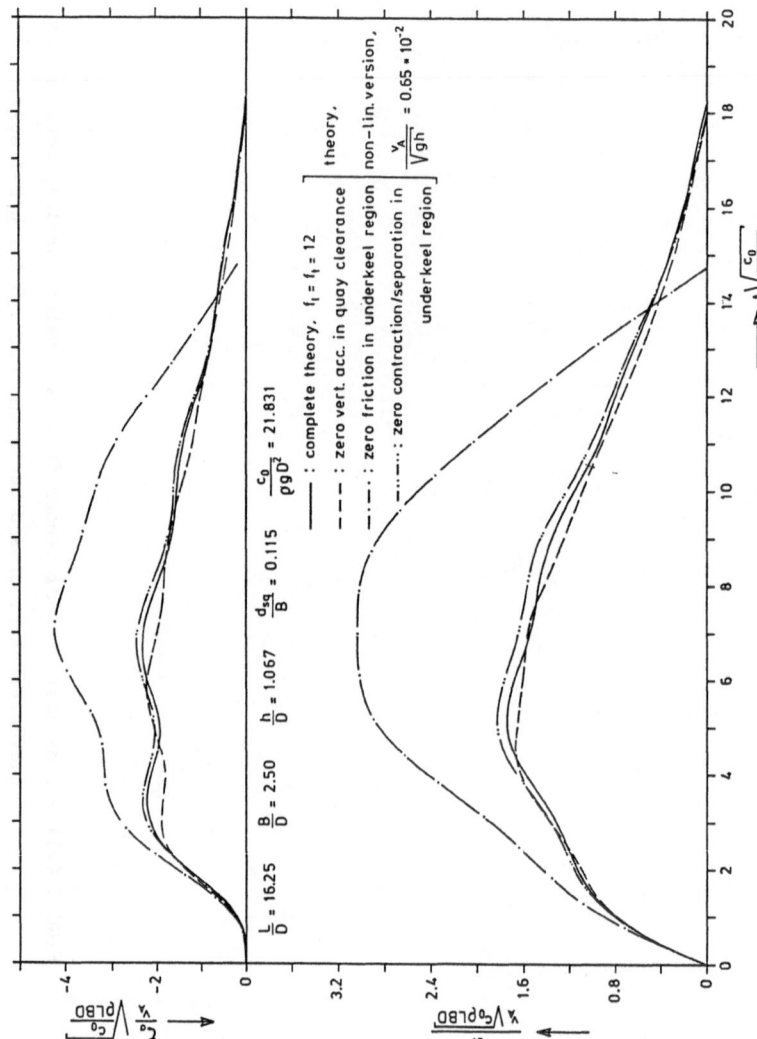

FIGURE 4.15 - Time histories of fender force and water-surface elevation in quay clearance: closed berth (direct T.D.A., h/D=1.067).

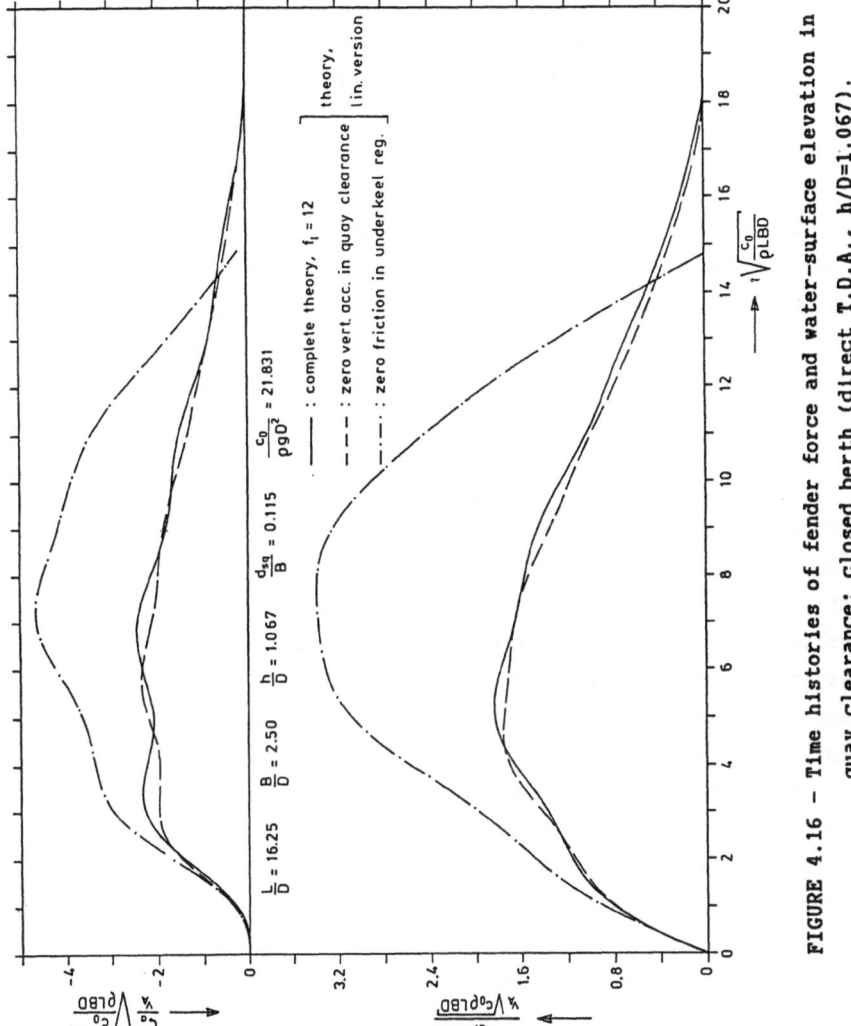

FIGURE 4.16 – Time histories of fender force and water-surface elevation in quay clearance: closed berth (direct T.D.A., h/D=1.067).

Elimination of $\zeta_a(t)$ from the linearized version of (4.17) and (4.18) yields a linear, homogeneous, ordinary differential equation of the fourth order with constant, real, positive coefficients. By considering the orders of magnitude of the four roots of its characteristic equation, approximate expressions can be derived for the two circular frequencies playing a part in the berthing-ship problem. One circular frequency corresponds with a period time approximately equalling twice the duration of the contact between ship and fender; the other can be discerned as a faster oscillation on the curves representing the time histories of $F_{2f}(t)$ and $\zeta_a(t)$. Further, calculation of the berthing operations with a better estimate for the initial condition $\zeta_a(0) \neq 0$ does not yield substantially differing results.

5. SUMMARY AND CONCLUSIONS

The description of a ship berthing to an open or a closed fender structure as well as the determination of the associated fender forces necessitate a time-domain approach, in which the fluid reactive forces are represented in an appropriate way and the remaining forces are taken into account over their entire time histories. The mathematical formulations presented satisfy these requirements and enable the inclusion of external forces of arbitrary nature; this implies that the fender may be damped, undamped, linear or non-linear. A practicable and sufficiently accurate foundation is provided for the description and determination of the relevant quantities figuring in the ship-berthing phenomenon.
The respective mathematical approaches are applied to the simplified case of a schematized ship, that on calm, shallow water at zero forward speed berthes to an open jetty-type berthing structure or a closed, straight, vertical quay-wall, each fitted with one undamped, (non-)linear fender without mass of its own. For verification experiments were carried out on a scale model.

In case of ship berthing to the open fender structure both centric and eccentric impacts are considered. This berthing situation is tackled by means of the i.r.f.-technique, which has the restrictions that the ship-fluid system is supposed to be linear and time-invariant; the fluid reactive forces then are described by way of the frequency-dependent hydrodynamic coefficients. Underkeel friction effects appear to be of secondary importance. Comparison of theory and experiment shows a good agreement. For a first estimation of the maximum value of the fender force use could be made of a long-wave approximation for the motion of the water.

In respect of ship berthing to the closed fender structure only centric impacts are dealt with. Two methods are used, namely the i.r.f.-technique and a direct T.D.A., both of them ensueing from the same mathematical model. Application of the i.r.f.-technique now leads to satisfactory results only if (linear) underkeel friction is incorporated in the hydrodynamic coefficients. The influence of the horizontal velocity in the quay clearance is small. An option of the direct T.D.A. is that the influence of non-linear (hydrodynamic) effects can be taken into account. Also this method presents a satisfactory agreement between theory and experiment, provided that the underkeel friction, at least, is estimated properly. In general, the influence of non-linearities is small. For both methods the vertical acceleration in the quay clearance plays a minor part.

For both berthing situations, the respective mathematical formulations presented are sufficiently accurate for the qualitative and quantitative description of the typical behaviour of a berthing ship as well as for the determination of the response of the fender(s). The calculated results are in satisfactory agreement with values obtained from measurements on a scale model, especially up to the point of time where the maximum value of the fender force is reached: leading slope and peak value of the fender force are predicted well. This applies for all water depths, fender stiffnesses (, initial clearances between ship and quay-wall) and lateral speeds of approach investigated.

Further, the assumption of linearity for the ship-fluid system in both berthing situations appears to yield a very well acceptable and practicable approximation for the quantitative analysis of transient motions of shiplike bodies.

Viscous effects of the fluid do not influence significantly the relevant quantities that play a part in the berthing phenomenon.

APPENDIX A. Expressions for hydrodynamic coefficients on horizontally unrestricted water

$$a_{22}(\omega) = 2\rho L \left(-\frac{1}{m_0^2} \frac{\hat{v}_b}{\omega \hat{x}_2} B_0' \sin(\theta)[\sinh(m_0 h)-\sinh\{m_0(h-D)\}] + \right.$$

$$\left. + \sum_{n=1}^{\infty} \frac{1}{m_n^2} \{A_n' + \frac{\hat{v}_b}{\omega \hat{x}_2} B_n' \cos(\theta)\}[\sin(m_n h)-\sin\{m_n(h-D)\}] \right) \quad ;$$

$$b_{22}(\omega) = 2\rho L \left(\frac{\omega}{m_0^2} \{A_0' + \frac{\hat{v}_b}{\omega \hat{x}_2} B_0' \cos(\theta)\}[\sinh(m_0 h)-\sinh\{m_0(h-D)\}] + \right.$$

$$\left. + \sum_{n=1}^{\infty} \frac{\omega}{m_n^2} \frac{\hat{v}_b}{\omega \hat{x}_2} B_n' \sin(\theta)[\sin(m_n h)-\sin\{m_n(h-D)\}] \right) \quad ;$$

m_0 = positive root of $\quad \omega^2 = gm_0 \tanh(m_0 h) \quad$,

m_n = positive roots of $\quad \omega^2 = -gm_n \tan(m_n h) \quad$, $m_1 < m_2 < \ldots < m_n < \ldots \quad$,

$$(n-\frac{1}{2})\pi < m_n h < n\pi \quad ;$$

$$A_0' = 2\frac{\sinh(m_0 h)-\sinh\{m_0(h-D)\}}{m_0 h+\sinh(m_0 h)\cosh(m_0 h)} \quad , \quad A_n' = 2\frac{\sin(m_n h)-\sin\{m_n(h-D)\}}{m_n h+\sin(m_n h)\cos(m_n h)} \quad ,$$

$$B_0' = \frac{2\sinh\{m_0(h-D)\}}{m_0 h+\sinh(m_0 h)\cosh(m_0 h)} \quad , \quad B_n' = \frac{2\sin\{m_n(h-D)\}}{m_n h+\sin(m_n h)\cos(m_n h)} \quad ,$$

$$A_0 = \hat{x}_2 A_0' \quad , \quad A_n = \hat{x}_2 A_n' \quad , \quad B_0 = \frac{\hat{v}_b}{\omega}B_0' \quad , \quad B_n = \frac{\hat{v}_b}{\omega}B_n' \quad ;$$

$$\tan(\theta) = \frac{-a_0(b_n+c)+a_n b_0}{-a_0 b_0-a_n(b_n+c)} \quad ,$$

$$\frac{\hat{v}_b}{\omega \hat{x}_2} = \frac{-a_0}{b_0 \cos(\theta)+(b_n+c)\sin(\theta)} = \frac{a_n}{b_0 \sin(\theta)-(b_n+c)\cos(\theta)} \quad ,$$

$$a_0 = \frac{2A_0'}{m_0^2} \sinh\{m_0(h-D)\} \quad , \qquad a_n = \sum_{n=1}^{\infty} \frac{2A_n'}{m_n^2} \sin\{m_n(h-D)\} \quad ,$$

$$b_0 = \frac{2B_0'}{m_0^2} \sinh\{m_0(h-D)\} \quad , \qquad b_n = \sum_{n=1}^{\infty} \frac{2B_n'}{m_n^2} \sin\{m_n(h-D)\} \quad ,$$

$$c = B(h-D) \quad ;$$

$$a_{22}(0) = 2\rho L \left\{ \frac{BD^2}{2(h-D)} + 2 \frac{h^4}{(h-D)^2} \sum_{n=1}^{\infty} \frac{1}{(n\pi)^3} \sin^2(n\pi\frac{h-D}{h}) \right\} \quad ,$$

$$a_{22}(\infty) = 4\rho L h^2 \left\{ \sum_{n=1}^{\infty} \{(-1)^n + P_n\}^2 Q_n^{-3} + \right.$$

$$\left. - \left[\sum_{n=1}^{\infty} \{(-1)^n + P_n\} P_n Q_n^{-3} \right]^2 \left[\frac{B(h-D)}{4h^2} + \sum_{n=1}^{\infty} P_n^2 Q_n^{-3} \right]^{-1} \right\} \quad ,$$

$$P_n = \sin(\frac{h-D}{h} Q_n) \quad , \quad Q_n = (2n-1)\frac{\pi}{2} \quad ,$$

$$b_{22}(0) = 0 \quad , \quad b_{22}(\infty) = 0 \quad ;$$

$$a_{66}(\omega) = \frac{1}{12} L^2 a_{22}(\omega) \quad , \qquad b_{66}(\omega) = \frac{1}{12} L^2 b_{22}(\omega) \quad .$$

APPENDIX B. Expressions for hydrodynamic sway coefficients near a vertical
        wall

$$a_{22}(\omega) = -\rho LBD - \rho\omega^{-2} \mathrm{Re}[\hat{f}_2(\hat{p x}_2)^{-1}] \quad ,$$

$$b_{22}(\omega) = \rho\omega^{-1} \mathrm{Im}[\hat{f}_2(\hat{p x}_2)^{-1}] \quad ,$$

$$\hat{f}_2(\hat{p x}_2)^{-1} = (-LBD + \frac{1}{6}f_{w_a} LD^3 d_{sq}^{-1})\omega^2 - DLTRP^{-1} +$$

$$- 2f_{u_a} DR(rQ)^{-1}(gD d_{sq} c_w^{-1} + \frac{1}{6} f_{w_a} i\omega D^2 - TQP^{-1})\tanh(\frac{1}{2}rL) +$$

$$+ \frac{1}{2}\alpha_1 c_w Bh^{-1}\{i\omega L(d_{sq} RP^{-1} + h) + 2f_{u_a} d_{sq} R(rQ)^{-1}(-i\omega QP^{-1} + D)\tanh(\frac{1}{2}rL)\} \quad ,$$

$$P = -\omega^2 \{ d_{sq}(h-D)^{-1} + f_{w_a} DB^{-1} \} + i\omega g d_{sq}(Bc_w)^{-1} + gB^{-1} + i\omega\alpha_1 g d_{sq}\{ c_w(h-D) \}^{-1} \quad ,$$

$$Q = i\omega D \{ d_{sq}(h-D)^{-1} + \tfrac{1}{2}f_{w_a} DB^{-1} \} + gDd_{sq}(Bc_w)^{-1} + \alpha_1 gDd_{sq}\{ c_w(h-D) \}^{-1} \quad ,$$

$$R = \omega^2 D \{ (h-D)^{-1} + \tfrac{1}{2}f_{w_a} D(Bd_{sq})^{-1} \} - i\omega\alpha_1 g \{ c_w(h-D) \}^{-1} \{ D + \tfrac{1}{2}(h-D) \} \quad ,$$

$$S = (g - \tfrac{1}{2}f_{w_a} D\omega^2) Q(i\omega P)^{-1} - \tfrac{1}{6}f_{w_a} D^2 \quad ,$$

$$T = g(1 + i\omega d_{sq} c_w^{-1}) - \tfrac{1}{2}f_{w_a} \omega^2 D \quad ,$$

$$r = S^{-\frac{1}{2}} \quad ,$$

$$\alpha_1 = 2\gamma c_w \{ \rho g(h-D) \}^{-1} \quad , \quad \gamma = \rho\sqrt{\upsilon\omega} \quad , \quad c_w = \sqrt{gh} \quad ,$$

$$f_{w_a} = \begin{array}{l} 0 \;\; : \;\; \text{zero} \\ +1 \;\; : \;\; \text{non-zero} \end{array} \Bigg\} \; \text{vertical acceleration in region a} \quad ,$$

$$f_{u_a} = \begin{array}{l} 0 \;\; : \;\; \text{zero} \\ +1 \;\; : \;\; \text{non-zero} \end{array} \Bigg\} \; \text{horizontal velocity in region a} \quad ,$$

$$a_{22}(0) = \rho LBD \{ D(h-D)^{-1} + \tfrac{1}{3}f_{w_a} D^2 (Bd_{sq})^{-1} \} \quad ,$$

$$b_{22}(0) = 0 \quad .$$

APPENDIX C. Time functions figuring in direct T.D.A.

$$g_1(t) = (d_{sq} + x_2)(h-D)^{-1} + f_{w_a} DB^{-1} \quad ,$$

$$g_2(t) = (d_{sq} + x_2)(h-D)^{-1} + \tfrac{1}{2}f_{w_a} DB^{-1} \quad ,$$

$$g_3(t) = D(h-D)^{-1} + \tfrac{1}{2}f_{w_a} D^2 \{ B(d_{sq} + x_2) \}^{-1} \quad ,$$

$$g_4(t) = 1 + D(h-D)^{-1} + \tfrac{1}{3}f_{w_a} D^2 \{ B(d_{sq} + x_2) \}^{-1} \quad ,$$

$$P_0(t) = - \{g(d_{sq}+ x_2)(Bc_w)^{-1} + \dot{x}_2(h-D)^{-1}\}\dot{\zeta}_a - gB^{-1}\zeta_a + \tfrac{1}{2}g_5(t)(\dot{x}_2)^2 +$$

$$-(h-D)^{-1}\{X(t)+Y(t)\} - \tfrac{1}{2}\xi B^{-1}(h-D)^{-2}Z^2(t)\,\mathrm{sgn}\{Z(t)\} \quad ,$$

$$Q_0(t) = -\dot{x}_2\dot{\zeta}_a(h-D)^{-1} + \tfrac{1}{3}g_5(t)(\dot{x}_2)^2 - (h-D)^{-1}\{X(t)+Y(t)\} - D^{-1}X(t) +$$

$$- \tfrac{1}{2}\xi B^{-1}(h-D)^{-2}Z^2(t)\,\mathrm{sgn}\{Z(t)\} + (\rho LBD)^{-1}f_2(t) \quad ,$$

$$g_5(t) = f_{w_a}D^2 B^{-1}(d_{sq}+x_2)^{-2} \quad ,$$

$$Z(t) = (d_{sq}+x_2)\dot{\zeta}_a + h\dot{x}_2 \quad , \quad \xi = 1.44 \quad ,$$

$$X(t) = 6f_1 \upsilon(h-D)^{-2}Z(t)$$

$$Y(t) = 6f_1 \upsilon(h-D)^{-2}\{Z(t)-(h-D)\dot{x}_2\}$$

$$\left.\right\} \quad \text{for } Re \le 2300 \quad ,$$

$$X(t) = 0.03326f_t \upsilon^{1/4}(h-D)^{-2}|Z(t)|^{7/4}\,\mathrm{sgn}\{Z(t)\}$$

$$Y(t) = 0.03326f_t \upsilon^{1/4}(h-D)^{-2}|Z(t)-(h-D)\dot{x}_2|^{7/4}\,\mathrm{sgn}\{Z(t)-(h-D)\dot{x}_2\}$$

$$\left.\right\} \quad \text{for } Re > 2300,$$

$$Re = \upsilon^{-1}|2Z(t) - (h-D)\dot{x}_2| \quad ,$$

f = multiplication factor for friction effect (standard value $\equiv$ 1).

The terms containing X(t) and Y(t) represent the underkeel friction effect. The transition between laminar and turbulent underkeel flow takes place at a Reynolds number Re = 2300, Re defined as $Re = 2\,\upsilon^{-1}|\bar{v}_b - \tfrac{1}{2}\dot{x}_2|(h-D)$.

APPENDIX D. References

1. Rupert, D.
   'Zur Bemessung und Konstruktion von Fendern und Dalben', Mitteilungen des Franzius-Instituts für Wasserbau und Küsteningenieurwesen der Technischen Universität Hannover, H. 44, Hannover 1976, pp. 112-288.
2. Report of the International Commission for Improving the Design of Fender Systems, P.I.A.N.C., suppl. to Bull. No. 45, 1984.

154

3. Pagès, M.

'Etude mécanique du choc se produisant lors de l'accostage d'un navire
à un quai', Annales des Ponts et Chaussées, 122e année, mars-avril 1952,
pp. 205-217.

4. Eggink, A.

XVIIIth Intern. Navigation Congress, P.I.A.N.C., Rome 1953, S.II-2, pp.
167-187.

5. Grim, O.

'Das Schiff und der Dalben', Schiff und Hafen, H.9, 1955, pp. 535-545.

6. Abbett, W. / Levington, Z.

'Design and construction of terminals for large ships', XXth Intern.
Navigation Congress, P.I.A.N.C., Baltimore 1961, S.II-1, pp. 307-328.

7. Gillespie, J.H.H. / Palmer, J.E.G. / e.a.

XXth Intern. Navigation Congress, Baltimore 1961, S.II-1, pp. 89-118.

8. Greco, L. / Gangemi, F. / Pezza, V.

XXth Intern. Navigation Congress, Baltimore 1961, S.II-1, pp. 131-155.

9. Woodruff, G.B.

'Berthing and mooring forces', Journ. of the Waterways and Harbors
Division, Proc. A.S.C.E., no. WW1, Vol. 88, Febr. 1962, pp. 71-82.
Discussion: idem, no. WW3, Vol. 88, Aug. 1962, pp. 189-193.

10. Saurin, B.F.

'Berthing forces of large tankers', Proc. of the Sixth World Petroleum
Congress, Frankfurt/Main, June 1963, Section VII, Paper 10, pp. 63-73.

11. Vasco Costa, F.

'The berthing ship. The effect of impact on the design of fenders and
other structures', The Dock & Harbour Authority. Vol. XLV, nos. 523,
524, 525, May, June, July 1964, pp. 22-26, 49-52, 90-94.

12. Proc. of the Nato Advanced Study Institute on 'Analytical treatment of
problems of berthing and mooring ships', Lisbon, July 1965 (publ.
A.S.C.E., New York, 1970).

13. Giraudet, P.

'Recherches expérimentales sur l'énergie d'accostage des navires',
Annales des Ponts et Chaussées, 136e année, no. II, mars-avril 1966,
pp. 103-127.

14. Tyrrell, B.G.

'Mooring dolphins. Evaluation of forces and principles of design', The Dock & Harbour Authority, Vol. XLVII, Nr. 550, August 1966, pp. 115-120; Nr. 551, Sept. 1966, pp. 161-166.

15. Vasco Costa, F.
    'Berthing manoeuvres of large ships', The Dock & Harbour Authority, Vol. XLVII, no. 569, March 1968, pp. 351-358.

16. Lee, T.T.
    'Design criteria recommended for marine fender systems', Proc. 11th Conf. on Coastal Engineering, London, Sept. 1968 (publ. A.S.C.E., 1969, part 3, pp. 1159-1184).

17. Nagai, S. / Oda, K. / Shigedo, M.
    'Impacts exerted on the dolphins of sea-berths by roll, sway and drift of supertankers subjected to waves and swells', XXIInd Intern. Navigation Congress, P.I.A.N.C., Paris 1969, S.II-3, pp. 63-90.

18. Reese, L.C. / O'Neill, M.W. / Radhakrishnan, N.
    'Rational design concept for breasting dolphins', Journ. of the Waterways and Harbors Division, Proc. A.S.C.E., no. WW2, Vol. 96, May 1970, pp. 433-450.

19. Shu-t'ien Li / Venkataswamy Ramakrishnan
    'Ultimate energy design of prestressed concrete fender piling', Journ. of the Waterways, Harbors and Coastal Engineering Division, Proc. A.S.C.E., no. WW4, Vol. 97, Nov. 1971, pp. 647-662.

20. Komatsu, S. / Salman, A.H.
    'Dynamic response of the ship and the berthing fender system after impact', Trans. Japanese Soc. of Civil Engineers, Vol. 4, 1972, pp. 18-19.

21. Papers presented at Nato Advanced Study Institute on 'Analytical treatment of problems in the berthing and mooring of ships', Wallingford (U.K.), May 1973 (publ. Hydraulics Research Station, Wallingford, U.K.).

22. Sakharov, S.M. / Voronin, P.P. / e.a.
    XXIIIrd Intern. Navigation Congress, P.I.A.N.C., Ottawa 1973, S.II-1, pp. 289-311.

23. Taubert, A.
    'Belastung von Bauwerken durch Schiffsstosz', HANSA, 110, 1973, No. 21, pp. 1864-1868.

24. Dubois, J. / Langlet, M.

156

'Etudes relatives aux conditions d'accostage et d'amarrage au terminal d'Antifer', 6th Int. Harbour Congress, Antwerpen, May 1974, pp. 2.29/1-2.29/6.

25. Lee, T.T. / Nagai, S. / Oda, K.
'On the determination of impact forces, mooring forces and motions of supertankers at marine terminal', 7th Annual Offshore Technology Conference, Houston, Texas, May 1975, Paper nr. OTC 2211, pp. 661-678.

26. Fischer, J.
'Conception of the fendering systems for the (very) large ships berthing', 24th Intern. Navigation Congress, P.I.A.N.C., Leningrad 1977, S.II-4, pp. 15-31.

27. Wirsbitzki, B.
'Criteria for economical design of fender systems', 24th Intern. Navigation Congress, P.I.A.N.C., Leningrad 1977, S.II-4, pp. 33-46.

28. Patrick, J.G.
'The design of jetties for large ships', Trans. Inst. Mar. Engrs. (C), Vol. 92, Conf. no. 6, 1980, Paper C48, pp. 19-24.

29. Nikerov, P.S. / Goryunov, B.F. / e.a.
'Improving the methods of determining the loads applied by berthing ships, the effect of wave disturbance, and examination of flexible fender systems', XXVth Intern. Navigation Congress, P.I.A.N.C., Edinburgh 1981; Inland & Maritime Waterways & Ports, Design-Construction-Operation, S.II: Maritime Ports and Seaways, Vol. I, Pergamon Press, Oxford, 1981, pp. 195-208.

30. Callet, P.
XVIIIth Intern. Navigation Congress, P.I.A.N.C., Rome 1953, S.II-Q.2, pp. 87-109.

31. Visioli, F. / Gleieses, M. / e.a.
XVIIIth Intern. Navigation Congress, P.I.A.N.C., Rome 1953, S.II-2, pp. 143-165.

32. Girgrah, M.
'Practical aspects of dock fender design', 24th Intern. Navigation Congress, P.I.A.N.C., Leningrad 1977, S.II-4, pp. 5-14.

33. Brolsma, J.U. / Hirs, J.A. / Langeveld, J.M.
'On fender design and berthing velocities', 24th Intern. Navigation Congress, P.I.A.N.C., Leningrad 1977, S.II-4, pp. 87-100.

34. Piaseckyj, P.J.

'Fender design in North America for large ships .....', 24th Intern. Navigation Congress, P.I.A.N.C., Leningrad 1977, S.II-4, pp. 133-145.

35. Leclercq, R.
'Résultats d'essais sur modèles réduits de dérive latérale des navires', Annales des Ponts et Chaussées, 130e année, mars-avril 1960, pp. 181-215.

36. Deschennes, H. / Dubois, J.
XXth Intern. Navigation Congress, P.I.A.N.C., Baltimore 1961, S.II-1, pp. 65-88.

37. Blok, J.J. / Dekker, J.N.
'On hydrodynamic aspects of ship collision with rigid or non-rigid structures', 11th Annual Offshore Technol. Conf., Houston, 1979, Proc. Vol. IV, Paper OTC 3664, pp. 2683-2697.

38. Yu, Y.S. / Ursell, F.
'Surface waves generated by an oscillating circular cylinder on water of finite depth: theory and experiment', Journ. of Fluid Mechanics, Vol. 11, 1961, pp. 529-551.

39. Newman, J.N.
'The damping of an oscillating ellipsoid near a free surface', Journ. of Ship Research, Vol. 5, No. 3, Dec. 1961, pp. 44-58.

40. Joosen, W.P.A.
'Slender body theory for an oscillating ship at forward speed', Proc. 5th O.N.R. Symp. on Naval Hydrodynamics, Bergen, 1964, pp. 167-183.

41. Newman, J.N. / Tuck, E.O.
'Current progress in the slender body theory for ship motions', Proc. 5th O.N.R. Symp. on Naval Hydrodynamics, Bergen, 1964, pp. 129-166.

42. Leeuwen, G. van
'The lateral damping and added mass of a horizontally oscillating ship model', Rep. No. 65 S, Netherlands Research Centre T.N.O. for Shipbuilding and Navigation, Delft, Dec. 1964.

43. Vugts, J.H.
'The hydrodynamic coefficients for swaying, heaving and rolling cylinders in a free surface', Report No. 112 S, May 1968, Netherlands Ship Research Center T.N.O., Shipbuilding Department, Delft.

44. Tasai, F. / Kim, C.H.

158

'Effect of shallow water on the natural period of heave', Reports of
Research Institute for Applied Mechanics, Kyushu University, Vol. XVI,
No. 54, 1968, pp. 223-229.

45. Kim, C.H.

'Hydrodynamic forces and moments for heaving, swaying and rolling
cylinders on water of finite depth', Journ. of Ship Research, Vol. 13,
No. 3, June 1969, pp. 137-154.

46. Vugts, J.H.

'The hydrodynamic forces and ship motions in waves', Thesis Delft
University of Technology, Uitgeverij Waltman, Delft, 1970.

47. Garrison, C.J.

'Hydrodynamics of large objects in the sea. Part I – Hydrodynamic
analysis', Journ. of Hydronautics, Vol. 8, No. 1, January 1974, pp. 5-
12.

48. Keil, H.

'Die hydrodynamische Kräfte bei der periodischen Bewegung
zweidimensionaler Körper an der Oberfläche flacher Gewässer', Institut
für Schiffbau der Universität Hamburg, Bericht Nr. 305, Februar 1974.

49. Garrison, C.J.

'Dynamic response of floating bodies', 6th Annual Offshore Technology
Conference, Houston, Texas, May 1974, Paper no. OTC 2067, pp. 365-377.

50. Gerritsma, J. / Beukelman, W. / Glansdorp, C.G.

'The effects of beam on the hydrodynamic characteristics of ship
hulls', Proc. 10th O.N.R. Symp. on Naval Hydrodynamics, June 1974, pp.
3-33.

51. Newton, R.E.

'Finite element analysis of two-dimensional added mass and damping',
Chapter 11 in 'Finite elements in fluids, Vol. 1, Viscous flow and
hydrodynamics' (ed. by R.H. Gallagher, J.T. Oden, C. Taylor, O.C.
Zienkiewicz), John Wiley & Sons, London, 1975.

52. Visser, W. / Wilt, M. van der

'A numerical approach to the study of irregular ship motions', Chapter
12 in 'Finite elements in fluids, Vol. 1, Viscous flow and
hydrodynamics' (ed. R.H. Gallagher, J.T. Oden, C. Taylor,
O.C. Zienkiewicz), John   Wiley & Sons, London 1975.

53. Chung, Y.K. / Coleman, M.I.

'Hydrodynamic forces and moments for oscillatory cylinders', Proc.
Civil Engineering in the Oceans III, June 1975, Univ. of Delaware,
Newark, Vol. 2 (publ. A.S.C.E., New York, 1975).

54. Fontijn, H.L.

'An approximative method for the determination of the hydrodynamic
coefficients of a ship in case of swaying and yawing on shallow water',
Communications on Hydraulics, Report no. 75-4, Dept. of Civil
Engineering, Delft University of Technology, Delft, The Netherlands,
1975.

55. Keil, H.

'Hydrodynamische Masse und Dämpfungskonstante tauchender Zylinder auf
flachem Wasser', Schiffstechnik, Bd. 23, 1976, pp. 186-188.

56. Ursell, F.

'On the virtual-mass and damping coefficients for long waves in water
of finite depth', Journ. of Fluid Mech., Vol. 76, part 1, 1976, pp. 17-
28.

57. Rhodes-Robinson, P.F.

'Note on the long-wave limit of the virtual-mass coefficient for a
half-immersed circular cylinder heaving on water of finite depth',
Journ. of Fluid Mech., Vol. 76, part 1, 1976, pp. 29-33.

58. Oortmerssen, G. van

'The motions of a moored ship in waves', Thesis Delft University of
Technology, H. Veenman en Zonen n.v., Wageningen, 1976.

59. Takaki, M.

'On the hydrodynamic forces and moments acting on the two-dimensional
bodies oscillating in shallow water', Reports of Research Institute for
Applied Mechanics, Vol. XXIV, No. 78, 1977, Kyushu University, Japan.

60. Kan, M.

'The added mass coefficient of a cylinder oscillating in shallow water
in the limit k → 0 and k → ∞ ', Papers of Ship Research Institute, No.
52, May 1977, Tokyo, Japan.

61. Plotkin, A.

'Heave and pitch motions in shallow water including the effect of
forward speed', Journ. of Fluid Mechanics, Vol. 80, part 3, 1977, pp.
433-441.

62. Chung, J.S.

'Forces on submerged cylinders oscillating near a free surface', Journ.
of Hydronautics, Vol. 11, No. 3, July 1977, pp. 100-106.

63. Keuning, J.A. / Beukelman, W.
'Hydrodynamic coefficients of rectangular barges in shallow water',
BOSS '79, 2nd Int. Conf. Behavior Off-Shore Struct., Imp. Coll. London,
Aug. 1979, Cranfield, BHRA Fluid Eng., 1979, Vol. 2, Paper 55, pp. 105-
124.

64. Sayer, P.
'The long-wave behaviour of the virtual mass in water of finite depth',
Proc. R. Soc. London, A, 372, 1980, pp. 65-91.

65. Sayer, P.
'An integral-equation method for determining the fluid motion due to a
cylinder heaving on water of finite depth', Proc. R. Soc. London, A,
372, 1980, pp. 93-110.

66. Fontijn, H.L.
'Ship berthing to a vertical quay-wall: fender forces and ship motion',
Communications on Hydraulics, Rep. no. 83-4, Dept. of Civil
Engineering, Delft University of Technology, Delft, The Netherlands,
1983.

67. Ogilvie, T.F.
'Recent progress toward the understanding and prediction of ship
motions', Proc. 5th O.N.R. Symp. on Naval Hydrodynamics, Bergen, 1964,
pp. 3-80.

68. Cummins, W.E.
'The impulse response function and ship motions', Schiffstechnik, 9, H.
47, 1962, pp. 101-109.

69. Tick, L.J.
'Differential equations with frequency-dependent coefficients', Journ.
of Ship Research, Techn. Note, Vol. 3, Nr. 2, Oct. 1959, pp. 45-46.

70. Oortmerssen, G. van
'The berthing of a large tanker to a jetty', 6th Annual Offshore
Technology Conference, Houston, Texas, May 1974, Paper nr. OTC 2100,
pp. 665-676.

71. Fontijn, H.L.
'Impact forces on berthing facilities resulting from moving ships', 6th
Int. Harbour Congress, Antwerpen, May 1974, Section 2.24, pp.
2.24.1/2.24.6.

72. Fontijn, H.L.
    'Forces on berthing structures from moving ships', Proc. XVIIth
    I.A.H.R. Congress, Baden-Baden, Aug. 1977, Paper C16, pp. 119-126.
73. Fontijn, H.L.
    'The berthing ship problem: forces on berthing structures from moving
    ships', Report no. 78-2, Communications on Hydraulics, Dept. of Civil
    Engineering, Delft University of Technology, Delft, The Netherlands,
    1978.
74. Fontijn, H.L.
    'The berthing of a ship to a jetty', Journal of the Waterway, Port,
    Coastal and Ocean Division, A.S.C.E., Vol. 106, No. WW2, Proc. paper
    15407, May 1980, pp. 239-259.
    Errata in Vol. 106, No. WW4, Nov. 1980, pp. 509-510.
75. Fontijn, H.L.
    'On the determination of berthing forces', Int. Conf. on Numerical and
    Hydraulic Modelling of Ports and Harbours, Birmingham, England, April
    1985, Paper G1, pp. 187-193.
76. Fontijn, H.L. / Kalkwijk, J.P.Th.
    'Prediction of fender loads at a closed berthing structure', Proc.
    Inst. Civ. Engrs., Part 2, Vol. 81, Dec. 1986, Paper 9036, pp. 511-534.
77. Wunsch, G.
    'Moderne Systemtheorie', Akademische Verlagsgesellschaft, Geest &
    Portig K.G., Leipzig 1962.
78. Wunsch, G.
    'Systemanalyse', B.1 1969, B.2 1970, Verlag Technik, Berlin.
79. Newman, J.N.
    'Some hydrodynamic aspects of ship maneuverability', Proc. 6th O.N.R.
    Symp. on Naval Hydrodynamics, Washington 1966, pp. 203-237.
80. Papoulis, A.
    'The Fourier integral and its applications', McGraw-Hill Book Company,
    Inc., New York, 1962.
81. Hwei P. Hsu
    'Fourier analysis', revised edition, Simon and Schuster, New York,
    1970.
82. Lighthill, M.J.
    'Fourier analysis and generalized functions', Cambridge University
    Press, 1970.

162

83. Kuipers, L. / Timman, R., eds.
'Handbook of mathematics - Chapter XIII, J.W. Cohen: The Laplace transform', Intern. series of monographs in pure and applied mathematics, Vol. 99, Oxford, Pergamon Press, 1969.

84. Todd, F.H.
'Ship hull vibration', Edward Arnold (Publishers) Ltd, London, 1961.

85. Koch, J.J.
'Eine experimentelle Methode zur Bestimmung der reduzierten Masse des mitschwingenden Wassers bei Schiffsschwingungen', Ingenieur-Archiv, IV. Band, 2. Heft, 1933, pp. 103-109.

86. Newman, J.N.
'The exciting forces on fixed bodies in waves', Journ. of Ship Research, Vol. 6, No. 3, December 1962, pp. 10-17.

87. Whittaker, E.T. / Watson, G.N.
'A course of modern analysis', 4th edition, Cambridge University Press, 1965.

88. Kotik, J. / Mangulis, V.
'On the Kramers-Kronig relations for ship motions', International Shipbuilding Progress, Vol. 9, Nr. 97, Sept. 1962, pp. 361-368.

89. Kotik, J. / Lurye, J.
'Some topics in the theory of coupled ship motions', Proc. 5th Symp. on Naval Hydrodynamics (O.N.R.), Bergen, 1964, pp. 407-424.

90. Lamb, H.
'Hydrodynamics', Cambridge University Press, Cambridge, 6th ed., 1932.

91. Schlichting, H.
'Boundary-layer theory', McGraw-Hill Book Company, New York, 6th ed., 1968.

92. Abramowitz, M. / Stegun, I.A.
'Handbook of mathematical functions', Ch. 25, Dover Publications, Inc., New York.

APPENDIX E. Notation

Symbols not included in the list below are only used at a specific place and are explained where they occur.

$\hat{a}$ amplitude of harmonically oscillating sway motion;

$a(t)$	additional acceleration in $ox_1x_2x_3$;
$a(\omega)$	hydrodynamic coefficient of mass term;
$b(\omega)$	hydrodynamic coefficient of damping force;
B	beam of ship;
c	hydrostatic restoring coefficient;
c_0	spring rate of linear fender;
c_1,c_2	respective spring rates of linear springs forming non-linear fender;
c_w	velocity of propagation of long wave;
d_{sc}	distance at rest between linear spring elements of non-linear fender;
d_{sq}	initial distance between ship and quay-wall;
D	draught of ship;
e_0	initial distance of line of action of fender to G;
f	multiplication factor for underkeel friction effect (standard value $\equiv$ 1;
$f(t)$	(external) forcing function (upon ship); hydrodynamic force; excitation of ship-fluid system (input signal); function of t;
f_u,f_w	switch parameter for horizontal velocity and vertical acceleration, respectively,
$F(t)$	reaction force (in fender);
$F(\omega)$	Fourier transform of f(t);
$\mathscr{F}\{f(t)\}$	Fourier transform of f(t);
g	acceleration due to gravity;
G	centre of gravity of ship;
$Gxyz$	moving ship-fixed right-handed Cartesian co-ordinate system with origin G;
h	water depth at rest;
$H(s)$	unilateral Laplace transform of k(t)= general transfer function;
$H(i\omega)$	non-generalized f.r.f.;
i	$\sqrt{-1}$;
$Im[...]$	imaginary part of;
$k(t)$	i.r.f.;
$K(\omega)$	Fourier transform of k(t)= harmonic transfer function for force excitation = generalized f.r.f.;
L	length of ship;
$\mathscr{L}\{f(t)\}$	unilateral Laplace transform of f(t);

m   generalized mass of ship;

M_0 reduced or effective mass of ship for horizontal motion;

$ox_1x_2x_3$ right-handed Cartesian co-ordinate system with origin o, parallel to $OX_1X_2X_3$ and translating with V; in case of rotation additional forces have to be introduced;

$ox_1x_2x_3$ co-ordinate system identical to and coinciding with $ox_1x_2x_3$, provided the latter system does not rotate;

$OX_1X_2X_3$ right-handed space-fixed Cartesian co-ordinate system with origin O;

p   fluid pressure;

R   fluid region coinciding with region c;

Re   Reynolds number;

Re[...]   real part of ...;

$R(\omega)$ harmonic transfer function for motion excitation;

s,s_1 complex variable, certain complex number, respectively;

t   time co-ordinate;

u(t)   response of ship-fluid system (output signal);

u,v,w   fluid velocity in x_1-,x_2-,x_3-direction, respectively;

$U(\omega)$ Fourier transform of x(t);

v_A (constant) lateral speed of approach of ship towards berth;

V   constant speed of ship;

x(t)   mode of motion c.q. motion variable of ship;

α dimensionless friction coefficient; $k(\infty)$;

γ proportionality coefficient for viscous shear stress;

$\delta(\omega)$ delta function or Dirac function;

δ_{jk} Kronecker delta;

Δt time increment; time step;

ΔX deflexion (of fender);

ζ elevation of water surface (with respect to h);

η height of long wave (with respect to h);

θ phase shift;

λ real part of s;

μ contraction coefficient;

υ kinematic viscosity of fluid;

ξ general head loss coefficient;

ξ_e,ξ_o head loss coefficient due to contraction at entrance and expansion at outlet, respectively;

m	generalized mass of ship;
M_0	reduced or effective mass of ship for horizontal motion;
$ox_1 x_2 x_3$	right-handed Cartesian co-ordinate system with origin o, parallel to $OX_1 X_2 X_3$ and translating with V; in case of rotation additional forces have to be introduced;
$ox_1 x_2 x_3$	co-ordinate system identical to and coinciding with $ox_1 x_2 x_3$, provided the latter system does not rotate;
$OX_1 X_2 X_3$	right-handed space-fixed Cartesian co-ordinate system with origin O;
p	fluid pressure;
R	fluid region coinciding with region c;
Re	Reynolds number;
Re[...]	real part of ...;
$R(\omega)$	harmonic transfer function for motion excitation;
s, s_1	complex variable, certain complex number, respectively;
t	time co-ordinate;
$u(t)$	response of ship-fluid system (output signal);
u,v,w	fluid velocity in x_1-,x_2-,x_3-direction, respectively;
$U(\omega)$	Fourier transform of $\dot{x}(t)$;
v_A	(constant) lateral speed of approach of ship towards berth;
V	constant speed of ship;
$x(t)$	mode of motion c.q. motion variable of ship;
$\propto$	dimensionless friction coefficient; $k(\infty)$;
γ	proportionality coefficient for viscous shear stress;
$\delta(\omega)$	delta function or Dirac function;
δ_{jk}	Kronecker delta;
Δt	time increment; time step;
ΔX	deflexion (of fender);
ζ	elevation of water surface (with respect to h);
η	height of long wave (with respect to h);
θ	phase shift;
λ	real part of s;
μ	contraction coefficient;
υ	kinematic viscosity of fluid;
ξ	general head loss coefficient;
ξ_e, ξ_o	head loss coefficient due to contraction at entrance and expansion at outlet, respectively;

ρ	specific mass density of fluid;
τ	integration variable (time);
Φ	velocity potential;
$\Psi(t)$	angle of rotation of ship in horizontal plane;
Ψ_0	amplitude of harmonically oscillating yaw motion;
ω	(circular) frequency; imaginary part of s.

Subscripts

a,b,c,d,	region a,b,c,d, respectively;
f	fender;
G	ship's centre of gravity;
i,j,k	direction or degree-of-freedom in Cartesian co-ordinate system, $(i,j,k)= 1,2,6$;
kc	keel clearance;
l,t	laminar or turbulent flow regime, respectively;
m,n	running index (real, positive integer);
o	origin of $ox_1 x_2 x_3$;
r	due to rotation of $ox_1 x_2 x_3$;
s	ship

Superscript

*	reduced version of its original.

General conventions and abbreviations

A bar over a quantity means 'average value of'.

A circumflex over a quantity means 'amplitude of'.

f.r.f.	frequency response function;
i.r.f.	impulse response function;
T.D.A.	time-domain approach.

USE OF STATISTICAL DATA AND METHODS IN IMPACT EVALUATION
OF FENDER DESIGN

PER TRYDE, Associate professor and I. A. Svendsen, Professor*
Institute of Hydrodynamics and Hydraulic Engineering
The Technical University of Denmark, Lyngby
* Now at the University of Delaware.
  The investigation was made by professor Svendsen.

1. INTRODUCTION
    Although it may be possible to derive analytical equations
to solve the problem in question one is often faced with
the fact that insufficient information is available to do
so. In solving engineering problems it is found appropriate
to make use of statistical data obtained from prototype
measurements of the berthing impact or the fender energy
E_f. As the statistical design method is based on measurements
of the energies actually absorbed in fenders at existing
terminals, the data automatically include the effect of impact
velocity, eccentricity, hydrodynamic mass, etc.
Thus, one of the advantages of the statistical design method
is that it shortcuts directly to the actual problem: The
berthing impact energy.(as stated in ref.4)
The philosophy behind the statistical approach is the determination of the risk level which can be accepted in the case
in question. How often will certain damage to the ship and
the berth be accepted? How often can a disturbance in the
operation of the berth be tolerated? In other words a problem
which may be solved by a cost/benefit analysis.

2. THE STATISTICAL DISTRIBUTION OF FENDER ENERGIES
    A large number of berthing impact energies have been re-
corded at several places in Europe. The measurements that
have been used in the analysis were taken at:

Rotterdam BP berth 1	461	arrivals
Rotterdam BP berth 2	526	-
BP berth Kent, U.K.	578	-
BP berth Finnart, U.K.	270	-
Gulfhavn Stigsnæs, Denmark	149	-
Prøvestenen, Copenhagen, Denmark	201	-
Esso berth Kalundborg, Denmark	166	-
Wilhelmshafen NWO-berth BRD	659	-
Torshamnen, Gothenburg, Sweden	1916	-

Because the measurements at Prøvestenen showed so much smaller impact energies, they have been left out of the analysis,
due to the use of some special large gravity fenders which
made the pilots feel uneasy during berthing.

E. Bratteland (ed.), Advances in Berthing and Mooring of Ships and Offshore Structures, 167–172.
© 1988 by Kluwer Academic Publishers.

168

To be able to use the data for prediction of energies at
a new berth, however, it must be known how the different
conditions at different locations influence the statistical
properties of the fender energy distribution. Such information can be obtained by correlating the measured energies
with the values of the important parameters as e.g. ship
size, current velocity and direction, and wind conditions.
Svendsen (1970) attempted this and found that the most
important parameters are the size of the ship and the current conditions.

3. VARIATION OF THE IMPACT ENERGY WITH THE SHIP'S DISPLACE-
MENT AND CURRENT CONDITIONS

It was found that with sufficiently observations at hand
all cumulated frequency distributions may be approximated
by a logarithmic-normal distribution.
Hence in a logarithmic-normal plot the distribution can be
represented by a straight line, and it is therefore uniquely
specified by two points (Fig. 1). In the following, the
E_f-values $E_{f,50}$ and $E_{f,5}$ corresponding to 50% and 5% probability of exceedance respectively, are used to describe the
cumulated frequency distributions. The following analysis
concentrates on the variation of the impact energies with
the displacement of the arriving ships and the general current conditions.
On the basis of the available material the harbours mentioned can be classified in three groups:
A: Harbours at well protected locations with no currents,
no waves, and sufficient tug boat assistance available (see
Fig. 2)
B: Harbours at locations well protected against waves but
exposed to moderate (2-4 knots) currents running parallel
to the orientation of the berth.(Fig.3).
C: Harbours at well protected locations exposed to currents
running in directions significantly different from the orientation of the berth. (example is Kent)

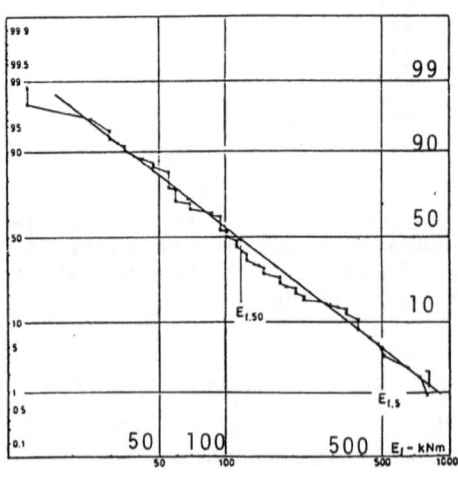

FIGURE 1.
Statistical distribution
of observed fender energy.

The reason for the strong influence of the currents is of course that the current forces on a berthing ship may be very large. As long as the axis of the ship is parallel with the current, the forces are moderate and may be balanced by the engines and the tug boats, and the pilot is able to reduce the impact velocity as the displacement increases. However, if an angle occurs between the axis of the ship and the current direction, very large transverse forces may develop on the ship, which has to be held in position relative to the berth against the current. The idea behind the three groups of harbours is that the manoeuvres for ships of the same sizes should be similar for berths in the same catagory with the same energy of the ship and the same position relative to the pier. In the following the energy E_f is shown in Figure 2, 3 and 4 as a function of the displacement of the arriving ships. The lines drawn in the figures correspond to the analytical expressions also given in the figures. It should be emphasized at this point, however, that since these lines are based on results from only a few locations, they can only represent a qualitative trend in the variation from place to place, as it is also indicated by the relatively crude division of harbours into the three groups, A, B and C. Hence they can be expected to change as further measurements become available.

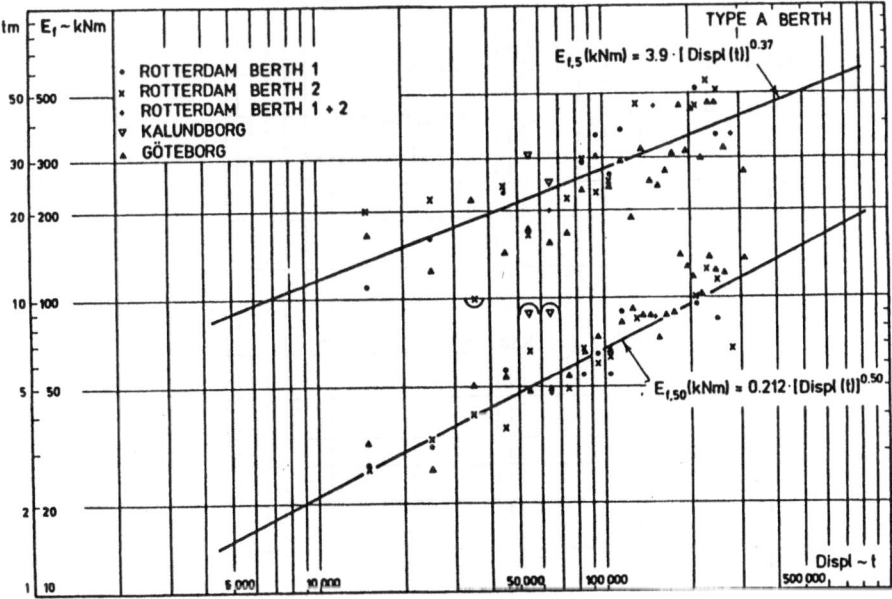

FIGURE 2. Type A berth.

170

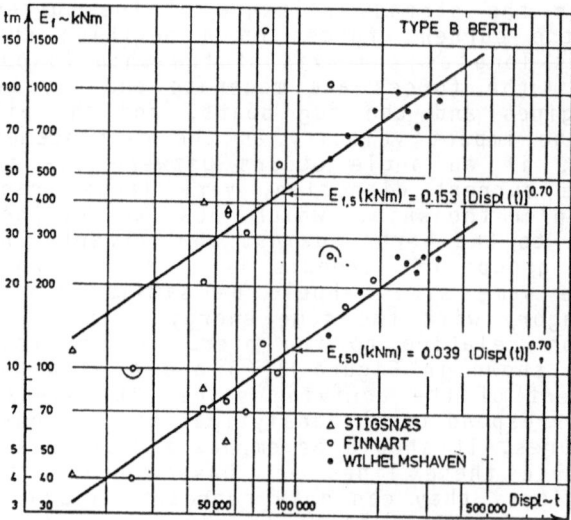

FIGURE 3. Type B berth.

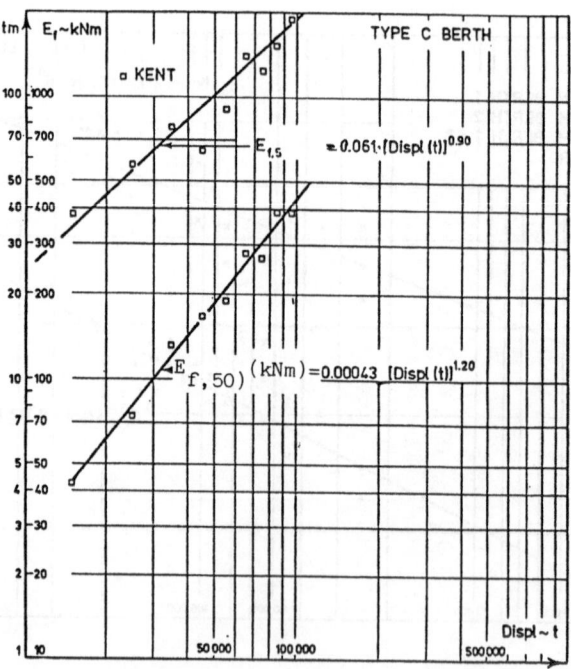

FIGURE 4. Type C berth.

4. EVALUATION OF THE DESIGN IMPACT ENERGY AND
FINAL CONCLUSIONS

The statistical behavior of the berthing impact energies implies that regardless of which design value is choosen for the impact energy E_f there is a risk, R, that it is exceeded. The risk that a choosen value E_f is exceeded at least once during a number N of arrivals, may be determined as the complement of the probability that E_f is not exceeded at all during these arrivals.

The result may be expressed by the relation:

$$R = 1 - e^{-N/T}$$

where e = 2.71828 and T is defined as the return period for the chosen value E_f, i.e.:

$$T(E_f) = 1/Q(E_f)$$

where $Q(E_f)$ is the probability that E_f is exceeded.

The statistical method is a practical tool for the design engineer. However the reliability of the method is depending on the number of measurements available. Harbour authorities should remember to report all arrivals.

The statistical nature also implies that it cannot be stated when the design load will be exceeded.

The extended abstract given here is partly extracted directly from ref. 4 : The Statistical Approach pg.20 prepared by professor I.A.Svendsen. It is strongly recommended to read this article in full.

REFERENCES

1. Balfour, J.S. Feben & D.L. Martin (1980), Fendering re-
   quirements/design, fender impact criteria. Ports '80.
   ASCE Speciality Conference.
2. Benjamin, J.R. & C.A. Cornell (1970), Probability , Statistics, and Decisions for Civil Engineers. McGraw Hill,
   New York.
3. Borgmann L.A. (1963), Risk Criteria, Proc. ASCE, 89, WW3-,
   paper 3607.
4. Permanent International Association of Navigation Congresses:
   Report of the International Commision for improving the
   design of fender systems. Suppl. to bulletin no. 45
   (1984)
5. Rogstad, G. & P. Leimdörfer (1961), Skarvikshamnen - the
   new oil port of Gothenburg. 20th International Nav.Con-
   gress, S II - S I, Baltimore.
6. Svendsen, I.A. (1370). Measurement of impact energies
   on fenders. The Dock and Harbour Authority, Vol. 51, no.
   599 & 600.

WIND AND WIND FLUCTUATIONS

Donald H. Lenschow
National Center for Atmospheric Research
Boulder, Colorado 80307-3000
USA

1. INTRODUCTION

Wind is one of the fundamental aspects of man's environment. In many ways, it is vital to our existence: it diffuses and removes local concentrations of pollutants and transports substances vital to life; it redistributes latent and sensible energy which fuel our weather systems; and historically it has been important as an energy source, both on land and sea. At the same time, its capacity for destruction is immense. Hurricanes, tornadoes, and föhn winds are some of the more obvious manifestations of its destructive fury, but even more commonplace storms can inflict considerable damage if the design or construction of structures does not adequately take into account the forces inflicted by the wind. In the following paragraphs, I shall describe the forces that produce and dissipate winds, the character of its essential turbulence, its variation with height and wave number, its frequency distribution, and its effects on structures.

2. DYNAMICS OF THE MEAN WIND

Wind is basically the response of the fluid surrounding our earth to forces whose origin is ultimately the differential heating and cooling of the earth and its atmosphere. This difference results from a combination of variations in solar insolation, albedo (i.e., the ratio of the incoming radiation that is reflected to the amount that is incident on a surface), infrared absorption and emission by both the atmosphere and the earth's surface, horizontal transport in both the atmosphere and the oceans, and phase changes (i.e., evaporation and condensation of water vapor).

One of the fundamental characteristics of wind is the vast range of scales over which wind variations are important—from planetary waves with wavelengths of 10^4 km to turbulence dissipation scales of a few millimeters. Because of the nonlinearity of the equations that describe air motion, every scale of motion affects every other scale. Furthermore, small changes in the initial conditions of the equations of motion do not necessarily result in small changes in the predicted flow. For these reasons,

E. Bratteland (ed.), Advances in Berthing and Mooring of Ships and Offshore Structures, 173–186.
© 1988 by Kluwer Academic Publishers.

prediction of flows is particularly difficult at intermediate scales where many of the weather phemonena of importance for everyday activities occur (*e.g.* tropical and extratropical cyclones, hurricanes, squall lines and thunderstorms). Despite these difficulties, useful simplifications to the equations of motion can be made in specific situations in order to gain a better understanding of the behavior of mean flows and their relation to the forces driving them, as well as the dissipative and diffusive effects of turbulence.

2.1. Equations of Motion

The fundamental equations of motion that apply to the earth's atmosphere are called the *Navier-Stokes equations*. For mean horizontal flow of a horizontally homogeneous fluid in a planetary atmosphere, these equations can be simplified to (*e.g.* Dutton, 1976)

$$\frac{\partial U}{\partial t} = -\frac{1}{\rho}\frac{\partial P}{\partial x} + fV + \frac{1}{\rho}\frac{\partial \tau_z}{\partial z} \tag{1}$$

$$\frac{\partial V}{\partial t} = -\frac{1}{\rho}\frac{\partial P}{\partial y} - fU + \frac{1}{\rho}\frac{\partial \tau_y}{\partial z}, \tag{2}$$

where U and V are the mean wind components in the x and y directions, respectively, ρ is density, P is pressure, τ_z and τ_y are stress components, and $f \equiv 2\Omega \sin\phi$ is the coriolis parameter. The coriolis force is an apparent force due to the rotation of the earth, with angular velocity $\Omega = 7.29 \times 10^{-5}\,\text{s}^{-1}$, and ϕ is latitude. If we consider only steady state flow, and define the geostrophic wind components $U_g \equiv -\frac{1}{\rho f}\frac{\partial P}{\partial y}$ and $V_g \equiv \frac{1}{\rho f}\frac{\partial P}{\partial x}$, (1) and (2) are reduced to

$$-f(V - V_g) - \frac{\partial \tau_z}{\partial z} = 0 \tag{3}$$

$$f(U - U_g) - \frac{\partial \tau_y}{\partial z} = 0. \tag{4}$$

In a turbulent atmosphere, the horizontal stress components result from vertical transport of momentum by turbulent eddies. To incorporate these turbulence effects into the equations of motion, we use the so-called Reynolds decomposition to divide a velocity component (as well as other variables) $\mathcal{U}$ into its mean U and departures from the mean u,

$$\mathcal{U} = U + u. \tag{5}$$

Introducing this decomposition into the full equations of motion (*e.g.* Businger, 1982), the stress components can be expressed as $\tau_z = -\rho\overline{uw}$ and $\tau_y = -\rho\overline{vw}$: that is, as the vertical turbulent flux of the fluctuations in the horizontal velocity components. The overbar denotes an average over a time or length long enough to assure a statistically reliable measure of the flux. Often, the vertical flux of an atmospheric property is parameterized by an eddy mixing process such that the flux of the property is proportional to its vertical gradient,

$$\overline{uw} = -K\frac{\partial U}{\partial z}, \qquad \overline{vw} = -K\frac{\partial V}{\partial z}. \tag{6}$$

This closure leads to the relations

$$-f(V - V_g) - \frac{\partial}{\partial z}\left(K\frac{\partial U}{\partial z}\right) = 0 \tag{7}$$

$$f(U - U_g) - \frac{\partial}{\partial z}\left(K\frac{\partial V}{\partial z}\right) = 0. \tag{8}$$

Solutions to these equations, with the boundary conditions of zero wind at the surface and wind equal to the geostrophic wind for $z \to \infty$ have been obtained for a variety of formulations for K, as well as the geostrophic wind components U_g and V_g as functions of height. In the simplest case, if we choose the x-axis to be parallel to the geostrophic wind (*i.e.* $U_g = G$ and $V_g = 0$), and let K and G be constant with height, we obtain the Ekman spiral (*e.g.* Businger, 1982):

$$U = G(1 - e^{-az}\cos az) \tag{9}$$

$$V = Ge^{-az}\sin az, \tag{10}$$

where $a = [f/(2K)]^{1/2}$. This was first used by Ekman (1905) to describe wind-driven currents in the upper ocean; qualitatively it also often describes flow in the lower part of the atmosphere—sometimes referred to as the Ekman Layer. A schematic example of a boundary layer wind profile is shown in Fig. 1.

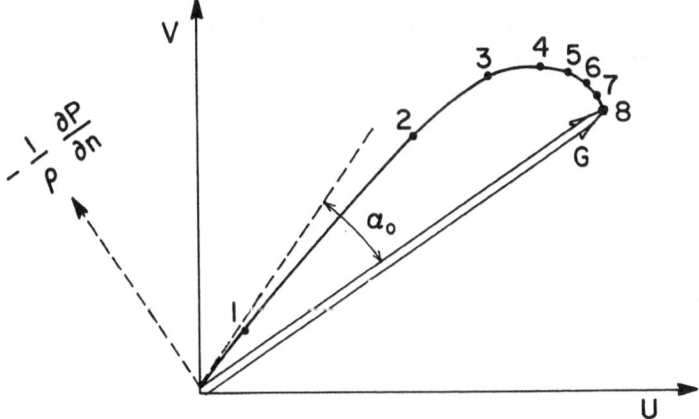

Figure 1. An example of a northern hemisphere boundary layer wind profile in a barotropic atmosphere (i.e., no change in the horizontal pressure gradient with height). The numbers indicate approximately equal height increments starting from

the surface and extending through the top of the boundary layer, where the wind speed becomes geostrophic; $G = \sqrt{U_g^2 + V_g^2}$; α_0 is the angle between the surface wind and the geostrophic wind G; and $\frac{\partial P}{\partial n}$ is the horizontal pressure gradient normal to the isobars.

2.2. Thermal Wind

In general, the geostrophic wind is not constant with height. Horizontal temperature gradients are ubiquitous throughout the atmosphere; their effect on the mean wind profile, specified in terms of the mean horizontal wind shear components (*e.g.* Dutton, 1976), can be expressed as

$$\frac{\partial U_g}{\partial z} = -\frac{g}{fT}\frac{\partial T}{\partial y}, \qquad \frac{\partial V_g}{\partial z} = \frac{g}{fT}\frac{\partial T}{\partial x}, \qquad (11)$$

where T is temperature and g is gravitational acceleration. The thermal wind is an important aspect, both of the large-scale circulation of the atmosphere as well as smaller scale phemonena such as land and sea breeze circulations, low-level jets and wintertime cold air outbreaks that can strongly affect the offshore environment.

Figure 2 shows an example of wind and temperature measurements along an airplane flight track over the East China Sea during a wintertime cold air outbreak. The temperature gradient is towards the north. Figure 3 shows that the wind measured from a rawindsonde released from a ship in the vicinity of the airplane flight path is out of the north below about 2.5 km. Above this, the wind shifts abruptly; the U component increases to about 22 m s^{-1}, while the V component remains nearly constant at about 10 m s^{-1}. This is opposite to the change in wind direction predicted by (9) and (10), and plotted in Fig. 1. It illustrates the important role that horizontal temperature gradients can play in determining the mean wind and its variation with height. Integrating (11) from the surface to 2.5 km, and using a horizontal temperature gradient estimated from the aircraft measurements of $\partial T/\partial y = -0.025$ K km^{-1} and $\partial T/\partial x = 0$, we obtain a change in U_g of $\Delta U_g \simeq 22$ m s^{-1}, which agrees well with the observed change in U.

Another example of a significant thermal wind effect in a marine environment is the observed diurnal variation in wind velocity along coasts. This results from daytime solar heating of the land which heats the overlying boundary layer. Beardsley, *et al.* (1987), for example, documented this phemonenon along the Northern California coast, where the amplitude of the diurnal oscillation in the along-shore component of the mean wind on the coast averages nearly 10 m s^{-1} in early summer.

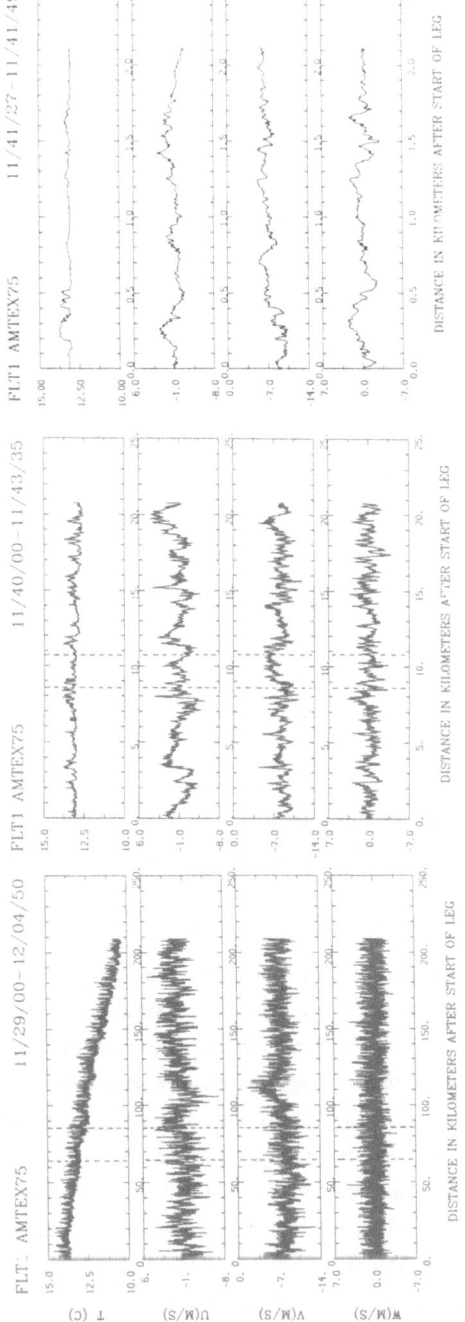

Figure 2. Wind velocity components and air temperature measured from an aircraft flying on a northwesterly track from Naha, Okinawa, over the East China Sea at 150 m height during the Air Mass Transformation Experiment (AMTEX). The flight occurred on 15 February 1975 during a cold air outbreak. Variables are sampled every 0.05 s; at an airspeed of 100 m s^{-1}, this corresponds to a distance of 5 m. The outputs are low-pass filtered at 10 Hz. The variables between the dashed lines at 64 and 86 km on the left panel are expanded and plotted on the middle panel. Similarly, the variables between 8.6 and 10.7 km on the middle panel are expanded and plotted on the right panel.

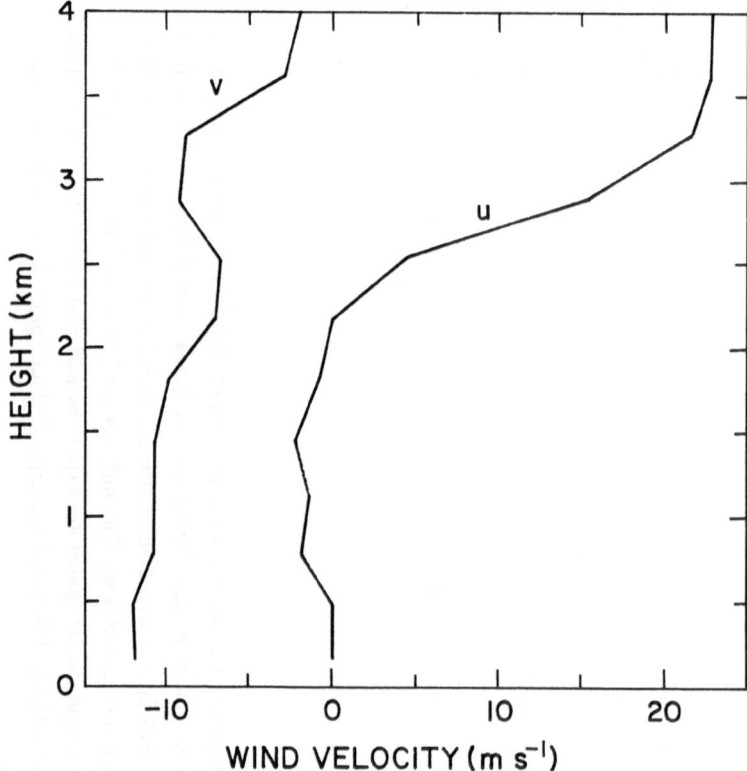

Figure 3. Wind profile obtained from a rawindsonde over the East China Sea. The balloon was released from the Research Vessel Keifu Maru at 14:30 LT in the vicinity of the airplane used to obtain the measurements shown in Fig. 2.

3. TURBULENCE IN THE ATMOSPHERIC BOUNDARY LAYER

The atmospheric boundary layer (ABL) is that region of the atmosphere which, because of turbulence, interacts with the earth's surface on a time scale of a few hours or less. Typically over the ocean, its depth is a few hundred meters to a kilometer or more. The sources of turbulence energy are wind shear, generated by drag at the surface and horizontal temperature gradients; and buoyancy, generated by surface heating and evaporation, and by radiative cooling in the upper part of cloud-capped boundary layers. The lowest ten meters or so of the ABL, where the fluxes of temperature and momentum vary by <10% of their surface values and change in mean wind direction with height is negligible, is called the surface layer. This is the layer within which most ships and offshore structures are contained, and therefore, for most engineering applications, we need be directly concerned only with surface layer structure. Nevertheless, the surface layer is an integral part of the ABL, and generalizations of its structure depend upon the structure of the rest of the ABL. Detailed expositions of the mean and turbulence structure of the ABL have been written by Businger (1982), Panofsky and Dutton (1984), Arya (1982), and Jensen and Busch (1982).

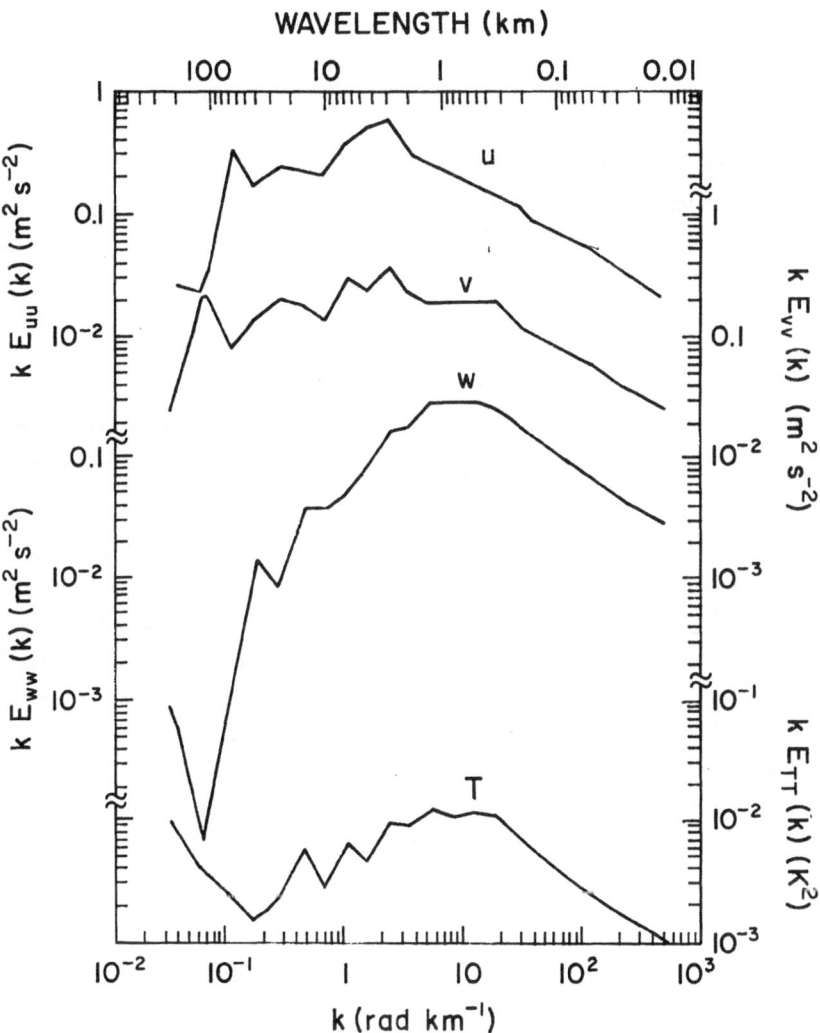

Figure 4. Wavenumber times the spectra of the three wind components and temperature for the time series plotted in Fig. 2.

Although an exact definition of turbulence is difficult to formulate, in practice it is usually easily recognized because of its chaotic and highly diffusive nature, its smoothly varying spectral density, and its intermittency. Examples of turbulence variables are shown in Fig. 2. We see in the top panel that on the large scale (>10 km) the variations in the horizontal velocity components are larger than the vertical velocity. On scales <10 km, the velocity components look similar, but temperature variations are positively skewed.

We can get a better idea of the contributions of different scales of these variables to their total variance by computing their spectral densities, or spectra. This is essentially a decomposition of a time series into variance density as a function of frequency (Hz) [or wavenumber (rad/km) or wavelength (km^{-1})]. Figure 4 shows plots of wavenumber times the spectra for the velocity components and temperature shown in Fig. 2. They confirm the observation that at large wavenumbers the velocity components are nearly identical, but at small wavenumbers, the spectrum of vertical velocity is much less than the spectra of horizontal velocity.

In Fig. 2, we can see correlation between $\mathcal{W}$ and $\mathcal{T}$, and $\mathcal{W}$ and $\mathcal{V}$ on scales of a few hundred meters to a few kilometers by a careful examination of the time series in the middle panel. The averages of the products of these two sets of variables are the vertical temperature flux and the northward component of the stress, respectively.

3.1 Vertical Gradients and Turbulence Statistics

Exact solutions to the equations of motion for a turbulent fluid have not been forthcoming because of their nonlinearity, so other means have been developed to understand the structure of the ABL. Dimensional analysis provides an extremely useful framework for generalizing observational results, and has led to the development of many empirically determined relations which can then be used generally for predicting ABL behavior. Since both mechanical (shearing) and buoyant forces are important sources of turbulence energy in the surface layer, a basic parameter for specifying surface layer structure is their ratio. The parameter usually used for this purpose is the Monin-Obukhov (1954) length,

$$L = -\frac{u_*^3}{k\frac{g}{T}(\overline{wT_v})_0},\tag{12}$$

where $\rho u_*^2 = \sqrt{\tau_{x0}^2 + \tau_{y0}^2}$, the zero subscript denotes a surface value, T_v is virtual temperature (the temperature of dry air having the same density and pressure as the moist air; virtual temperature flux is proportional to buoyancy flux) and k is von Kármán's constant ($\simeq 0.4$). Surface layer variables are usually specified as functions of the dimensionless ratio z/L. Over the ocean, the temperature difference between the water and the air, and consequently the buoyancy flux, is generally smaller than over land, which means that, at a given height, the marine surface layer is generally closer to neutral than the surface layer over land. If $z/L < 0$, the buoyancy flux generates turbulence and the surface layer is unstably stratified; if $z/L > 0$, the buoyancy flux acts to suppress turbulence and the surface layer is stably stratified. At $z/L = 0$, the buoyancy flux is zero and the surface layer is neutrally stratified. Thus, if we are close enough to the surface that $z \ll |L|$, the neutral surface layer formulations can be used.

Universal forms for vertical gradients and turbulence statistics have been obtained empirically as functions of z/L. For example, the expression for vertical wind shear is

$$\frac{\partial U}{\partial z} = \frac{u_*}{kz}\phi_m\left(\frac{z}{L}\right),\qquad(13)$$

where $\phi_m(z/L)$ is an empirically determined function; commonly used examples are (Businger et al., 1971)

$$\phi_m(z/L) = \left(1 - 16\frac{z}{L}\right)^{-1/4}, \quad z/L < 0 \qquad(14)$$

$$\phi_m(z/L) = 1 + 5\frac{z}{L}, \quad z/L > 0. \qquad(15)$$

These expressions can be integrated (from a lower boundary condition of $z = z_0$, where z_0 is the surface roughness length) to obtain the wind profile. As can be easily seen, the neutral case gives the well-known logarithmic wind profile. Similarly, an expression that is used for the standard deviation of vertical velocity in an unstably stratified surface layer is (Hicks, 1981)

$$\sigma_w/u_* = 1.25(1 - 2z/L)^{1/3}. \qquad(16)$$

On the other hand, Hicks (1981) has found that the normalized standard deviations for the horizontal wind components are ~ 3 for moderate or greater instability; that is, they are independent of z/L. Panofsky et al. (1977) suggest that they are functions of z_i/L; i.e. that the relevant length scale is not the height above the surface, but is z_i, depth of the ABL.

Often it is not feasible to measure the friction velocity u_* by direct measurements of the velocity fluctuations. Fortunately, over the ocean, u_* can be determined with an accuracy acceptable for most purposes from a measurement of the mean wind through the relation

$$u_*^2 = C_D U_{10}^2, \qquad(17)$$

where C_D is the drag coefficient for a mean wind measurement at 10 m height. For neutral stability, large fetch and steady-state conditions, this coefficient has been given by Large and Pond (1981) as

$$10^3 C_D = \begin{cases} 1.2 & 4 \leq U_{10} < 11\text{ms}^{-1} \\ 0.49 + 0.065 U_{10} & 11 \leq U_{10} \leq 25\text{ms}^{-1}. \end{cases} \quad (18)$$

The drag coefficient increases somewhat for unstable stratification.

Spectra of the velocity components have also been categorized as a function of z/L, but with somewhat less certainty. In particular, for unstable stratification, the low frequency part of all the velocity component spectra seems to scale with the depth of the ABL (*e.g.*, Panofsky and Dutton, 1984). A normalized longitudinal velocity spectrum is shown in Fig. 5.

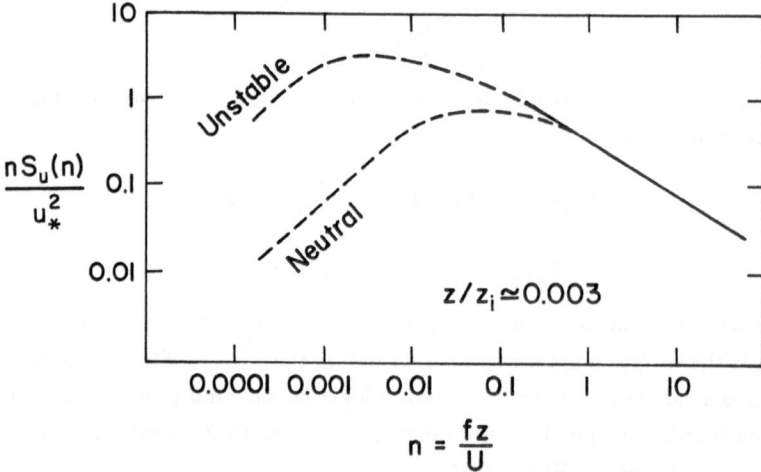

Figure 5. Normalized longitudinal velocity spectrum versus normalized frequency n, where f is frequency, at a particular normalized height above the surface showing the increase in variance and integral scale of the turbulence for unstable stratification.

3.2 Wind Distributions

Monin-Obukhov scaling can be used to obtain various turbulence statistics for wind fluctuations in conditions of steady-state flow over horizontally homogeneous surfaces. In this situation, for most practical considerations, the probability density function of fluctuations in a wind component measured as a function of time or space can be assumed to scale with the friction velocity u_* and z/L, and have a Gaussian or normal distribution,

$$p(u) = \frac{1}{\sqrt{2\pi}\sigma_u} \exp(-u^2/2\sigma_u^2). \quad (19)$$

From a structural engineering standpoint, however, a more important consideration may be the distribution of the wind speed over long time periods which would include synoptic variability (*i.e.*, wind variations contributed by quasi-periodic passage of cyclonic disturbances) as well as occasional intense storms. Such a probability density function must start at $p(0) = 0$, rise to at least one maximum and then decrease to zero for large wind speed $S = \sqrt{\mathcal{U}^2 + \mathcal{V}^2}$. A generalized two-parameter distribution function that has these properties and has been successful in representing the mean wind speed distribution (independent of direction) in a variety of situations is the Weibull distribution,

$$p(S) = \frac{k}{c} \left(\frac{S}{c}\right)^{k-1} e^{-(S/c)^k}. \tag{20}$$

Typical values of the parameters used in this distribution are $k \simeq 2$ [for $k = 2$, (20) is also known as a Rayleigh distribution] and $c \simeq 5$ m s^{-1} (Davenport, 1977). A convenient way to test the hypothesis that wind speed follows a Weibull distribution is to plot its exceedance frequency

$$P(S \geq S_x) \equiv \int_{S_x}^{\infty} p(S')dS' = \exp[-(S_x/c)^k] \tag{21}$$

on a graph whose abscissa and ordinate are the functions $\ln S$ and $\ln\{-\ln[P(S \geq S_x)]\}$, respectively. A Weibull distribution yields a straight line on such a plot. A schematic example is shown in Fig. 6. Due to the relative infrequency and significantly different dynamics of such anomalous phemonena as hurricanes and tornadoes, this distribution may not adequately incorporate their effects (Davenport, 1982).

4. EFFECTS OF WINDS ON STRUCTURES

Since structures typically have resonant response modes, the frequency response of the aerodynamic forcing, which is related to the square of the wind speed integrated over the area of the structure, can be useful in estimating structural response to winds. This is a complicated problem, however, since the structure itself strongly affects the wind field. Furthermore, the integration over the structure depends upon the transverse coherence between the velocity components. Therefore, in practice, various approximations are used to simplify the analysis. For example, often overall drag and lift coefficients, defined by

$$C_D = \frac{F_D}{\frac{1}{2}A\rho U^2}, \quad C_L = \frac{F_L}{\frac{1}{2}A\rho U^2}, \tag{22}$$

where F_D and F_L are the overall drag (parallel to the flow) and lift (perpendicular to the flow) forces exerted on the structure by the wind, respectively, and A is the cross-sectional area of the structure, are used to parameterize the force exerted by the wind. These coefficients are likely to be functions of the wind direction. Values of C_D range from a few tenths up to a value somewhat greater than one, depending

Figure 6. A schematic example of exceedance frequency of wind speed. The functions used to calculate spacing on the ordinate and abscissa, which are defined in the text, result in a straight line for a variable with a Weibull distribution (19). In this case, the coefficients are $c \simeq 8.2$ knots and $k \simeq 1.8$.

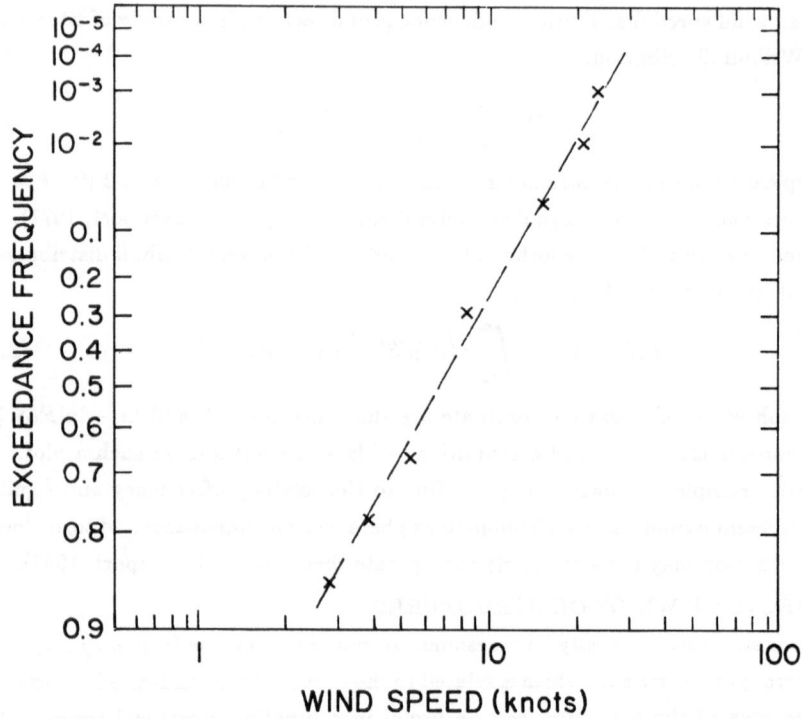

on the shape and size of the structure.

If we consider also response of structures to turbulence, the ratio of the size of the turbulence eddies to the size of the structure is an important parameter. If, for example, the structure is small compared to the dominant wavelength of the turbulence being considered, the entire area of the structure will respond uniformly to the gust; on the other hand, if the structure is large compared to the turbulence eddy, the response of the structure, which is related to the integral of the gust over the surface of the structure, will be diminished as the spatial correlation of the gust decreases across the face of the structure.

A common measure of this spatial correlation is the spectral coherence function, defined by

$$Coh_{12}(f, D) = \frac{Co_{12}^2(f, D) + Quad_{12}^2(f, D)}{S_1(f, D)S_2(f, D)}, \tag{23}$$

where $Co_{12}(f, D)$ and $Quad_{12}(f, D)$ are the cospectrum and quadrature spectrum, respectively, D is the spatial separation between variables 1 and 2, and $S_1(f, D)$ and $S_2(f, D)$ are the spectra of the two variables used in computing the coherence. The cospectrum and quadrature spectrum are the analogous transformation of the covariance of two variables into the frequency or wavenumber domain. Davenport (1961) first proposed that the coherence of the longitudinal wind component is a function only of D/λ, where λ is wavelength, and that for vertical separation this function is an exponential:

$$Coh(D/\lambda) = \exp(-aD/\lambda). \tag{24}$$

This exponential form has been used for lateral coherences as well. Generally, the decay parameter a has been found to vary strongly as a function of stability (larger for stable conditions; smaller for unstable conditions), but for both lateral and vertical separations, a is of order ten. Further discussion of experimental values for a, as well as a theoretical determination of lateral coherence functions in isotropic turbulence, which has been found to vary somewhat from the exponential form, is given by Kristensen and Jensen (1979). We can see from (24), however, that for $\lambda \simeq 100$ m, which is equivalent to $f \simeq 0.2$ Hz at a wind speed of 5 m s^{-1}, the coherence is already very small at a separation distance of 20 m. Therefore, the wind forces on a structure of this size would not be as large as would be predicted by measurements of the instantaneous point-measured wind speed fluctuations at this frequency. This reduction in force has been parameterized by Davenport (1977) with what he calls an aerodynamic admittance function. On the basis of both theoretical and empirical evidence, he shows that this function is equal to one for $D/\lambda \to 0$, is about 0.2 to 0.3 for $D/\lambda \simeq 1$, and continues to decrease as D/λ increases.

REFERENCES

Arya, S. P., 1982: Atmospheric boundary layers over homogeneous terrain. In *Engineering Meteorology*, edited by E.J. Plate, Elsevier, 233–267.

Beardsley, R. C., C. Dorman, C. Friehe, L. Rosenfeld and C. Winant, 1987: Local atmospheric forcing during CODE 1 and CODE 2. Part 1: A description of the marine boundary layer and atmospheric conditions over a northern California upwelling region. *J. Geophys. Res.*, 92, C2. 1467–1488.

Businger, J. A., 1971: Flux profile relationships in the atmospheric surface layer. *J. Atmos. Sci.*, 25, 181–189.

Businger, J. A., 1982: Equations and concepts. In *Atmospheric Turbulence and Air Pollution Modeling*, edited by F.T.M. Nieuwstadt and H. van Dop, D. Reidel Publishing Company, 1–36.

Davenport, A. G., 1961: The spectrum of horizontal gustiness near the ground in high winds. *Quart. J. Roy. Meteor. Soc.*, 87, 194–211.

Davenport, A. G., 1977: Wind structure and wind climate. In *Safety of Structures Under Dynamic Loading*, edited by I. Holand, D. Kavlie, G. Moe and R. Sigbjörnsson, Norwegian Institute of Technology, Trondheim, Norway, 209–284.

Davenport, A. G., 1982: The interaction of wind and structures. In *Engineering Meteorology*, edited by E. J. Plate, Elsevier, 527–572.

Dutton, J. A., 1976: *The Ceaseless Wind*. McGraw-Hill Book Company, 579 pp.

Ekman, V. W., 1905: On the influence of the earth's rotation on ocean currents. *Arkiv. Math. Astron. O. Fysik*, 2, 11.

Hicks, B. B., 1981: An examination of turbulence statistics in the surface boundary layer. *Boundary Layer Meteorol.*, 21, 389–402.

Kristensen. L., and N. O. Jensen, 1979: Lateral coherence in isotropic turbulence and in the natural wind. *Boundary Layer Meteorol.*, 17, 353–373.

Jensen, N.O. and N.E. Busch, 1982: Atmospheric turbulence. In *Engineering Meteorology*, edited by E.J. Plate, Elsevier, 179–231.

Large, W. G. and S. Pond, 1981: Open ocean momentum flux measurements in moderate to strong winds. *J. Phys. Oceanogr.*, 11, 324–336.

Monin, A. S., and A. M. Obukhov, 1954: Basic laws of turbulent mixing in the ground layer of the atmosphere. *Trans. Geophys. Inst. Akad. Nauk USSR*, 151, 163–187.

Panofsky, H. A., H. Tennekes, D. H. Lenschow and J. C. Wyngaard, 1977: The characteristics of turbulent velocity components in the surface layer under convective conditions. *Boundary Layer Meteorol.*, 11, 355–361.

Panofsky, H. A., and J. A. Dutton, 1984: *Atmospheric Turbulence*. Wiley and Sons, 239 pp.

COMPUTER MODELLING OF WAVE PROPAGATION IN COASTAL AREAS AND HARBOURS

MARTIN MATHIESEN

Norwegian Hydrotechnical Laboratory, Trondheim, Norway

1. INTRODUCTION

Most of the computer wave models used in port and harbour studies have been developed over the past 15 - 20 years but the basic ideas behind some of the models can be traced back a 100 years or so. It is now fair to say that the use of numerical methods makes it possible to give a good description of ocean waves as they travel inshore: along the shipping lane, at the harbour entrance and within the harbour itself. As a result, by use of such models we can study most aspects of wave motions related to port and harbour design, i.e. both short and long period wave motions which effect the behaviour of moored ships.

In this presentation we review some of the computer wave models most commonly used in port and harbour studies today. These are models based on the refraction equations, the combined refraction-diffraction equation or the mild-slope equation and the Boussinesq equations. The model equations are stated without derivation but the basic assumptions behind them are given and the numerical solution techniques are outlined.

2. REFRACTION

In most harbour studies performed, we generally have a fairly good knowledge of the offshore wave conditions, either as a result of wave measurements or from hindcast wave studies. However, as the waves travel inshore, they are modified due to a varying bathymetry and coastal currents; i.e. wave refraction effects. As a result, the wave conditions along the shipping lane towards the harbour may change considerably.

Wave refraction theory and the ray tracing method were introduced in coastal engineering 40 years ago by Munk and Traylor [1947]. Even though the theory of wave refraction has advanced considerably since then, most of today's wave refraction computer models are still based on some kind of ray tracing technique.

Here we only give a brief introduction to the theory of wave refraction. For a more detailed description we refer to Phillips [1977, pp. 23-25,59-60,179-183]. We assume that the waves can be described within the framework of linear wave theory, that is;

$$ak \ll 1 \quad \text{and} \quad a/h \ll (kh)^2 \quad \text{when} \quad kh \ll 1$$

where a is the wave amplitude, $k = |\underline{k}|$ is the wave number and h is the water depth. The wavelength is defined by $L = 2\pi/k$. We further assume that the relative change in water depth over wavelength is small;

$$|\nabla h|/kh \ll 1$$

187

E. Bratteland (ed.), Advances in Berthing and Mooring of Ships and Offshore Structures, 187–197.
© 1988 by Kluwer Academic Publishers.

and that the relative change in the current velocity $\underline{u}$ over wavelength and period is small;

$$|\nabla\underline{u}|/ku \ll 1 \qquad \text{and} \qquad |\partial\underline{u}/\partial t|/\omega u \ll 1$$

where ω is the absolute angular wave frequency. The wave period is defined by $T = 2\pi/\omega$. Finally, we assume that the relative change in wave amplitude over wavelength is small;

$$|\nabla a|/ak \ll 1$$

The results obtained are the ray equations which read;

$$d\underline{r}/dt = \underline{c}_g + \underline{u} \qquad\qquad (1)$$

$$d\omega/dt = \underline{k}.(\partial\underline{u}/\partial t) \qquad\qquad (2)$$

$$d\underline{k}/dt = -(\partial\sigma/\partial h)\nabla h - (\nabla\underline{u}).\underline{k} \qquad\qquad (3)$$

$$d(S(\underline{k})/\sigma)/dt = 0 \qquad\qquad (4)$$

The rays $\underline{r} = \underline{r}(t)$ are parametric curves which can conveniently be interpreted as streamlines of wave energy transport. The other variables are the relative group velocity $\underline{c}_g$, the intrinsic angular frequency σ and the wave number spectrum $S(\underline{k})$.

When the current field is assumed steady in time, we see from (2) that the absolute frequency $f = \omega/2\pi$ is constant along a ray. It is then convenient to use the directional wave spectrum $S(f,\theta)$ obtained from $S(\underline{k})$ through the relation $S(\underline{k}) = J(f,\theta) S(f,\theta)$ where θ is the wave direction and $J(f,\theta)$ is the Jacobian determinant of the mapping from wave number to frequency direction domain. Consequently (4) can be restated in the form:

$$d(J(f,\theta) S(f,\theta)/\sigma)dt = 0 \qquad\qquad (5)$$

The Jacobian is given by;

$$J(f,\theta) = |\partial\omega/\partial\underline{k} \times \partial\theta/\partial\underline{k}|/2\pi \qquad\qquad (6)$$

which after insertion for the partial derivatives may be rewritten in the form:

$$J(f,\theta) = \underline{c}_g.(\underline{c}_g + \underline{u})/2\pi kc_g \qquad\qquad (7)$$

The ray equations have to be solved together with the Doppler shift equation;

$$\omega = \sigma + \underline{k}.\underline{u} \qquad\qquad (8)$$

and the dispersion relation;

$$\sigma^2 = gk \tanh(kh) \qquad\qquad (9)$$

where g is the acceleration due to gravity.

When solving these equations we advocate using the backward ray tracing technique pioneered by Abernethy and Gilbert [1975], here generalized to include current refraction. This technique does not make direct use of the local ray separation distance as is the case for most other ray tracing techniques.

When using the backward ray tracing method, rays are traced in the reverse direction from points in the inshore area outward to the offshore area where the waves do not refract. It follows from (5) that the inshore spectrum S can be determined in terms of any choice of offshore spectrum model S_0 by;

$$S(f,\theta) = [J_0(f,\theta_0)\sigma/J(f,\theta)\sigma_0] \, S_0(f,\theta_0) \tag{10}$$

where the zero-index refers to offshore values. The only purpose of the ray calculations is then to establish the relation $\theta_0 = \theta_0(f,\theta)$ between the inshore and the offshore wave direction. The numerical task, therefore, is to calculate the ray paths. This can be done either by direct integration of (1) and (3), or through the calculation of ray curvature. For an illustration of model use we refer to Mathiesen [1987] who has applied the model to the study of wave propagation through a current whirl.

The backward ray tracing described above can be extended to include the effect of wave breaking. From the basic form of the spectral balance equation (4), which states that in the case of no dissipation $S(\underline{k})/\sigma$ is constant along a ray, the effect of wave breaking can be modelled by imposing a maximum allowed value of $S(\underline{k})$ along each ray. This maximum value can be determined from a modified version of Phillips' [1985] equilibrium spectrum form;

$$S(\underline{k}) = \beta[\cos(2\theta)]^p (u_*/c)k^{-4}$$

where β and p are parameters ($\beta \approx 0.06$ and $p \approx 0.5$), u_* is the friction velocity and c the phase velocity ($c = \omega/k$). The limiting spectral density is determined at the point along the ray where the wave number attains its maximum.

3. REFRACTION - DIFFRACTION
3.1. Mathematical formulation

In areas where there is a strong focusing of waves or where the waves are scattered due to the presence of obstacles or solid boundaries, the relative variation in wave height may be large, i.e. $|\nabla a|/ka \approx O(1)$. Consequently, the refraction model cannot be used to study these wave phenomena. The reason is that the model does not allow for the change in wave direction due to local variations in wave amplitude and a more general wave model is needed.

The combined refraction-diffraction equation or the mild-slope equation, was first derived by Berkhoff [1972]. The basic assumptions are the same as for the refraction approximation except that the relative change in wave amplitude is no longer assumed to be small. Here we shall consider a generalized version of Berkhoff's equation, due to Chen [1986], that also includes the effect of bottom friction and absorption at lateral boundaries.

In his derivation of the generalized equation Chen [1984,1986] essentially follows the arguments of Smith and Sprinks [1975] except for the addition of a bottom stress term to the momentum equation. The resulting equation is:

$$\nabla \cdot (\lambda cc_g \nabla\varphi) + (\omega^2 c/c_g)\varphi = 0 \tag{1}$$

where

∇ : horizontal gradient operator ($\nabla = (\partial x, \partial y)$)
λ : friction factor ($\lambda = 1$ gives Berkhoff's equation)
c : phase speed ($c = \omega/k$)
c_g : group speed ($c_g = \partial\omega/\partial k$)
φ : two-dimensional velocity potential ($\underline{u} = \lambda f (\nabla\varphi)\exp(-i\omega t)$)
i : imaginary unit ($i^2 = -1$)
f : z - dependancy of wave field ($f = \cosh(k(z + h))/\cosh(kh)$)
k : wave number ($\omega^2 = gk \tanh(kh)$)
g : acceleration of gravity
h : depth

The friction factor is given by:

$$\lambda = 1 / [1 + i\beta a \exp(i\gamma)/(h \sinh(kh))]$$

where

β : friction coefficient ($\beta \approx 0.1$)
a : wave amplitude
γ : phase angle ($\gamma = -\pi/4$)

Absorption or reflection at closed boundaries is modelled by:

$$\partial\varphi/\partial n - \alpha\varphi = 0 \tag{2}$$

where

$$\alpha = ik (1 - K_r) / (1 + K_r)$$

K_r is the reflection coefficient and n represents the normal unit vector pointing out from the area modelled.

For the scattered wave field φ^s a radiation condition is imposed at infinity:

$$\lim_{r\to\infty} [\sqrt{r} (\partial/\partial r - ik) \varphi^s] = 0 \tag{3}$$

where r is the radial distance.

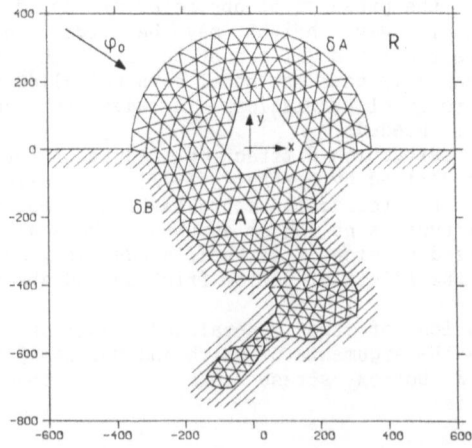

Figure 1. Finite element model with definitions of symbols used in text.

The boundary value problem (1)-(3) is solved using a hybrid element method analogous to the method originally developed by Chen and Mei [1972]. The water domain is divided into two regions, a finite domain A and an infinite domain R, see Figure 1. A conventional finite element technique is used for the region A, while an analytical solution with unknown coefficients is used to describe the wave field in region R.

The variational principle established for the boundary problem requires that the following functional Π is stationary with respect to arbitrary first variations of the potentials φ and φ_R in the near and far region respectively;

$$\Pi(\varphi, \varphi_R) = \iint_A (1/2)[\lambda cc_g(\nabla\varphi)^2 - \omega^2 c_g\varphi^2/c] \, dA - \int_{\delta B} (1/2)\alpha\lambda cc_g\varphi^2 \, dl \qquad (4)$$

$$+ \int_{\delta A} \lambda cc_g[(1/2)(\varphi_R - \varphi_0) - (\varphi - \varphi_0)](\partial\varphi_R/\partial n_A) \, dl$$

$$- \int_{\delta A} (1/2)\lambda cc_g(\varphi_R - \varphi_0)(\partial\varphi_0/\partial n_A) \, dl$$

where φ_0 is the potential for the incoming waves and n_A represents the unit normal outward from region A. For a more detailed description of the system of equations and the solution technique, we refer to Chen [1984, 1985] and to Mei [1983, pp. 168-181]. An alternative method of derivation of (4) for $\lambda = 1$, i.e. no bottom friction, is given by Behrendt and Jonsson [1984].

The numerical method developed can be used to study the response of harbours to both short and long period waves. However, for open harbours for which it may be necessary to model a large part of the coastal area, the study of short waves, i.e waves of periods 7 - 25 s, can be rather costly.

A possible solution to this problem is to use the parabolic approximation to (1) which was derived by Radder [1979] and also by Lozano and Liu [1980]. Radder [1979] obtains the parabolic approximation using a splitting technique whereby the wave field is split into transmitted and reflected components. Neglecting the reflected field, the equation for the transmitted field reads as follows;

$$\partial\varphi/\partial x = [ik - (2kcc_g)^{-1}\{\partial(kcc_g)/\partial x - i\partial(cc_g\partial)/\partial y^2\}] \, \varphi \qquad (5)$$

where x is the main direction of wave advance. The reduction in computer cost is obtained because the parabolic equation permits a numerical solution using a marching method in the direction of wave advance.

The hybrid finite element method based on (4) is probably most effective in the study of harbour response to long period waves, i.e. waves of period 1 - 10 mins. In this case, because the diameter of the finite elements can be made large, the number of elements is reduced and the computer time involved will be moderate at most.

In the study of harbour resonance, the emphasis has mostly been on harbour plan form and the effect of a varying bathymetry largely ignored. However, during the last decade, this has changed considerably. Papers by Raichlen et al. [1983], Liu [1983,1986] and also by Tinti [1980] and Noiseux [1983] clearly demonstrate that changes in water depth can have a marked influence on the amplification of wave amplitude at resonance. In general, harbour resonance is most likely to effect harbours that are long

192

and narrow, and then even more so if there is a marked increase in depth
near the harbour entrance.

The relative effect of dissipation due to bottom friction and boundary
absorption is most clearly noted at frequencies close to the frequencies
of resonance where the radiation losses are small. However, as long as the
dissipative losses are moderate, the frequencies of resonance are only
slightly reduced as compared with the nondissipative case, see Shaw and
Lai [1974].

In some harbours the effect of entrance loss can be important. In an
analytical study Ünlüata and Mei [1975] show that this loss depends on a
factor proportional to $f(a/h)(kb)^{-2}$ where f is the friction coefficient, a
the wave amplitude, h the water depth, k the wave number and b the width
of the breakwater gap. In practice, the effect of entrance loss is only
important for the lowest mode of resonance (longest waves), and then only
when the harbour entrance is narrow.

3.2. Application of model

The harbour at Sklinna, a group of islands off the coast of Central
Norway, is mainly used during the seasonal fisheries but also serves as
harbour of refuge for smaller vessels. The semi-enclosed harbour basin
between Heimøy and Hansholmen is protected by a breakwater to the
northeast as shown in figure 2. However, the harbour does not provide
adequate shelter for incoming ocean waves from the southwest and it is
planned to construct a new breakwater to remedy this situation.

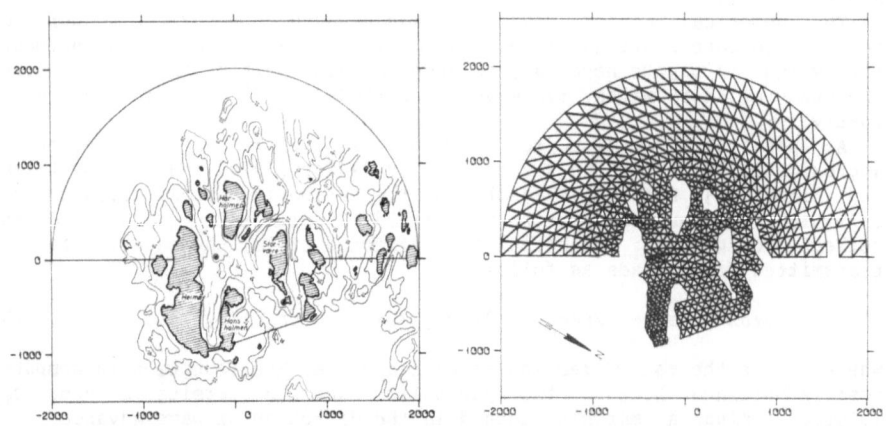

Figure 2. Map of Sklinna islands.   Figure 3. Finite element grid model.

Data from wave measurements, see figure 4 and 5, show the occurrance of
long period waves with a dominant period of about 2 minutes and wave
heights close to 1.0 m in the inner part of the harbour. It is desirable,
therefore, to make a choice of breakwater alignment so that the long
period wave motions within the harbour can also be reduced.

For the numerical wave study we have constructed a finite element grid
model as shown in figure 3. The water depth is variable within the grid
but constant (50 m) outside. Reflection coefficients at island and
breakwater boundaries are set to 0.99.

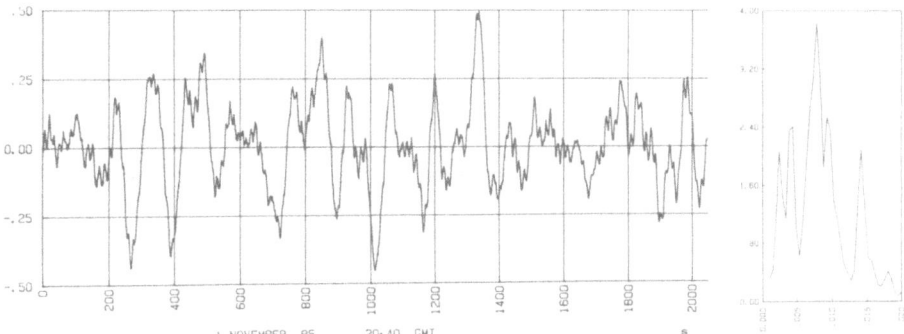

Figure 4. Time series of surface elevations.

Figure 5. Spectrum of time series in figure 4.

Computed amplification factors, defined as the ratio between local wave height and incoming wave height, are presented in figure 6. The similarity with the spectrum of the measured wave data is striking: The four main peaks in the spectrum are seen to correspond to local amplification maxima, i.e. resonance peaks. The computed velocity field and isolines for the amplification factors for incoming waves of 125 s period and 5 cm wave height (at 50 m depth) from the west are presented in figure 8.

Harbour resonance can be reduced by placing a breakwater along a nodal line for the standing wave motion, that is where the velocity oscillations are the largest and the amplifications factors are small. However, along the nodal line close to the middle of the harbour basin the water depths are close to 30 m and a breakwater here would be prohibitively expensive. Along the next nodal line the best choice of breakwater alignment seems to be from the southwest of Heimøy to the skerry between Heimøy and Hårholmen. Results from the calculation for this alternative are presented in figures 7 and 9. We see that the wave amplification at resonance within the harbour basin is reduced considerably (to about 1/3). Consequently, we have found a breakwater alignment that provides adequate shelter for the incoming short period ocean waves as well as reduction of the long period wave motions within the harbour.

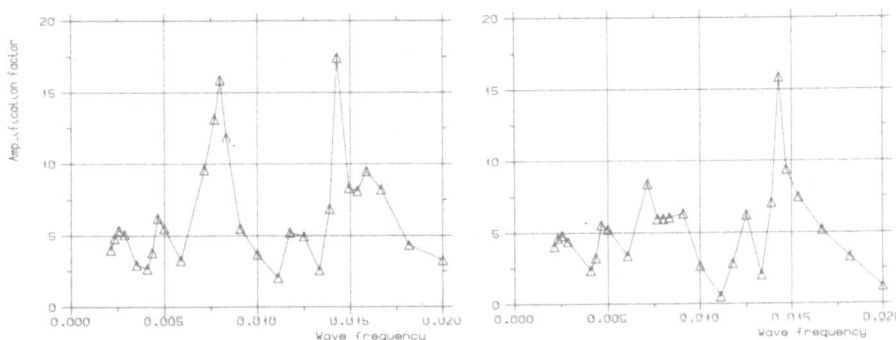

Figure 6. Computed amplification factors for existing situation.

Figure 7. Computed amplification factors with new outer breakwater.

194

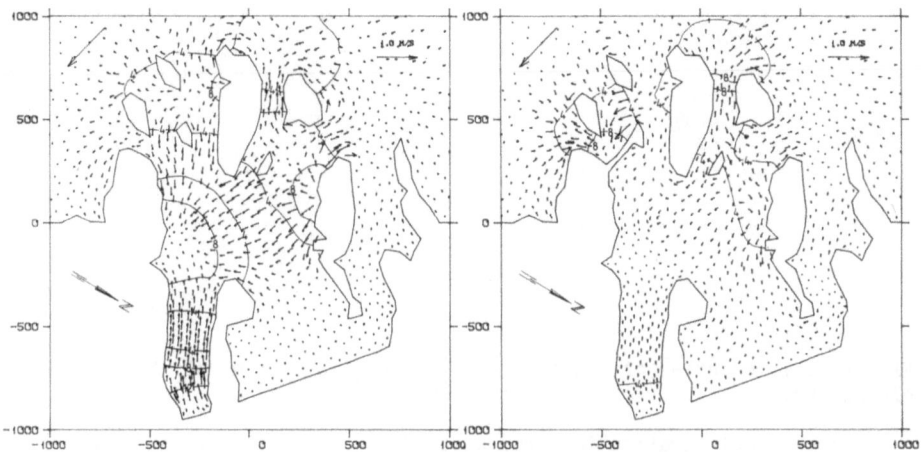

Figure 8. Computed velocity field for existing situation.

Figure 9. Computed velocity field with new outer breakwater.

4. BOUSSINESQ-EQUATIONS

In the study of harbour response to short period waves, the assumption of linear wave theory, a/h << (kh)^2 when kh << 1, turns out to be very restrictive for many coastal harbours. An alternative wave model is needed. One such model is provided by the Boussinesq-equations which holds for weakly nonlinear waves of moderate wavelength.

We start by defining two parameters α = a/h and β = (kh)^2, both assumed to be small and of the same order:

$$\alpha << 1 \qquad \beta << 1 \qquad \alpha \approx \beta$$

For a derivation of the Boussinesq-equations we refer to Peregrine [1967] who first derived these equations for the case of a variable depth. The equations read:

$$\partial \zeta / \partial t + \nabla . [(h + \zeta) \underline{u}] = 0 \tag{1}$$

$$\partial \underline{u} / \partial t + \underline{u} . \nabla \underline{u} + g \nabla \zeta - \partial \{(h/2) \nabla [\nabla . (h \underline{u})] - (h^2/6) \nabla [\nabla . \underline{u}] \} / \partial t = 0 \tag{2}$$

where ζ is the surface elevation, $\underline{u}$ the depth-averaged velocity, h the water depth, g acceleration of gravity and ∇ = ($\partial / \partial x, \partial / \partial y$) the horizontal gradient operator.

The Boussinesq-equations form a fairly general set of equations in wave theory. The following equations are obtained as limiting cases:

Airy's theory of very long waves : α = O(1), $\beta \rightarrow$ o
Linear waves with weak dispersion : $\alpha << \beta$
Linear shallow water waves : $\alpha \rightarrow 0, \beta \rightarrow 0$

The equations (1) and (2) have to be solved together with the boundary conditions:

$$P \partial \zeta / \partial t - \underline{n} . (\partial \underline{u} / \partial t) = P \partial \zeta_G / \partial t - \underline{n} . (\partial \underline{u}_G / \partial t) \tag{3}$$

$$Q \ \partial\zeta/\partial t + \underline{n}.(\partial\underline{u}/\partial t) = -Q \ \nabla.[(h + \zeta)\underline{u}] - g\underline{n}.\nabla\zeta - \underline{n}.[\underline{u}.\nabla\underline{u}]$$

$$+ (h/2) \ \underline{n}.\nabla[\nabla(h\partial\underline{u}/\partial t)] - (h^2/6)\underline{n}.\nabla[\nabla.(\partial\underline{u}/\partial t)] \quad (4)$$

$$\underline{s}.(\partial\underline{u}/\partial t) = - g\underline{s}.\nabla\zeta - \underline{s}.[\underline{u}.\nabla\underline{u}]$$

$$+ (h/2) \ \underline{s}.\nabla[\nabla(h\partial\underline{u}/\partial t)] - (h^2/6)\underline{s}.\nabla[\nabla.(\partial\underline{u}/\partial t)] \quad (5)$$

where ζ_G and $\underline{u}_G$ defines a "wave generator" at the open boundary (zero at closed boundary), $\underline{n}$ and $\underline{s}$ are unit normal (outwards) and tangent vectors at the boundary. The parameters P and Q are given by;

$$P = [g/h(1 + k^2h^2/3)]^{1/2} \ (1 - K_r) \ / \ (1 + K_r)$$

$$Q = [g/h]^{1/2}$$

where K_r is the reflection coefficient.

Numerical methods for the solution of the Boussinesq-equations have been presented by various authors, see Abbott et al. [1978], Håugel [1980] and Rottmann-Söde et al. [1985]. These authors all use finite difference techniques, but the methods of solution do otherwise differ. The choices of boundary conditions are also different from described above.

An alternative "semi-discrete" finite difference method is presented here: The Boussinesq-equations and the associated boundary value problem involve only first derivatives with respect to time. By approximating the spatial derivatives (using a 9-point molecule for internal points) the system of partial differential equations (1)-(5) can be "reduced" to a system of ordinary differential equations of the form:

$$M \ d\underline{z}/dt = \underline{F}(\underline{z},t) \quad (6)$$

where M is a known matrix, $\underline{z}$ is a vector representing the unknown variables ζ and $\underline{u}$ at all grid points, and $\underline{F}$ is a known vector function. The matrix M is independent of $\underline{z}$ and t and consequently the LU-decomposition of M has to be performed only once to obtain:

$$d\underline{z}/dt = M^{-1}\underline{F}(\underline{z},t)$$
$$\quad (7)$$

which is integrated using a third order Runge-Kutta method.

In practical applications of the model results are generally presented in terms of time series plots of surface elevations, contour plots of significant wave heights, contour and perspective plots of instantaneous surface elevations and plots of instantaneous velocity fields.

For an illustration of model use we refer to Warren et al. [1985] who present results from a study of wave penetration into Valencia harbour.

ACKNOWLEDGEMENT
The author is indepted to H. S. Chen and the Coastal Engineering Research Centre, Vicksburg, Mississippi, USA for making their refraction-diffraction computer algorithm available for the study of long period waves in Sklinna harbour.

REFERENCES

ABBOTT, M. B., H. M. PETERSEN, O. SKOVGAARD, On the numerical modelling of short waves in shallow water, Journal of hydraulic research, Vol 16, no 3, pp. 173-204, 1978.

ABERNETHY, C. L. and G. GILBERT, Refraction of wave spectra. Report INT 117, Hydraulics Research Station, Wallingford, England, 1975.

BEHRENDT, L. and I. JONSSON, The physical basis of the mild-slope wave equation, Proceedings of 19th international conference on coastal engineering, Houston, Texas, 1984.

BERKHOFF, J. C. W. , Computation of combined refraction - diffraction, Proceedings of 13th international conference on coastal engineering, Vancouver, B. C., Canada, 1972.

CHEN, H. S., Hybrid element modelling of harbor resonance. 4th international conference on applied numerical modelling, 1984.

CHEN, H. S., Effects of bottom friction and boundary absorption on water wave scattering. Applied ocean research, vol 8, no. 2, pp. 99-104, 1986.

CHEN, H. S. and C. C. MEI, Oscillations and wave forces in an offshore harbor. Ralph M. Parsons Laboratory report no. 190, MIT, 1974.

HAUGEL, A., Adaption of tidal numerical models to shallow water wave problems, Proceedings 17th conference on coastal engineering, Sydney, Australia, 1980.

LIU, P. L.-F., Effects of the continental shelf on harbor resonance, in Tsunamis - Their science and engineering, edited by K. Iida and T. Kawasaki, pp. 359-385, Terra scientific publishing company, Tokyo, 1983.

LIU, P. L.-F., Effects of depth discontinuity on harbor oscillations, Coastal engineering, Vol 10, pp. 395-404, 1986.

LOZANO, C. J. and P. L.-F. LIU, Refraction-diffraction model for linear surface waves, Journal of fluid mechanics, Vol 101, pp. 705-720, 1980.

MATHIESEN, M., Wave refraction by a current whirl, Journal of geophysical research, Vol 92, no. C4, pp. 3905-3912, 1987.

MEI, C. C., The applied dynamics of ocean surface waves, pp. 740, John Wiley & Sons, Inc., 1983.

MUNK, W. H. and M. A. TRAYLOR, Refraction of ocean waves, Journal of geology, No 55, pp. 1-26, 1947.

NOISEUX, C. F., Resonance in open harbours, Journal of fluid mechanics, Vol 126, pp. 219-235, 1983.

PEREGRINE, D. H., Long waves on a beach, Journal of fluid mechanics, Vol 27, pp. 815-827, 1967.

PHILLIPS, O. M., The dynamics of the upper ocean. Second edition, pp. 336, Cambridge University Press, 1977.

PHILLIPS, O. M., Spectral and statistical properties of the equilibrium range in wind-generated gravity waves, Journal of fluid mechanics, vol 156, pp. 505-531, 1985.

RADDER, A. C., On the parabolic equation method for water-wave propagation, Journal of fluid mechanics, Vol 95, pp. 159-176, 1979.

RAICHLEN, F., T. G. LEPELLETIER and C. K. TAM, The excitation of harbours by tsunamis, in Tsunamis - Their science and engineering, edited by K. Iida and T. Kawasaki, pp. 359-385, Terra scientific publishing company, Tokyo, 1983.

ROTTMANN-SÖDE, W., H. SCHAPER and W. ZIELKE, Two numerical wave models for harbours, International conference on numerical and hydraulic modelling of ports and harbours, Paper K3, pp. 285-293, 1985.

SHAW, R. P., and C.-K. LAI, Channel friction and slope effects on harbor resonance, ASCE, Journal of the waterways, harbors and coastal engineering division, Vol 100, no WW3, pp. 205-215, 1974.

SMITH, R. and T. SPRINKS, T., Scattering of surface waves by a conical island, Journal of fluid mechanics, Vol 72, pp. 373-384, 1975.

TINTI, S., Response of a harbour opened to a sea of variable depth, Pageoph. Vol 118, pp. 783-795, 1980.

ÜNLÜATA, Ü and C. C. MEI, Effects of entrance loss on harbor oscillations, ASCE, Journal of the waterways, harbors and coastal division, Vol 101, no WW2, pp. 161-180, 1975.

WARREN, R., J. LARSEN and P. E. MADSEN, Application of short wave numerical models to harbour design and future development of the model, International conference on numerical and hydraulic modelling of ports and harbours, Paper L1, pp. 303-308, 1985.

FIELD OBSERVATIONS OF SHIP BEHAVIOUR AT BERTH.

GISLI VIGGOSSON M.Sc. Senior Res. Eng.
Icelandic Harbour Authority.

ABSTRACT.
The knowledge of ship behaviour at berth is of fundamental importance for designing an improvement of harbours. For smaller harbours like fishery harbours the only tool available sofar have been hydraulic wave disturbance tests with moored vessels. These tests have not been compared with field measurements until such measurements were made as a part of a joint Nordic research project.
The purpose of that project was to establish criteria for acceptable ship movements in harbours for working and safe mooring conditions.
This paper deals mainly with the Icelandic part of this project and describes:

-    The general description of ship behaviour at berth.
-    The instruments which were developed for ship motion measurements.
-    The field measurements and interviews.
-    The analysis of field data.
-    The comparison between measurements in field and model.
-    The definitions of the working and safe mooring conditions.
-    The criteria for acceptable ship movements in harbours are based on results of the joint Nordic project.

The Icelandic field instrumentation included instrument for recording of ship movements in six degrees of freedom. The instrumentation was truck-mounted, with the truck placed in a fixed position on the quay during the observations.
During February, March, October and November 1983 a total of 24 measurements were undertaken from 8 ships.
In Reykjavik harbour ship movements were measured for two identical container vessels of Lo-Lo and Ro-Ro type (Loa = 105.6 m). A separated study of the parameters affecting the efficiency of loading and unloading was made.
In Akranes harbour measurements were done on a cargo vessel (Loa = 78.5 m) under critical working conditions. The movements of two travlers (Loa = 35.2 m and 50.05 m) were observed under mooring condition at berth.

E. Bratteland (ed.), Advances in Berthing and Mooring of Ships and Offshore Structures, 198–264.
© 1988 by Kluwer Academic Publishers.

In Thorlakhöfn harbour the movements of trawler (Loa = 47.0 m),
one 200 grt vessel (Loa = 36.3 m) and a 30 grt boat (Loa =17.4
m) were measured. The trawler was observed under both working
and safe mooring conditions, but the others were measured
under safe mooring conditions.
A captain or another experienced person was interviewed during the measurements. The purpose of the questionnaire was
to correlate the measured movements of the ship to the experience of the captain.He estimated the wind velocity,wave heights
and long period waves together with the six movements of the
ship. Information on the harbour, the berth and the fenders
as well as the moorings were collected. He was asked about
his experiences with the ship at that berth under working and
critical mooring conditions. Difficulties in maneuvering etc
related to the harbour, were also asked for and critical working and safe mooring conditions included in term of all six
movements of the ship and how often such conditions had occurred based on his experience.   35 questionnaires were
filled out during the interviews.
The paper shows the layout of the harbour with the moored
ship at berth. All related data are shown with a few comments
on the conditions of the ship. The other figures show time
series of all parameters and the results of these time series.
Some velocities are also indicated.

1. INTRODUCTION
The importance of fishery for Iceland is best described by the
fact that more than 70 percent of the country's exports are
fishery products.
Iceland has about 70 harbours mostly serving the fishing fleet.
They are spread along the entire coastline,with the exception
of the sandy region along the south coast.
The fishing fleet consists of almost 1000 ships of 10-1000 grt
and about 2500 fishing boats under 10 grt. Most Icelandic harbours are small, especially in relation to the wavelength they
are exposed to,(breakwater length versus wavelength).
Most Icelandic harbours are located in relatively protected
waters, like fiords. But even so the wave penetration from the
ocean may still be considerable, necessitating protective works
like breakwaters.
In many harbours the wave action causes long period waves, a
wave set-up inside the surf zone and a sediment transport on
sandy coasts.
Surges and wave action often make movements of ships exceeding
the acceptable limits for loading, unloading and mooring.
The research section of Icelandic Harbour Authority (IHA) runs
a hydraulic laboratory where model testing with irregular waves
is performed. Usually a model scale of 1:60 is used to under-
take both three dimensional wave disturbance tests with moored
ships and stability tests on rubble mound breakwaters. A selection of measured wave data is reproduced in the model. The
application of irregular waves in wave disturbance models and
stability tests necessitates better knowledge of prototype wave
climate, the prototype ship movements and the investigation of
breakwater material (quarry).

Several model tests per year have been performed since 1977
with moored ships in irregular waves. At the beginning the
goal was to improve the worst conditions to acceptable limits
for a moored ship during the time of wave measurements inside
and outside the harbour.
In the model the worst conditions were simulated by modelling
the ship, the moorings and the fender systems with the assistance of experienced captains and harbour personnels.
Based on the type and size of the Icelandic fleet tree types
of ship were chosen as a reference: A cargo vessel of 5000 DWT
(Loa = 95 m), a trawler of 900 DWT (Loa = 47.5 m) and a small
20 DWT fishing boat (Loa = 17 m).
During the first years of model testing with moored ships the
first set of criteria for safe mooring and working conditions
was established.
IHA is responsible for collecting wave data for the design,
construction and maintence of harbours. To fulfill these requirements the research section operates 8 waverider buoys and
5 pressure gauges.
The wave measuring program consists usually of one offshore
waverider and several waveriders near harbours with pressure
gauges inside harbours.
The purpose of measuring waves offshore, outside and inside the
harbours simultaneously, is to establish the wave energy transfer functions between these three locations. These transfer
functions are estimated by comparing rms-and spectrum values
at the same time.
Directional wave spectrum-refraction analysis have also been
used to estimate the offshore-inshore energy transfer functions,
and model characteristics.

2. SHIP BEHAVIOUR AT BERTH
2.1 General description of ship behaviour at berth.
The movements of a moored ship at berth can be described by
the three translatory movements, surge, sway and heave, and
the three movements of rotation pitch, roll and yaw.
The movements are defined as shown in fig. 2.1.

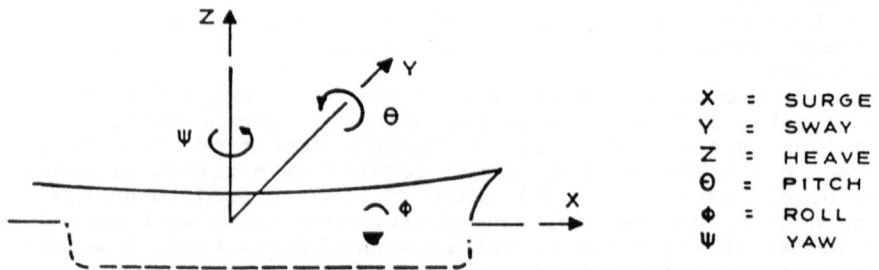

Fig. 2.1. Definition of Ship Movements.

The movements of a ship are combinations of these six movements
with few dominating movements.
The movements of rotation i.e. pitch, roll and yaw are almost
independant of the mooring system. But they depend on the resonance period of the ship, the wave spectrum, the berth angle
relative to waves, the reflection pattern for the waves and
the type of berth.
During loading and unloading roll causes the maximum vertical
'for crane on ship' operation, as the movements of the load in
the crane will be amplified due to the roll movement.
The movements of pitch cause the same type of problems as roll
but to a less extent.
The vertical displacement in each point of the ship is the sum
of the movement of heave, roll and pitch.
The translatory movements of a moored ship depend upon the type
of ship, the mooring and the fender system, the type of berth,
the wave spectrum, the berth angle in relation to waves and
the reflection pattern of the waves.
The mooring arrangements and the differences in spring constants of the mooring lines and the fenders cause a symmetry in
the system. The periods of movements in the horizontal plane
are therefore much longer than the wave period. When the mooring lines are slack or very elastic the movements of the ship
are due to the period of the long waves.
For the loading and unloading operation the most critical movements for a quay based crane are surge and sway.
The various types of movements cause different kinds of damages. Large surge displacements cause breaking of mooring lines
and collision with another ship moored aft or for. The yaw motion may result in damage due to collision between the ship and
the quay or another ship.
Where the tidal range is high the slackness of the mooring lines depend on the relative height of bow and stern versus the
quay. For larger ship the mooring lines are taut at high tide
but for smaller boat the mooring lines can be taut on low tide.
The following Fig. 2.2 presents schematically the relation between the various parameters which might cause problems for ship
behaviour at berth.

2.2 Mooring lines
The most common mooring lines used for mooring of larger fishing vessels are 56 mm (2") polypropylene. The elongation of
the line is ab. 18% at 50% of breaking strength, but for steel
wire the elongation is ab. 0.5% at 50% for the breaking strength. A new 56 mm polypropylene has 36 t breaking strength,
but after some use this strength is lowered to some 80% or 29t.
The spring stiffness is 8 t/m for 10 m lines at 50% of breaking strength.
Under heavy wave conditions the vessels are often moored with
3-4 mooring lines at each bollard also with different length
of lines. A steel wire with tails are also used.
In harbours associated with heavy wave agitation, smaller fishing vessels up to 30 grt use 40 mm polypropylene, but vessels
of 200 grt and larger use 56 mm mooring lines.

202

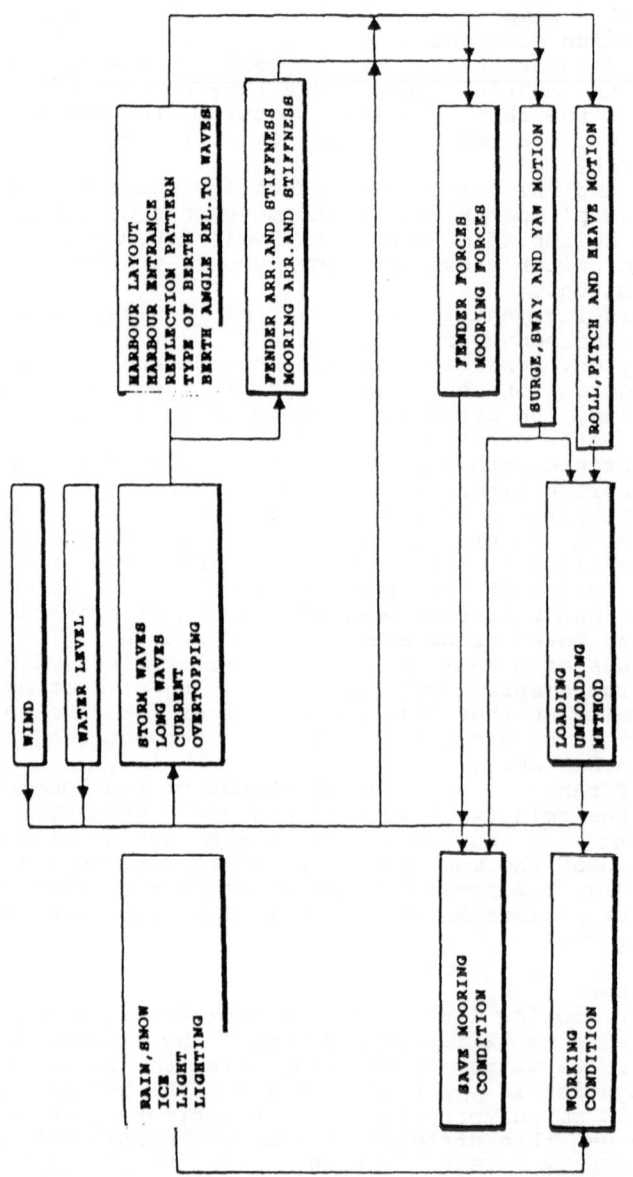

Fig. 2.2 Parameters of importance for ship behaviour at berth.

The total length of mooring lines is about 60 m for 35 grt vessels, 90 m for 200 grt and 120 m for trawlers.

2.3 Fenders.
In fishing harbours the fenders are usually a row of truck tires, except in Reykjavik harbour where ϕ 400 + ϕ 200 mm cylindrical fenders of Vredestein type of 4 m length are used.
The reaction force and the energy absorption for truck tires (1000 200 mm) are in average for 50% deflection: $P_{50\%}$ = 11 t and E = 0.6 tm with $P_{50\%}$ / E = 18, according to measurements made by the Iceland Harbour Authority.
For the cylindrical fender the respective values are $P_{50\%}$ =9.8 t, E = 0.52 tm with $P_{50\%}$ / E = 19.
For comparison a payload tire have for $P_{50\%}$ = 6.1 t, E = 0.62 tm with $P_{50\%}$ / E = 9.8.
The Reykjavik Harbour Authority has developed fenders of 6 truck tires connected together by two steel rod. They are placed horizontally at the berth. The reaction force and the energy absorption for this type of fenders are $P_{50\%}$ = 12.0 t, E = 3.3 tm with $P_{50\%}$ / E = 3.6, which is a soft fender.

2.4 The velocity of the ship movements.
Safe mooring at berth depend on the amplitudes and velocities of ship movements. The velocity and the mass of the ship express the dynamic forces which may cause damage to the berth and/or the vessel.
When a vessel moves along the quay the kinetic energy of the vessel will be absorbed by the elastic potential energy of the mooring lines, the friction forces between the ship hull and the fenders and the fluid dissipation.
When a vessel moves perpendicular to the quay, (sway) the kinetic energy will be absorbed by the elastic potential energy of the fenders.
A vessel of 1000 to 5000 DWT, moored at a berth with 56 mm polypropylene and with fenders, which consist of row of truck tires, provides a simple case of these relations.
The spring stiffness of the mooring lines along the quay are 30 to 60 t/m for 50 % elongation and a tire has a energy absorption of 0.5 tm for 50% compression. Assuming the spring stiffness (k) of 40 t/m with 50% of breaking strength (F) of 40 t, the amplitude (X) of the movements (F=k.x) is 1.0 m and the elastic potential energy (E = $\frac{1}{2}$ kx) is 20 tm. Most of this energy converts to kinetic energy of the vessel, at least when the vessel is not in contact with the rubber fenders. For vessel of size 1000 DWT the corresponding velocity will be ab. 0.60 m/s (E = $\frac{1}{2}$ mv^2) and for 5000 DWT the velocity will be ab. 0.3 m/s.
For the sway movement the acceptable velocity is almost identical for the same size of vessel, assuming that 60% of the length of the vessel is in contact with the quay under mooring conditions. Disregarding the added mass the kinetic energy for sway is 20 tm. For a velocity of 0.6 m/s the absorption of potential energy of the tires per meter is 20/0.6 · 60 = 0.56 tm/m for a 1000 DWT vessel. For a 5000 DWT vessel a velocity of 0,3 m/s will provide the same absorption of potential energy.

2.5 Conventional methods to reduce ship movements.
In fishing harbours where the wave agitation is high several
methods are used to reduce the movements both under working
condition and under safe stay at berth.
These methods can be one or a combination of the following:
-    Tauting the mooring lines.
-    Propeller operation keep spring lines taut.
-    Bow and stern propelles press the ship to the berth.
-    The wind force the ship toward the berth.
-    The wind force the ship away from the berth.
-    Balance tanks reduse the roll.
-    The ship turns few degrees from vertical away from the
     berth to reduce roll.
-    Storm moorings, breast lines are used.
-    Winches hold the mooring lines taut.
-    Moorings lines and fenders are of non recoiling type.

3. SHIP MOVEMENTS IN HARBOURS - A JOINT PROJECT BETWEEN THE
   NORDIC COUNTRIES
From 1981 to 1986 a joint research project has been made by
governmental and research institutes from the nordic countries
with an economic support from the Nordic Council.
The nordic countries involved were Denmark, Finland, Faroe
Islands, Iceland, Norway and Sweden. Finland did not partici-
pate technically, but did give economic support. The project
was initated and coordinated by Iceland.
The purpose of the project was to establish criteria for acc-
eptable ship movements in relation to the following situations:
-    Loading and unloading (working conditions).
-    Ship moored at berth (safe mooring conditions at berth)
In the proposal forming the basic for application for economic
support from the Nordic Council the background for the project
is described as follows:
      Modern transport by ship requires minimum time for
      loading and unloading in ports and at marine terminals.
      This requirement is often restricted by ship movements
      at quay. If too large ship movements are experienced,
      damage to the ships and port installastions may occur.
      In recent years changes in the cargo-handling methods
      have been introduced which have resulted in changes of
      ships and port installations. All these changes have
      large economic consequences.
      The above mentioned changes involve mainly the contain-
      erization of general cargo and changes of methods of
      handling fish in fishing ports, (the latter is of great
      importance in the nordic countries that all have large
      fishery fleets.)
      To minimize future expenses it is required to establish
      pertinent criteria for acceptable ship movements.
      Smaller ship movements can be decreased by different
      means, especially by the extension of protecting break-
      waters which is often costly.

The project included the following four phases:
-    Literature study.

- Pre-study in each country to identify ports with ship movement problems that could be selected for field measurements of ship movements etc.
- Field measurements of ship movements and simultaneous interviews with captains and port personnel. The field measurements of ship movements were made in Denmark, Faroe Island, Iceland, Norway and Sweden. These measurements were made with two different types of equipments developed in Norway and Iceland.
- Supplementary studies comprising interviews with port masters and operators etc. and comparison with results from existing hydraulic investigations of the same ports. At occasions, studies of container operations were also made.

The results of the project is summariced in a final report. NET-PROJEKT, SKIBSBEVEGELSER I HAVNE, nov. 1986 (with summary in English). The detailed results of these measurements are described in separate national reports prepared by the institutions involved in the projects. These national reports are written in the Scandinavian languages.
This paper is based on the Icelandic report: NET-PROJEKT, SKIPSBEVEGELSER I HAVNER, islandsk delraport, oktober 1986, and above mentioned final report.

4. FIELD MEASUREMENTS
4.1 A Brief description of the Icelandic system for measuring ship movements.

The ship movements are detected by setting up two coordinate systems, one at a reference point on the quay (point 1) and another at the shipside (point 2). By measuring the proper values the six fundamental movements of the ship can be calculated mathematically.
A special device was developed to measure the vector between the two coordinate systems directly. This is done simply by holding a wire straight between the origins of the two coordinate systems. By measuring two angles at point 1 and the length of the wire, the vector at point 2 is fixed. At point 2 the ships inclination and two perpendicular angles to reference point 1 are measured, and by these six measurements the ship movements can be calculated.
The following assumptions are made during the calculations:
A) The coordinate system at point 1 has Y-axis vertical and X-axis horizontal.
B) The X-axis at point 1 and 2 are made parallel by calculating the mean of the horizontal angles at each point and cancelling the difference. The same holds true for the Y-axis.
c) The pitch movements were measured directly while the other movements are calculated from the measured angles and the length of the wire.

The device is mainly a hydraulic winch which produces around 10 kg tension on the wire, which goes over a pulley with a potentiometer and then through a universal joint with two angle meters.

The wire ends at a universal joint which holds two angle meters and an inclinometer and this is all clamped to the shipside with four strong magnets.
The winch is powered by a 24 volt battery.
The winch is installed in a van, parked at the shipside. The
van also contains the data logging equipment, in our case, a
HP-9825 computer, scanner, digital voltmeter and power supply
for the sensors.
Wind velocity and direction meter is mounted on top of the
ship. Four load cells measure the tension on the mooring
lines. A pressure sensor, installed underneath the ship, and
waverider outside the harbour measure the waves.
Deflection of the fenders (truck tires) were not measured.
Each channel is scanned twice a second for a total of 2048
samples and then stored on a cassette.
The principle of the instruments are shown on fig. 4.1 - 4.2
and in photos, fig. 4.3 - 4.4.

4.2 The methods of analysis.
The time series have been analysed according to zero-crossing
methods and spectral analysis using Fast Fourier Transforma-
tions (FFT). The floating point of the ship is used as a reference in the presentation of time series of ship movements.
The computer program for calculating the time series was made
by Mr. J. Thomsen, M.Sc. Head of Dept Faroe Island Harbour
Authority.

5. ENVIRONMENTAL DATA
Waves: The purpose in measuring waves both offshore and outside and inside the harbour simultaneously, is to establish
the wave energy transfer functions between these three locations and to relate the ship movements to the offshore wave
statistics. The prototype measurements were made in three
harbours: Reykjavik, Akranes and Thorlakshöfn.
The wave climate in these harbours is dominated by swells,
windwaves, and long period waves.
For swells the refraction coefficients are 0.22 for Akranes,
0.14 for Reykjavik and 0.5 to 0.8 for Thorlakshöfn based on
wave measurements and refraction calculations.
Waverider was installed offshore 8 miles west of the coast at
80 m waterdepth. (64° 05 N. 23° 12 W) (fig. 5.1).
Based on measured and hindcasted data of the significant wave
heights the long term wave heights statistics were determined
for this location (3.hours duration).

Return period in years	1/24	1/10	1	5	10	50
H_s (m)	8,6	9,3	10,5	11,4	12,2	13,3
T_p (sec)	13,0	14,0	15,0	16,0	18,0	20,0

Wind: The long term wind statistics in this area is estimated:

Return period in years	1	5	10	50
Knots	60	62	66	78

In coastal area the wind velocity is expected to be about
10% higher.

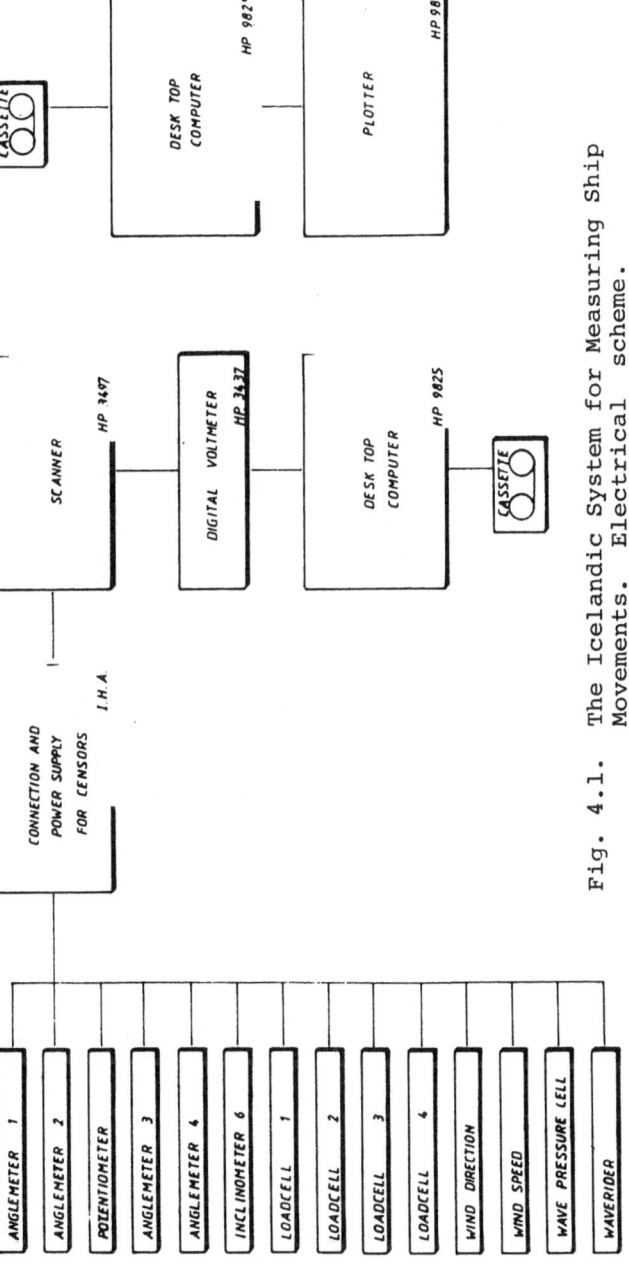

Fig. 4.1. The Icelandic System for Measuring Ship Movements. Electrical scheme.

208

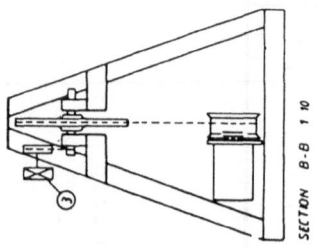

SECTION B-B 1:10

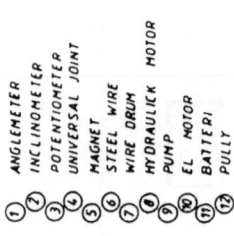

1 ANGLEMETER
2 INCLINOMETER
3 POTENTIOMETER
4 UNIVERSAL JOINT
5 MAGNET
6 STEEL WIRE
7 WIRE DRUM
8 HYDRAULICK MOTOR
9 PUMP
10 EL MOTOR
11 BATTERI
12 PULLY

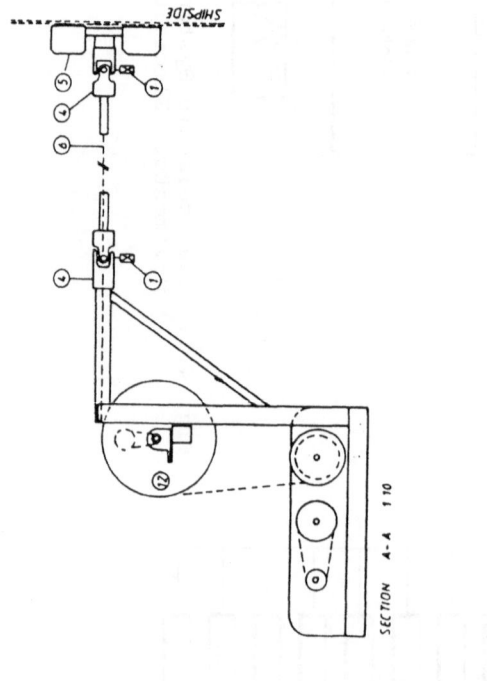

SECTION A-A 1:10

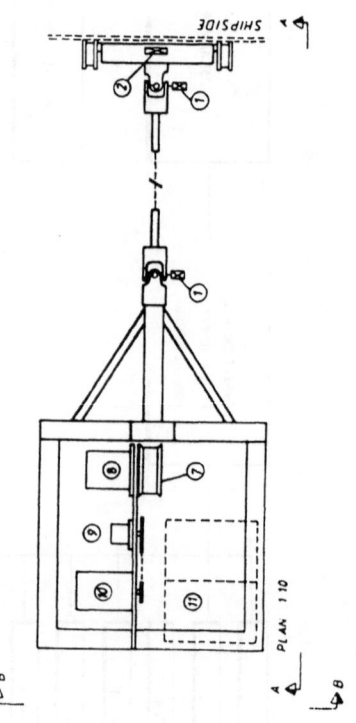

PLAN 1:10

Fig. 4.2.   The Icelandic System for Measuring Ship
            Movements.   Hydraulic wince.

Fig. 4.3 The van parked at the shipside.

210

Fig. 4.4   A closer look at the tensionwire system and
the hydraulic winche.

Tidal range: The tidal range in these harbours is:

	Akranes	Thorlákshöfn	Reykjavik
MHWS	4,17	3,15	4,14
MHWN	2,98	2,23	2.98
MSL	2,18	1,64	2,16
MLWN	1,36	1,06	1,36
MLWS	0,17	0,14	0,17

Fig. 5.1 shows the location of the three harbours and the off-shore Waveriders. The Waveriders was installed about 600m out-side Akranes harbour, about 800 m outside Thorlakshöfn harbour and approx. 800 m outside Reykjavik harbour.

6. FIELD MEASUREMENTS AND INTERVIEWS
6.1 Field measurement
During February, March,October and November 1983 a total of 24 measurements were undertaken from 8 ships in three harbours. The significant wave height offshore exceeded 10 m twice during these measurements and 8 m seven times out of a total of 24 measurements.
In Reykjavik harbour ship movements were measured for two identical container vessels of Lo-Lo and Ro-Ro type (Loa = 105.6 m). The wave situation offshore exceeded once a situation with a return period one per year. A separate study of the parameters affecting the efficiency of loading and unloading was made.
In Akranes harbour ship measurements were undertaken on a freighter(Loa = 78.5 m) under critical working condition. The movements of two trawlers (Loa = 35.2 m and 50.05 m) were recorded under mooring condition at berth. During one situation offshore waves reached the once per year occurrance.
In Thorlákshöfn harbour the movements of a trawler (Loa = 47.7 m), a 200 grt boat (Loa 36.3 m) and a 30 grt boat (Loa = 17.4 m) were observed under working conditions.
The mooring systems varied from taut moorings by winches in Reykjavik harbour to taut and slack moorings in the other harbours.

6.2 Interviews
A captain or another experienced person was interviewed during the measurements. The purpose of the questionnaire was to correlate the measured movements of the ship to the experience of the captain. He estimated the wind velocity, waveheights and long period waves together with the six movements of the ship. Information on the harbour, the berth and the fenders as well as the moorings were collected. He was asked about his experiences with the ship at that berth under working and critical mooring conditions.
Difficulties in manoeuvering etc. related to the harbour, were also asked for and critical working and safe mooring conditions included in term of all six movements of the ship and how often such conditions had occurred based on his experience. The 35 questionnaires were filled out during the interviews.

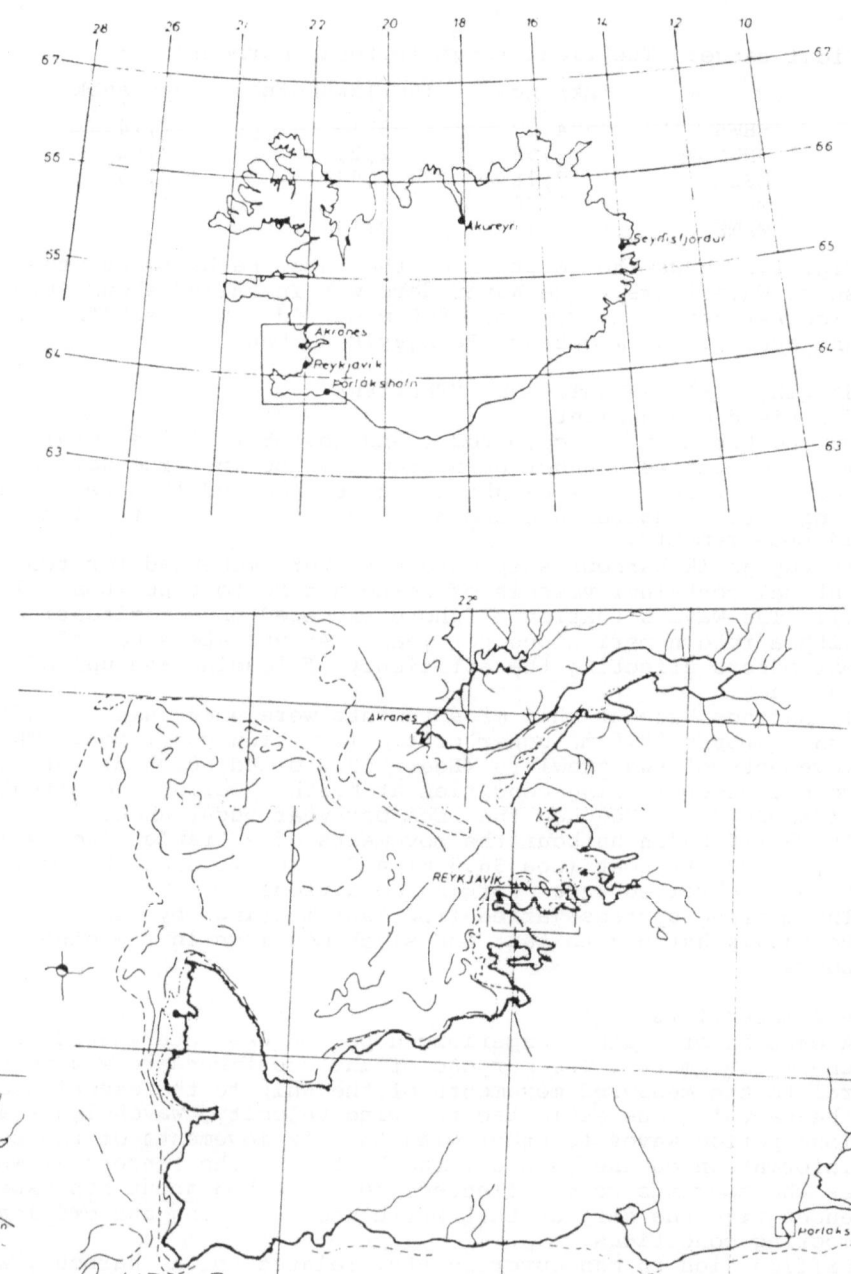

Fig. 5.1   Locations of Prototype Measurements

6.3 Presentation of ship movements data.
The paper shows the layout of the three harbours with the moored ship at berth. All related data are shown with a few comments on the conditions of the ship. The other figures show
the time series of all parameters and the analysed results
from these time series. In some cases figures of velocities
are also shown.
From a total of 24 obervations 7 examples are presented in this
paper:
Arnarfell is a cargo vessel of length (Loa = 78,5 m).The plan
of Akranes harbour and the mooring arrangement is shown in fig.
6.1. The harbour is located in a small bay. The berths are
vertical and the wave reflection pattern inside the harbour
are complicated. Before the measurements the mooring arrangements were carefully recorded. The definitions are given in
the list of symbols.
The measurements took place during critical loading of herr-
ing-drums by the crane on the vessel. The effectiveness of
the loading operation was reduced by the ship movements to some
extent.
The significant wave height offshore during the measurements
was H_s = 6,3 m and at the berthed ship H_s = 0,22 m, T_p = 13,4s.
The average wind velocity was 15 m/s with the maximum of 27,5
m/s and wind direction 194 deg in relation to the ship. The
tidal elevation was + 2,6 m.
Inspection of the time series, Fig. 6.2, show long periodic
oscillations both in the time series of the wind velocity and
the wave pressure underneath the ship. Likewise wave-grouping
are pronounced. As could be expected the time series of the
wave pressure and the heave are similar with the same long periodic trends. The planar motions of surge and yaw are similar
with strong correlation with the mooring forces. The mooring
forces are dominating with peaked forces and slack moorings in
between. The spectrum analysis, Fig. 6.3, show that heave,
pitch and roll have energy in the same frequency domain as the
wind waves.
The maximum peak to peak values (H_{max}) was recorded as surge
1,67m, sway 1,72m, roll 2,56 deg and yaw 4,01deg. The maximum
velocities were 0.16m/s for heave, surge and sway and 1.08deg/s
for roll. The maximum mooring forces was 8.03 t.   Some velocities of the time series are shown in Fig. 6.4.
Óskar Magnusson is a trawler of length (Loa = 50.1 m). The
mooring arrangement is shown in Fig. 6.5.
Unloading operation of fish took place during the measurement.
In such adverse weather with heavy swells the captain preferred to use elastic mooring lines, Fig. 6.6, by using only one
line per bollard which caused large displacements of the
vessel. When the unloading operation was finished, the number
of mooring lines per bollard were increased to two or three.
The significant wave height offshore during the measurements
was H_s = 8,9 m and at ship side H_s = 0,27 m, T_p = 13,4 s. The
average wind velocity was 25.6 m/s with a maximum of 39 m/s and
a wind direction 263 deg in relation to the ship. The tidal
elevation was + 3.4 m.
In Thorlakshöfn harbour three vessels were measured at the same
place during the same day.

JÓN VÍDALÍN is a trawler of Loa = 47.7 m, fig. 6.8. The wave height offshore was H_s = 5.4 m and the average wind velocity 19.2 m/s, with a maximum of 27.5 m/s. and wind direction 3.5 deg in relation to the ship. The tide was + 1.4 m.
This example shows how the vessel movements are dependant on whether the mooring lines are slack or taut. The time series shown in fig. 6.9 illustrate the movements and mooring forces in a situation with slack moorings.
The H_{max} values, fig. 6.10, of surge was 3.07 m, sway 1,03 m and roll 4,0 deg. The maximum mooring forces is 16.1 t.
The time series shown in fig. 6.11 show the movements and mooring forces 50 min later. The mooring are now taut and loading/unloading can take place.
The H_{max} value, fig. 6.12, of surge was now 1.19 m, sway 0.65 m and roll 3.39 deg. The maximum mooring force was 11.8 t.
Some velocities of time series are shown in fig. 6.13.

JÓHANN GÍSLASON is a fishing boat (Loa = 36.3 m), fig. 6.14.
The wave height offshore was H_s = 3,4 m and the average wind velocity is 25.3 m/s, with a maximum velocity of 43.5 m/s and wind direction 3.5 deg. The tide was + 3.0 m.
This example shows vessel movements with normal mooring system under loading/unloading operation and the movements when the mooring lines are kept taut by propelled operation.
Using propelled operation to decrease the movements under loading/unloading operation is commom under adverse weather conditions.
The time series shown in fig.6.15 show the movement and mooring forces with normal mooring system. The H_{max} values, fig. 6.16, of surge is 0.84 m, sway 0.49m, yaw 3.25 deg., roll 3.26 deg. and the maximum force is 3.04 t.
The time series show in fig. 6.17, illustrates the movements and mooring forces 30 min later. The mooring system is the same but the movements are almost stopped by propelled operation of half engine force.
The H_{max} values, fig. 6.18, of surge was 0.16m, sway 0.26 m, yaw 1.02 deg., roll 3.16 deg. and the maximum force is 4.46 t.

JULÍUS is a fishing boat Loa = 17.4 m, fig. 6.19. The wave height offshore was H_s = 3.7 m and the average wind velocity 16 m/s with maximum velocity 22.1 m/s and wind direction 3.5 deg. The tide was + 3.2 m.
This example fig. 6.20, demonstrates movement of small vessel under normal wave situation.
The H_{max} value, fig. 6.21, of surge was 0.32 m, sway 0.71 m, yaw 6.35 deg. and roll 2.34 deg. The mooring forces were not recorded. The maximum force was estimated within 1 t.

ALAFOSS is a container vessel of Lo-Lo and Ro-Ro type (Loa = 105,6 m) fig. 6.22. The wave height offshore was H_s = 9.8 m and the average wind velocity 10.3 m/s, with a maximum velocity of 23.8 m/s and wind direction 273 deg in relation to the ship. The tidal elevation was + 3.80 m.

The example on fig. 6.23, illustrates movements of a vessel
with a wave direction almost perpendicular to the vessel with
all mooring lines taut by mooring winches. It is interesting
to note that almost all movements are harmonics as shown by
the spectrum (fig. 6.24) due to the taut moorings.
A loading operation by cranes on the quay took place under the
measurements.
The H_{max} value, fig. 6.24, of surge was 0.16 m, sway 0.62 m,
yaw 1.11 deg., roll 2.01 deg. and the maximum force was 7.2 t.
Some velocity of time series are shown in fig. 6.25.

7. ANALYSIS OF FIELD DATA
Evaluation of the ship movements is based on maximum values.
Fig. 7.1 shows all maximum translatory and rotatory movements
as a function of the ship length.
Fig. 7.2 shows the correlation between estimated and measured
maximum ship movements. The correlation is better than expected
as the captains are inside the ship during the interviews.
These findings and others caused that the results of the interviews have been used widely during recent years to seak to define the problems involved in the respective harbours.
Fig. 7.3 to 7.5 show plotting of all Hmax values in relation
to the significant wave height offshore. At the top the wave
angle with the berth is shown. On the left side of the figure
movements dependent on mooring lines (surge, sway and yaw) are
shown.
In Reykjavik harbour two situations are shown where the parameters were estimated. One situation when the effect of ship
movement was measured on container handling rate, and another
situation when the windwave were estimated with a return period of once a year. The combination of sway and yaw velocity
were estimated to 80 cm/sec perpendicular to the berth in that
situation. The plottings are described as a working or mooring conditions.
Fig. 7.6 to 7.8 show plotting of maximum velocity of ship movements in relation to the significant wave height offshore.
Fig. 7.9 to 7.11 are plottings of average wind velocity and
directions and maximum mooring forces.
Fig. 7.12 to 7.14 show plotting of the significant wave height
outside the harbour and at the shipside related to the waves
offshore and plotting of corresponding significant wave heights
outside the harbour and at the shipside.

8.COMPARISON OF RESULTS FROM MODEL TESTS AND FIELD MEASUREMENTS
Comparison between results from measurements of ship movements
in models and in the field were made for the three harbours.
Wave disturbance tests with moored ship in irregular waves were
performed in scale 1:60 for Akranes harbour during 1977 - 1978
and for Reykjavik harbour in scale 1:100 during 1976. Both models were run by the hydraulic laboratory of the Iceland Harbour Authority in Reykjavik.
Thorlakshöfn harbour was tested in scale 1:100 through model-
test in Denmark 1973 to 1974 by the Danish Hydraulic Institute.
The proposed improvement based on these model tests have been
carried out by building two breakwaters in Thorlakshöfn, one

breakwater in Akranes harbour and a container berth in Reykjavik Harbour.
Figs. 8.1 to 8.3 show   comparison of some results from measurements of ship movements, mooring forces and significant
wave heights in models and in nature. Only related data are
shown such as waves measured at the same location, ship location and size are similar.
In the model tests only dominating ship movements are measured
as sway in Reykjavik harbour, surge in Akranes harbour and
Thorlakshöfn harbour.
Inspection of these plottings show that hydraulic wave disturbance model is able to reproduce the prototype conditions in
a reliable way, provided the model conditions are properly
reproduced.

9. WORKING AND SAFE MOORING CONDITIONS - THE MAIN RESULTS OF
   THE JOINT PROJECT BETWEEN THE NORDIC COUNTRIES.
9.1 Discussion of loading/unloading and safe mooring criteria.
The acceptable ship movements during loading/unloading operations are to some extent determined by local traditions. Generally, this acceptance does not consider possible reduced cost
effectiveness. In relation to optimization of harbour layouts,
however, it is necessary to take these factors into consideration.
The choise of criteria depends on different parameters, such
as:
-    Loading/unloading methods, - gear, personnel.
-    Vessel type, type of goods.
-    Characteristics of the vessel movements.
Therefore more detailed criteria relating to vessel type and
loading/unloading methods have been established. For safe stay
at berth the basic criteria is that no damage occurs to ship
or quay.
The interacting forces between quay and ship depend on the
following factors:
ENVIRONMENTAL FACTORS
-    Waves
-    Winds
-    Currents

SHIP AND HARBOUR CONDITIONS
-    Size of ship
-    Characteristics of vessel movements
-    Type of quay
-    Mooring and fender system

9.2 Results of field measurements.
The results from nearly all measurements of ship movements
have been related to length of vessel (Loa). They are shown
in fig. 9.1.   Distinction was made between condition where
it was possible to load/unload and where the movements were
excessive for operations but still allowed the ship to stay
at the berth.

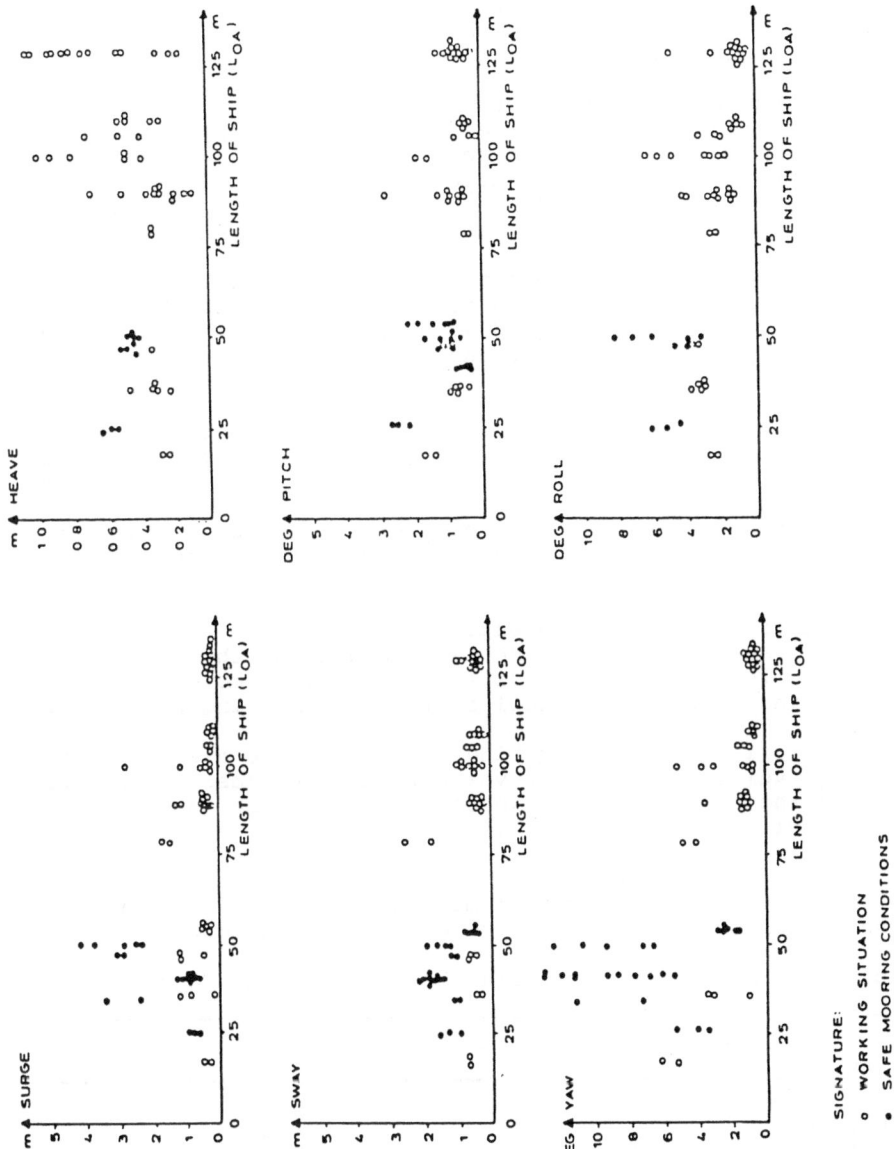

Dif. 9.1. Ship movements in relation to lenth of ship

9.3 Definition of movement criteria
With reference to the conditions described above, the following definitions of movement criteria have been selected.

Interrupted working situation:
This situation is characterized by movements, causing a substantial reduction of the effectiveness of the loading/unload-
ing operations. For this critical movement, analysis of
frequency of occurrence were undertaken. Results depend to a
certain extent upon type of vessel and loading/unloading operations.
The investigations carried out show that a harbour in which
operations have to be stopped for a total of less than 1 week/
year may normally be regarded as having acceptable conditions
for loading/unloading operations.

Safe stay at berth:
This condition is limited by the largest movements for which
no damage occurs to vessel or quay, provided that the vessel
is reasonably well moored and the quay is well equipped with
fenders.
For acceptable condition in a harbour, attempts should be made
to keep the frequency of occurrrence of this situation less
than once a year.

9.4 Criteria for ship movements
Table 9.1 and 9.2 present the main results of the joint Nordic
project given as criteria for acceptable ship movements for
the working situation as well as for safe mooring conditions
at berth.
The critierisia for safe berthing situations are also given as
critical velocities, as the velocity and the mass of the vesse
express the dynamic forces which may damage the quay and/or
the vessel.

9.5 Criteria for wave height
Wave height criteria for safe mooring conditions at berth were
also established. Table 9.3 presents the main result.

Table 9.3 Wave Height Criteria for Safe Mooring Conditions.

Ship type	L_{oa} (m)	H_s (m)
Open boats	5-12	0.20
Other boats	5-12	0.30
Small fishing vessels	15-30	0.30
Coasters (<2000 DWT)		0.45

H_s is the significant wave height equal to the mean
value of the highest third of the waves in a wave train.
This criterion is valid only if wind waves are causing the dis
turbance. In harbours where seiching/long periodic waves are
significant, the indicated wave height will not be a criterion
for acceptable conditions.

CRITERIA FOR SHIP MOVEMENTS.

Maximum values during Working Conditions.

Table 9.1 Criteria for ship movements (loading/unloading operations). The movements are maximum peak-peak.

Type of vessel	Surge (m)	Sway (m)	Heave (m)	Yaw (deg)	Pitch (deg)	Roll (deg)
Fishing vessels [1] (L_{oa} =25-60 m)						
LO-LO	1.0-1.5	1.0-1.5	0.4-0.6	3-5	4	3-5
Elevator crane	0.15	0.15				1.5
Suction pump	2.0-3.0	1.5-2.0				
Freighters, Coasters [1] (L_{oa} = 60-130 m)						
Crane on the vess.	1.0-2.0	1.2-1.5	0.6-1.0	1-3	1-2	2-3
Crane on the quay	1.0-2.0	1.2-1.5	0.8-1.2	2-4	1-2	3-5
Ferries [2] (L_{oa} =100-150 m)		0.8	1.0	1.0	1.0	2.0
Container Vessels [1] (L_{oa} =100-200 m)						
90-100% efficincy	0.6-1.0	0.6-0.8	0.6-0.9	0.5	1.5	3.0
50% efficiency	2.0	2.0	1.2	1.5	2.5	6.0

1) Frequency of these movements should be less than 1 week/year (2% of time)

2) Frequency of these movements should be less than 3 hours/year (0.3% of time).

CRITERIA FOR SHIP MOVEMENTS.

Maximum values for Safe Mooring Conditions at Berth.

Table 9.2 Criteria for vessel movements for safe mooring conditions at berth. The movements are peak-peak values. For the berth to be acceptable, the frequency of these movements should be less than 3 hours/year.

Type of vessel	Surge (m)	Sway (m)	Heave (m)	Yaw (deg)	Pitch (deg)	Roll (deg)
Fishing vessel (L_{oa} = 25-60 m)						
Movement	1.2-1.5	1.0-2.0	0.6-1.0	6	4	8
Freighters/coasters (L_{oa} = 60-120 m)						
Movement	1.0-2.0	1.5-2.0	1.0-1.5	3-5	2-3	6
Velocity						
Size of vessel						
about 1000 DWT	0.6 m/s	0.6 m/s		2.0 deg/s	2.0 deg/s	
about 2000 DWT	0.4 m/s	0.4 m/s		1.5 deg/s	1.5 deg/s	
about 5000 DWT	0.3 m/s	0.3 m/s		1.0 deg/s	1.0 deg/s	

Acknowledgements. The author would like to thank his colleuagues at the Icelandic Harbour Authority Mr.S.Sigurdarson Civ. Eng. and Mr.G.Sch.Tryggvason Civ.Eng. for their time and effort, which made this lecutrue possible.
Likewise I would like to use this opportunity also to express my very best of thanks to all my colleagues involved in the Nordic Joint Research project for very interesting and profitalbe cooperation.

Notation. The following symbols are used in this paper:
Notation of mooring system, fig. 6.1, 6.5, 6.8, 6.14 and 6.22:

1 - 4	:	mooring lines with load celles. The number refers to the load cells.
4 60 L = 7,0	:	four mooring lines of diameter 60mm, each of lenght 7.0 m.
H	:	Vertical height between the bollard and the locks at the ship
D_f	:	Draught at bow
D_a	:	Draught at stern
min	:	Minimum displacement
max	:	Maximum displacement
G1,G2,G3,	:	Coordinates of the floating point of the ship related to the measuring point at the ship side.

The parameters presented in the figures resulting from the analysis are defined as:

H	:	The double amplitude
T	:	The period
T_s	:	The significant period
T_z	:	The mean period
T_{mo1}	:	The mean period
T_{mo2}	:	The mean period
T_{p1}	:	The peak period for period longer than 20 sec.
T_{p2}	:	The peak period for period shorter than 20 sec.
H_s	:	The significant wave heigh
H_{max}	:	The maximum double amplitude
Max	:	The maximum differences between highest and lowest values expected for: maximum forces, maximum wind velocity and maximum water pressure
H_{4rms}	:	The 4 times rms-values
Mt	:	The mean value of the time series expected for mooring forces at Mt represent the average time between peak forces
So	:	The peak energy top of the spectrum (m^2/s, O^2/s)
S_p	:	The maximum velocity of the time series (m/s, $^0/s$)

T_s, T_z, H_s, Max Mt, are determined by the zero-crossing method, while T_{mo1}, T_{mo2}, T_{p1}, T_{p2}, H4rms, So, Sp, are determined by the spectral analysis.
To estimate the significant wave height the H4rms for the wave pressure series are multiplied with the transfer function.

222

REFERENCES

Gísli Viggósson, Guðjón Sch. Tryggvason, Sigurður Sigurðarson. 1986. **Skipsbevegelser i havner.** (Ship Movements in Harbours), NET- project, Icelandic National Report. Icelandic Harbour Authority (IHA).

NET- project, 1986. **Skibsbevægelser i havne.** (Ship Movements in Harbours). Joint Nordic Research Project.

NET- project, 1982 **Questionnarie for captains under field observation.**

Gísli Viggósson, Gestur Gunnarsson, Steingrímur Arason, 1978 **Akraneshöfn, skyrsla um gagnasöfnun og líkantilraunir.** (Akranes Harbour Hydraulic Model Tests) IHA.

Torben Sörensen, Helge Gravesen, 1974. Thorlakshöfn Harbour Project. **Report og Hydraulic Model Tests.** Danish Hydraulic Institute.

Gísli Viggósson, Jónas Elíasson, Hannes Valdimarsson, 1977 **Sundahöfn. Líkantilraunir undirstöðugögn.** (Reykjavík Harbour, Hydraulic Model Tests) Reykjavík Harbour.

Gísli Viggósson, Guðjón Sch. Tryggvason, Sigurður Sigurðarson, 1987. **Wave Measurements in Iceland.** To appear in Journal of Coastal Research.

Gísli Viggósson, Sigurður Sigurðarson 1987. **Mat á ölduhreyfingu og skipshreyfingum við stækkun Kleppsbakka.** (Extension of Kleppsbakki. A studdy of Wave- and Ship Movements.) Reykjavík Harbour.

Per Bruun, 1984. **Mooring and fendering, rational principles in design.** The Dock and Harbour Authority. Vol 64. No. 758 1984.

Per Bruun, 1985. **Fendering.** Dock and Harbour Authority, Vol LXU No. 767 march 1985.

Danish Hydraulic Institute, 1986. **Skibsbevægelser i havne** (Ship Movements in Harbours), NET-project, Danish National Report.

Landsverkfröðingurin, 1986. **Skibsbevægelser i havne** (Ship Movements in Harbours), NET-project, Faroese National Report.

Kystdirektoratet, 1986. **Skibsbevægelser i havne** (Ship Movements in Harbours), NET-project, Norwegian National Report.

SSPA Maritime Research and Consulting, 1986. **Skibsbevægelser i havne** (Ship Movements in Harbours), NET-project, Swedish National Report.

223

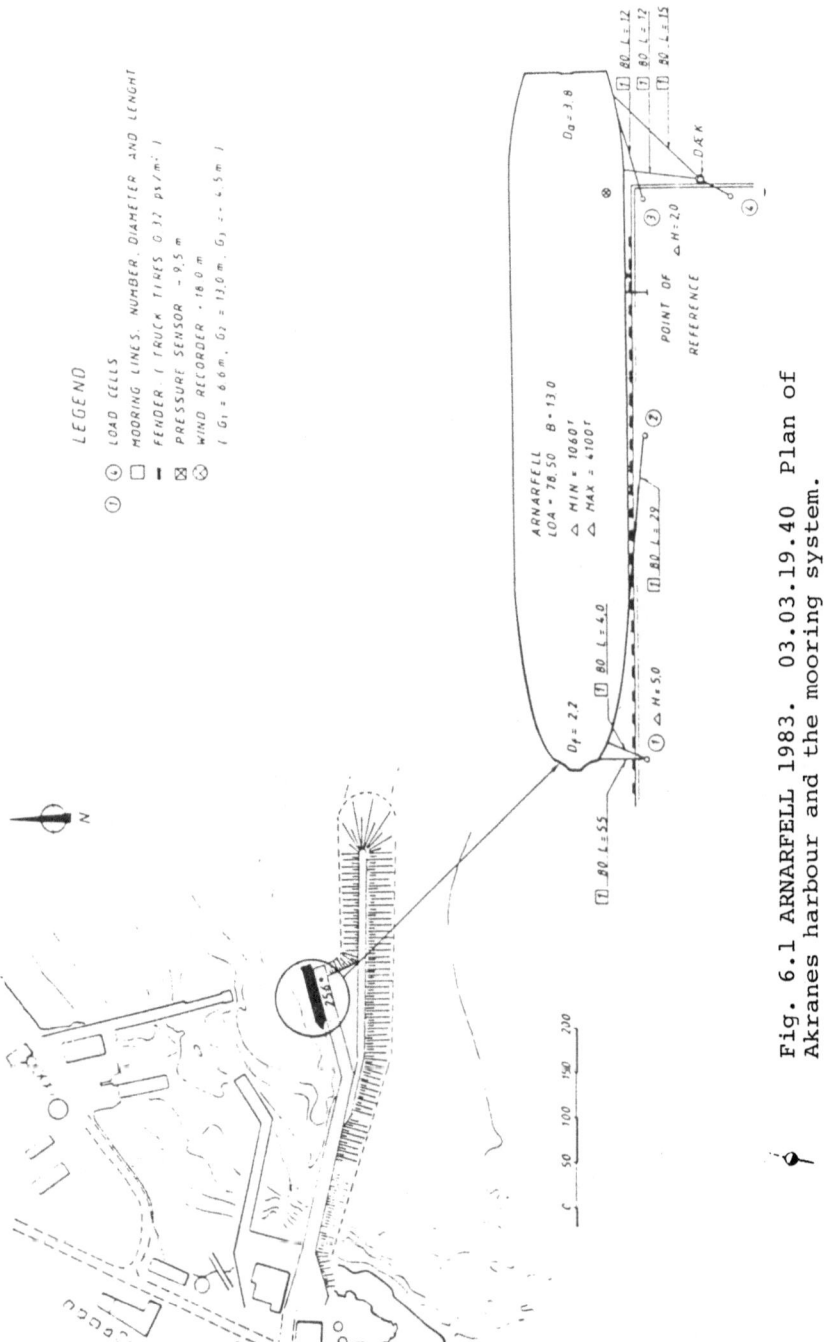

Fig. 6.1 ARNARFELL 1983. 03.03.19.40 Plan of
Akranes harbour and the mooring system.

224

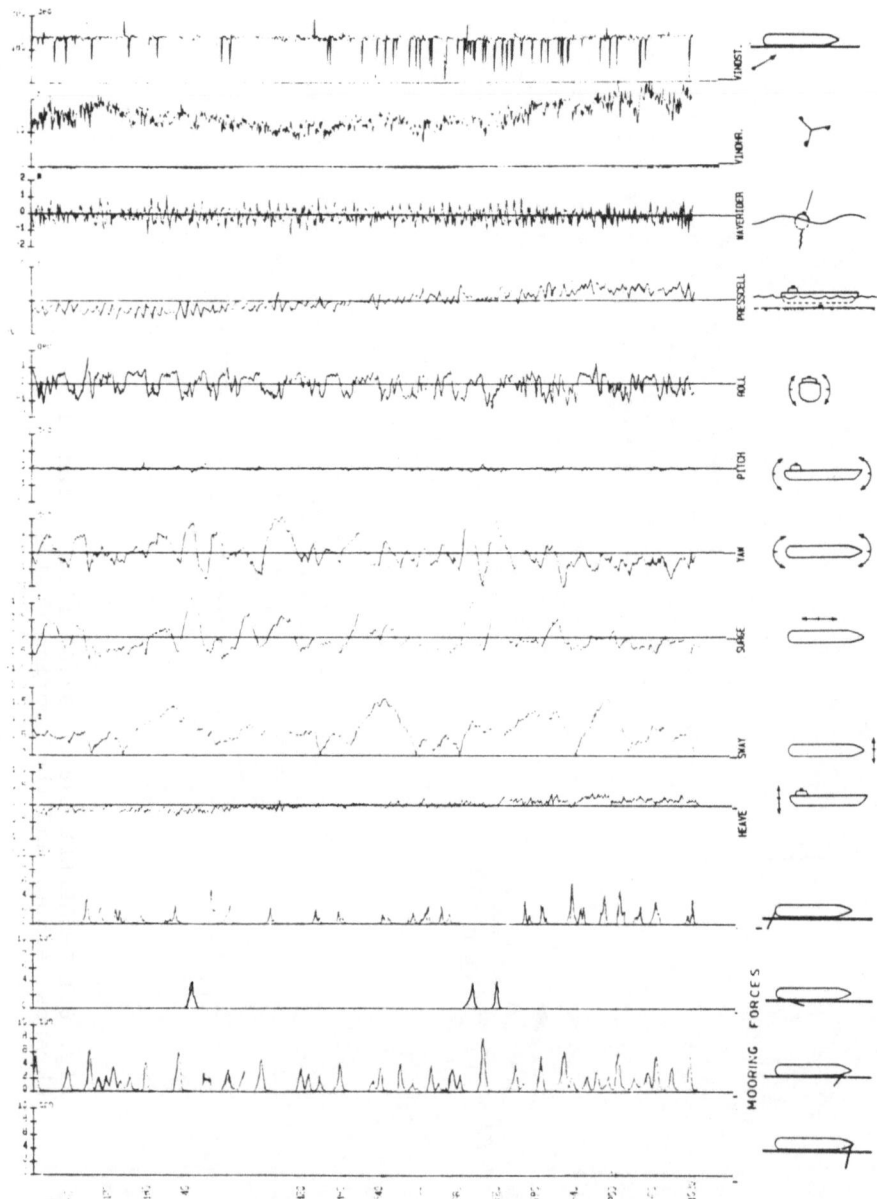

Fig.6.2. ARNERFELL 1983.03.03.18.40 Time series of ship
movement 0 critical loading operation.

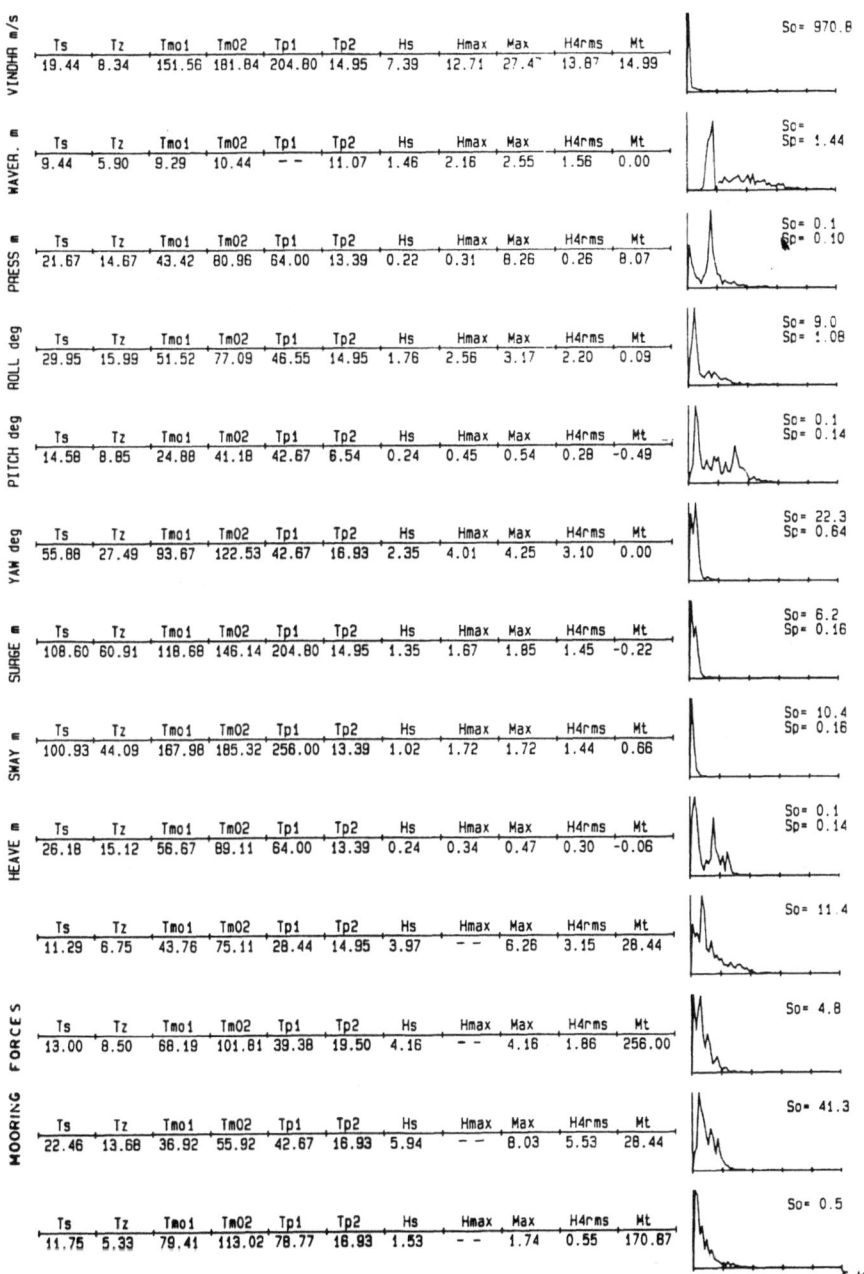

	Ts	Tz	Tmo1	Tm02	Tp1	Tp2	Hs	Hmax	Max	H4rms	Mt
VINDHR m/s	19.44	8.34	151.56	181.84	204.80	14.95	7.39	12.71	27.4⁻	13.87	14.99
WAVER. m	9.44	5.90	9.29	10.44	- -	11.07	1.46	2.16	2.55	1.56	0.00
PRESS m	21.67	14.67	43.42	80.96	64.00	13.39	0.22	0.31	8.26	0.26	8.07
ROLL deg	29.95	15.99	51.52	77.09	46.55	14.95	1.76	2.56	3.17	2.20	0.09
PITCH deg	14.58	8.85	24.88	41.18	42.67	6.54	0.24	0.45	0.54	0.28	-0.49
YAW deg	55.88	27.49	93.87	122.53	42.67	16.93	2.35	4.01	4.25	3.10	0.00
SURGE m	108.60	60.91	118.68	146.14	204.80	14.95	1.35	1.67	1.85	1.45	-0.22
SWAY m	100.93	44.09	167.98	185.32	256.00	13.39	1.02	1.72	1.72	1.44	0.66
HEAVE m	26.18	15.12	56.67	89.11	64.00	13.39	0.24	0.34	0.47	0.30	-0.06
	11.29	6.75	43.76	75.11	28.44	14.95	3.97	- -	6.26	3.15	28.44
FORCE S	13.00	8.50	68.19	101.81	39.38	19.50	4.16	- -	4.16	1.86	256.00
MOORING	22.46	13.68	36.92	55.92	42.67	16.93	5.94	- -	8.03	5.53	28.44
	11.75	5.33	79.41	113.02	78.77	16.93	1.53	- -	1.74	0.55	170.87

So= 970.8

So=
Sp= 1.44

So= 0.1
Sp= 0.10

So= 9.0
Sp= 1.08

So= 0.1
Sp= 0.14

So= 22.3
Sp= 0.64

So= 6.2
Sp= 0.16

So= 10.4
Sp= 0.16

So= 0.1
Sp= 0.14

So= 11.4

So= 4.8

So= 41.3

So= 0.5

.5 Hz

Fig.6.3. ARNARFELL 1983.03.03.18.40
Analysis of time series.

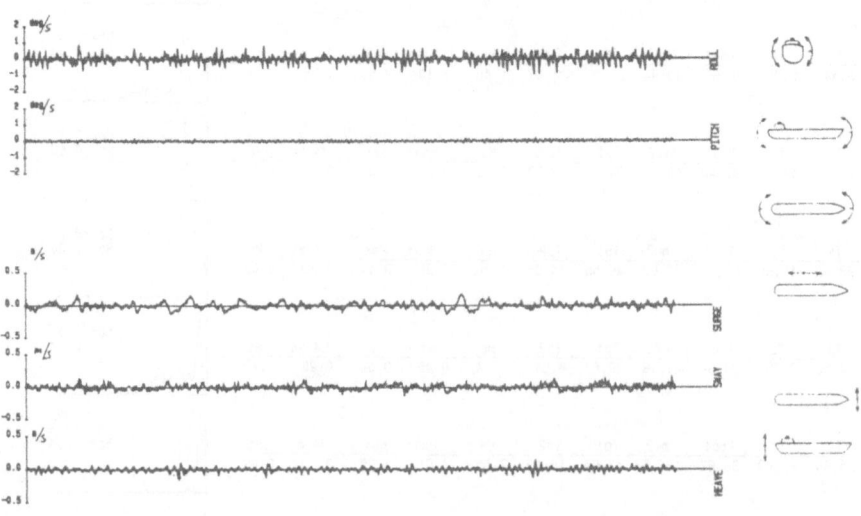

Fig. 6.4   ARNARFELL   1983. 03.03.18.40
Some velocities of the time series.

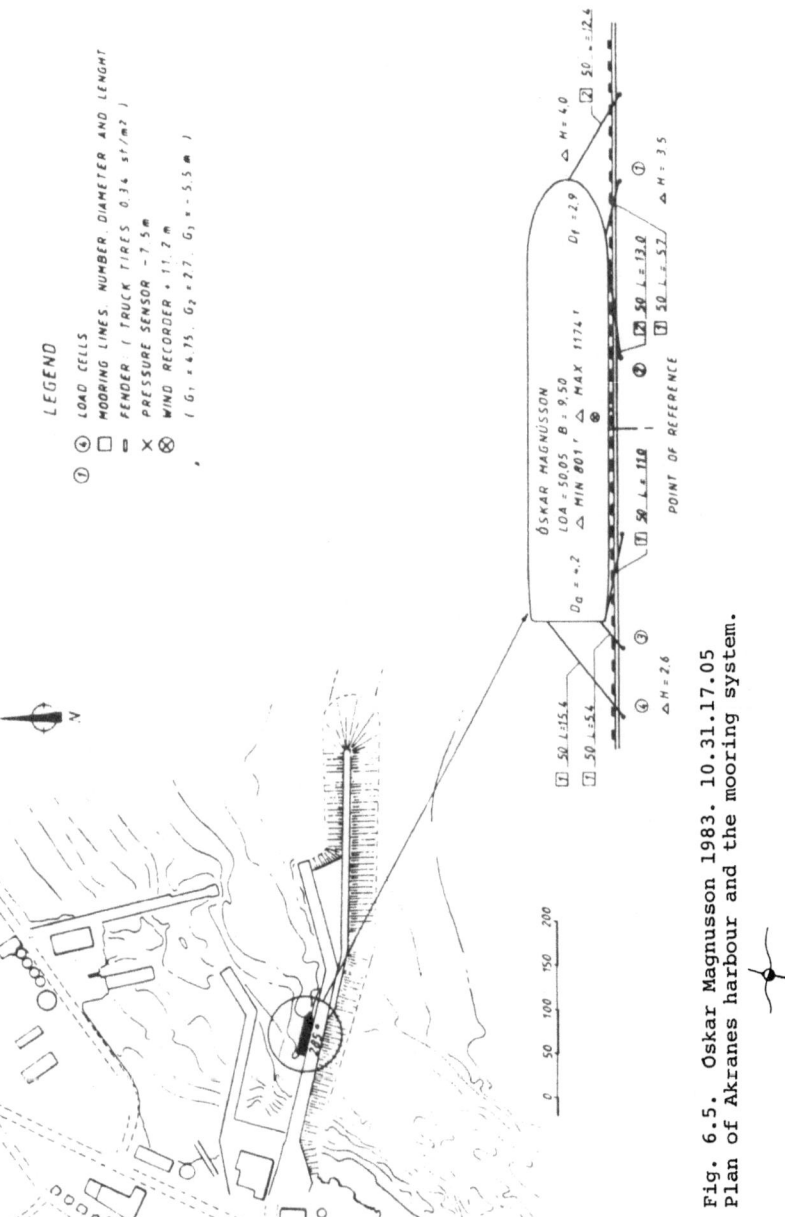

Fig. 6.5. Oskar Magnusson 1983. 10.31.17.05
Plan of Akranes harbour and the mooring system.

228

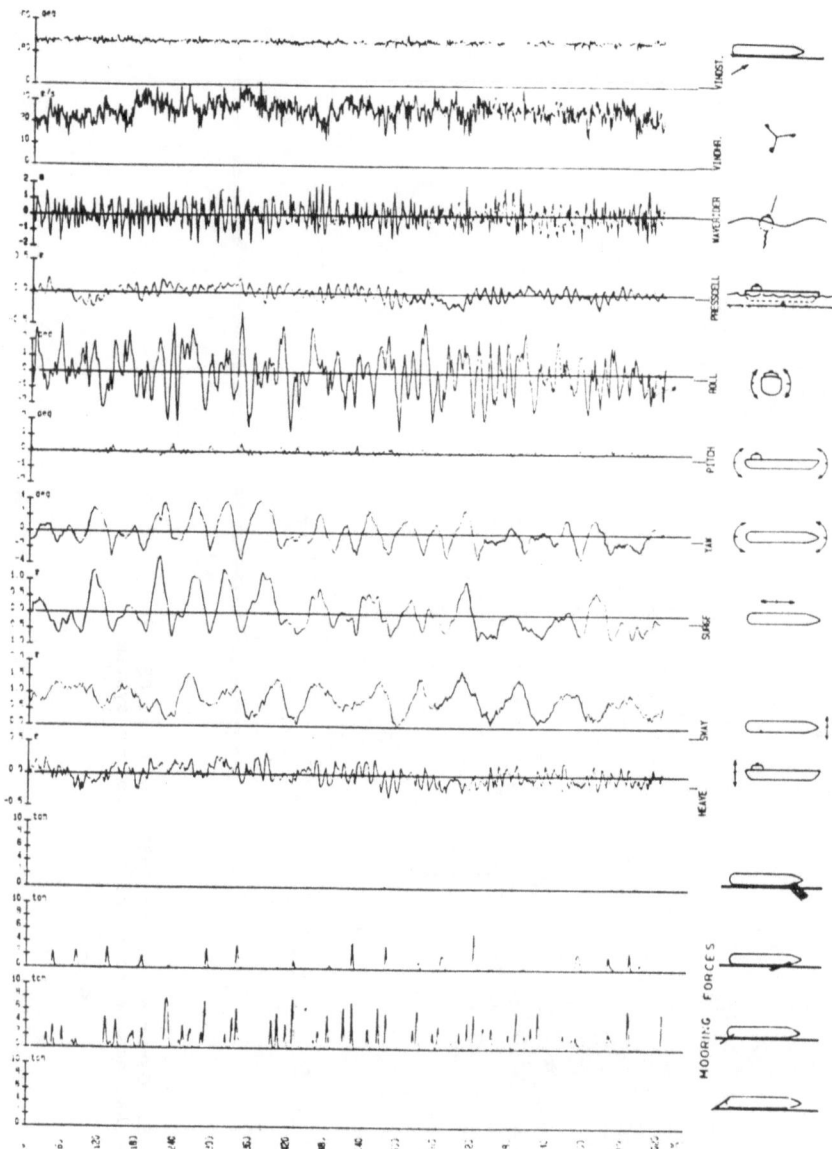

Fig. 6.6 ÓSKAR MAGNÚSSON 1983. 10.31.17.05
Time series of thip movements, elastic mooring.

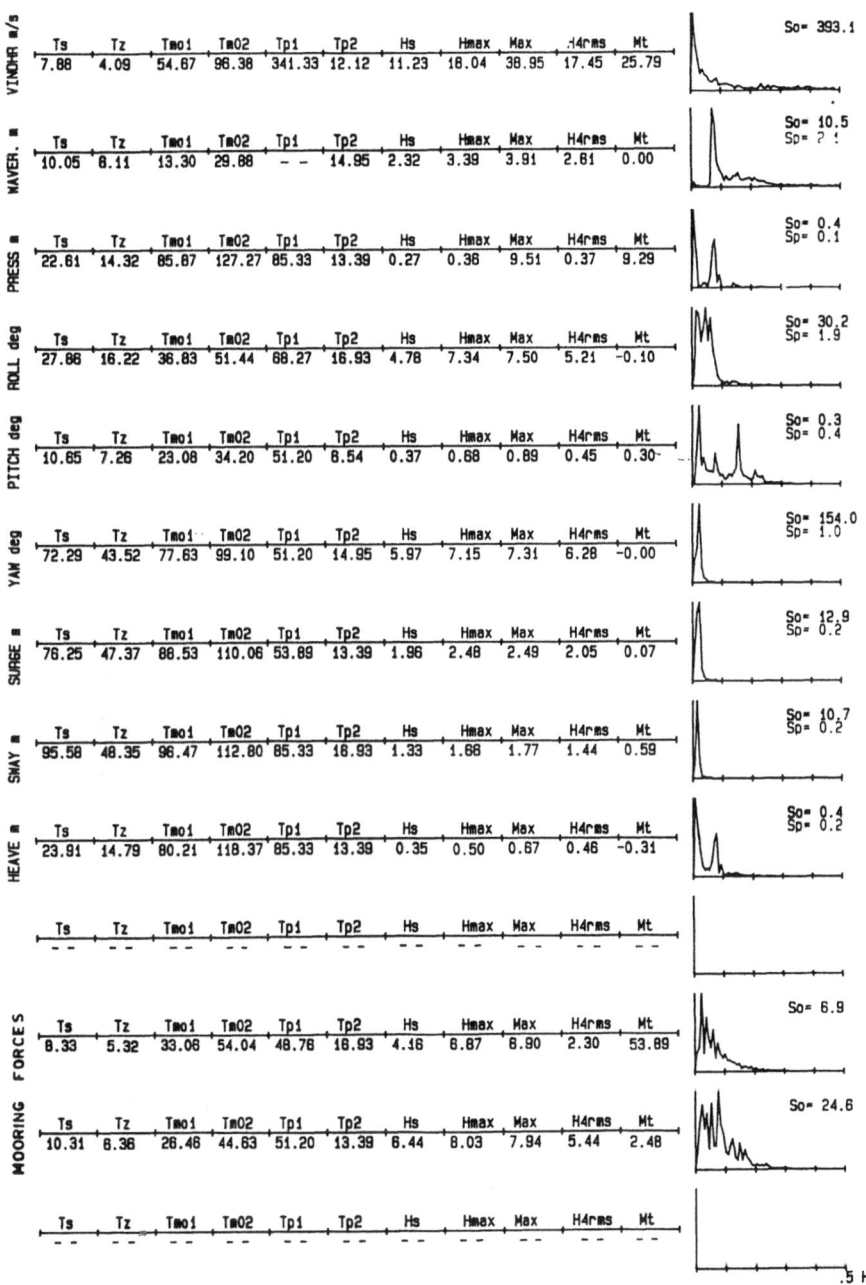

Fig. 6.7 ÓSKAR MAGNUSSON 1983. 10.31.17.05
Analysis of time series.

230

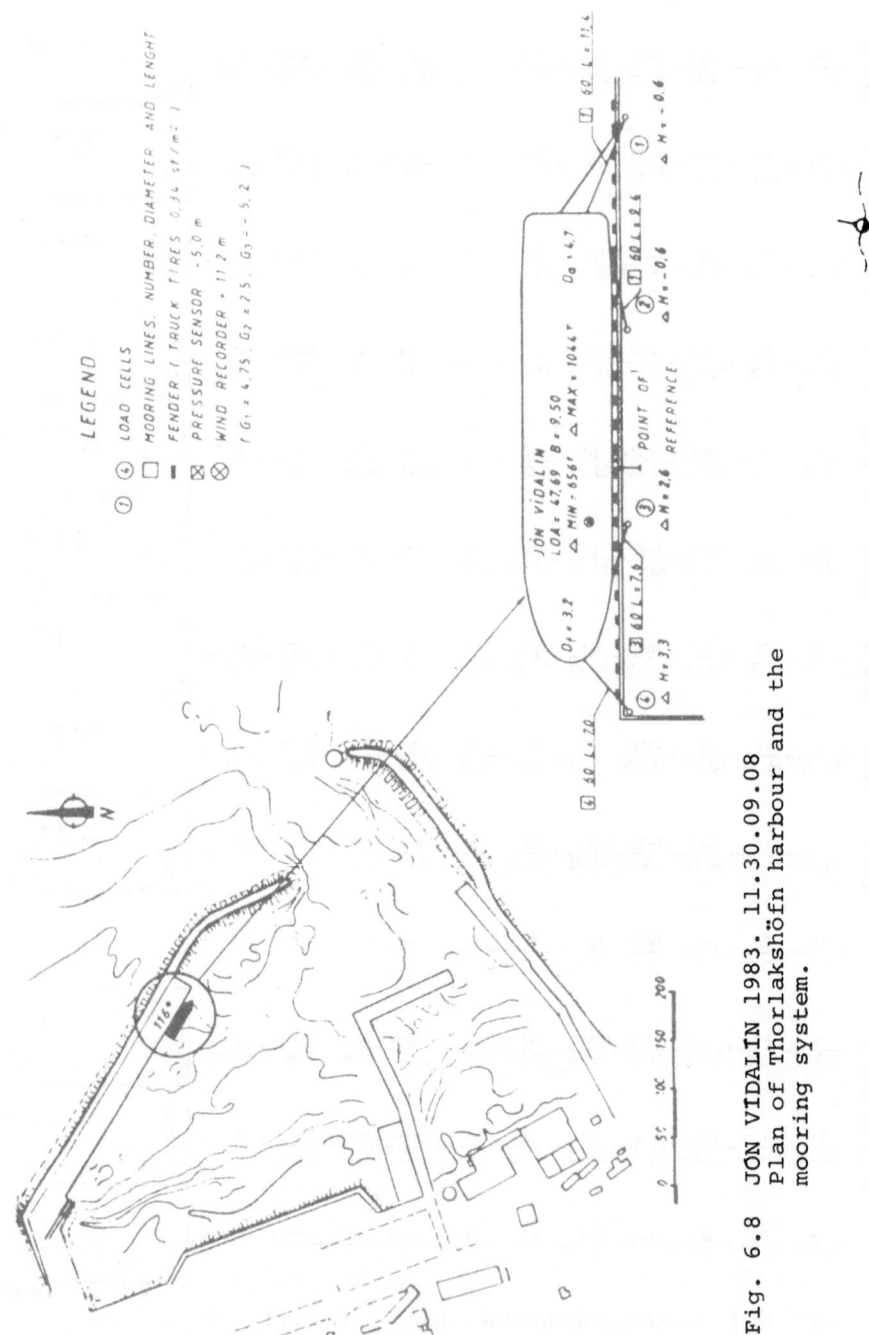

Fig. 6.8 JÓN VIDALIN 1983.11.30.09.08
Plan of Thorlakshöfn harbour and the
mooring system.

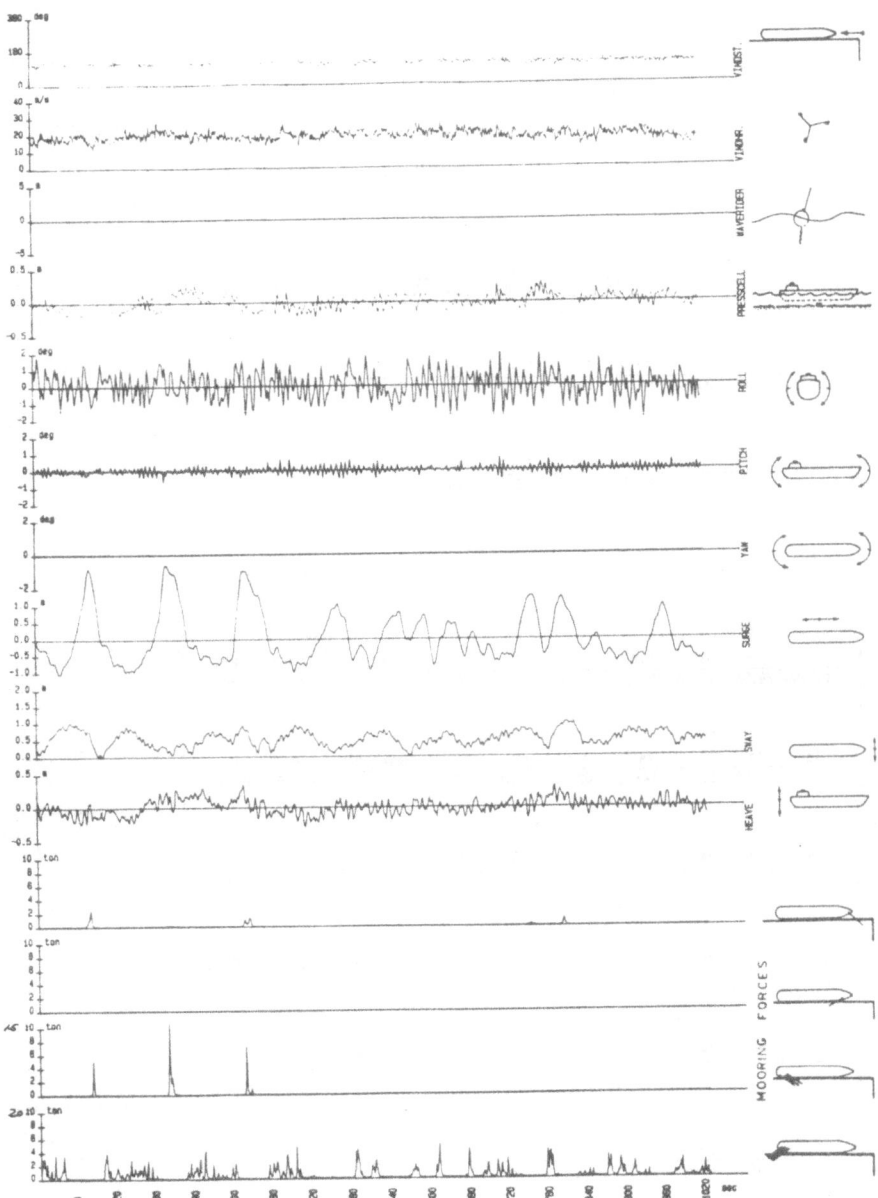

Fig. 6.9 JON VIDALIN 1983. 11.30.09.08
Time series of ship movements, slack
moorings.

232

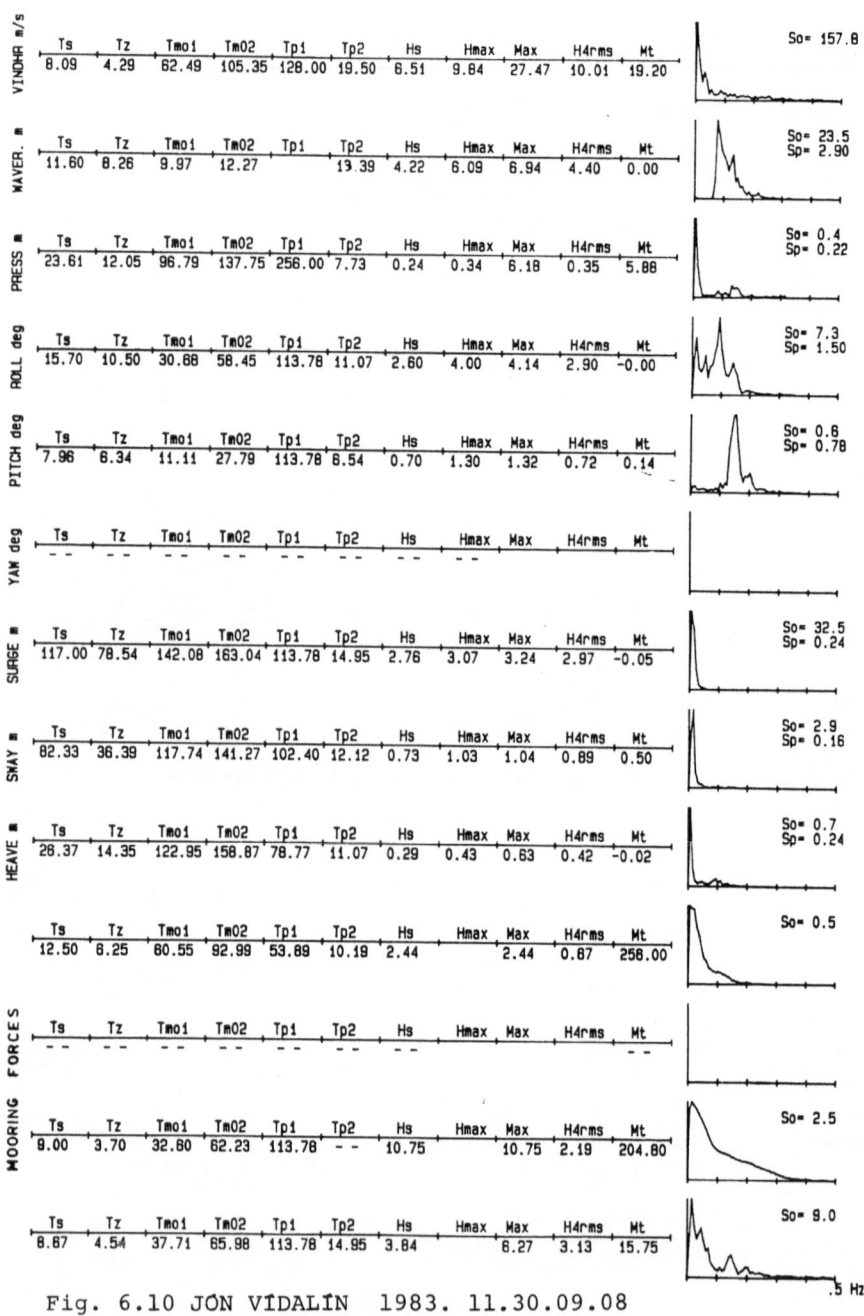

	Ts	Tz	Tmo1	Tm02	Tp1	Tp2	Hs	Hmax	Max	H4rms	Mt	
VINDHR m/s	8.09	4.29	62.49	105.35	128.00	19.50	6.51	9.84	27.47	10.01	19.20	So= 157.8
WAVER. m	11.60	8.26	9.97	12.27		13.39	4.22	6.09	6.94	4.40	0.00	So= 23.5 Sp= 2.90
PRESS m	23.61	12.05	96.79	137.75	256.00	7.73	0.24	0.34	6.18	0.35	5.88	So= 0.4 Sp= 0.22
ROLL deg	15.70	10.50	30.88	58.45	113.78	11.07	2.60	4.00	4.14	2.90	-0.00	So= 7.3 Sp= 1.50
PITCH deg	7.96	8.34	11.11	27.79	113.78	8.54	0.70	1.30	1.32	0.72	0.14	So= 0.6 Sp= 0.78
YAW deg	--	--	--	--	--	--	--	--			--	
SURGE m	117.00	78.54	142.08	163.04	113.78	14.95	2.76	3.07	3.24	2.97	-0.05	So= 32.5 Sp= 0.24
SWAY m	82.33	36.39	117.74	141.27	102.40	12.12	0.73	1.03	1.04	0.89	0.50	So= 2.9 Sp= 0.16
HEAVE m	28.37	14.35	122.95	158.87	78.77	11.07	0.29	0.43	0.63	0.42	-0.02	So= 0.7 Sp= 0.24
	12.50	8.25	60.55	92.99	53.89	10.19	2.44		2.44	0.87	258.00	So= 0.5
MOORING FORCES	--	--	--	--	--	--	--				--	
	9.00	3.70	32.80	62.23	113.78	-- --	10.75		10.75	2.19	204.80	So= 2.5
	8.87	4.54	37.71	65.98	113.78	14.95	3.84		6.27	3.13	15.75	So= 9.0

.5 Hz

Fig. 6.10 JÓN VÍDALÍN   1983. 11.30.09.08
Analysis of time series.

233

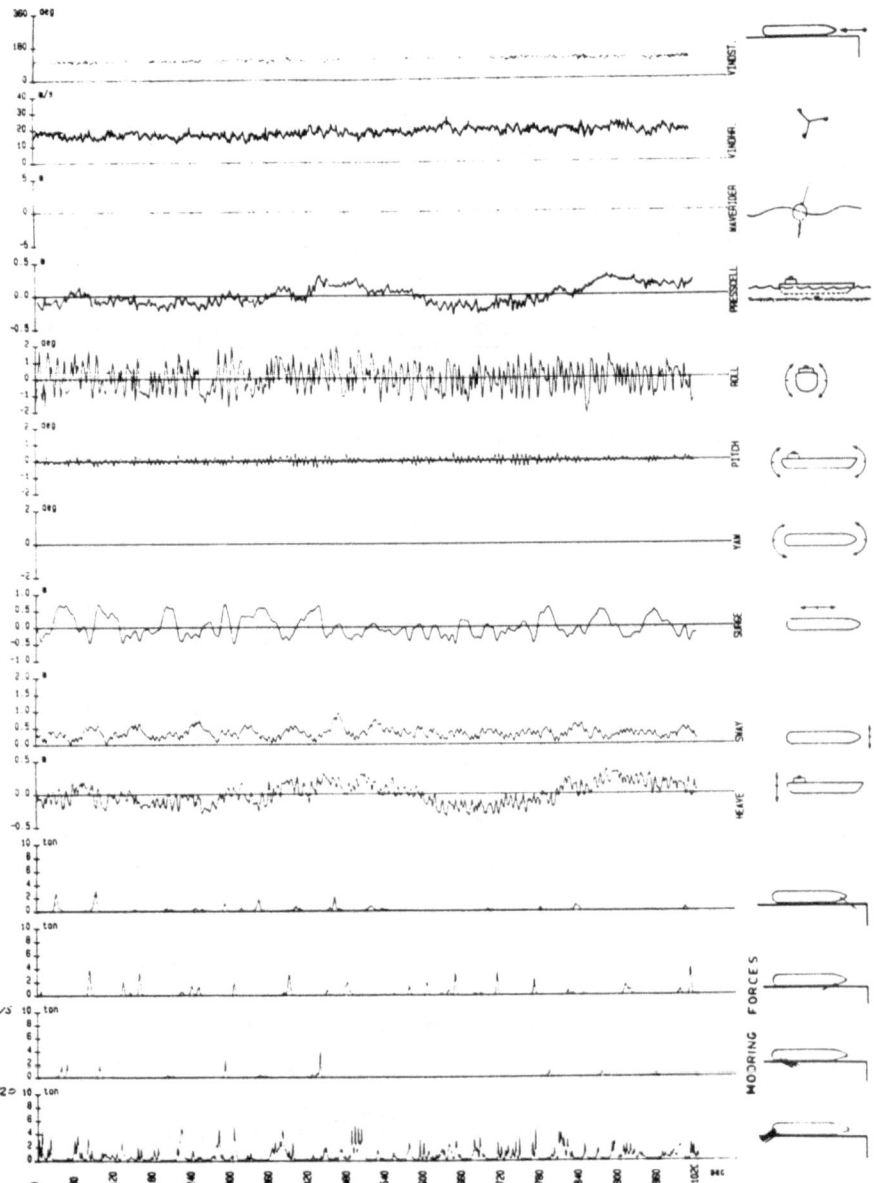

Fig. 6.11   JÓN VÍDALÍN 1983. 11.30.09.48
Time series of ship movements,
taut moorings.

234

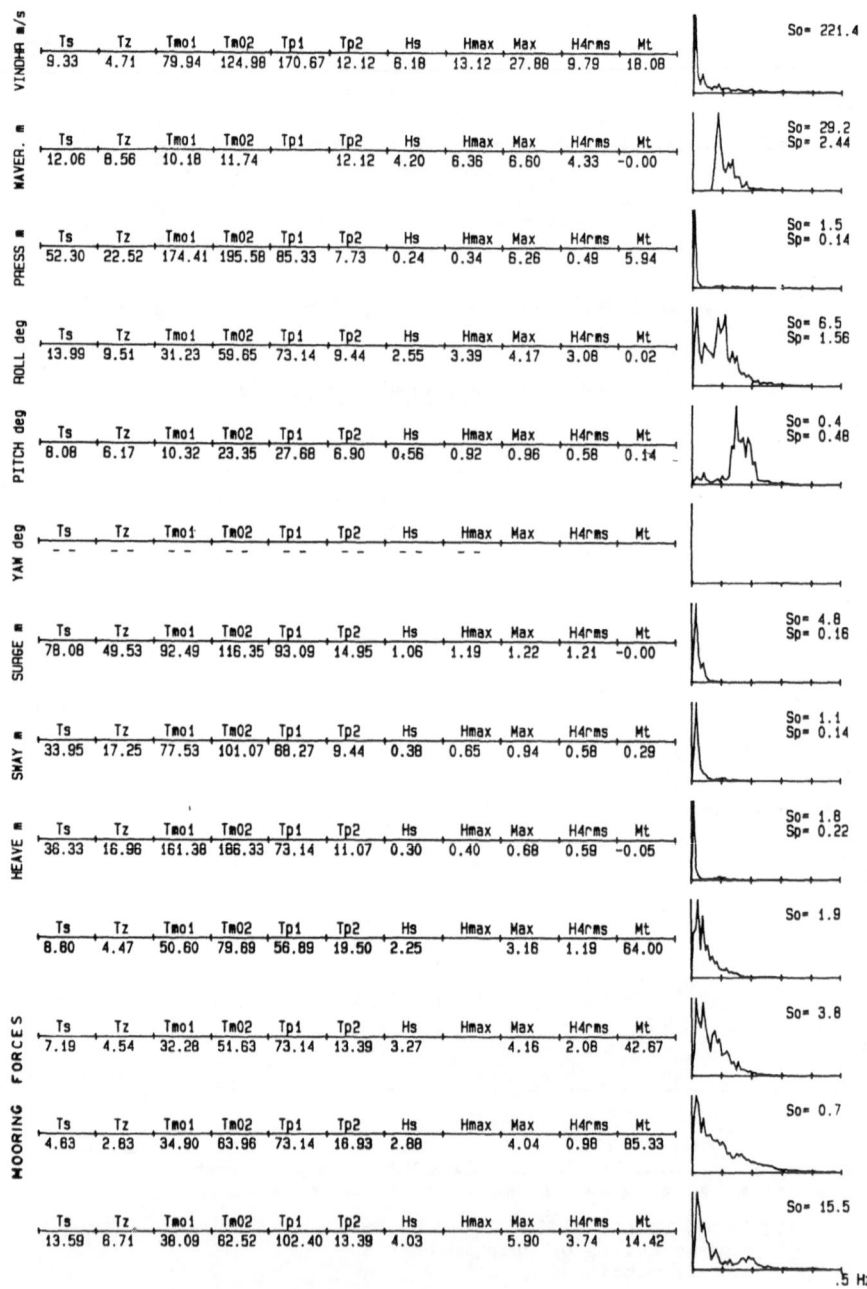

	Ts	Tz	Tmo1	Tm02	Tp1	Tp2	Hs	Hmax	Max	H4rms	Mt	
VINDHR m/s	9.33	4.71	79.94	124.98	170.67	12.12	6.18	13.12	27.88	9.79	18.08	So= 221.4
MAVER. m	12.06	8.56	10.18	11.74		12.12	4.20	6.36	6.60	4.33	-0.00	So= 29.2 / Sp= 2.44
PRESS m	52.30	22.52	174.41	195.58	85.33	7.73	0.24	0.34	6.26	0.49	5.94	So= 1.5 / Sp= 0.14
ROLL deg	13.99	9.51	31.23	59.65	73.14	9.44	2.55	3.39	4.17	3.08	0.02	So= 6.5 / Sp= 1.56
PITCH deg	8.08	6.17	10.32	23.35	27.68	6.90	0.56	0.92	0.96	0.58	0.14	So= 0.4 / Sp= 0.48
YAW deg	-	-	-	-	-	-	-	-	-	-	-	
SURGE m	78.08	49.53	92.49	116.35	93.09	14.95	1.06	1.19	1.22	1.21	-0.00	So= 4.8 / Sp= 0.16
SWAY m	33.95	17.25	77.53	101.07	88.27	9.44	0.38	0.65	0.94	0.58	0.29	So= 1.1 / Sp= 0.14
HEAVE m	36.33	16.96	161.38	186.33	73.14	11.07	0.30	0.40	0.68	0.59	-0.05	So= 1.8 / Sp= 0.22
	8.80	4.47	50.60	79.69	56.89	19.50	2.25		3.16	1.19	64.00	So= 1.9
FORCES	7.19	4.54	32.28	51.63	73.14	13.39	3.27		4.16	2.08	42.67	So= 3.8
MOORING	4.63	2.83	34.90	63.96	73.14	16.93	2.88		4.04	0.98	85.33	So= 0.7
	13.59	8.71	38.09	62.52	102.40	13.39	4.03		5.90	3.74	14.42	So= 15.5

.5 Hz

Fig. 6.12   JON VIDALIN   1983. 11.30.09.48
            Analysis of time series.

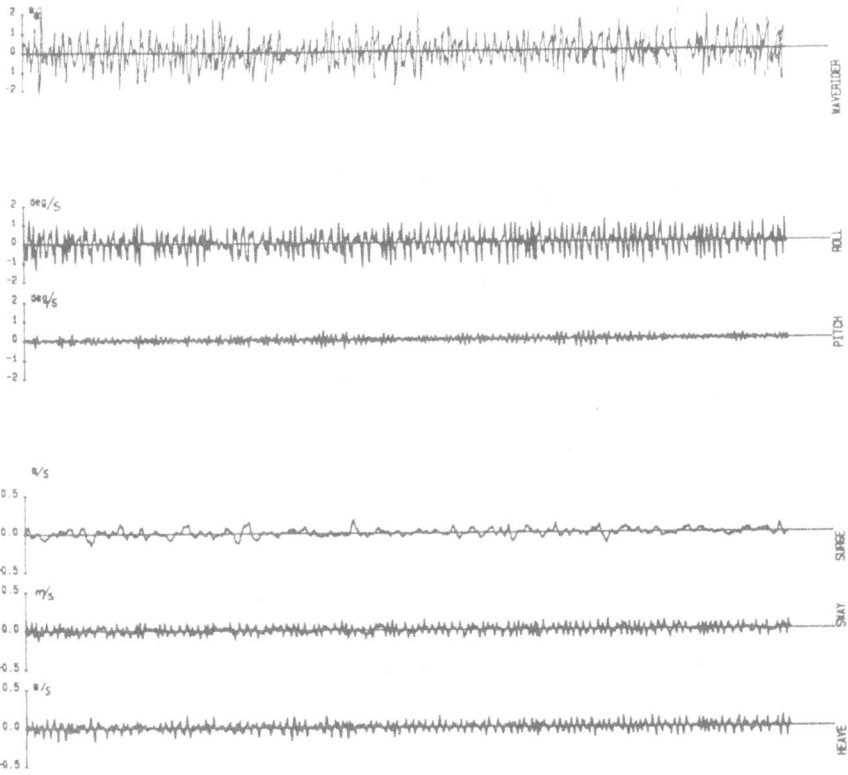

Fig. 6.13   JÓN VIDALÍN   1983. 11.30.09.48
            Some velocities of time series.

236

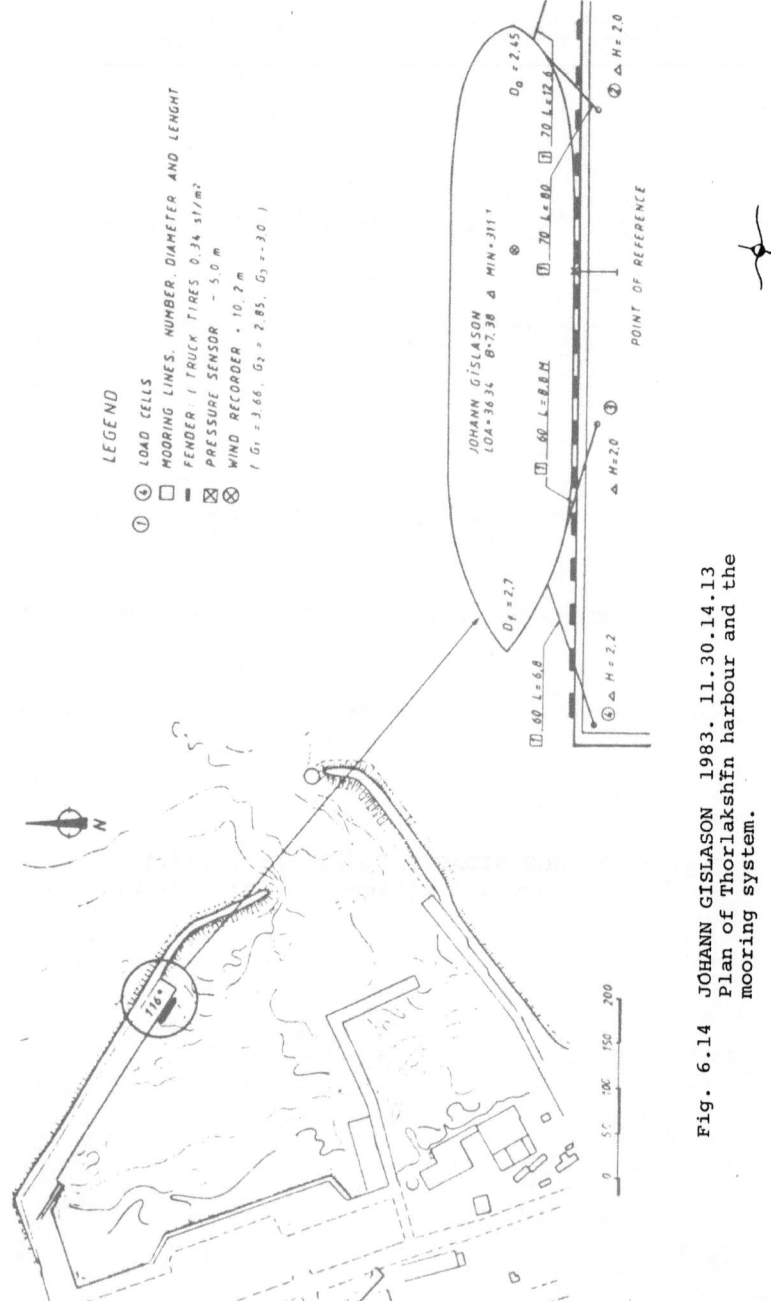

Fig. 6.14 JOHANN GISLASON 1983. 11.30.14.13
Plan of Thorlakshfn harbour and the
mooring system.

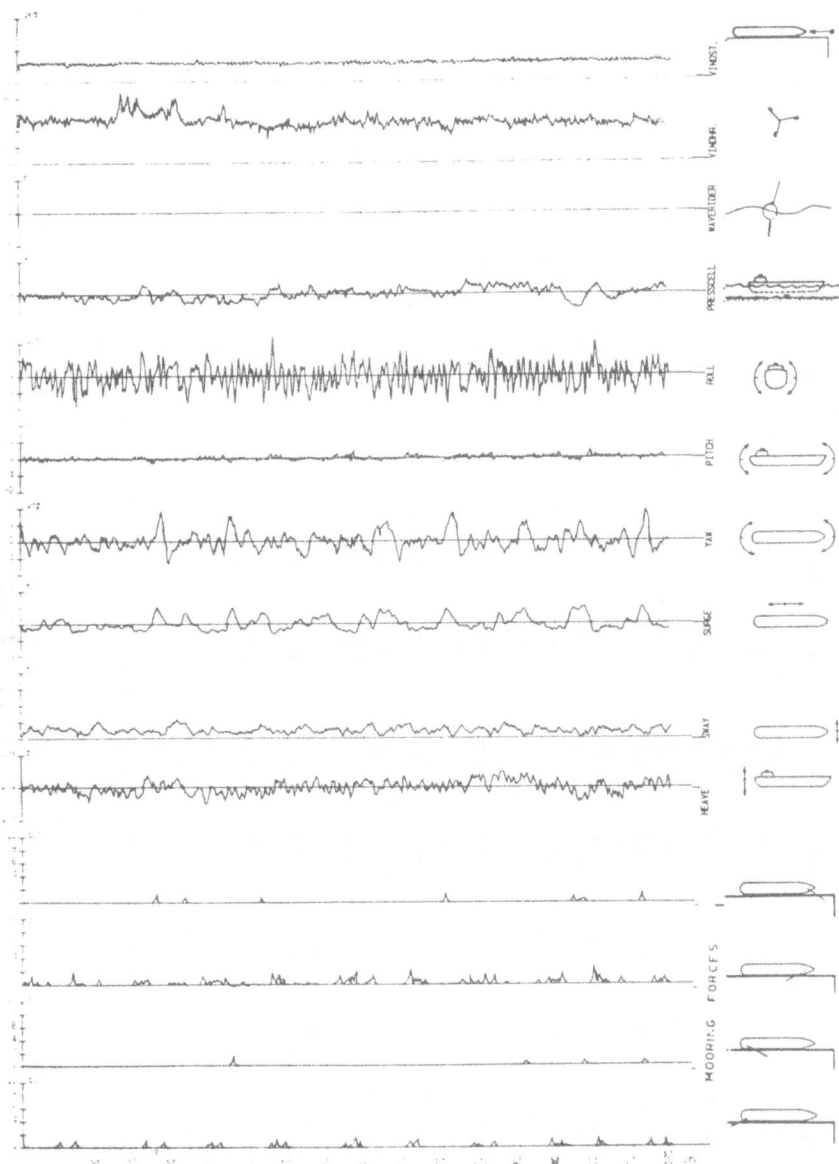

Fig. 6.15 JÓHANN GÍSLASON 1983. 11.30.14.13
Time series of ship movements, normal
moorings under loading/unloading operation.

238

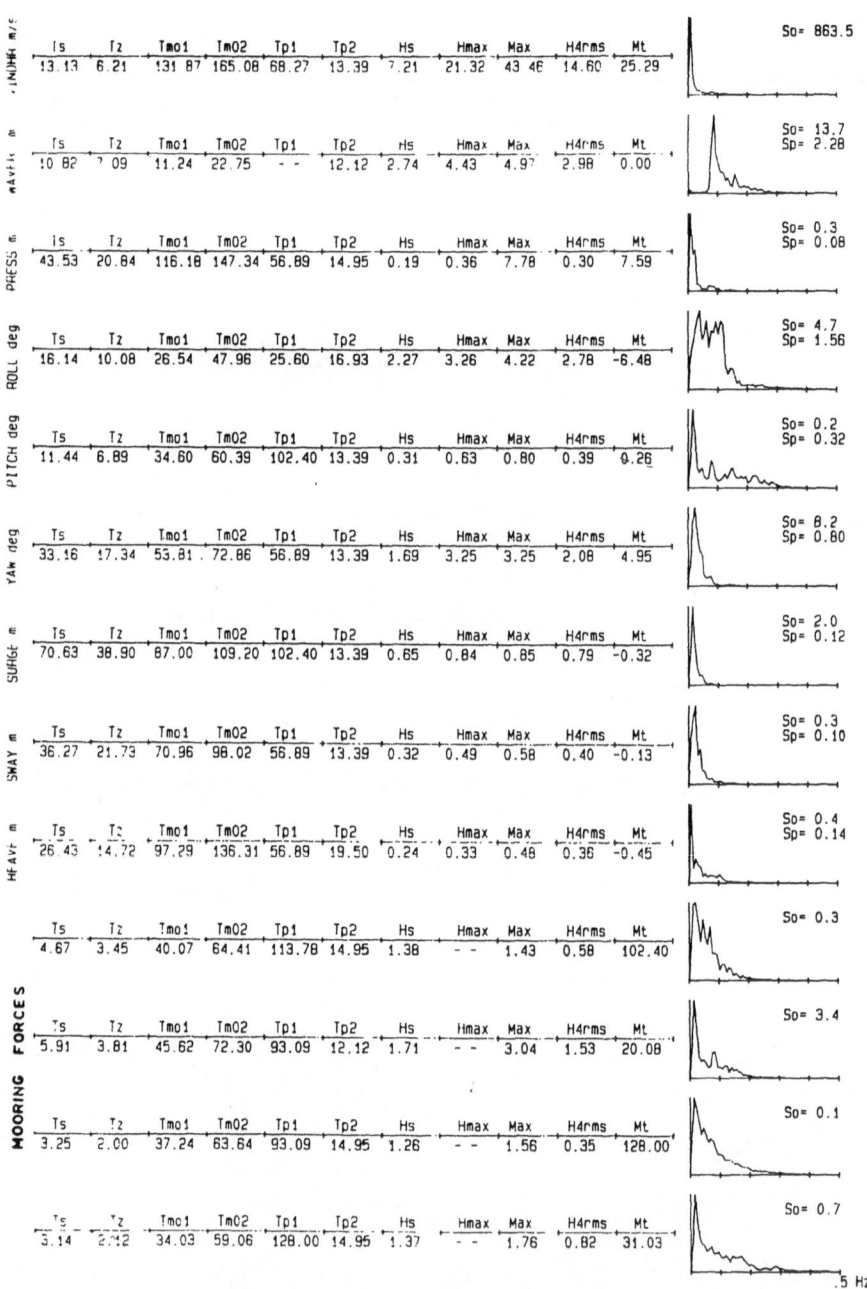

Fig. 6.16   JÓHANN GÍSLASON 1983. 11.30.14.13
            Analysis of time series.

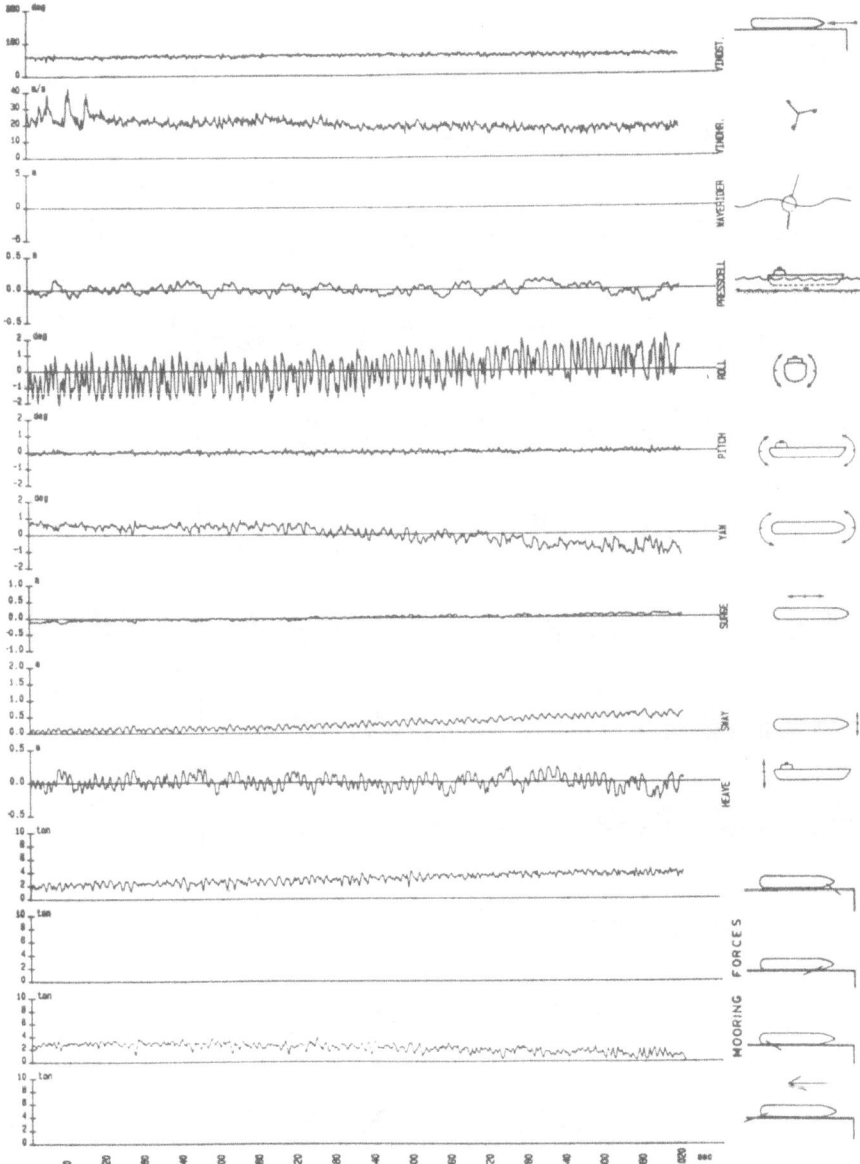

Fig. 6.17   JOHANN GISLASON   1983. 11.30.14.13
Time series of ship movements, propeller
operation by using haf machine force to
decrease ship movements under working
conditions.

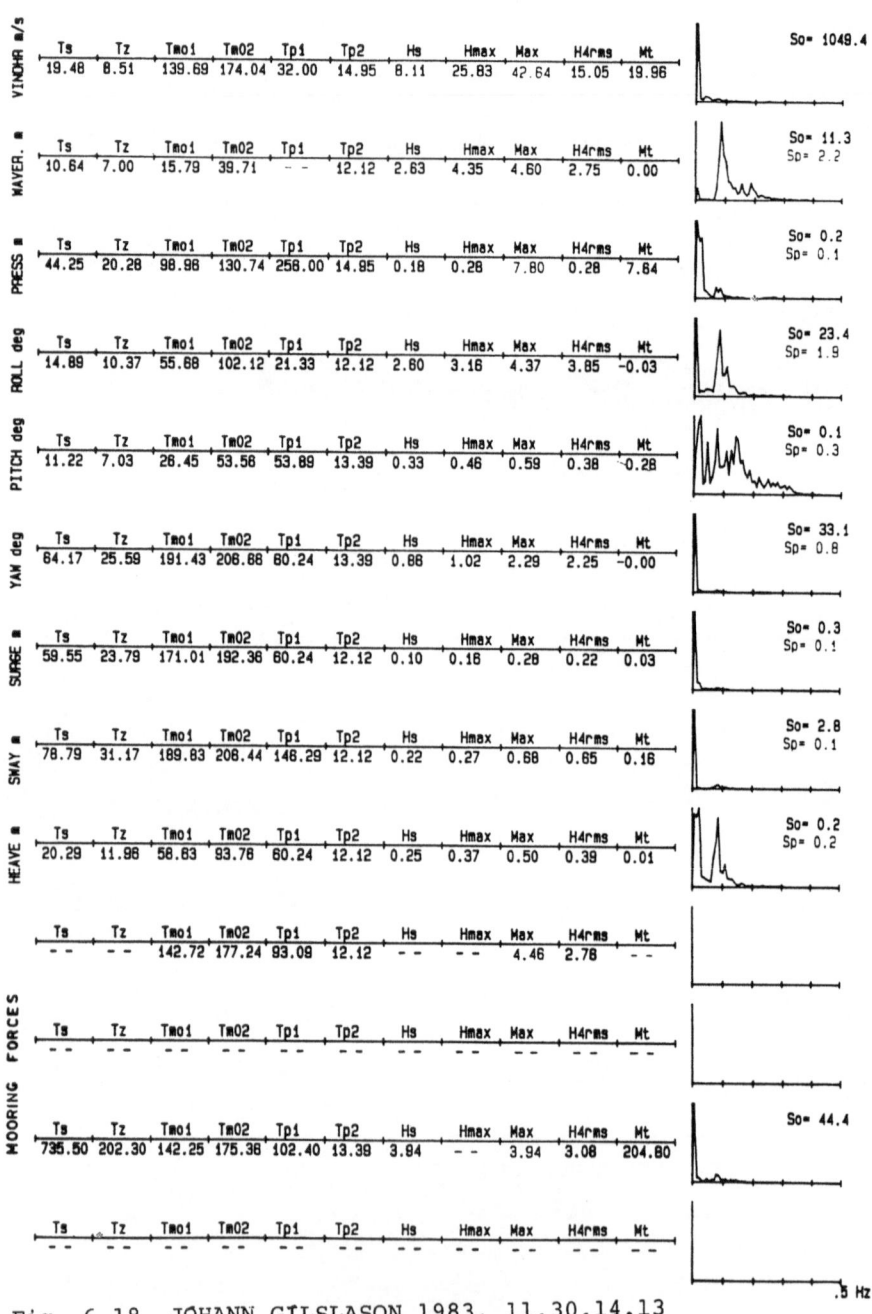

	Ts	Tz	Tmo1	Tm02	Tp1	Tp2	Hs	Hmax	Max	H4rms	Mt	
VINDHR a/s	19.48	8.51	139.69	174.04	32.00	14.95	8.11	25.83	42.64	15.05	19.96	So= 1049.4
MAYER. m	10.64	7.00	15.79	39.71	--	12.12	2.63	4.35	4.60	2.75	0.00	So= 11.3 / Sp= 2.2
PRESS m	44.25	20.28	98.96	130.74	258.00	14.95	0.18	0.28	7.80	0.28	7.64	So= 0.2 / Sp= 0.1
ROLL deg	14.89	10.37	55.68	102.12	21.33	12.12	2.60	3.16	4.37	3.85	-0.03	So= 23.4 / Sp= 1.9
PITCH deg	11.22	7.03	26.45	53.58	53.89	13.39	0.33	0.46	0.59	0.38	-0.28	So= 0.1 / Sp= 0.3
YAW deg	64.17	25.59	191.43	206.88	80.24	13.39	0.86	1.02	2.29	2.25	-0.00	So= 33.1 / Sp= 0.8
SURGE m	59.55	23.79	171.01	192.36	80.24	12.12	0.10	0.16	0.28	0.22	0.03	So= 0.3 / Sp= 0.1
SWAY m	78.79	31.17	189.83	206.44	146.29	12.12	0.22	0.27	0.68	0.65	0.16	So= 2.8 / Sp= 0.1
HEAVE m	20.29	11.98	58.63	93.76	80.24	12.12	0.25	0.37	0.50	0.39	0.01	So= 0.2 / Sp= 0.2
MOORING FORCES	--	--	142.72	177.24	93.09	12.12	--	--	4.46	2.78	--	
	--	--	--	--	--	--	--	--	--	--	--	
	735.50	202.30	142.25	175.38	102.40	13.39	3.94	--	3.94	3.08	204.80	So= 44.4
	--	--	--	--	--	--	--	--	--	--	--	

Fig. 6.18   JOHANN GILSLASON 1983. 11.30.14.13
Analysis of time series.

241

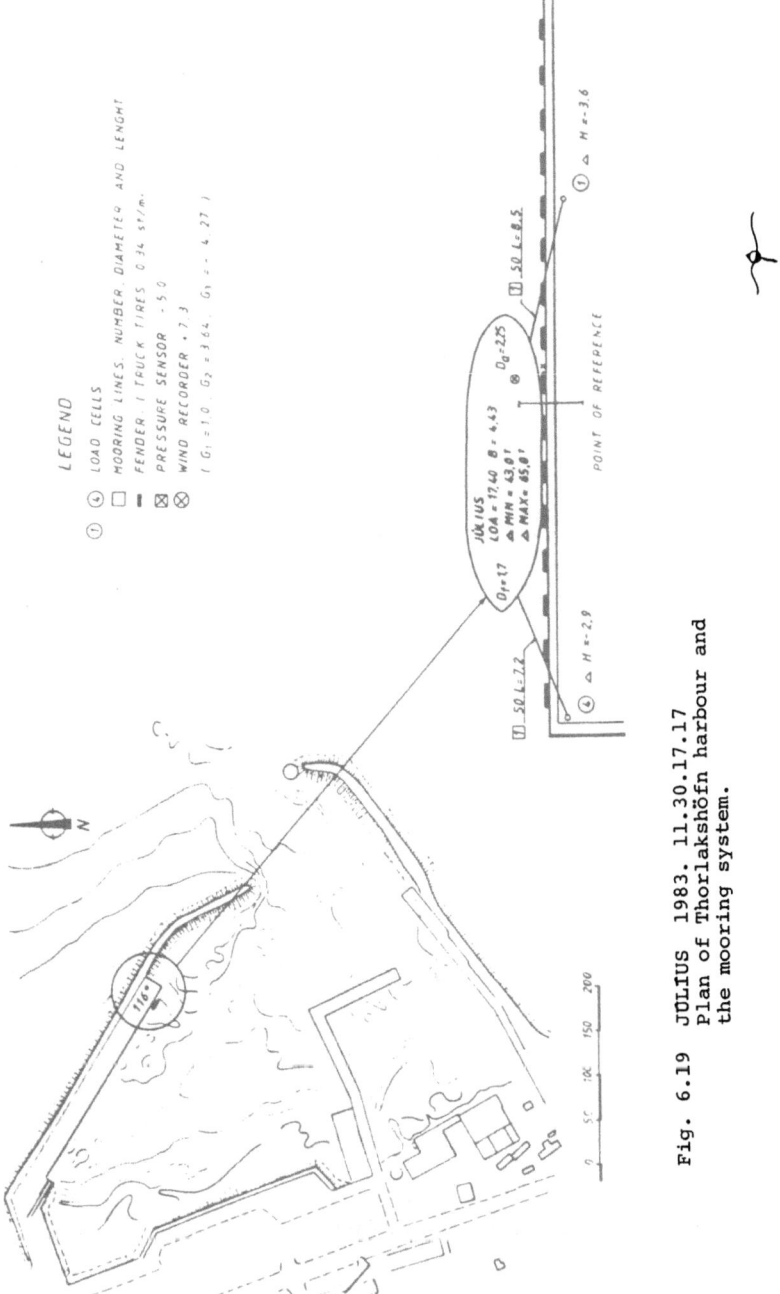

Fig. 6.19 JÚLIUS 1983. 11.30.17.17
Plan of Thorlakshöfn harbour and
the mooring system.

242

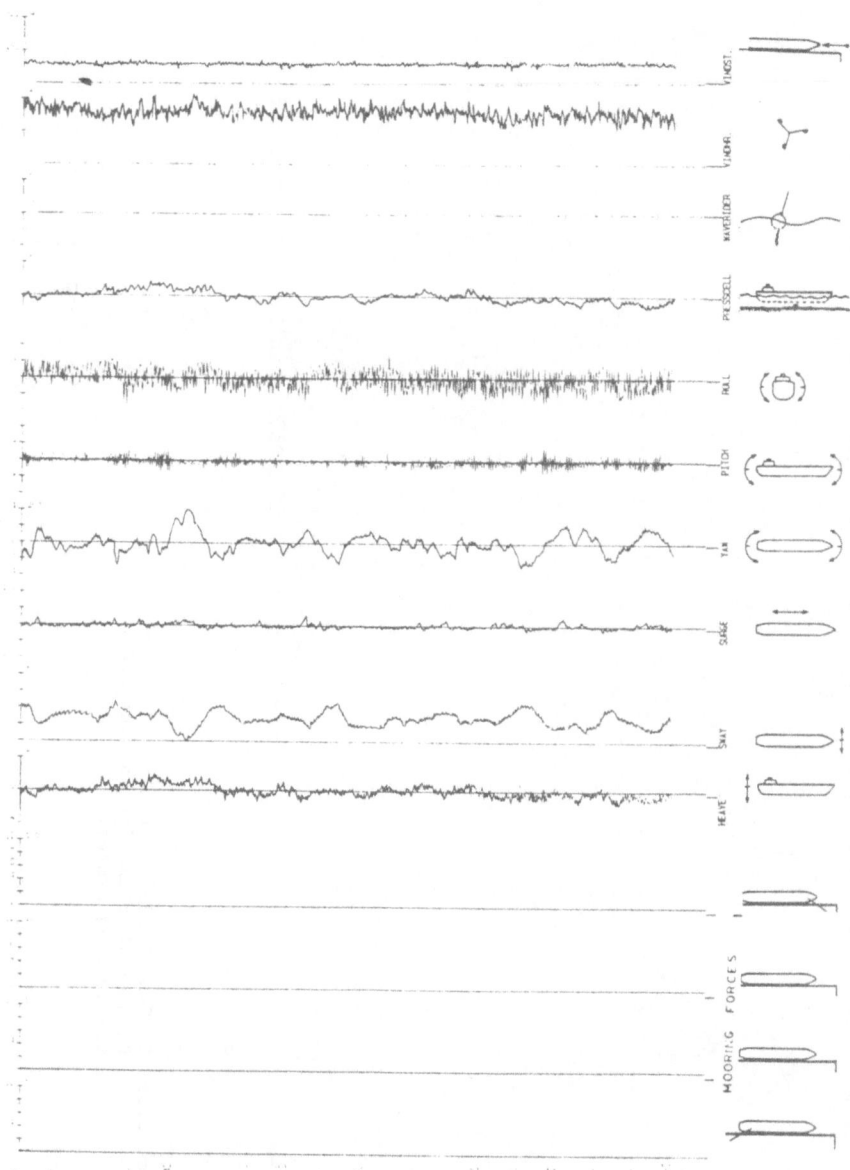

Fig. 6.20 JULIUS 1983. 11.30.17.17
Times series of ship movements.

243

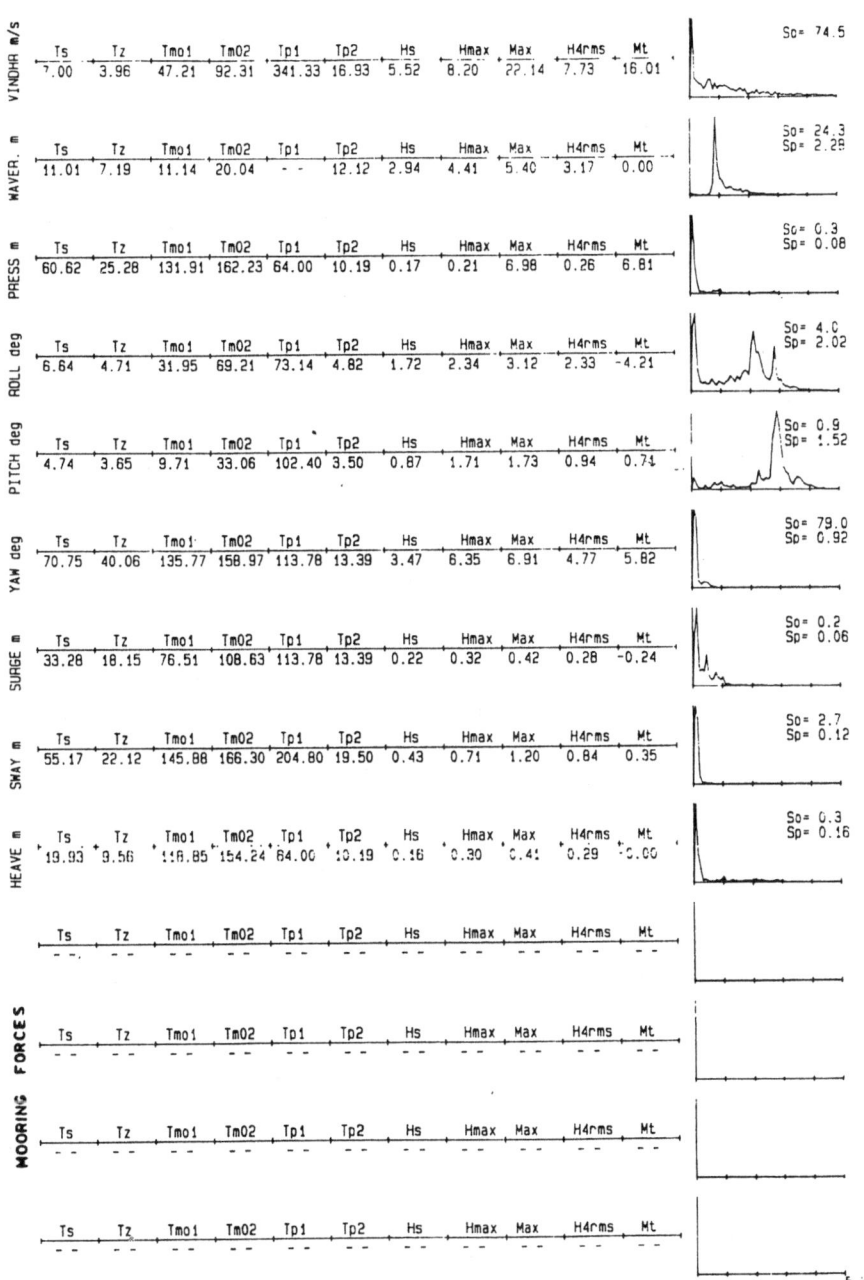

	Ts	Tz	Tmo1	Tm02	Tp1	Tp2	Hs	Hmax	Max	H4rms	Mt	So= 74.5
VINDHA m/s	7.00	3.96	47.21	92.31	341.33	16.93	5.52	8.20	22.14	7.73	16.01	
MAVER. m	Ts 11.01	Tz 7.19	Tmo1 11.14	Tm02 20.04	Tp1 - -	Tp2 12.12	Hs 2.94	Hmax 4.41	Max 5.40	H4rms 3.17	Mt 0.00	So= 24.3 / Sp= 2.28
PRESS m	Ts 60.62	Tz 25.28	Tmo1 131.91	Tm02 162.23	Tp1 64.00	Tp2 10.19	Hs 0.17	Hmax 0.21	Max 6.98	H4rms 0.26	Mt 6.81	So= 0.3 / Sp= 0.08
ROLL deg	Ts 6.64	Tz 4.71	Tmo1 31.95	Tm02 69.21	Tp1 73.14	Tp2 4.82	Hs 1.72	Hmax 2.34	Max 3.12	H4rms 2.33	Mt -4.21	So= 4.0 / Sp= 2.02
PITCH deg	Ts 4.74	Tz 3.65	Tmo1 9.71	Tm02 33.06	Tp1 102.40	Tp2 3.50	Hs 0.87	Hmax 1.71	Max 1.73	H4rms 0.94	Mt 0.74	So= 0.9 / Sp= 1.52
YAW deg	Ts 70.75	Tz 40.06	Tmo1 135.77	Tm02 158.97	Tp1 113.78	Tp2 13.39	Hs 3.47	Hmax 6.35	Max 6.91	H4rms 4.77	Mt 5.82	So= 79.0 / Sp= 0.92
SURGE m	Ts 33.28	Tz 18.15	Tmo1 76.51	Tm02 108.63	Tp1 113.78	Tp2 13.39	Hs 0.22	Hmax 0.32	Max 0.42	H4rms 0.28	Mt -0.24	So= 0.2 / Sp= 0.06
SWAY m	Ts 55.17	Tz 22.12	Tmo1 145.88	Tm02 166.30	Tp1 204.80	Tp2 19.50	Hs 0.43	Hmax 0.71	Max 1.20	H4rms 0.84	Mt 0.35	So= 2.7 / Sp= 0.12
HEAVE m	Ts 19.93	Tz 9.56	Tmo1 116.85	Tm02 154.24	Tp1 64.00	Tp2 10.19	Hs 0.18	Hmax 0.30	Max 0.41	H4rms 0.29	Mt -0.00	So= 0.3 / Sp= 0.16

Fig. 6.21 JULIUS 1983. 11.30.17.17
Analysis of time series.

244

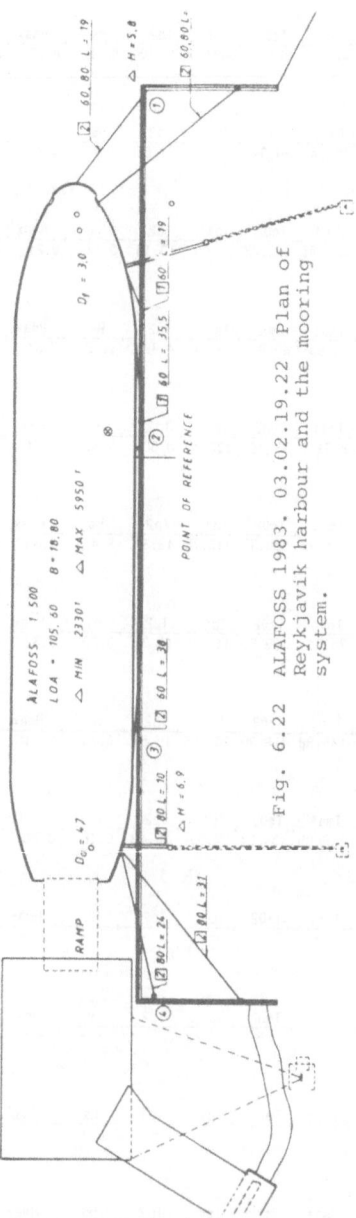

LEGEND

① LOAD CELLS
⊕ MOORING LINES. NUMBER. DIAMETER AND LENGHT
□ FENDER CYLINDRICAL ⌀ 400 - ⌀ 200 m⁻ 14 m m
■ PRESSURE SENSOR - 10 cm
⊠ WIND RECORDER - 135 m
⊗

(G₁ = 9.4, G₂ = 19 T   G₃ = - 2.4 T)

Fig. 6.22 ALAFOSS 1983. 03.02.19.22 Plan of Reykjavik harbour and the mooring system.

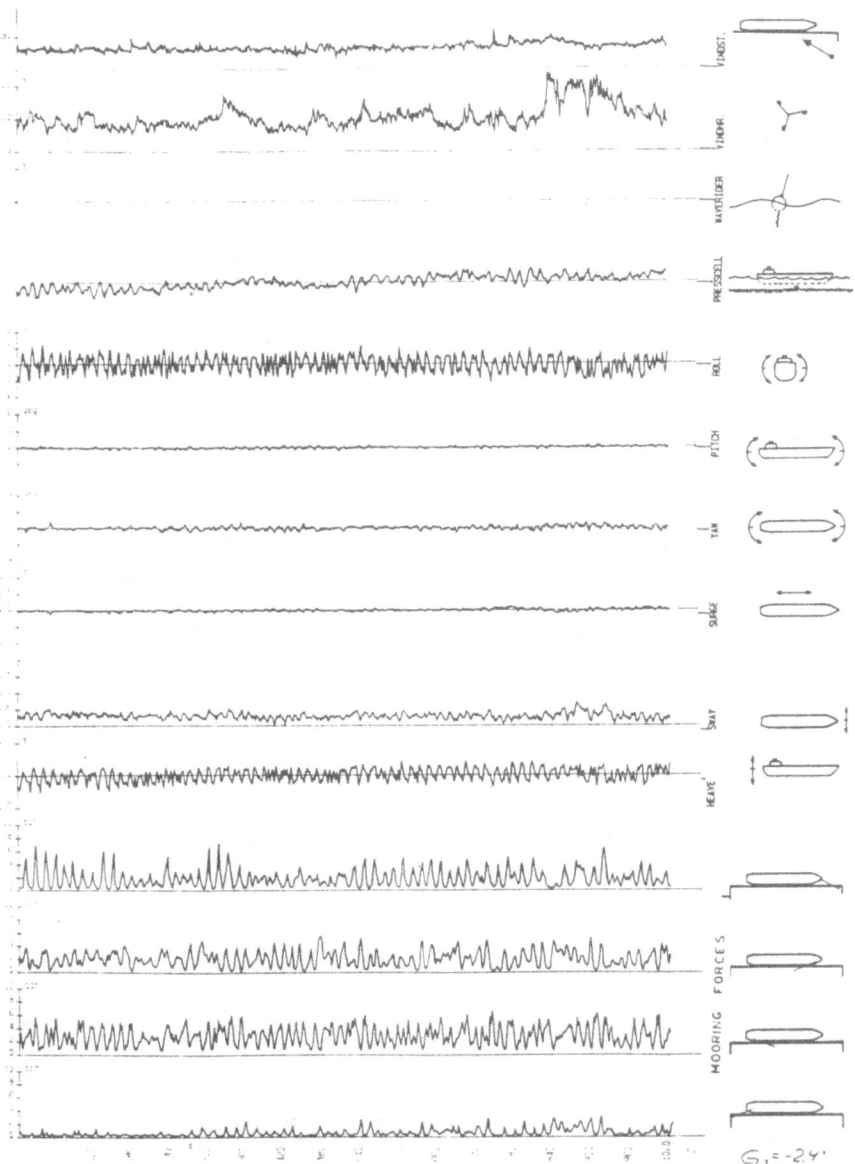

Fig. 6.23   ALAFOSS 1983. 03.02.19.22
            Time series of ship movements, taut
            moorings by winches.

246

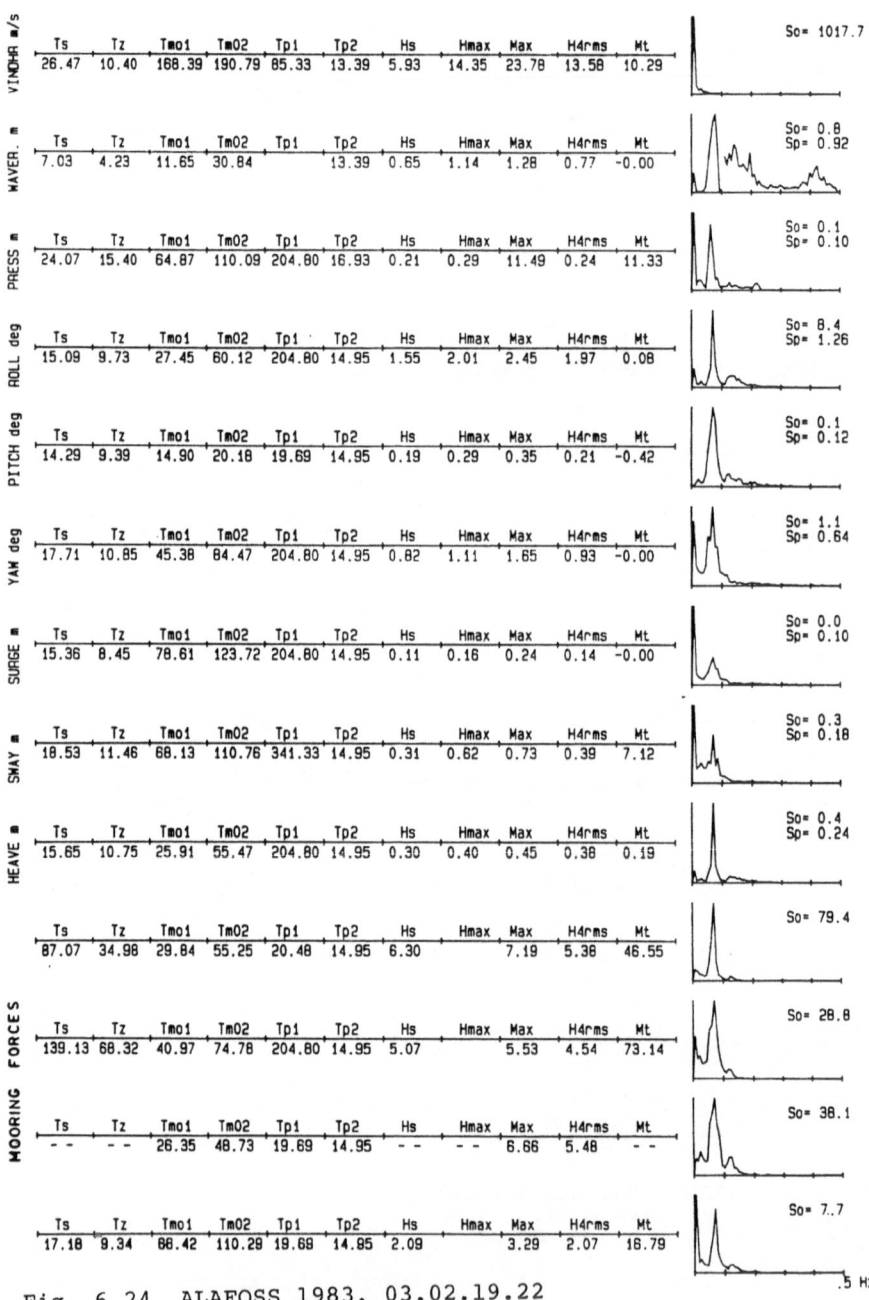

	Ts	Tz	Tmo1	TmO2	Tp1	Tp2	Hs	Hmax	Max	H4rms	Mt	
VINDHR m/s	26.47	10.40	168.39	190.79	85.33	13.39	5.93	14.35	23.78	13.58	10.29	So= 1017.7
MAVER. m	7.03	4.23	11.65	30.84		13.39	0.65	1.14	1.28	0.77	-0.00	So= 0.8 Sp= 0.92
PRESS m	24.07	15.40	64.87	110.09	204.80	16.93	0.21	0.29	11.49	0.24	11.33	So= 0.1 Sp= 0.10
ROLL deg	15.09	9.73	27.45	60.12	204.80	14.95	1.55	2.01	2.45	1.97	0.08	So= 8.4 Sp= 1.26
PITCH deg	14.29	9.39	14.90	20.18	19.69	14.95	0.19	0.29	0.35	0.21	-0.42	So= 0.1 Sp= 0.12
YAW deg	17.71	10.85	45.38	84.47	204.80	14.95	0.82	1.11	1.65	0.93	-0.00	So= 1.1 Sp= 0.64
SURGE m	15.36	8.45	78.61	123.72	204.80	14.95	0.11	0.16	0.24	0.14	-0.00	So= 0.0 Sp= 0.10
SWAY m	18.53	11.46	68.13	110.76	341.33	14.95	0.31	0.62	0.73	0.39	7.12	So= 0.3 Sp= 0.18
HEAVE m	15.65	10.75	25.91	55.47	204.80	14.95	0.30	0.40	0.45	0.38	0.19	So= 0.4 Sp= 0.24
	87.07	34.98	29.84	55.25	20.48	14.95	6.30		7.19	5.38	46.55	So= 79.4
MOORING FORCES	139.13	68.32	40.97	74.78	204.80	14.95	5.07		5.53	4.54	73.14	So= 28.8
	- -	- -	26.35	48.73	19.69	14.95	- -	- -	6.66	5.48	- -	So= 38.1
	17.18	9.34	68.42	110.28	19.68	14.95	2.09		3.29	2.07	18.79	So= 7.7

.5 Hz

Fig. 6.24   ALAFOSS 1983. 03.02.19.22
Analysis of time series.

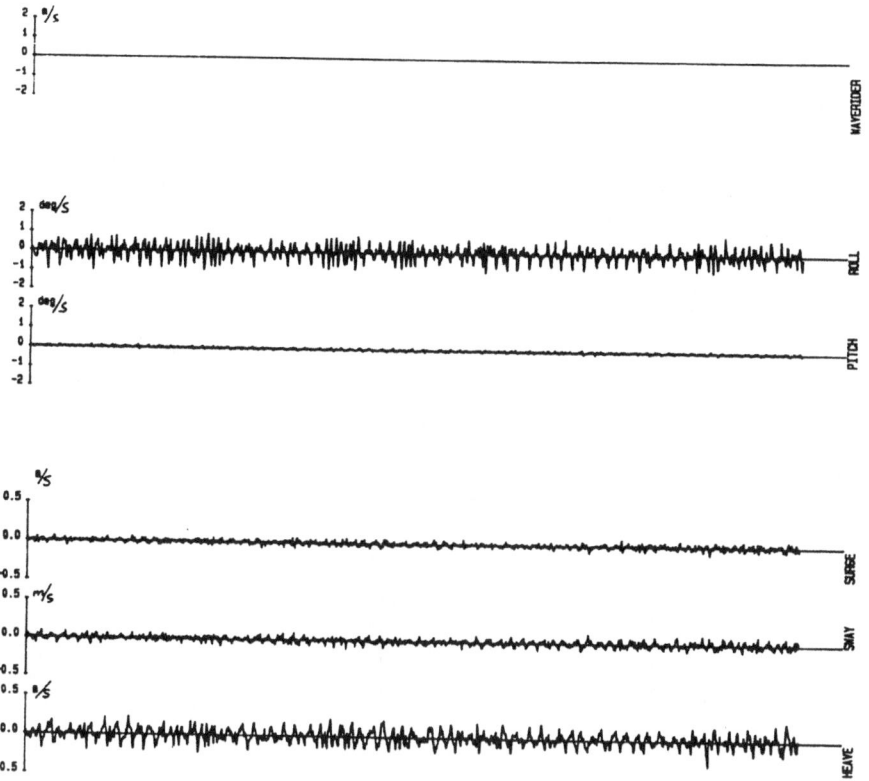

Fig. 6.25   ALAFOSS 1983. 03.02.19.22
           Some velocities of time series.

248

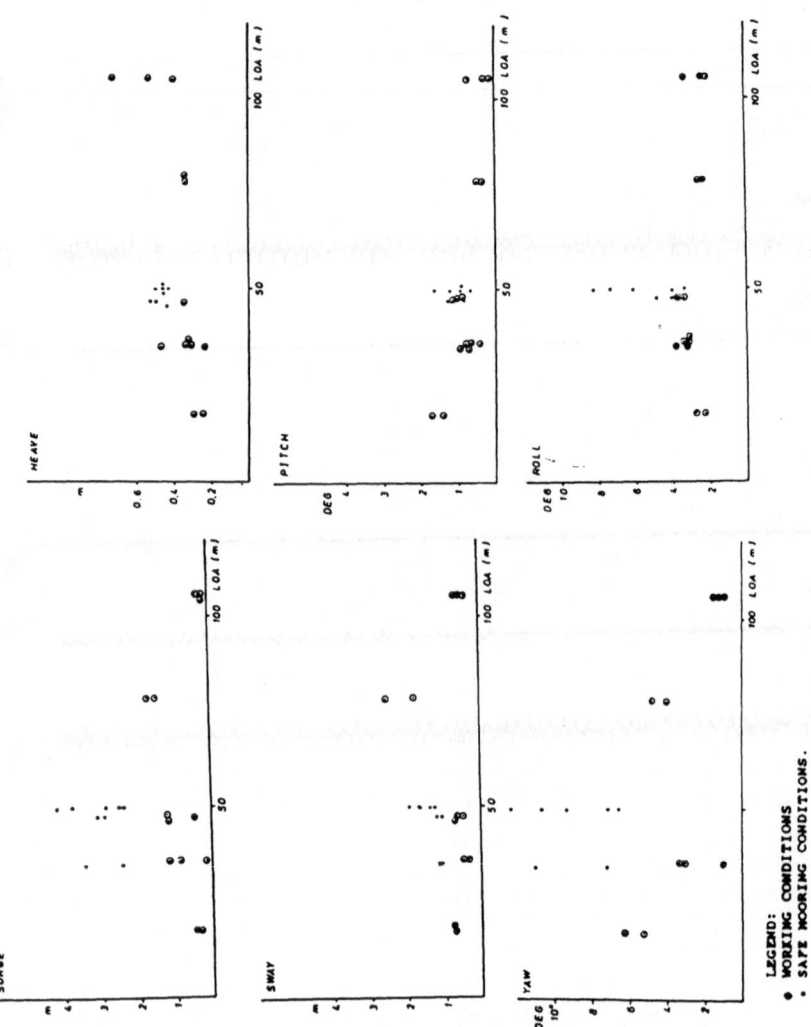

Fig. 7.1   Maximum translatory and rotatory movements
as a function of the ship length.

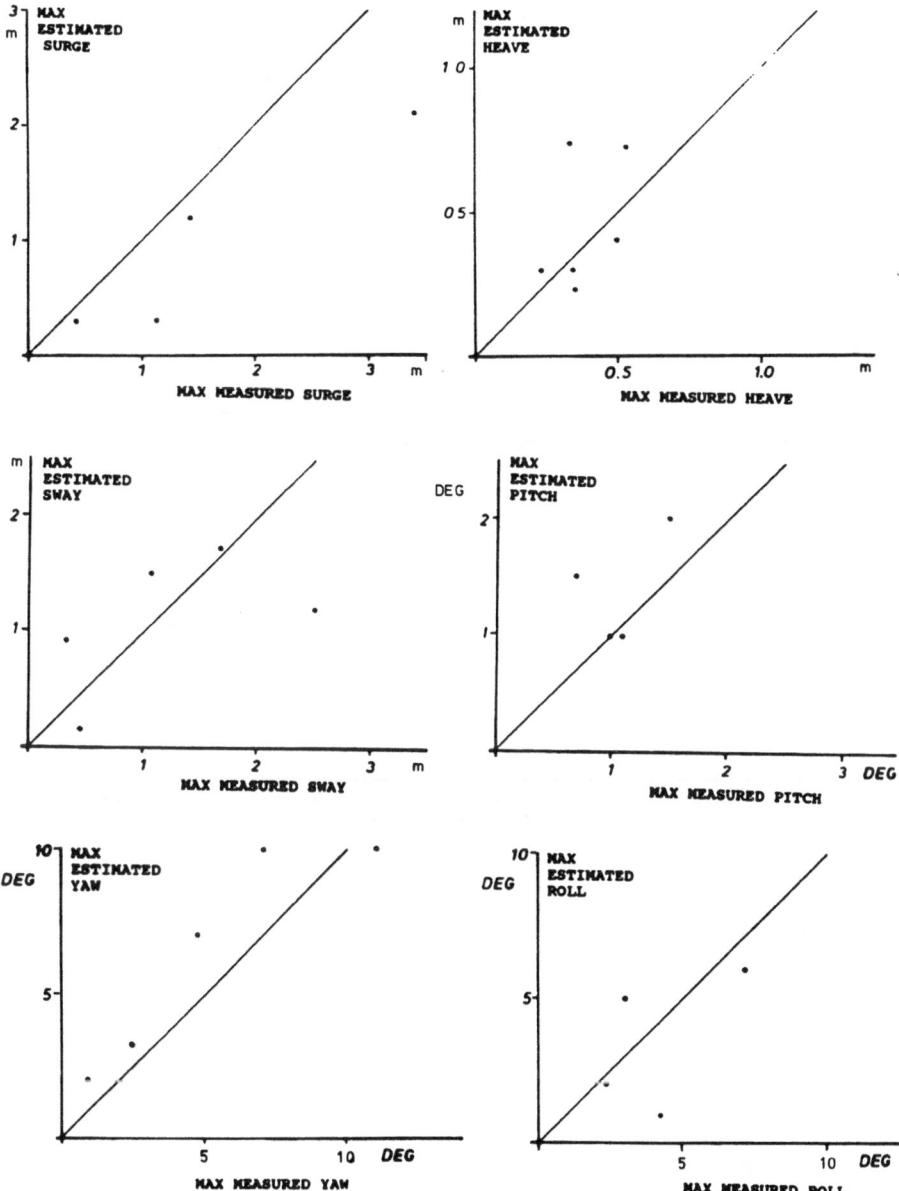

Fig. 7.2 Correlation between estimated and measured maximum
ship movements. The estimation was done by the
captains during field measurements.

250

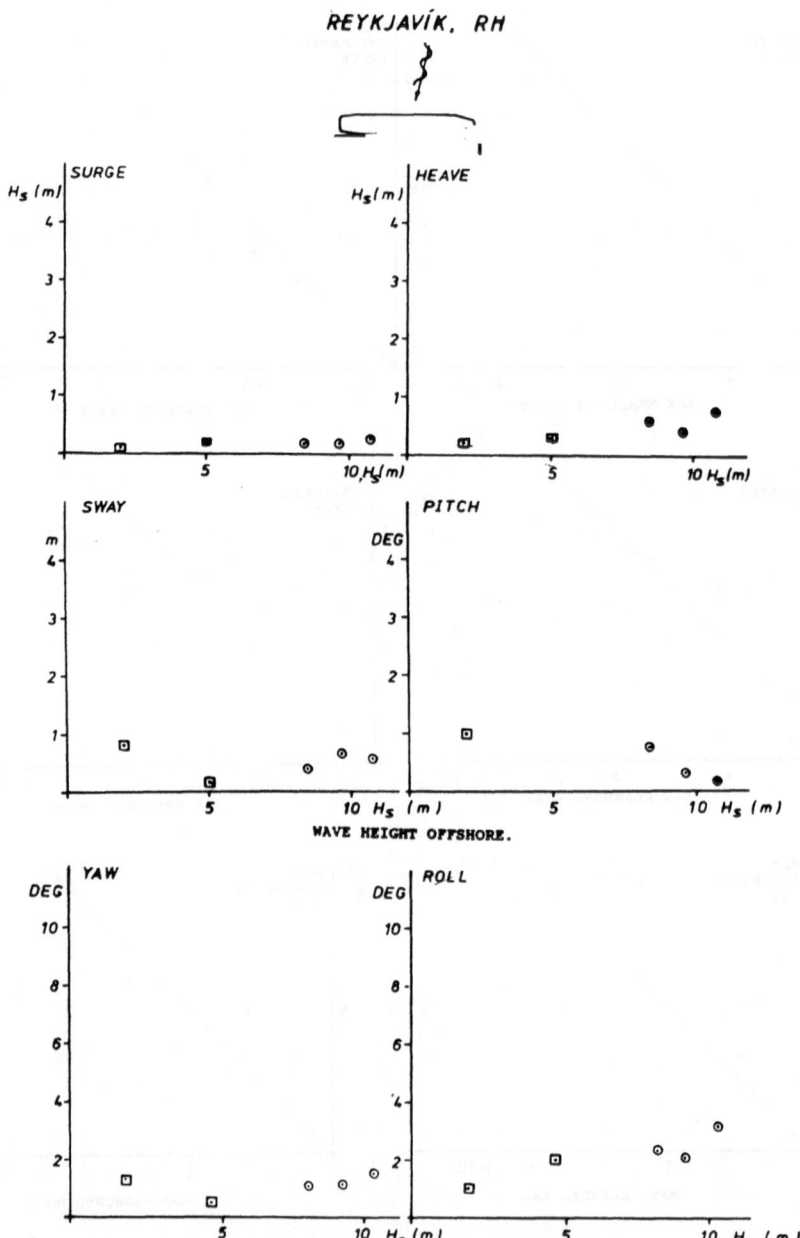

Fig. 7.3   Hmax values in relation to significant wave height
          offshore.

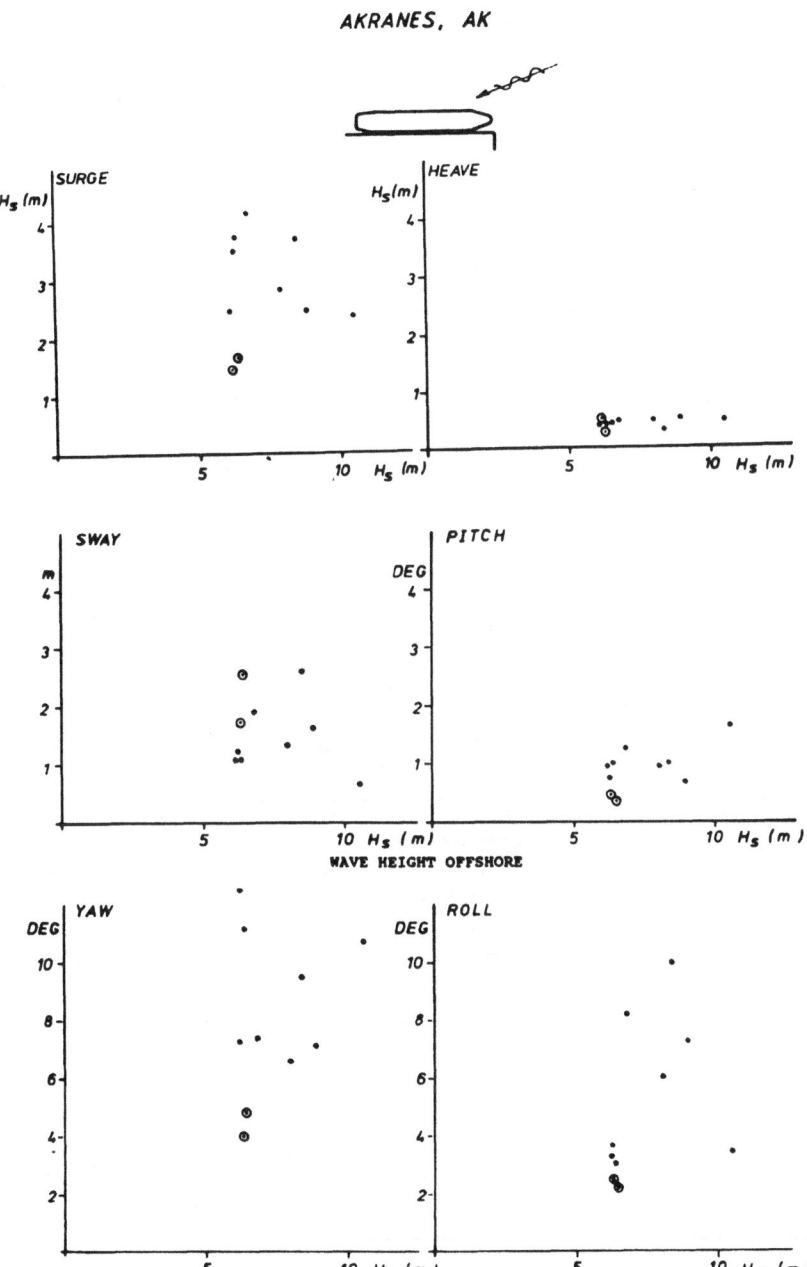

AKRANES, AK

Fig. 7.4 Hmax values in relation to significant wave
height offshore.

252

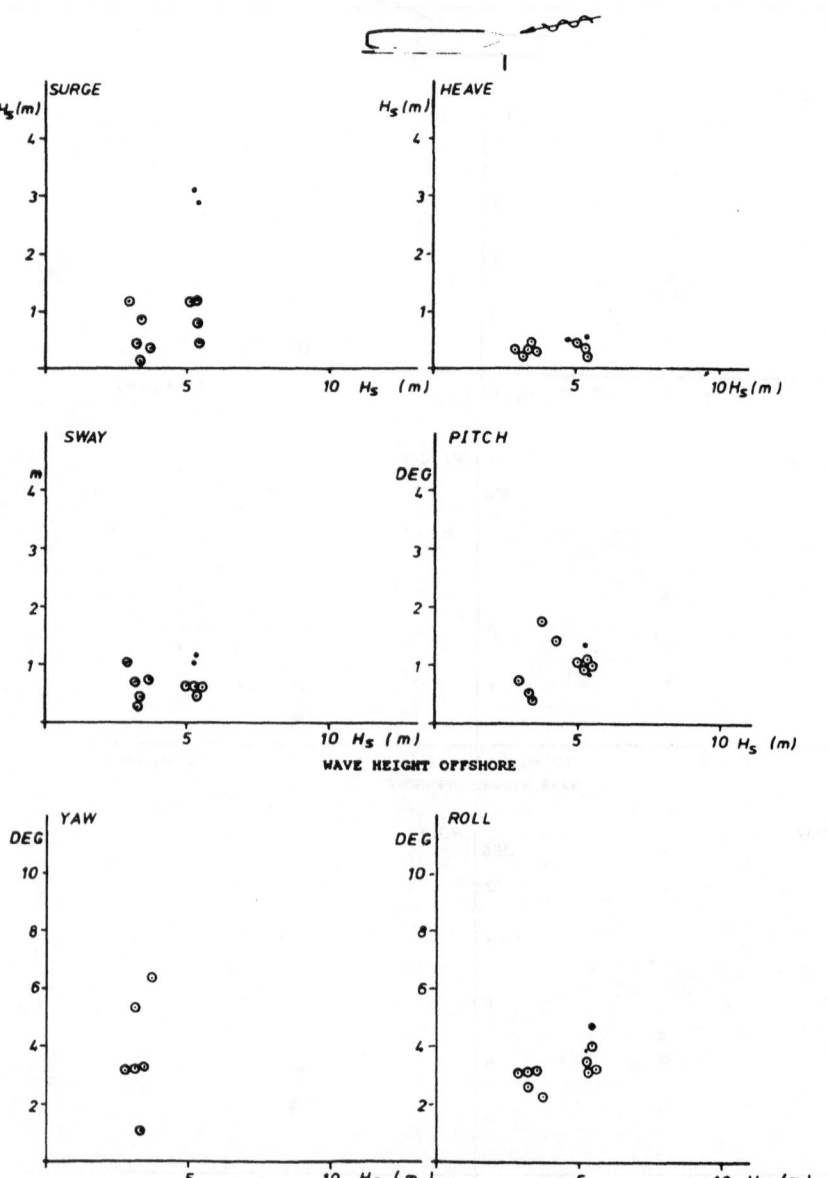

Fig. 7.5 Hmax values in relation to significant
wave height offshore.

253

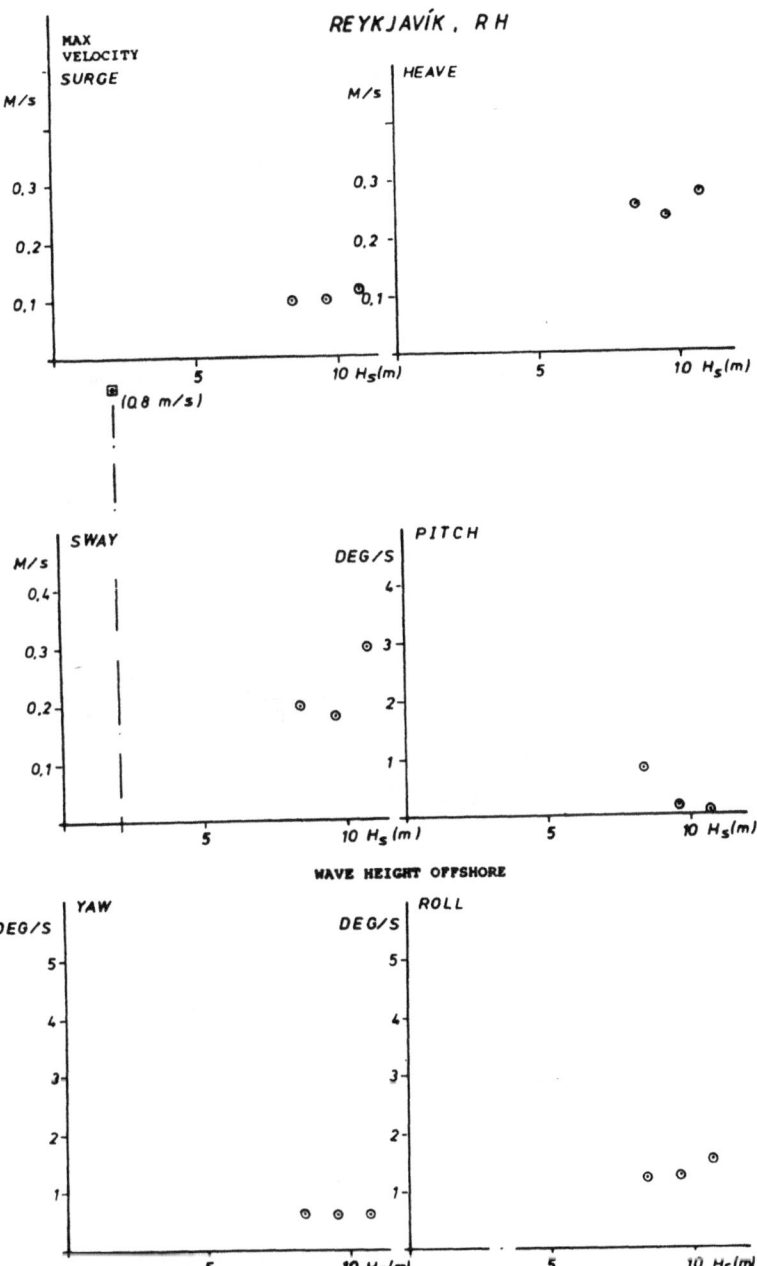

Fig. 7.6 Maximum velocity of ship movements in relation
to significant wave height offshore.

254

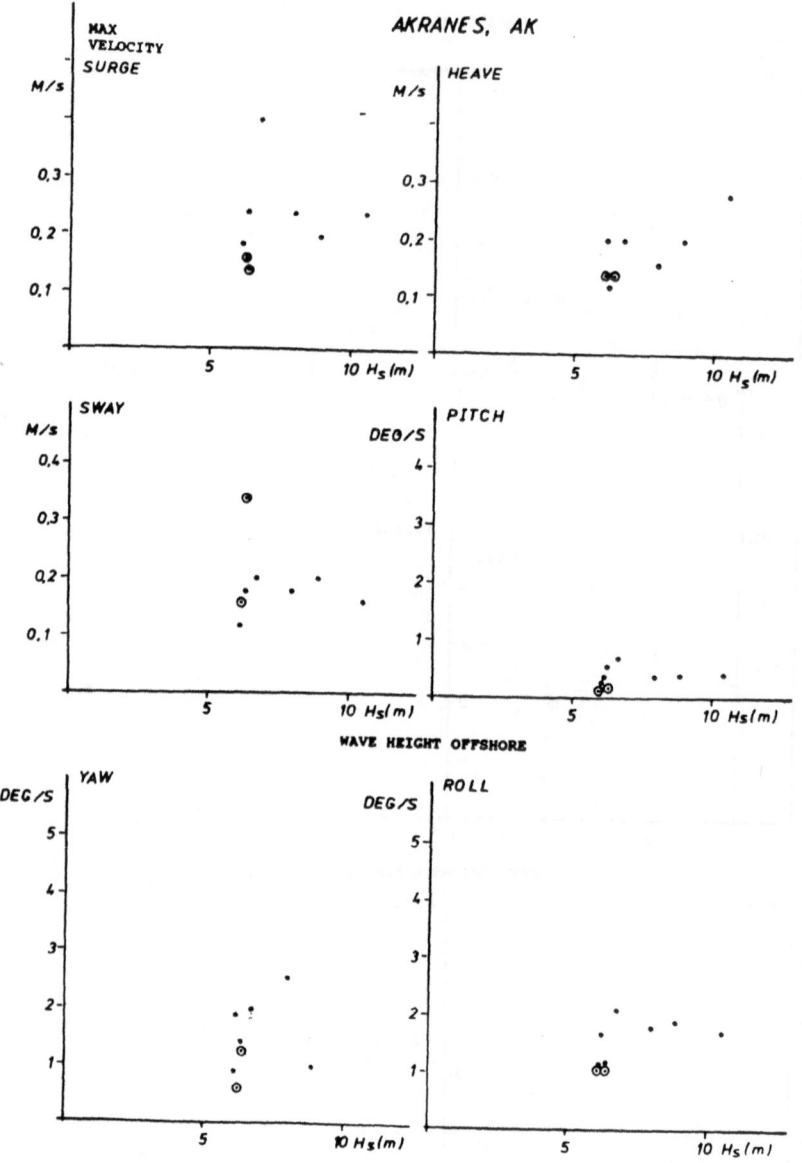

Fig. 7.7   Maximum velocity of ship movements in relation
to significant wave height offshore.

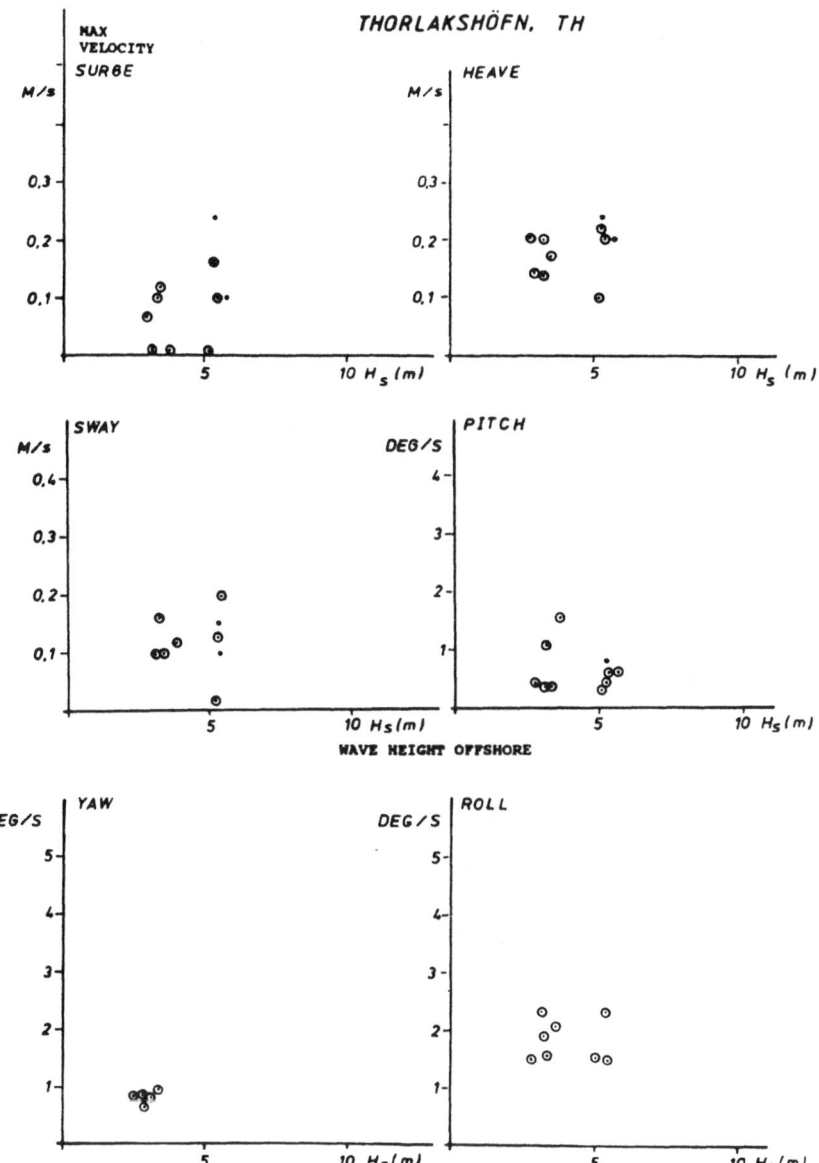

Fig. 7.8 Maximum velocity of ship movements in relation to significant wave height offshore.

256

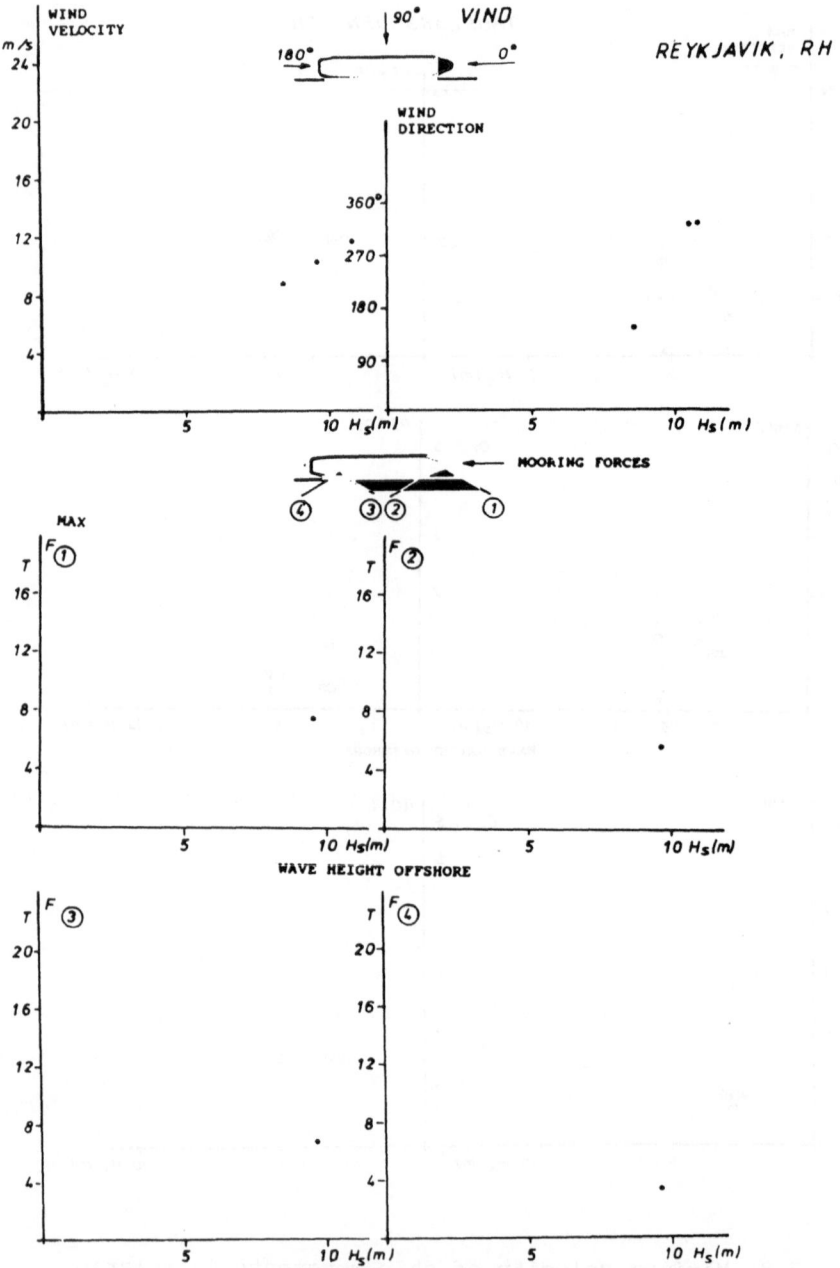

Fig. 7.9 Average wind velocity, wind direction and
maximum mooring forces in relation to
wave height offshore.

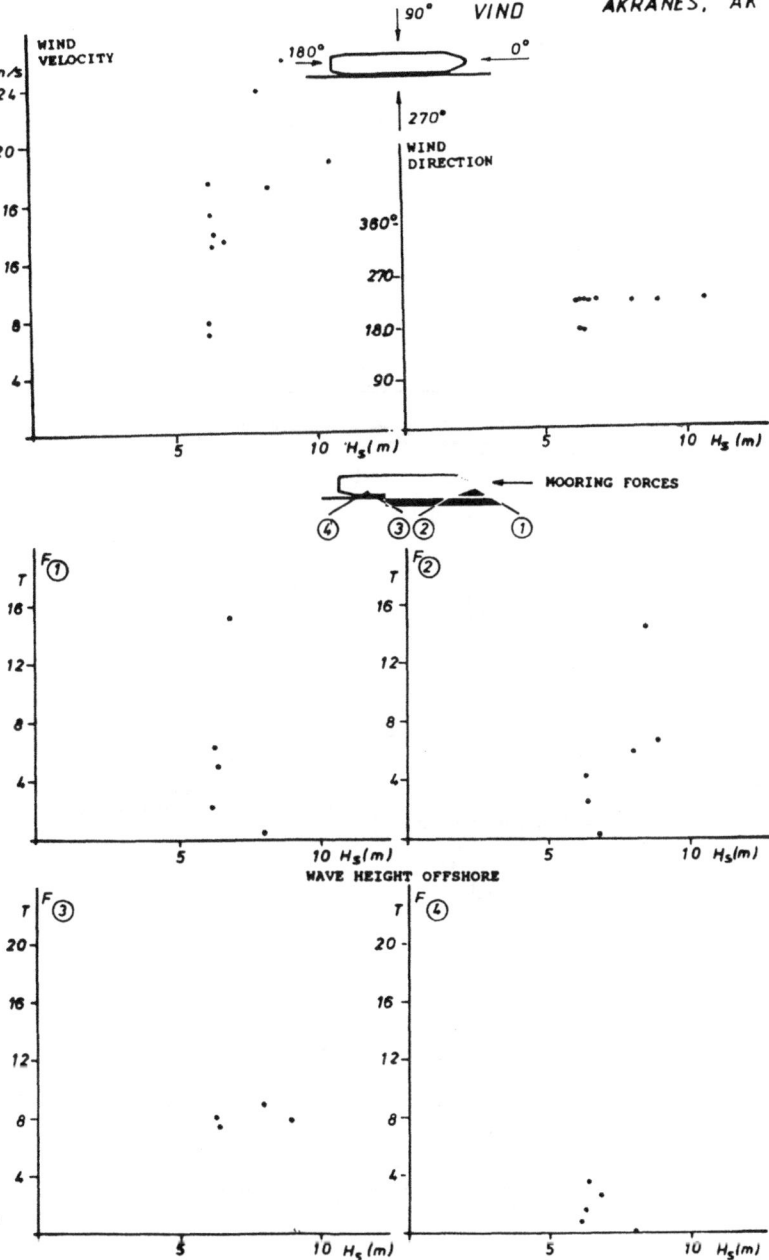

Fig. 7.10 Average wind velocity, wind direction and maximum mooring forces in relation to wave height offshore.

258

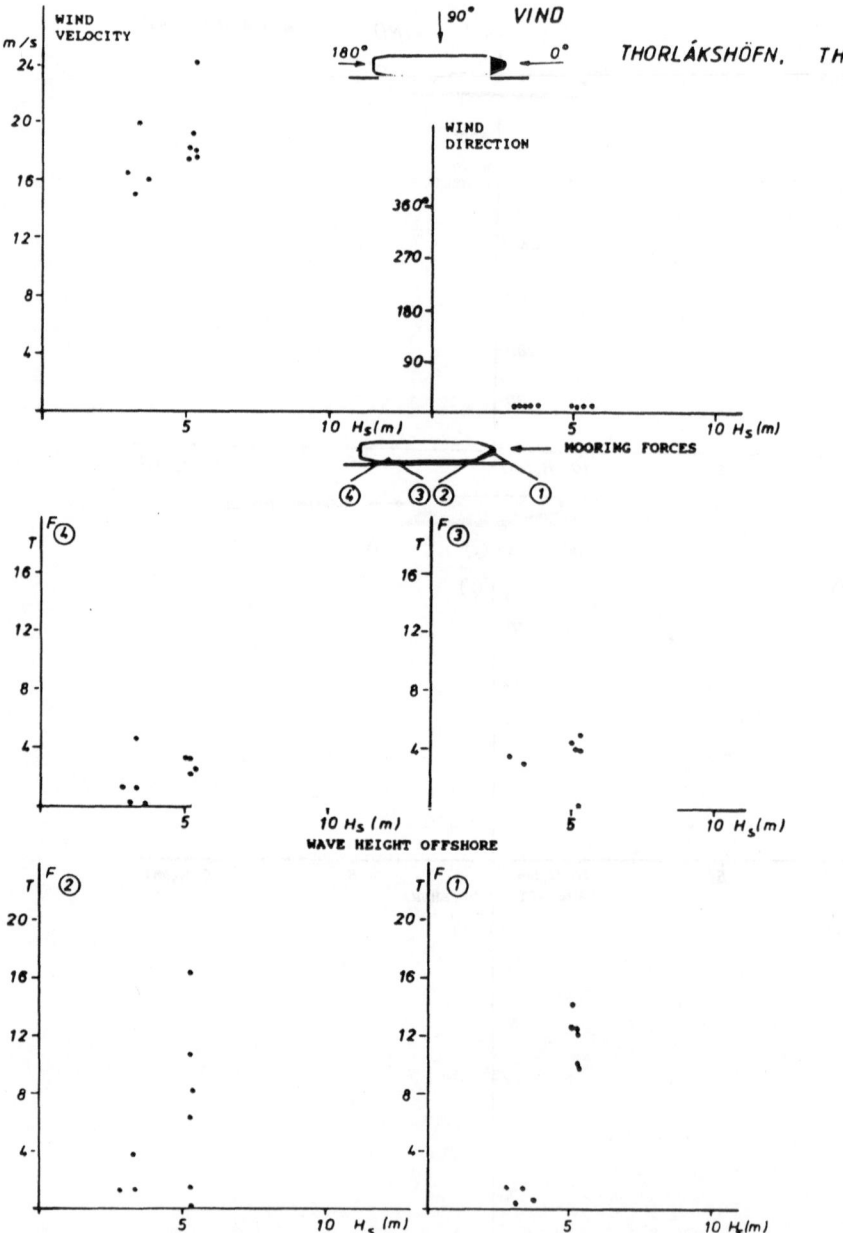

Fig. 7.11 Average wind velocity, wind direction and
          maximum mooring forces in relation to wave
          height offshore.

259

REYKJAVÍK

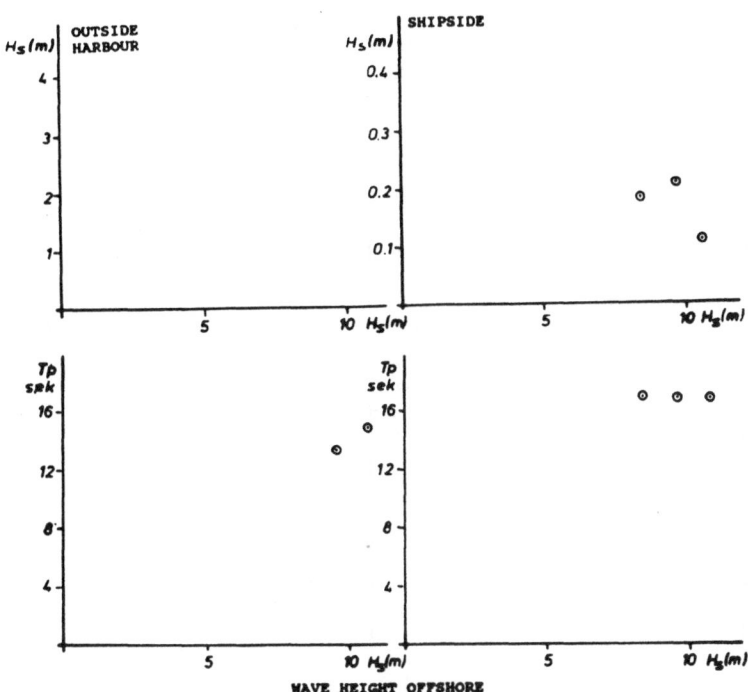

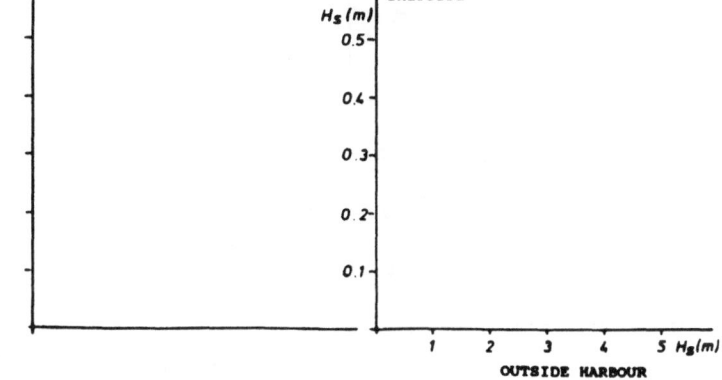

Fig. 7.12 Siqnificant wave height and wave period outside
the harbour and at the shipside in relation to
wave height offshore.

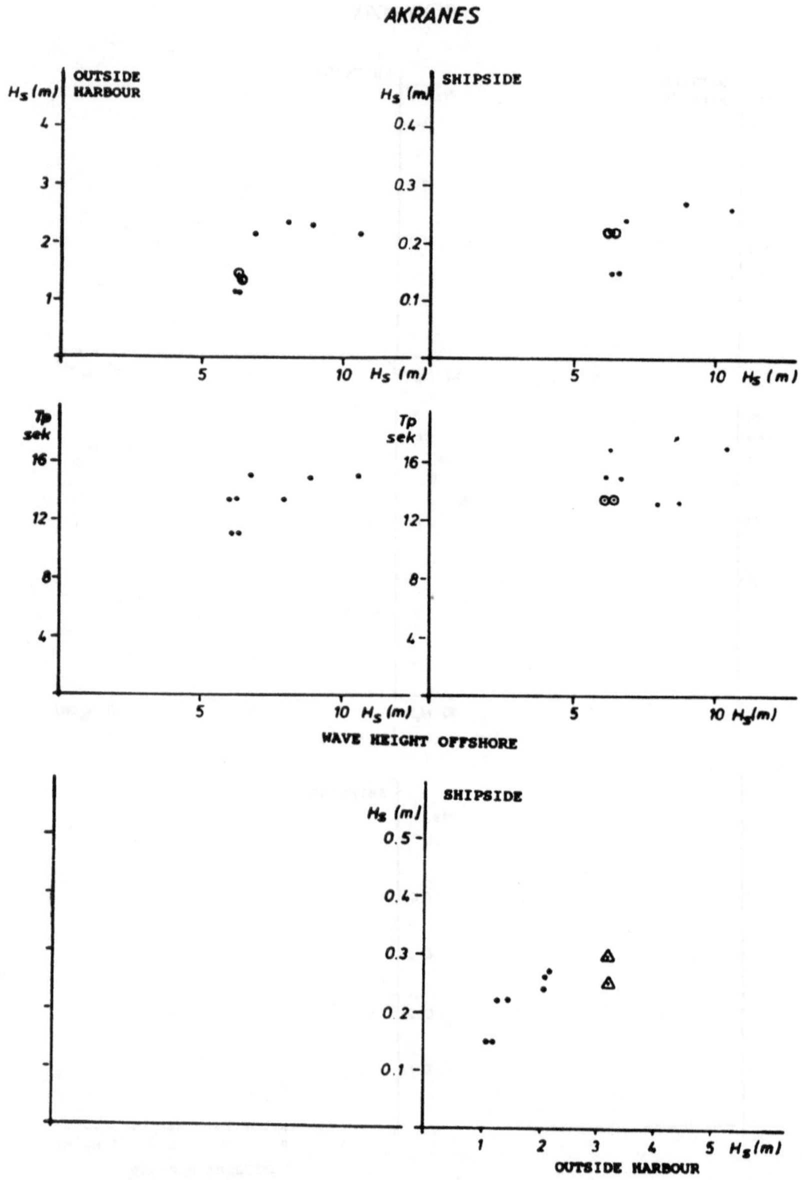

260

Fig. 7.13 Significant wave height and wave period outside
the harbour and at the shipside in relation to
wave height offshore.

THORLÁKSHÖFN

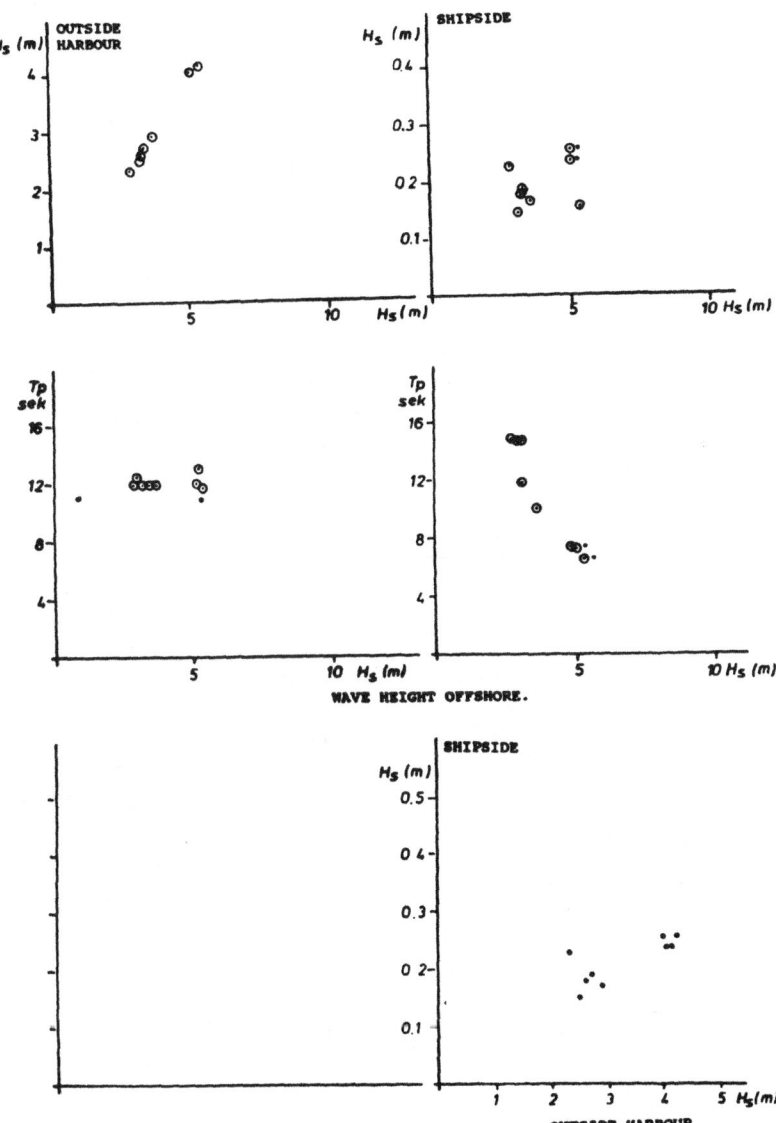

Fig. 7.14 Significant wave height and wave period outside
the harbour and at the shipside in relation to
wave height offshore.

262

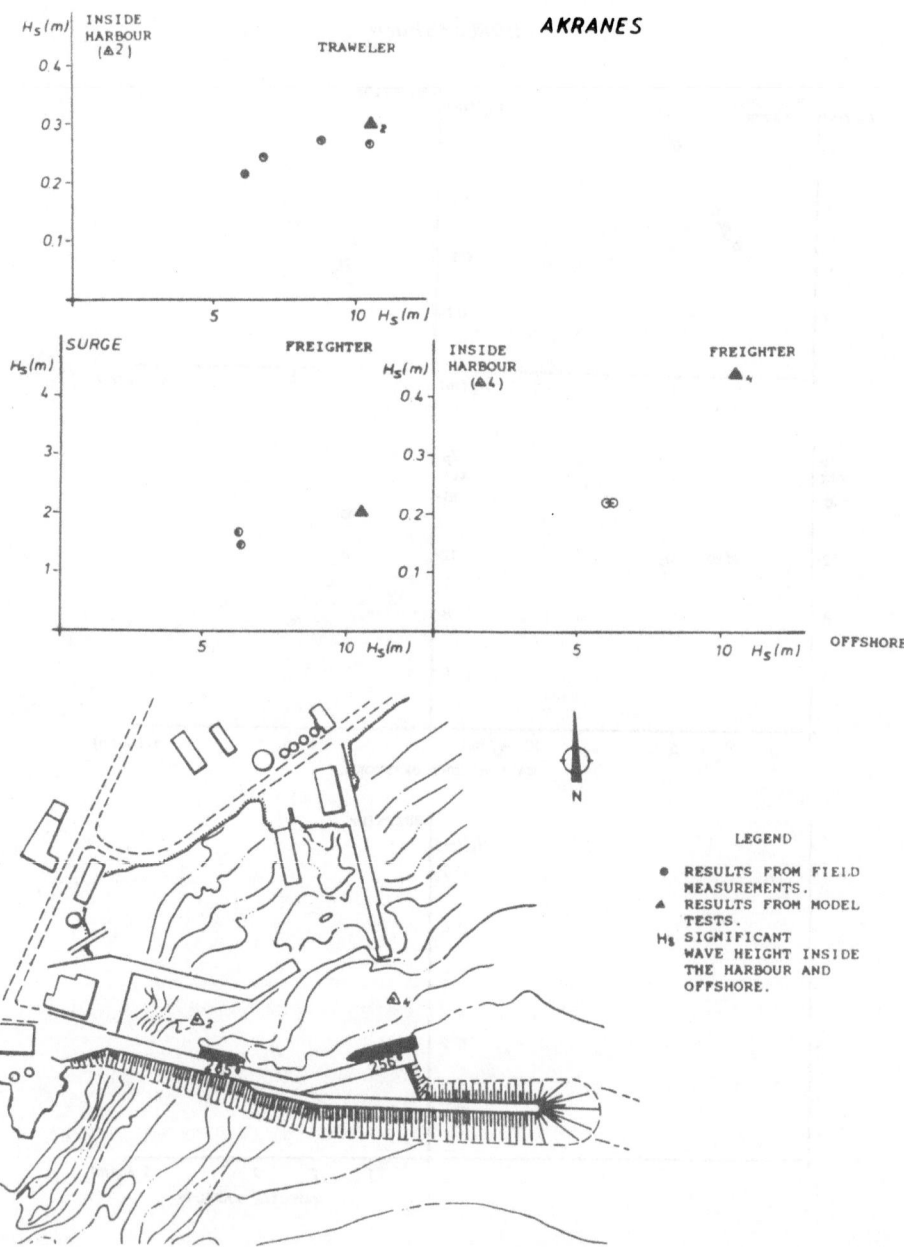

Fig. 8.1 Akranes harbour, comparison of results from
measurements in model and in nature in realation
to wave height offshore.

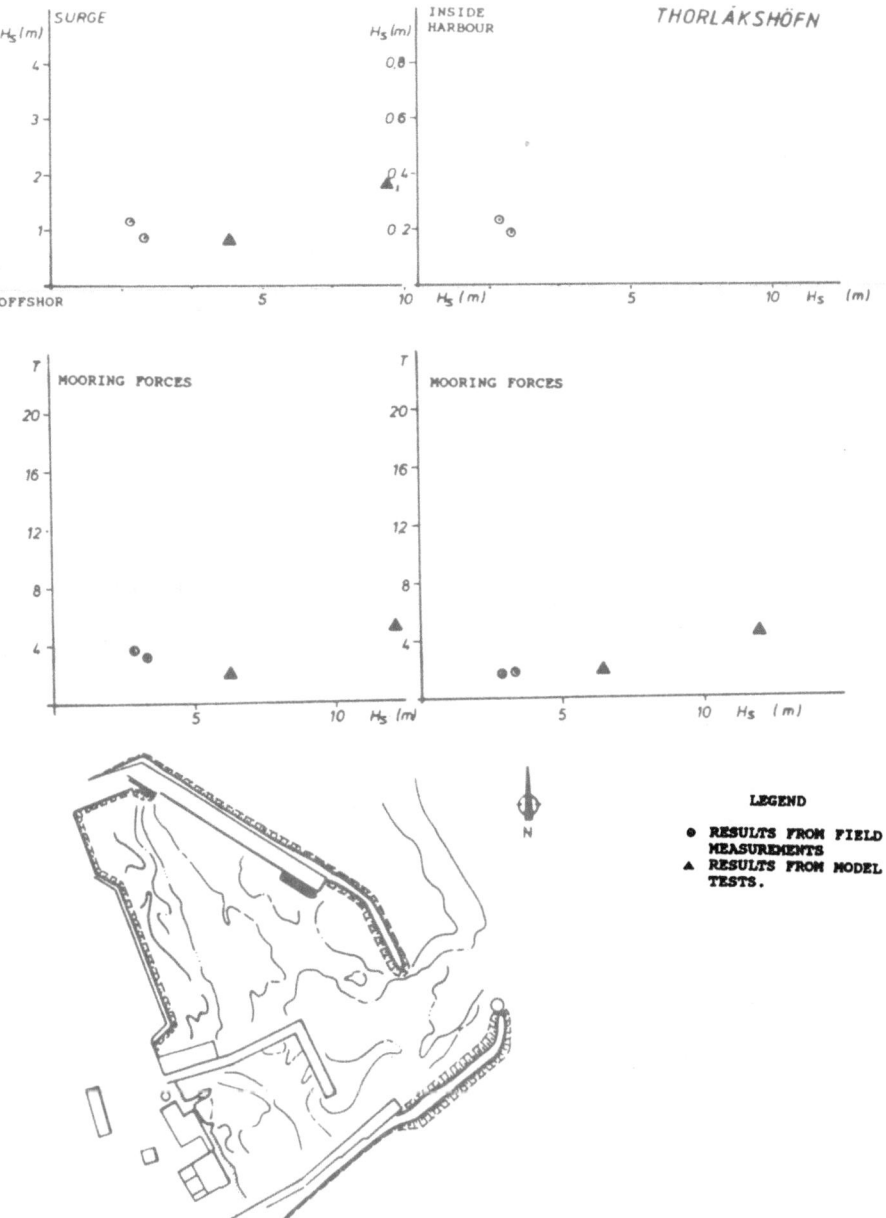

Fig. 8.2 Thorlakshöfn harbour, comparison of results from
measurements of ship movements, and mooring forces
in model and in nature in relation to wave height
offshore in nature.

264

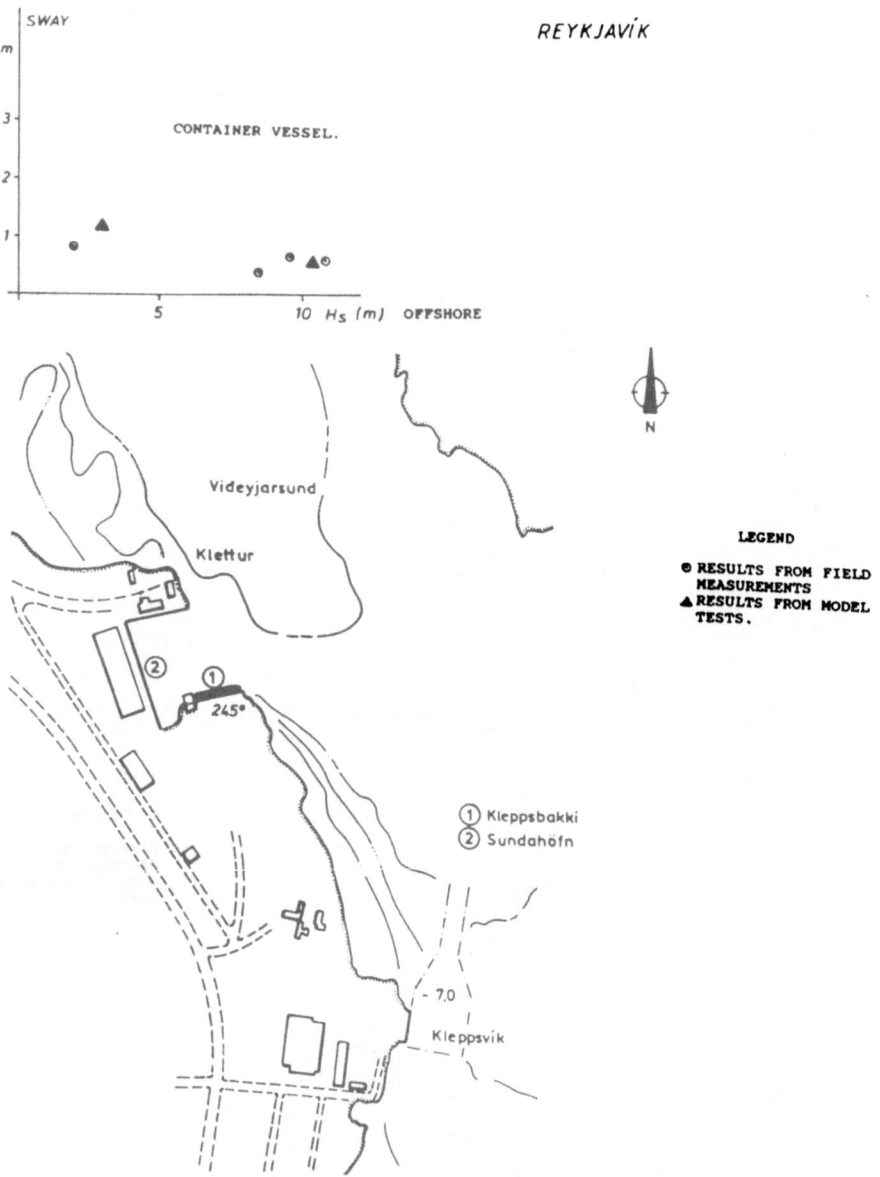

Fig. 8.3 Reykjavik harbour, comparison of results from
measurements of ship movements, in models and in
nature in relation to wave height offshore.

FORCES RELATED TO MOTIONS OF MOORED SHIPS/
ANALYTICAL METHODS OF MOORED SHIP MOTIONS

GERARD VAN OORTMERSSEN

MARITIME RESEARCH INSTITUTE NETHERLANDS

1. INTRODUCTION
The behaviour of moored ships in harbours, coastal areas or offshore is extremely complex, due to nonlinearities and dynamic effects. It is of vital importance to have methods available by which the behaviour of the moored ship can be predicted in the design stage, in order to determine the extreme loads in the worst conditions and to assess the operational limitations of the mooring system. In the past, the only way by which accurate estimates of the motion behaviour and mooring loads could be obtained was to perform model tests.
During the last decade, computer simulations became feasible due to the development of large powerful computers and because much progress has been made in developing mathematical models. Although model tests are still often used for a final design check, computer simulations are of growing importance in an early design stage to quickly assess the behaviour of a large number of design alternatives.
In addition, computer simulations can be a valuable tool for operators of mooring terminals. Actual operational conditions often differ from those assumed during the design stage. Modern hardware and software enable the prediction of the actual behaviour of a moored ship on short term. Thus the operator will be in a position to make better decisions, resulting in improved performance and reduced risk of calamities.

2. CHARACTERISTICS OF MOORED SHIP BEHAVIOUR
A moored ship may be regarded as a mass-spring system. The mass of the ship and the load-excursion characteristics of the ship are such, that the natural frequencies of the horizontal motions are very low, much lower than the frequency of normal waves. Typically the natural period of horizontal motions of a moored ship are in the order of 60 - 200 seconds. As a result, the motions of moored floating structures generally contain three components:
- mean excursions due to the mean wind, current and wave forces
- motions at wave frequencies
- low frequency motions around the natural frequency of the moored ship.
The resonant low frequency motions can have various causes:
- the non-linear elasticity of the mooring system may induce so-called subharmonic response, as will be discussed in Section 10;

265

E. Bratteland (ed.), Advances in Berthing and Mooring of Ships and Offshore Structures, 265–281.

- wind speed variations may in some cases cause low frequency
  motions;
- harbour basin resonances or seiches may occur in certain
  harbours and if the frequencies of such long standing waves
  are near the natural frequencies of the moored ship these
  may directly excite resonant ship motions;
- in many cases second order wave drift forces are an important source for low frequency motions. Wave drift forces are
  associated with wave grouping effects, and their frequency
  equals the difference of frequencies of individual wave components.

The magnitude of the low frequency excitation is generally
small, in particular when compared with the magnitude of the
"normal" wave forces, but since the fluid damping at the
natural frequency is usually very small, a small force may
still result in resonant motions of large amplitude.

The forces acting on the moored ship may be summarized as
follows:

- aerodynamic forces (wind forces)
- forces exerted by the moving water particles against the
  ship's hull;
  the water motions may be due to current, waves or caused by
  passing ships. In addition, the motions of the moored ship
  cause water motions which in turn influences the motions of
  the ship;
- forces exerted by the mooring system, i.e. lines, fenders,
  chains.

Another way of subdividing the loads is in exciting forces,
which cause the ship to move, and in reaction forces, which are
the result of the motion of the moored ship. Within the linear
potential theory the superposition principle may be applied
which states that the total hydrodynamic forces on a moving
ship in waves consist of the wave forces on the restrained ship
plus the hydrodynamic forces on the moving ship in still water.

The various forces will be discussed in the following
sections.

3. WAVE LOADS, WAVE DRIFT FORCES
3.1. First order wave loads

The wave loads on a moored ship can be calculated adequately
by means of boundary integral equation methods. Such methods
describe the 3-dimensional flow around large bodies by means of
the linear potential theory. The wetted surface of the ship is
subdivided into a number of panels (see Figure 1), and on these
panels a singularity distribution is applied. The singularity
distribution is described by means of a Green's function, which
is a solution of the Laplace equation which satifies:

- the linearized boundary condition in the free surface;
- the kinematic boundary condition at the sea bed;
- the radiation condition at infinity.

By satisfying the proper boundary condition on each panel,
i.e. that of zero normal velocity, the magnitude of the singularity distribution on the panels is determined for a given
wave condition. Subsequently, the wave loads are found by integrating the linear hydrodynamic pressure over the wetted
surface.

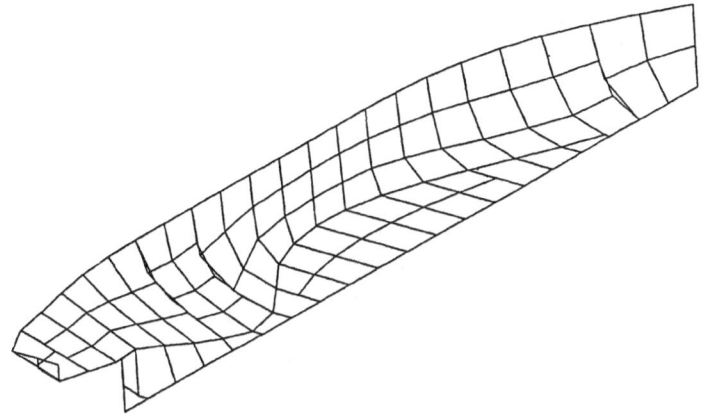

FIGURE 1. Facet distribution on a ship's hull.

The computations are carried out in the frequency domain, assuming simple harmonic waves. By repeating the calculations for
a number of frequencies, the information necessary for determining the wave loads in irregular seas can be obtained in the
form of wave load transfer functions.

Computer programs based on the boundary integral equation
method have been extensively validated and are found to yield
quite reliable results in general, Ref. [1] through [5].

3.2. Wave drift forces

The total wave load on a ship in irregular waves consists of
a part which oscillates with the wave frequency and a low frequency part, which oscillates with the frequency of the wave
groups. This so-called second order wave drift force has a
small magnitude but may induce resonant motions of moored
ships.

The magnitude of the wave drift force is proportional to the
square of the wave height. In regular waves, the wave drift
force is constant.

Wave drift forces can be estimated by integrating the second
order hydrodynamic pressures over the instantaneous wetted
surface of the ship (see Pinkster [6, 7]).

The total wave drift force consists of five components:
1. a component which depends on the relative wave elevation;
2. a component which depends on first order fluid velocities
   (the velocity squared term in Bernoulli's equation);
3. a component which is the product of the gradient of first
   order pressures and ship motions;
4. a componment which is a product of first order angular
   motions of the ship and inertia forces;

5. a contribution from second order potentials.

The first four components can be calculated from linear potential theory, and several computer programs based on the boundary integral equation method have been extended in order to yield wave drift force results [6, 7].

The fifth component is more difficult to obtain. This component is related to the second order wave which is locked to the wave groups, sometimes called set-down. This component is mainly important in higher waves in shallow water. Pinkster [7] has given an approximative method to estimate this force.

The low frequency drift force in irregular waves can be described by means of a quadratic transfer function, which takes the interaction between various frequency components into account. The nature of the quadratic transfer function can be illustrated as follows:

A regular wave group consisting of two wave components with frequency ω_1 and ω_2 and amplitude ζ_1 and ζ_2 is given by:

$$\zeta = \zeta_1 \sin(\omega_1 t + \varepsilon_1) + \zeta_2 \sin(\omega_2 t + \varepsilon_2) \tag{1}$$

The low frequency wave drift force in this group is as follows:

$$F^{(2)}(t) = P_{11}\zeta_1^2 + P_{22}\zeta_2^2 + (P_{12}+P_{21})\zeta_1\zeta_2 \cos\{(\omega_1-\omega_2)t+(\varepsilon_1-\varepsilon_2)\}+$$
$$+(Q_{12}+Q_{21})\zeta_1\zeta_2 \sin\{(\omega_1-\omega_2)t+(\varepsilon_1-\varepsilon_2)\} \tag{2}$$

P and Q are quadratic transfer functions:
$P_{12} = P_{21} = P(\omega_1, \omega_2)$ – in-phase component
$Q_{12} = -Q_{21} = Q(\omega_1, \omega_2)$ – quadratic component

The amplitude transfer is given by:

$$T(\omega_1,\omega_2) = \sqrt{P^2(\omega_1,\omega_2)+Q^2(\omega_1,\omega_2)}$$

The mean part of the drift force in an irregular sea with spectrum $S_\zeta(\omega)$ is:

$$\overline{F}^{(2)} = 2 \int_0^\infty S_\zeta(\omega)P(\omega,\omega) \, d\omega \tag{3}$$

and the slowly varying part is given by the spectrum:

$$S_F(\mu) = 8 \int_0^\infty S_\zeta(\omega+\mu) \, S_\zeta(\omega)T^2(\omega+\mu,\omega) \, d\omega \tag{4}$$

An example of a quadratic transfer function as obtained from theory and model tests is given in Figure 2.

The experimental results in this figure have been obtained from tests in irregular waves with a special set-up which allowed the ship model to move freely with the wave frequenies but which restrained the low frequency motions. From the measured force records the quadratic transfer function was obtained by applying cross-bi-spectral analysis techniques.

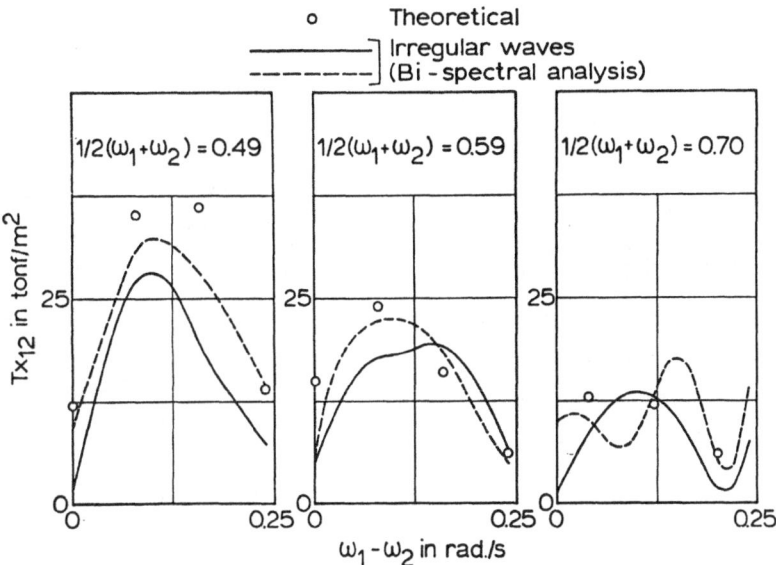

FIGURE 2. Quadratic transfer function of wave drift force (in longitudinal direction) as determined from potential theory and from model tests in irregular waves (two different wave spectra).

4. HYDRODYNAMIC FORCES, DAMPING
4.1. Hydrodynamic reaction forces

The motions of a moored ship induce water motions which in turn exert forces on the ship. These hydrodynamic reaction forces are usually described in terms of added mass' and damping, being the in-phase and quadrature components in case of harmonic oscillatory motions:

$$F = -\omega^2 a(\omega) x_a \sin \omega t + \omega b(\omega) x_a \cos \omega t \qquad (5)$$

where:

F = fluid reaction force
$a(\omega)$ = added mass coefficient
$b(\omega)$ = damping coefficient
x_a = amplitude of harmonic motion
ω = frequency of motion

The added mass and damping coefficients can be calculated by means of the boundary integral equation methods described in Section 3.1. The only difference with the wave load calculation is the formulation of the boundary condition on the ship's hull.

An example of the added mass and damping coefficients as obtained by means of linear potential theory and the correlation with experimental values is given in Figure 3.

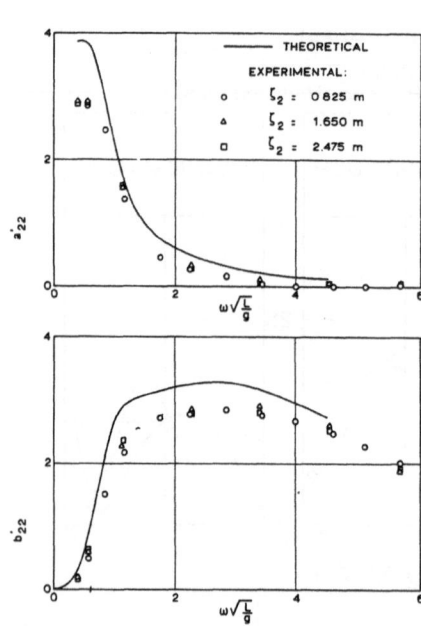

FIGURE 3. Added mass and
damping coefficient in
sway for a water depth
to draught ratio of 1.2.
Experimental values are
given for three values
of the motion amplitude.

FIGURE 4. Added mass and
damping coefficient in
sway for different water
depth to draught ratios δ.

The influence of the water depth on the added mass and damping is quite important as can be seen from Figure 4. From this
figure it also becomes apparent, that the frequency dependency
increases with decreasing water depth.

If a ship is moored against a quay, the presence of this quay
can greatly influence the hydrodynamic forces, as has been
shown in Ref. [8]. The effect of a (vertical) quay can be included in numerical calculations without much difficulties. An
example of the results is given in Figure 5.

When two or more ships are moored in close proximity, their
motions will introduce interaction forces on the other ships.
Boundary integral equation methods can be extended as to cover
these interaction effects as well [9, 10].

4.2. Damping of low frequency motions

The resonant type low frequency motions of moored ships are
to a large extent determined by the magnitude of the damping.
Since the linear potential theory damping discussed in Section
4.1 is quite small at the low frequencies around the natural
frequency of the moored ship, damping from other sources becomes very important. In this respect the damping due to viscous pressure drag and the so-called wave drift damping should
be mentioned.

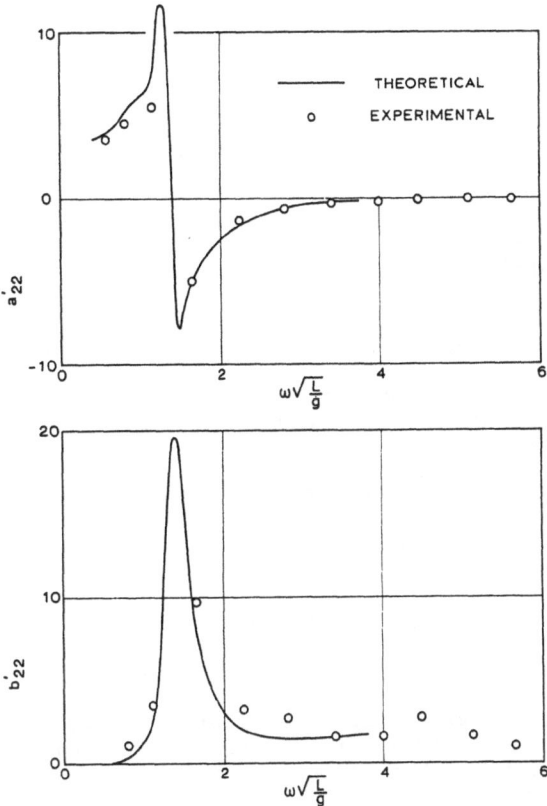

FIGURE 5.   Added mass and damping coefficient in sway: distance
between ship and quay 16.50 m, water depth to
draught ratio 1.2.

Damping due to viscous pressure drag or eddy making is of
particular importance for the sway and yaw motions and may be
estimated from the drag coefficient of the ship [11].

When a ship is moored in waves, wave drift damping occurs.
This damping is caused by changes in the drift forces due to
the slow motions of the vessel. The magnitude of this damping
force is proportional to the square of the wave height, and
therefore this type of damping is mainly of importance when
ships are moored at offshore locations in high seas. See Ref.
[12, 13, 14].

5. WIND AND CURRENT LOADS
Wind and current forces are of great importance for the
dimensions of mooring systems, since they form the larger part
of the steady forces that have to be restrained by the
moorings. Wind and current forces are mainly caused by viscous

pressure drag. Their magnitude may be estimated by the following empirical formula:

$$F_k = 1/2 \; \rho V^2 A_k C_{d_k} \; (\alpha) \tag{6}$$

in which:

F_k = force/moment in the k-mode (k=1,2 or 6)
ρ = fluid density
V = current/wind speed
A_k = projected area for force in the k-mode, as a function of current/wind direction
C_{d_k} = drag coefficient in k-mode

In the case of wind, C_d is usually obtained from wind tunnel tests. Data on wind loads on ships can be found in Ref. [15] through [22].

Current drag force coefficients are best obtained from model tests in large basins which allow relatively unrestricted flow in the case of oblique current. The influence of restricted water depth is extremely important with respect to current loads. Data on current drag force coefficients can be found in [17], [20] and [22]. In [20] wind and current force data are presented in a convenient form using Fourier series.

Current and wind are often considered as steady forces. It should be borne in mind, however, that depending on the location and the conditions considered, significant dynamic components can be present in both the direction and magnitude of wind and current. This results in dynamic components in the wind and current forces.

Wind and current may also contribute to the damping forces. This effect can be taken into account in simulation calculations by inserting the relative velocity between fluid and vessel in equation (6).

6. EFFECT OF COMBINED LOADS, WAVE DIRECTIONALITY

When designing mooring systems it is essential to take proper account of the combined actions of wind, current and waves. It is usually not possible to say beforehand which combinations of environmental conditions will yield the largest motions or the highest mooring loads, and therefore various situations have to be considered.

In addition, the possible effect of wave directionality has to be taken into consideration. Directional spreading of waves or bi-modal waves consisting of a combination of swell from one direction and wind waves from another may have a considerable effect on the behaviour of moored ships, especially in case of single point mooring systems.

Pinkster [23] investigated the effect of wave directionality on wave drift forces. Martinussen and Nielsen [24] have given a qualitative comparison of the response of SPM - turret- and catenary moored vessels in long crested and short crested seas.

7. LOADS DUE TO PASSING SHIPS

A ship which passes another ship which is moored at a berth, induces loads on the moored ship. The ensueing transient motions of the moored ship may lead to high peak loads in the mooring lines. Many instances are known where lines actually broke as a result of a ship passage. Loads due to passing ships are best determined by means of model tests. In the literature a very limited amount of data can be found. In [25] Remery presents the results of extensive measurements on model scale of the loads on a tanker induced by the passage of another tanker. Fig. 1 gives an example of his results. The loads induced by a passing ship on a moored vessel were found to be proportional to the square of the speed of the passing vessel.

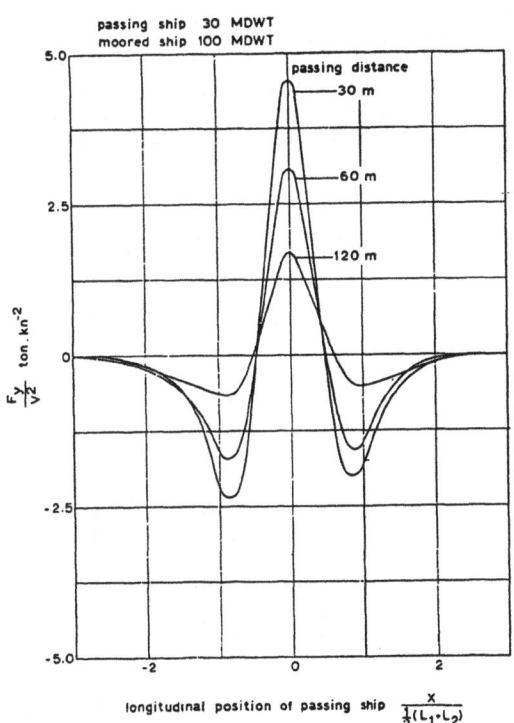

FIGURE 6. Lateral force on a captive 100 MDWT tanker due to the passage of a 30 MDWT tanker.

8. STATIC AND DYNAMIC MOORING LOADS
   It is usually assumed that dynamic behaviour of individual components of the mooring system will not influence the behaviour of the moored structure. Consequently, the mooring forces may be found from the instantaneous position of the chock locations and from the static load-excursion characteristics of the mooring system. Several studies, however, have indicated that dynamic effects may greatly influence the maximum tension in individual mooring lines or anchor chains (see for instance [26]). This is particularly the case for catenery

274

moorings in deeper water.

Analytical as well as numerical methods are in use to predict the dynamic behaviour. Under the assumption of linear dynamic behaviour, asymptotic solution techniques, perturbation methods and approximate formulae may be employed to predict the overall dynamic behaviour of mooring lines in the early design stages [27 - 30]. For more rigorous numerical analysis finite element and finite difference methods can be used, the latter method being the most efficient for mooring line analysis [31 - 34].

Finite difference methods are based on the so-called lumped mass method. Van den Boom [32] has clearly demonstrated the adequacy of this method, by comparing the results of calculations with model test results. Fig. 2 shows an example of his results for harmonic oscillations. He also showed that the dynamic behaviour of a mooring line under random excitation may increase both the effective mooring stiffness and the low frequency damping and thus affect the low frequency motions of the moored ship.

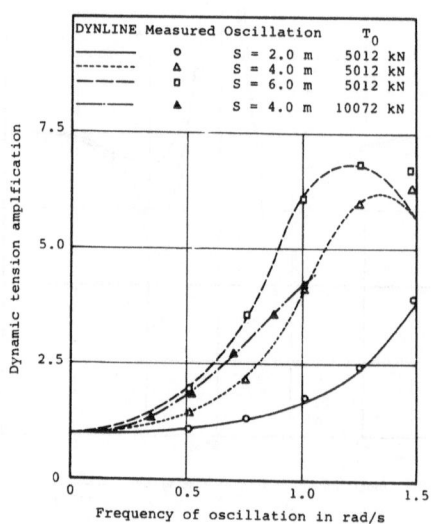

FIGURE 7. Results of harmonic oscillation tests for 152 mm chain at 150 m water depth.

9. GENERAL REVIEW OF ANALYTICAL METHODS OF MOORED SHIP MOTIONS

The motion response of moored ships can be studied by solving the equations of motion. The equations of motion are based upon Newton's law of dynamics, but in the formulation and solution technique various possibilities exist, each with its own approximations and limitations. Basically, there are two approaches possible:

1. solve the equations of motion in the frequency domain;
2. solution in the time domain.

The first, frequency domain analysis, is quite attractive since it provides statistical data for motions and mooring loads under random wave excitation in an efficient way. The main drawback is that the hydrodynamic reaction forces and elasticity characteristics of the moored system must be linear. Another disadvantage is that transient motions such as induced by passing ships can not be coped with.

Since the elasticity characteristics of most mooring systems are strongly non-linear, one has to take recourse to the time domain simulation technique. This technique will be further discussed in Section 10.

Simple analytical approaches can be very helpful to make rough estimates in the early design stages or in helping to explain the typical behaviour of moored ships. Assuming for instance constant added mass and zero damping, a moored ship can be described as a non-linear mass-spring system:

$$x + c_1 x + c_2 x^3 = \dot{F}_a \cos \omega t \tag{7}$$

By solving this so-called Duffing equation, it can be shown that the response of non-linear systems can contain components with frequencies lower than the frequency of excitation, so-called subharmonics [8]. Such an analysis underlines the importance of time domain simulation techniques which allow incorporation of the relevant non-linearities.

10. TIME DOMAIN ANALYSIS OF MOORED SHIPS

The motions of moored objects under the influence of arbitrary external loading can be determined from the solution of the following equations of motion in the time domain [35]:

$$\sum_{j=1}^{6} \left\{ (M_{kj} + m_{kj})x_j + \int_{-\infty}^{t} R_{kj}(t - \tau) \, \dot{x}_j(\tau) \, d\tau + C_{kj}x_j \right\} = F_k(t) \tag{8}$$

in which:
x_j = motion in j-direction
M_{kj} = inertia matrix
C_{kj} = matrix of hydrostatic restoring forces
R_{kj} = matrix of retardation functions
m_{kj} = added inertia matrix
$F_k(t)$ = arbitrary time varying external forces in the k-mode

The left-hand side of equation (8) contains the inertia forces and linear hydrodynamic reaction forces only. The right-hand side contains the external forces due to waves, wind and current, mooring loads and, if needed, non-linear drag terms and wave drift damping.

The fluid reaction forces are primarily obtained from potential theory as frequency dependent added mass and damping coefficients. Based on these coefficients the added inertia matrix m_{kj} and matrix of retardation functions R_{kj} of equation (8) are obtained from:

$$R_{kj}(t) = \frac{2}{\pi} \int_0^\infty b_{kj}(\omega) \cos \omega t \, d\omega \tag{9}$$

$$m_{kj} = a_{kj}(\omega*) + \frac{1}{\omega*} \int_0^\infty R_{kj}(t) \sin \omega*t \, dt \tag{10}$$

in which:

$a_{kj}(\omega), b_{kj}(\omega)$ = frequency dependent added mass and damping coefficient matrices

$\omega*$ = a constant frequency.

The advantage of describing the hydrodynamic reaction forces due to added mass and damping effects using retardation functions lies in the fact that arbitrary or transient motions can be accommodated correctly irrespective of the nature of the motions.

In order to be able to compute the first and second order wave loads on a vessel in the time domain for arbitrary wave conditions, use is made of the Volterra series formulation [36]. According to this formulation the total wave load including first and second order contributions follows from:

$$F_k(t) = \int_{-\infty}^{+\infty} h_k^{(1)}(\tau) \zeta(t - \tau) \, d\tau +$$
$$+ \int_{-\infty}^{+\infty} \int_{-\infty}^{+\infty} h_k^{(2)}(\tau_1 \tau_2) \zeta(t - \tau_1) \zeta(t - \tau_2) \, d\tau_1 d\tau_2 \tag{11}$$

in which:

$\zeta(t)$ = wave elevation time record

$h_k^{(1)}(\tau), h_k^{(2)}(\tau_1 \tau_2)$ = first and second order impulse response functions relating the elevation to the force in the k-mode.

The impulse response functions are found from:

$$h_k^{(1)}(\tau) = \frac{1}{2\pi} \int_{-\infty}^{+\infty} H_k^{(1)}(\omega) \, e^{i\omega\tau} d\omega \tag{12}$$

$$h_k^{(2)}(\tau_1 \tau_2) = \frac{1}{(2\pi)^2} \int_{-\infty}^{+\infty} \int_{-\infty}^{+\infty} H_k^{(2)}(\omega_1 \omega_2) \, e^{i(\omega_1 \tau_1 - \omega_2 \tau_2)} d\omega_1 d\omega_2 \tag{13}$$

in which $H_k^{(1)}(\omega)$ and $H_k^{(2)}(\omega_1 \omega_2)$ are the first order wave force transfer function and the second order drift force transfer function respectively.

The advantage of the Volterra series approach is that simulation computations can be carried out based on measured wave records.

The equations of motion in the time domain can be solved on a computer using numerical integration techniques. A flow diagram of a typical time domain computer simulation procedure is given in Fig. 8.

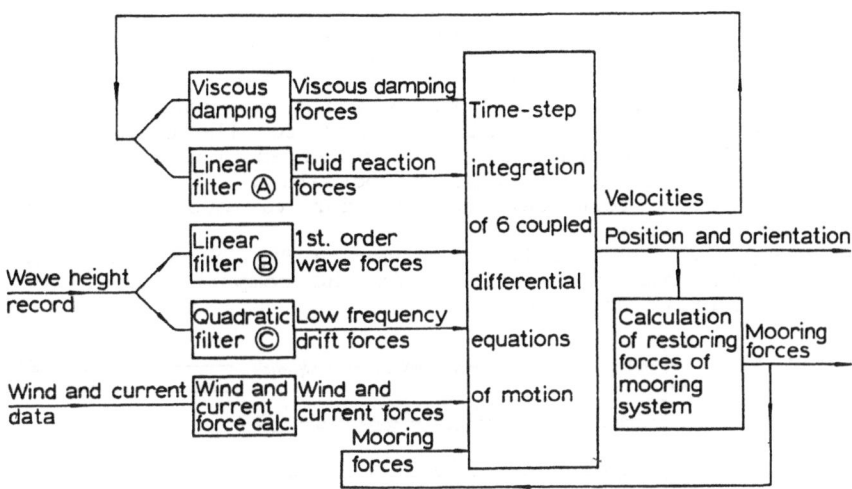

FIGURE 8. Block diagram of time domain simulation.

When carrying out time domain simulations one has to keep in mind that the results such as maximum motions and line loads depend on the particular realisation of the wave spectrum used as input and on the duration of the simulation. Pinkster and Wichers [37] investigated the statistical variance of results of simulations as a function of the simulation duration.

11. ANALYTICAL METHODS FOR SHIPS MOORED TO SINGLE POINT MOORINGS

The behaviour of a vessel moored to an SPM system in waves, wind and current is determined to a large extent by slow motions with large amplitude of the tanker in the horizontal plane, which are due to second order wave drift forces or instabilities. These large amplitude motions have a number of inconvenient consequencies for the mathematical model:

- the motions have to be described in a body-fixed system of coordinates, which leads to extra terms in the equations of motion in similarity to the equations of motion used for manoeuvring studies;
- the motions have a significant influence on the environmental loading; in particular, large heading changes influence the wave, wind and current loads, but also the effect of large displacements on the wave loads has to be taken into account;
- due to the large amplitude of the motions viscous damping of low frequency horizontal motions becomes important.

A more detailed description of the way in which these effects can be accounted for in the mathematical model is given in Ref. [38].

12. CORRELATION WITH MODEL TESTS/FULL SCALE

An extensive validation study was carried out for a large tanker moored to an offshore jetty [39]. Model tests were carried out in the Wave and Current Basin of MARIN, which measures 60 m x 40 m x 1 m. By using the random wave signal measured in the model basin as input for the computer simulation, a direct comparison of measured and computed records of motions and mooring loads became possible.

Correlations were carried out for various wave, wind and current conditions.
An example of the results is given in Fig. 9. The results generally confirmed the validity of the computer model for this kind of mooring situation.

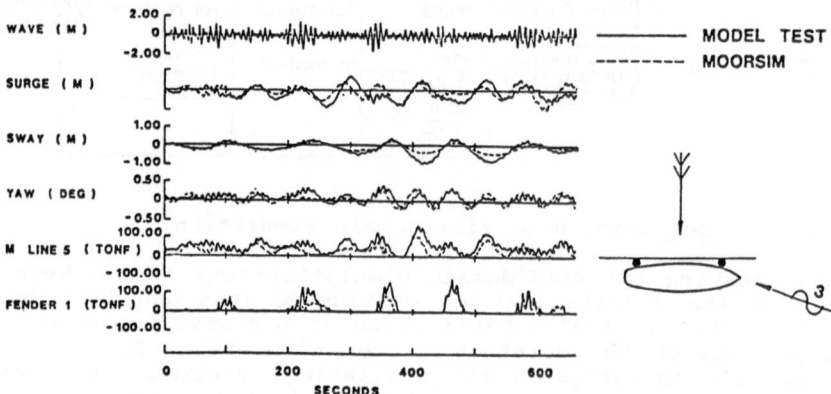

FIGURE 9. Comparison for bow quartering waves and beam wind.

Model tests and time domain simulations were carried out by Wichers and Van den Boom [38] for a tanker moored to a Single Point Mooring system. The wave, wind and current conditions were such, that they yielded stable motions with limited amplitudes. For these conditions, satisfactory results were obtained as can be seen from the sample in Fig. 10

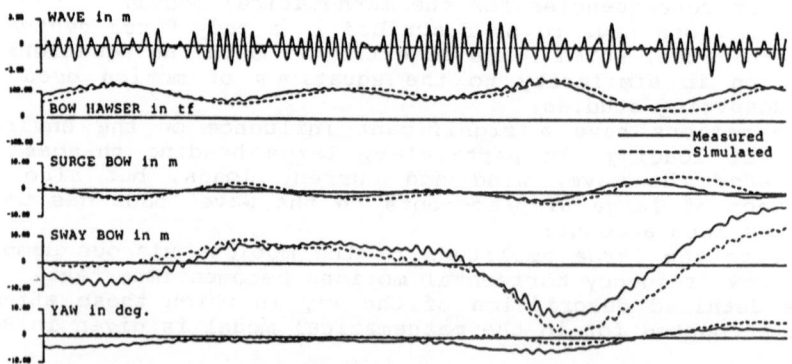

FIGURE 10. Records of motion and bow hawser force.

Data on full scale behaviour of moored ships is very scarce. Some recent experiences with Single Point Mooring systems have been reported in [40] through [42].

REFERENCES
1. Garrison CJ and Chow PY: Wave Forces on Submerged Bodies, ASCE, Waterways and Harbour Div. 98, 1972.
2. Oortmerssen G van: The Motions of a Ship in Shallow Water, Ocean Engineering, Pergamon Press, Vol. 3, 1976.
3. Faltinsen OM and Michelsen FC: Motions of Large Structures in Waves at Zero Froude Number, Proc. Symp. on the Dynamics of Marine Vehicles and Structures in Waves, Univ. College London, 1974.
4. Hogben N and Standing RG: Wave Loads on Large Bodies, Proc. Symp. on the Dynamics of Marine Vehicles and Structures in Waves, Univ. College Londen, 1974.
5. Løken AE and Olsen OA: Diffraction Theory and Statistical Methods to Predict Wave Induced Motions and Loads for Large Structures, OTC Paper 2502, Houston, 1976.
6. Pinkster JA: Mean and Low Frequency Wave Drifting Forces on Floating Structures, Ocean Engineering, Pergamon Press, October 1979.
7. Pinkster JA: Low Frequency Second Order Wave Exciting Forces on Floating Structures, MARIN Publication No. 650, 1980.
8. Oortmerssen G van: The Motions of a Moored Ship in Waves, NSMB Publication No. 510, 1976.
9. Oortmerssen G van: Hydrodynamic Interaction Between Two Structures, Floating in Waves, BOSS Conference, London, 1979.
10. Oortmerssen G van: Some Hydrodynamic Aspects of Multi-body Systems, Proc. International Symp. on Hydrodynamics in Ocean Engineering, Trondheim, 1981.
11. Saito K et al.: On the Low Frequency Damping Forces acting on a Moored Body in Waves, Journal of the Kansai Soc. of Naval Architects, No. 195, Japan 1984.
12. Wichers JEW, and Van Sluijs MF: The Influence of Waves on the Low Frequency Hydrodynamic Coefficients of Moored Vessels, OTC Paper 3625, Houston, 1979.
13. Wichers JEW: On the Low Frequency Surge Motions of Vessels Moored in high Seas, OTC Paper 4437, Houston, 1982.
14. Wichers JEW and Huijsmans RHM: On the Low Frequency Damping Forces Acting on Offshore Moored Vessels, OTC, Houston, 1984.
15. Gould RWF: The Estimation of Wind Loads on Ship Super-structures, Maritime Technology Monograph No. 8, RINA, London.
16. Norrbin NH: Fairway Design with Respect to Ship Dynamics and Operational Requirements, SSPA Research Report No. 102, 1986.
17. Palo PA: Steady Wind and Current Induced Loads on Moored Vessels, OTC Paper No. 4530, Houston, 1983.
18. Aage C: Wind Coefficients for Nine Model Ships, Hydro-and Aerodynamics Lab. Denmark, Report No. A3, 1971.
19. Isherwood MA: Wind Resistance of Merchant Ships, RINA, London, 1972.

20. OCIMF: Prediction of Wind and Current Loads on VLCC's, London, 1977.
21. Berlekom WB van: Wind Forces on Modern Ship Forms-effects on Performance, RINA, London, 1981.
22. Remery GFM and Oortmerssen G van: The Mean Wave, Wind and Current Forces on Offshore Structures and their Role in the Design of Mooring Systems, OTC Paper No. 1741, Houston, 1973.
23. Pinkster JA: Numerical Modelling of Directional Seas, Proc. Symp. on Description and Modelling of Directional Seas, Techn. Univ. Denmark, 1984.
24. Martinussen J and Nielsen JK: DP-assisted Turret Moored Production and Storage Facility in Short Crested Waves, Proc. 6th Offshore South East Asia Conf., 1986.
25. Remery GFM: Mooring Forces Induced by Passing Ships, OTC Paper No. 2066, Houston, 1974.
26. Sluijs MF van and Blok JJ: The Dynamic Behaviour of Mooring Lines, OTC Paper No. 2881, Houston, 1977.
27. Triantafyllou MS: Preliminary Design of Mooring Systems, Journal of Ship Research, Vol. 26, 1982.
28. Triantafyllou MS, Bliek A and Shin H: Dynamic Analysis as a Tool for Open-sea Mooring Design, SNAME Trans., Vol. 93, 1985.
29. Ractliffe AT: The Validity of Quasi-static and Approximate Formulae in the Context of Cable and Flexible Riser Dynamics, BOSS Conference, Delft, 1985.
30. Polderdijk SH: Response of Anchor Lines to Excitation at the Top, BOSS Conference, Delft, 1985.
31. Nakajima T, Motora S and Fujino M: On the Dynamic Analysis of Multi-component Mooring Lines, OTC Paper No. 4309, Houston, 1982.
32. Boom HJJ van den: Dynamic Behaviour of Mooring Lines, BOSS Conference, Delft, 1985.
33. Felippa CA: Finite Element Analysis of Three-dimensional Cable Structures, Computational Methods in Non-linear Mechanics, TICOM, Univ. of Texas, 1974.
34. Boom HJJ van den, Dekker JN and Elsäcker AW: Dynamic Aspects of Offshore Riser and Mooring Concepts, OTC Paper No. 5531, Houston, 1987.
35. Cummins WE: The Impulse Response Function and Ship Motions, DTMB Report No. 1661, Washington D.C. 1962.
36. Dalzell JF: Application of the Fundamental Polynomial Model to the Ship added Resistance Problem, 11th Symp. on Naval Hydrodynamics, UCL, London, 1976.
37. Pinkster JA and Wichers JEW: The Statistical Properties of Low Frequency Motions of Non-linearly Moored Tankers, OTC Paper No. 5457, Houston, 1987.
38. Wichers JEW and Boom HJJ van den: Simulation of the Behaviour of SPM-moored Vessels in Irregular Waves, Wind and Current, 2nd Int. Conf. Deep Offshore Technology, Malta, 1983.
39. Oortmerssen G van, Pinkster JA and Boom HJJ van den: Mathematical Simulation of the Behaviour of Ships Moored to Offshore Jetties, 19th ICCE, Houston, 1984.

40. Remery GFM and Stambouzos MA: Tanker Based Marginal Field Production: 8 Years Operational Experience, OTC Paper No. 4036, Houston, 1985.
41. Eppley DR and Van Berkel PB: 12 Month's Operational Experience With an FPSO, Handling the Production from 3 Fields Offshore Nigeria, OTC Paper No. 5491, Houston, 1987.
42. Grancini G, Jovenitti LM and Pastore P: Moored Tanker Behaviour in Crossed Sea; Field Experiences and Model Tests, Symp. on Description and Modelling of Directional Seas, Techn. Univ. of Denmark, 1984.

OPTIMIZATION AND SAFETY CONSIDERATIONS
IN THE DESIGN OF STATIONKEEPING SYSTEMS

I. J. Fylling,   MARINTEK A/S   NORWAY
G.O. Ottera,
F. Godtliebsen

ABSTRACT

Spread mooring anchor systems have been the dominating stationkeeping
alternative for floating drilling vessels and other large work vessels. The
use of thruster systems and dynamic positioning is, however, increasing,
both due to activity in greater water depths and limitations of anchor
system's capability, and due to increased reliability of automatic control
systems. The most important aspects of a comparative evaluation and selection of stationkeeping system are discussed.

Improved tools for response analysis and in particular long-term operation
simulation makes it possible to carry out optimization studies related to
operational aspects as well as direct capacity-load relations.

Existing rules and guidelines for design of offshore structures are not
necessarily calibrated to fit an optimized design procedure. It is emphasized that new design practices must be used with care in order to
increase, or at least not to reduce the total safety level.

1. SELECTION OF STATIONKEEPING SYSTEMS

The problem of selecting stationkeeping system is visualized in figure 1.
In the present paper we will discuss aspects related to the four boxes A-D.

In the present section we concentrate on operational aspects related to
selection of stationkeeping systems. The subsequent sections will deal in
some more detail with characteristic properties of environmental forces and
optimization potential.

Four different aspects of operational requirements that are of importance
in station keeping system selection are indicated in figure 1. Mobility,
regularity, position tolerances and heading requirements.

This paper was also presented at PRADS'87 - The Third International Symposium on Practical Design of Ships and Mobile Units, The Norwegian Institute of Technology, Trondheim, Norway, June 22-26, 1987.

E. Bratteland (ed.), Advances in Berthing and Mooring of Ships and Offshore Structures, 282–297.
© 1988 by Kluwer Academic Publishers.

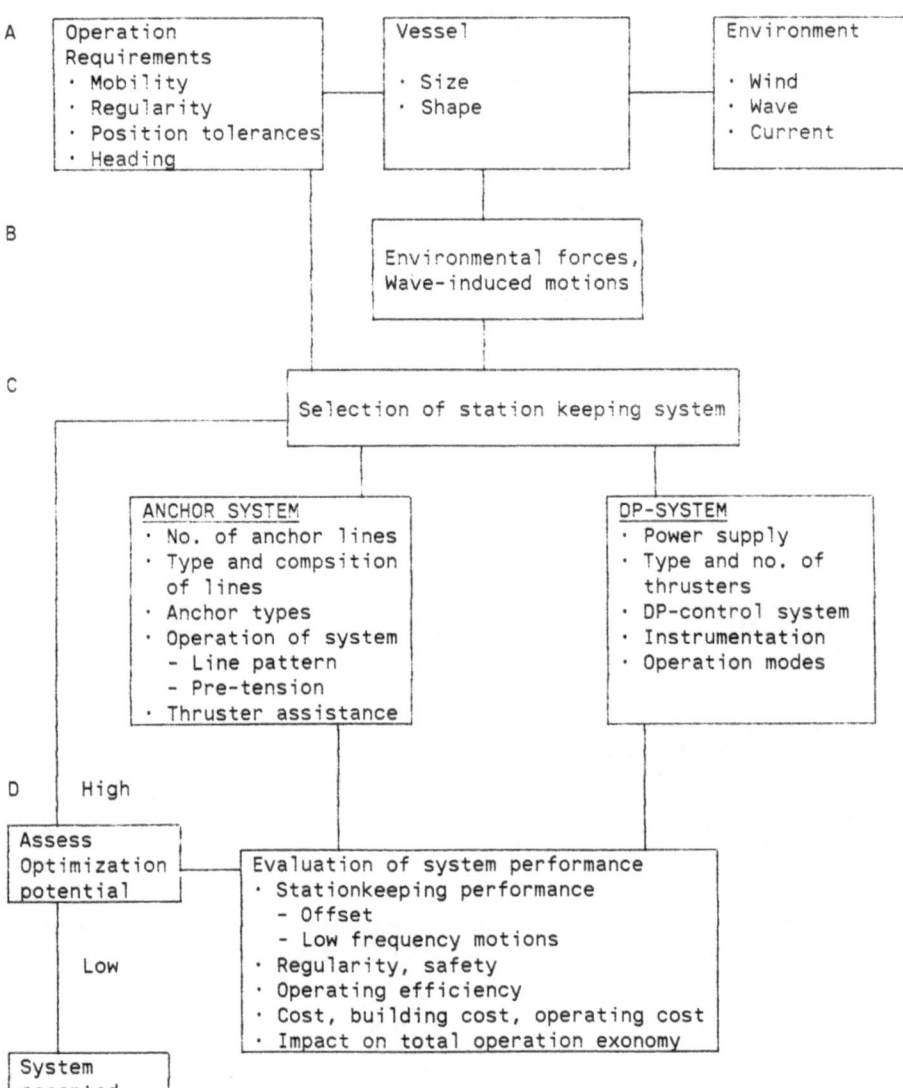

A   Operation Requirements
     · Mobility
     · Regularity
     · Position tolerances
     · Heading

Vessel
· Size
· Shape

Environment
· Wind
· Wave
· Current

B   Environmental forces, Wave-induced motions

C   Selection of station keeping system

ANCHOR SYSTEM
· No. of anchor lines
· Type and compsition of lines
· Anchor types
· Operation of system
  - Line pattern
  - Pre-tension
· Thruster assistance

DP-SYSTEM
· Power supply
· Type and no. of thrusters
· DP-control system
· Instrumentation
· Operation modes

D   High

Assess Optimization potential

Low

Evaluation of system performance
· Stationkeeping performance
  - Offset
  - Low frequency motions
· Regularity, safety
· Operating efficiency
· Cost, building cost, operating cost
· Impact on total operation exonomy

System accepted

Figure 1 Key elements in the design of a station keeping system.

Table 1 shows a qualitative evaluation of four broad categories of station
keeping systems related to these operational considerations.

Table 1   Basic requirements and qualitative properties of four categories
of station keeping systems.

Basic station keeping requirements	Spread mooring Anchor system	Single point mooring system	Turret anchor system	Dynamic Pos.
Mobility				
Fixed	0	0	0	0
Mobile	0	+/- 1)	0	+
Regularity				
Permanent	0	0	0	0
Disconnect allowed	0	+	0	+
Position tolerance				
Small	+	-	+	+
Wide	0	0	0	0
Heading				
Fixed	+	-	0	0
Limited	0	-	0	0
Free	0	+ 2)	+ 2)	+ 2)
Space restrictions in the sea or at the seafloor	-	0	-	+

```
0   - Neutral
+   - Advantageous system
-   - Disadvantageous system
1)  - The vessel itself is mobile but the station keeping system is sta-
      tionary
2)  - The advantage relates mainly to a ship-shaped type of vessel.
```

Mobility

Most types of offshore work vessels are designed to be mobile. Examples of
different requirements to mobility expressed in terms of time on location:

```
1)  Diving support vessels.        2-8 days    on location.
2)  Tankers loading offshore       3-5 days      "      "
3)  Crane barges                   1-4 months    "      "
4)  Drilling vessels               2-3 months    "      "
5)  Flotels, accommodation         6-12 months   "      "
6)  Production vessels             5-20 years    "      "
```

Except for the floating production vessel designed to stay for several
years at the same location, all other vessels must be characterized as
mobile. The time and cost for mobilization are more important for the 3
first categories than for the drilling vessels and accommodation vessels.

It is obvious that a DP-system is favourable with respect to mobility.
Single point mooring systems are normally fixed, but the vessel itself is
easily mobilized.

Regularity

The regularity requirement may be important with respect to capacity,
redundancy and reliability of the stationkeeping system.

In the case of an un-interruptible operation, the system must have capacity
for a survival condition, and the safety must correspond to the safety
required with respect to structural strength and stability. This may be the
case for a large capacity floating production system.

For interruptible operations one will normally want to hava a capacity
and/or redundancy of the station keeping system that corresponds to other
critical systems on the vessel, so that the station keeping system does not
contribute to weather related downtime.

It is very important to decide whether the vessel has to stay on location
in a 100 year storm or not. The 100 year mooring force may be 2-3 times as
large as the force that is exceeded 1 per cent of the time, and the last
per cent of regularity/availability may become very expensive.

The advantage of DP-systems and single point mooring system in this respect
is primarily related to the easy disconnect and mobilization.

Position tolerances and heading requirement

The single point mooring system can not compete with the other categories
in operations where small position tolerances are required. This type of
system will allow large motions and large heading variations. Wherever a
fixed heading is required, the spread mooring system is advantageous. If
the heading can be selected to minimize mooring forces, the other 3 alternatives are favourable, provided that the vessel has an elongated hull
shape so that there is a significant difference between forces in the
various directions.

Other considerations

In addition to the above, operational requirements, the chosen system must
be designed to function for a specified vessel in a specified environment.
This implies selection of holding force capacity, water depth range etc.
The subsequent sections will mainly deal with these aspects.

2. ENVIRONMENTAL FORCES

The main purpose of any stationkeeping system is to provide forces to coun-
teract the external environmental forces. These forces depend upon:
   .   Size and shape of the vessel.
   .   Wind- and current velocity and seastate.
   .   Directions of wind, waves and current.

Two particularly important aspects with respect to selection of sta-
tionkeeping system are:
   .   The degree of symmetry of the vessel.
   .   Whether the operation is indifferent to vessel heading or not.

A nearly rotation symmetric vessel will have no heading preference and is ideal for operations that require a stationary vessel in a directionally, homogeneous weather.

Semisubmersible platforms without pontoons (e.g. Pentagone-type) or with ring shaped pontoons are close to this ideal. A twin-pontoon semisubmersible may have of the order 25 per cent less forces in the longitudinal direction than transverse. Figure 2a shows external forces based on worst combination of wind, wave and current directions for a twin pontoon semisubmersible. Figures 2b and 2c shows corresponding results for two ship-shaped hulls, a tanker hull and a diving-support vessel.

If the vessels are required to maintain a fixed heading in critical operations, the holding force capacity of the stationkeeping system should be distributed around the vessel according to the shape of the dotted curves in figure 2.

If the vessel is allowed to weathervane or select an optimum heading with respect to stationkeeping forces (or motions), a ship-shaped or elongated hull shape will be preferable. According to figures 2b, 2c the force in the longitudinal direction is only 25-30 per cent of the transverse force on a ship-shaped hull, assuming worst relative directions of wind, waves and current.

In this case it is interesting to look at the distribution of heading directions. This distribution can be established by carrying out a long-term simulation, provided that time series of wind, wave and current are available for a period of several years.

Figure 3a shows distribution of equilibrium heading for single point moored tanker in the Troll field offshore Norway. The direction distribution information is important when the anchoring point has to be located near other fixed installations.

Figure 3b shows an example of distribution of external forces on a diving support vessel located at Haltenbanken in 2 different operation modes:
a)   Fixed heading.
b)   Free heading, selected to minimize stationkeeping forces.

The capacity requirement to the stationkeeping system will be more than twice as large for case a) than for case b).

The direction dependency of external forces that is not related to the vessel shape is also an important factor. For most types of stationary (bottom fixed) anchoring systems it will be favourable to utilize information about the directional in-homogeneity of the weather. An example is discussed in the chapter on line pattern optimization.

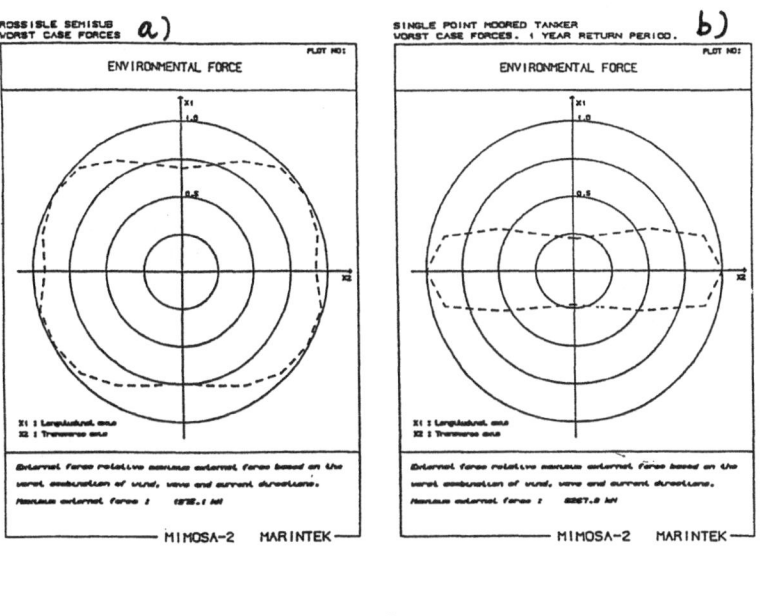

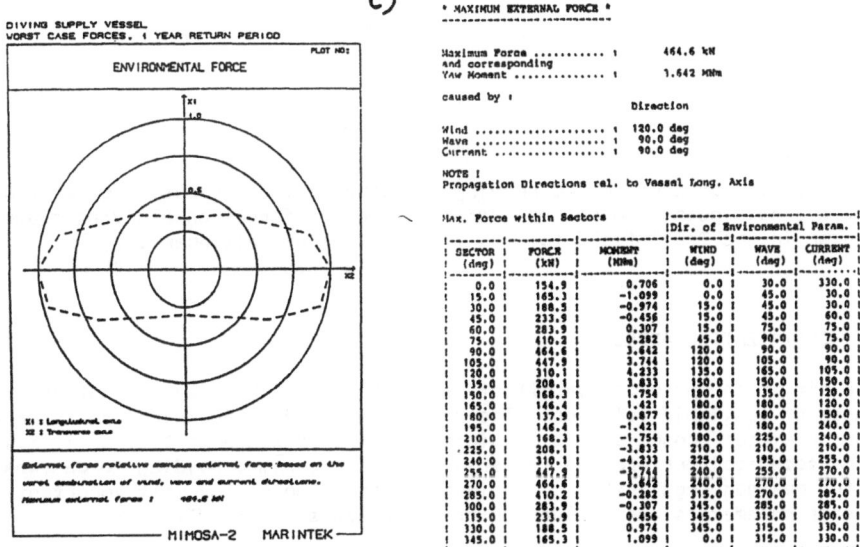

* MAXIMUM EXTERNAL FORCE *

Maximum Force .... : 464.6 kN
and corresponding
Yaw Moment .... : 1.642 MNm

caused by :

               Direction

Wind .... : 120.0 deg
Wave .... : 90.0 deg
Current .... : 90.0 deg

NOTE :
Propagation Directions rel. to Vessel Long. Axis

Max. Force within Sectors

			Dir. of Environmental Param.		
SECTOR (deg)	FORCE (kN)	MOMENT (MNm)	WIND (deg)	WAVE (deg)	CURRENT (deg)
0.0	154.9	0.706	0.0	30.0	330.0
15.0	165.3	-1.099	0.0	45.0	30.0
30.0	188.5	-0.974	15.0	45.0	30.0
45.0	233.9	-0.456	15.0	45.0	60.0
60.0	283.9	0.307	15.0	75.0	75.0
75.0	410.2	0.282	45.0	90.0	75.0
90.0	464.6	3.642	120.0	90.0	90.0
105.0	447.9	3.744	120.0	105.0	90.0
120.0	310.1	4.233	135.0	165.0	105.0
135.0	208.1	3.833	150.0	150.0	150.0
150.0	168.3	1.754	180.0	135.0	120.0
165.0	146.4	1.421	180.0	180.0	120.0
180.0	137.9	0.877	180.0	180.0	150.0
195.0	146.4	-1.421	180.0	180.0	240.0
210.0	168.3	-1.754	180.0	225.0	240.0
225.0	208.1	-3.833	210.0	210.0	210.0
240.0	310.1	-4.233	225.0	195.0	255.0
255.0	447.9	-3.744	240.0	255.0	270.0
270.0	464.6	-3.642	240.0	270.0	270.0
285.0	410.2	-0.282	315.0	270.0	285.0
300.0	283.9	-0.307	345.0	285.0	285.0
315.0	233.9	0.456	345.0	315.0	300.0
330.0	188.5	0.974	345.0	315.0	330.0
345.0	165.3	1.099	0.0	315.0	330.0

Figure 2
Worst-case environmental forces in various directions for 3 different
vessels. The dotted line shows maximum force in the various directions
relative to the largest one. The most unfavourable combinations of wind-
wave- and current directions are calculated automatically by the program.

288

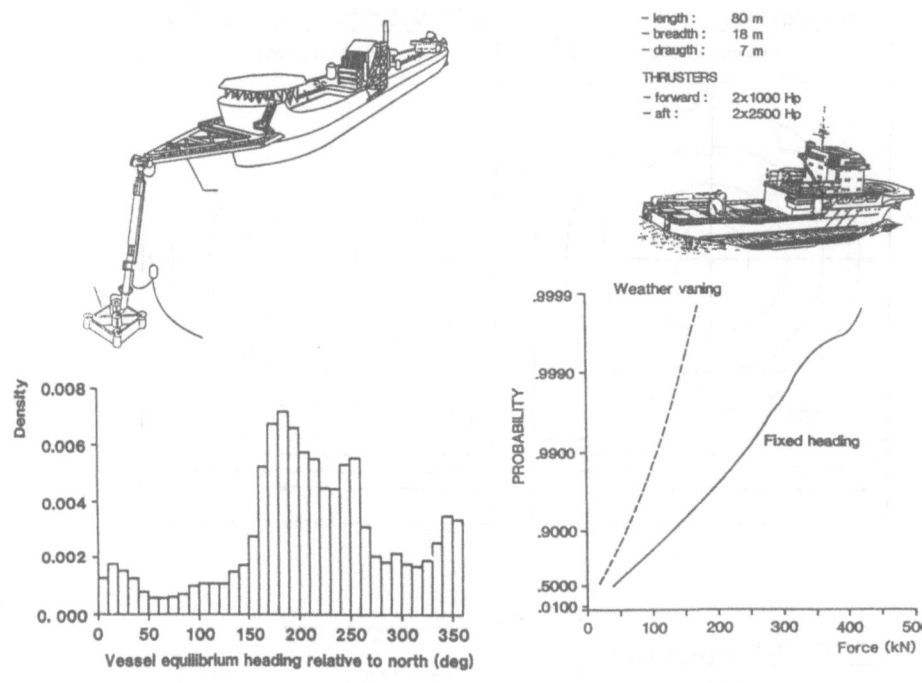

Figure 3a
Distribution of equilibrium heading
for a single point moored production
tanker in the Troll field (6)

Figure 3b
Long-term distribution of station-
keeping forces on a diving-support
vessel at Haltenbanken.
Ship-shaped hull, L = 80 m,
B = 18 m.

3. OPTIMIZATION ASPECTS

The aim of an optimization effort is to select the best system out from a
range of available solutions. The criterion for judging which one is best
can be either economical, e.g. minimum cost or maximum profit, or it may be
some technical performance criterion, such as maximum holding force capacity, maximum regularity, or minimum offset. Some of the most important
design parameters and evaluation criteria are shown in Table 2.

To the operator it will be most important to consider the total economy of
the operation. This will involve a comprehensive cost model that also may
include other equipment and related operations. The designer will normally
not have sufficient information to carry out an overall optimization of the
operation in which the platform/vessel is to be used.

Table 2 System features and economical optimization aspects.

Design parameters, performance parameters	Cost impact
Type, size and number of components.	Building cost.
Operation and manning requirement. Fuel consumption.	Operating cost.
Time for stationing. Time lost due to bad performance.	Efficiency related. Loss of income.

An example of a total optimization for one specified type of operations, namely drilling of exploratory wells, was presented in reference (1). Table 3 shows the result of an exploratory well cost comparison. Various anchor systems and dynamic positioning are compared. With the present assumptions the 12 line chain system is favourable. The fully redundant DP + anchor system gives nearly 4 per cent more expensive wells. A DP-system is equivalent with an ordinary chain- or wire system with thruster assistance. This comparison is based on cost structure data from the operator and economical assumptions with regard to capital recovery factors etc.

For the major part of the present discussion we will focus on the technical aspects that have primary interest for engineers and practical designers.

Table 3 Cost comparison of different stationkeeping systems for a drilling
platform. (Ref. /1/.
A - Anchor system, AAA - Automatic Anchor Assist, thrusters.
DP - Dynamic positioning.

Alternative	Weight (tonnes)	Vessel cost increase (Mill. USD)	Dayrate increase (Mill. USD)	Timesav. days/well t	Well Cost %
1. DP, 20000 kW 2 emergency anchor lines	1896	4.318	.0061	2.0	+0.9
2. AAA-chain, 8 lines, 8000 kW	2630	3.977	.0031	0.5	+1.1
3. AAA-chain, 8 lines, 1200 kW	3085	3.068	.0027	0.5	+0.9
4. DP+Chain, 8 lines DP 2000 kW, 8 thrus. A	3721	10.860 10.860	.0110 .0079	2.0 0	+3.7 +3.9
5. A-Chain + wire, 12 lines, no thrusters	3255	3.523	.0025	0	+1.5
6. A-Chain 12 lines, no thrusers	3788	0	.0000	0	0

4. OPTIMIZATION OF LINE COMPOSITION

The capacity and characteristics of a spread mooring system is a combined result of the individual line characteristics. For a given tension level the line characteristic is mainly governed by the geometrical shape of the line and to some extent by the elasticity. Figure 4 shows some examples of uniform and composite lines.

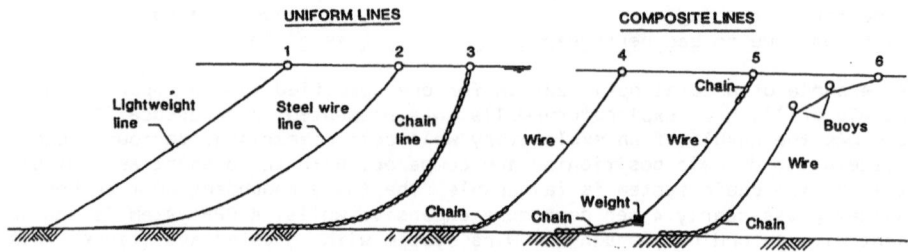

Figure 4 Alternatives for selecting anchor lines.

The uniform lines are normally preferred for mobile units in order to minimize equipment requirement in handling and installation. Further, steel wire rope is preferred for systems that have to be run more or less regularly (pipelaying vessels, crane barges, etc.), while chain systems are preferred on systems that are locked in normal operation (drilling platforms).

For deepwater systems and for permanent anchoring, the composite lines become more attractive because they can be designed to have superior characteristics. The combined chain-wire lines have been used on several drilling vessels, and deepwater floating production units are designed also with buoy system for very deep water, see ref. (3).

In order to illustrate the optimization potential of wire-chain combinations a result from (2) is included. This concerns a study of an anchoring system for a flotel that is to be connected by a gangway to a fixed platform.

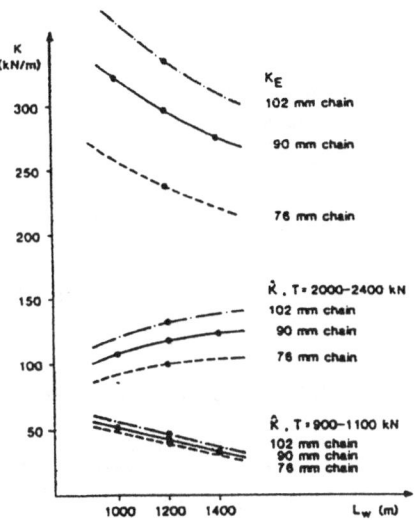

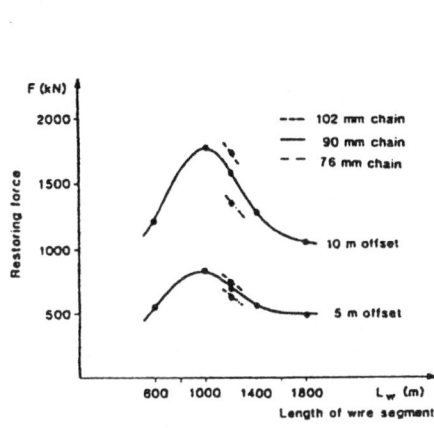

Figure 5 Comparison of stiffness
properties for chain-wire lines.
Various length of upper wire segment,
L_W, K_E = elastic stiffness
      K = $\Delta T/\Delta X$

Figure 6 Restoring force of
12-line composite system at 5 and
10 m offset as function of upper
wire segment length, L_W

All anchor lines were two-segment lines with wire as the upper segment and
chain as the lower segment. The diameter for wire is 3 3/4". The length of
wire and the chain diameter are varied. These systems all have identical
strength, determined by the upper wire diameter. The second most important
property is the stiffness. Three stiffness parameters of the individual
lines are displayed in Figure 5.

1) The secant stiffness for tension level 900-1000 kN (typical operation
    condition). A high value is preferable in order to limit offset
    variations with changing environment.

2) The secant stiffness for tension level 2000-2400 kN (typical extreme
    condition). A low value is preferable in order to limit the dynamic
    loading.

3) The elastic stiffness (independent of tension level). A low value is
    preferable in order to limit dynamic loading.

The figure indicates that the shortest range of wire lengths will be preferable for a system where high stiffness in the operating condition is

required. The reduction of stiffness for extreme tensions is also favourable. The drawback of reducing the wire length is that the elastic stiffness increases so that the dynamic effects may increase.

The combined characteristics of a 12-line system is evaluated in terms of restoring force at a specified offset distance in Figure 6. This parameter reflects the system's tolerance to changing intensity and direction of the weather. It appears that the system with 1000 m wire length has a favourable characteristic for this set of operating conditions. Increasing or decreasing the wire length 400 m from this optimum will give roughly 30 per cent increase of vessel offset for a given external force.

1000 m wire, 96 mm (3 3/4") and 810 m chain, 90 mm (3 1/2") was selected as the most favourable of the presently investigated systems. This conclusion was valid also after evaluation of dynamical response properties.

5. OPTIMIZATION OF LINE PATTERN

In a study of long-term extreme loads in anchor systems (4) it was demonstrated that in a symmetrical anchor system the degree of utilization of the lines may differ widely from one side of the vessel to the other.

The weather and the environmental loading may be very inhomogeneous with respect to direction. In order to clarify the consequences of this related to anchor system design, additional investigations have been undertaken where instrumental as well as hindcast weather data have been used in order to obtain representative long-term directional distribution. Some results are reported in (5).

Since the anchor line loading is the result of combined wind, wave and current forces, at least 7 statistical variables have to be included. These are:

- Wind velocity and direction.
- Current velocity and direction.
- Wave height, period and direction. (Assuming that waves can be described by a 2-parameter spectrum).

At present it is impractical to describe all these parameters statistically in terms of joint distributions.

A further complication is due to the fact that the resulting anchor line force can not be obtained from linear combinations of wind-, wave and current effects.

The most practical way to provide information on long-term distribution of this type of combined, non-linear load effects is to literally place the structure in the environment. In the computer the environment is represented by long time series of jointly occurring values of the seven parameters listed above (or more parameters if available). Values are stored for e.g. 3 or 6 hours intervals over a period of 3-30 years.

The vessel is represented by a mathematical model that describes the relations between the environmental parameters and the mooring forces. The calculation of forces and motions is carried out fast for each time interval, and in this way the instantaneous values of mooring forces, or any other critical response can be established for 3-6 hour intervals over a period of several years. This time series is then used to estimate extreme loads, and we have bypassed the problem of selecting an environmental design condition. In order to have a basis for selecting an optimum line pattern, the results from the long-term simulation are sorted into disjoint sets according to the direction of the resulting environmental forces, and we have a set of design loads instead of design conditions. Figure 7 gives an impression of how the external force varies with direction.

Now there are two ways to design an anchor system that is optimal with respect to capacity.

a) Use identical lines and adjust the capacity by the angle between the lines. Small angles in sectors with high forces. Larger angles in sectors with small forces.

b) Use equal angular spacing, and adjust the line dimension according to the force in each sector.

The first approach is the only feasible one for mobile units and probably also the most practical one for permanent anchorings.

Figure 8 shows a straight forward procedure for selecting an optimum line configuration.

Table 4 shows maximum line tension in the 8 lines for a symmetric line system and an optimized directional spreading of the anchor lines. The two anchor line patterns are shown in Figure 9. The ratio of maximum tension in highest and lowest loaded lines, all directions considered, is 1.84 for the symmetric and 1.19 for the "optimized" system.

Table 4 Maximum line tension for a symmetric and optimized anchor line pattern.

Line	1	2	3	4	5	6	7	8	Max ten Min ten
Symmetric Line angle (deg.)	22.5	67.5	112.5	157.5	202.3	247.5	242.5	337.5	-
Max tension (kN)	734	783	756	1121	1389	1353	1041	710	1.96
Optimized Line angle (deg.)	15.0	84.0	141.0	182.0	11.5	238.5	269.0	309.0	-
Max tension (kN)	887	910	952	1000	1064	1046	917	923	1.19

The load in the highest loaded line has been reduced by 23 per cent, from 1389 kN to 1066 kN.

This means that the capacity of the line system in this environment has increased significantly by the optimum directional spreading. Because of the discretization of the force and motion in the directions, and the fact that the wave induced motions are not taken into account in the one-step

294

optimization, the degree of utilization is still not equal for all lines. An iterative optimization procedure may improve the results further.

The present study has indicated that the directional variation of anchor system loading is considerable. This conclusion is expected to be qualitatively correct for areas not dominated by cyclone or hurricane storms. Optimization of anchor line pattern according to the actual environmental loading may contribute towards a more efficient use of available anchoring systems in different ways:

- Extreme loads may be reduced.
- System capacity in terms of environmental loading may be increased.
- Weight and size of the anchoring system may be reduced.

On the negative side it must be observed that:

- Reduction of loading in highest loaded line is obtained at the cost of increasing the load level in several of the other lines.
- A non-uniform distribution of the lines may have a negative impact on the redundancy of the system.
- The un-certainty of a directional extreme value estimate will be higher than the total extreme estimated from the same data base.

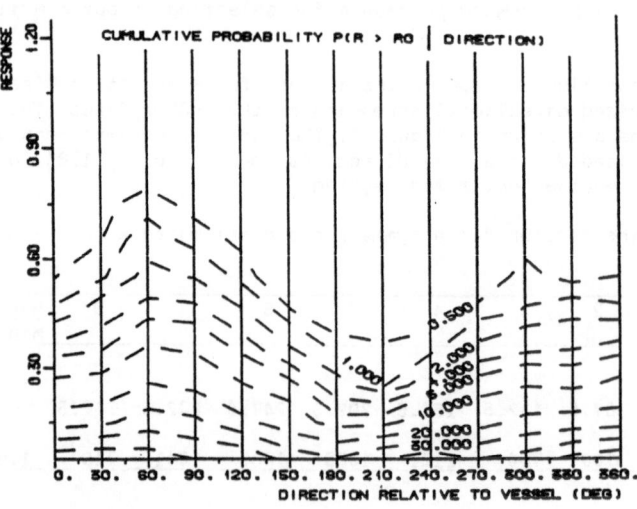

Figure 7

Equi-probability curves for conditional probability of level exceedance of external forces, in various directions.

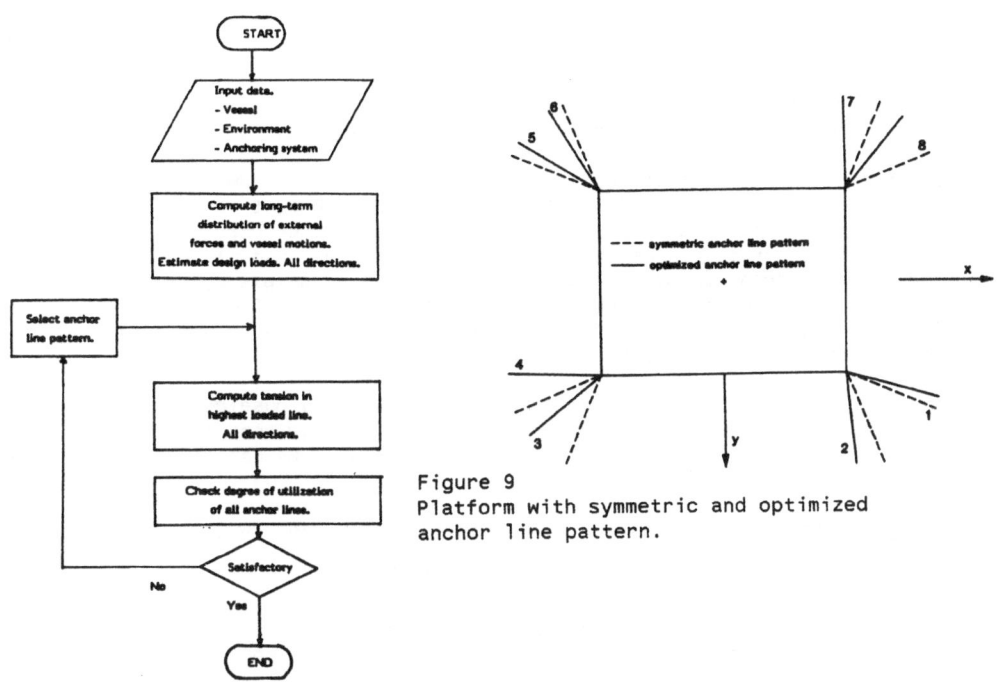

Figure 9
Platform with symmetric and optimized
anchor line pattern.

Figure 8
Computation tasks for selecting
optimum line pattern.

In order to use a directional scatter of design conditions in practice,
some precautions and further investigations should be undertaken:

- Fatigue load capacity and accidental load conditions must be included.
- Data bases for environmental conditions should be extended, particularly with regard to current velocities in the upper layers.
- Impact of higher degree of utilization in the form of damage probability and redundancy requirement should be evaluated. This is briefly
  discussed in the next section.

6. SAFETY AND RELIABILITY RELATED TO OPTIMIZATION OF ANCHOR SYSTEMS.

The design of stationkeeping systems has developed mainly empirically, and
attempts to build rules or design guidelines through the last decade have
encountered several problems. Norwegian authorities that are concerned with
the safety of offshore structures and operations have tried 2 different
approaches:

The Norwegian Maritime Directorate (NMD) has developed a set of rules and
guidelines based on experience with spread-mooring anchoring systems for
semisubmersible drilling platforms and flotels. A design verification based

on simple models is accepted. The safety factors are then serving as a practical calibration in order to obtain a reasonable fit between empirical practice and simple, straight forward analysis. One weakness of this approach is that when new types of systems are designed, or new operations are introduced, it may be a laborious task to establish equivalent safety factors and design analysis requirements to such systems.

A different approach is attempted by the Norwegian Petroleum Directorate (NPD). The established partial safety factor approach that is applied for rigid support structures is extended and attempted to be used on floating and compliant structures as well. The theoretical load and strength requirements appear to be well defined and are in principle system independent in this approach, provided that all loads and load effects are quantified. The practical problems in defining design conditions and analysis requirements may, however, be considerable.

Both of the above codes and most comparable codes only pose requirements to local loads and not failure probability of the total system. Thus, this type of codes will have to be calibrated for the various categories of systems.

As a practical example consider a 10 line anchor system. Three hypothetical versions of the system have been specified:

System A: Symmetric. Return period of critical design load: 100 years in highest loaded line. Unspecified in all the other lines. ("Critical design load" here means a load equal to the breaking strength divided by material utilization factor).'

System B: Optimized system. Return period of critical design loads: 100 years in each of the 10 lines.

System C: Optimized system. Return period of critical design loads: 1000 years in each of the 10 lines. 100 years for the total system.

A straight forward design check on total load level or highest loaded component according to literal interpretation of present rules will not show any difference between the cases A and B.

If the environment is homogeneous, systems A and B are equivalent and system C is more safe with respect to risk of overloading.

However, if the environment is inhomogeneous, as the case is on the Norwegian continental shelf, system A is obviously safer than B. If one weather direction is predominant with regard to extreme loads, systems A and C become equivalent, and system B will have less safety.

Although this problem is inherent in all structural design it is particularly simple to demonstrate and analyze for a discrete anchor system, and the effect of optimization is considerable as indicated in the previous section. The consequences of this type of optimization in terms of statistical uncertainty, utilization of environmental data and extrapolation to low probability levels are discussed in ref. (8).

This discussion emphasizes the need to adjust design requirements to the design method. Efforts to optimize the design may change the safety level, not only for empirical and simple design codes, but also for more advanced design codes.

7. REFERENCES

(1) Haslum, K., Fylling, I.J.: "Design of semisubmersible drilling units, main parameter selection", Second International Marine Systems Design Conference, Lyngby 1985.

(2) Falkenberg, E., Lie, H.: "Positioning of flotels in deep water", Automation for Safety in Shipping and Offshore Petroleum Operations, Trondheim, June 1985.

(3) Filson, J.J.: "Floater key to deepwater future" Offshore Vol 47, No. 1, January 1987.

(4) Ormberg, H., Fylling, I.J.: "Extreme loads in anchor systems for longterm operation". 4th Offshore Mechanics and Arctic Engineering symposium, Dallas 1985.

(5) Lie, H., Fylling, I.J.: "Mooring system design, Aspects of environmental loading and mooring system optimization potential", Offshore and Arctic Frontiers, OMAE specialty Symposium, New Orleans, 1986.

(6) Ottera, G.O., Fylling, I.J.: "Long-term behaviour of a production tanker in single point mooring". Offshore Mechanics and Arctic Engineering Symposium, Dallas, 1985.

(7) Ottera, G.O.: "Posisjonering av skip, retningsavhengighet og posisjoneringssystem" Offshore Marine Operasjoner, Norske Sivilingeniørers Forening, Storefjell/Oslo 1985.

(8) Godtliebsen, F., et al.: "Reliability and safety related to optimization of anchor systems", Report no 511024.00.01, Marintek, Trondheim 1987.

SIMULATION OF MOTIONS OF BERTHED VESSELS - SIMPLIFIED SIMULATION MODEL.

Ivar J. Fylling and Frank Andersson.

MARINTEK - The Norwegian Marine Technology Research Institute A/S
Trondheim, Norway.

1. INTRODUCTION.

A simulation model for analyses of motions of berted vessels is
described. The simulation is performed as a time domain analyses.

The simulation program called MOSSI-Q is developed at MARINTEK, and has
been used as a tool in parameter studies in the design of loading
terminals for oil and gas tankers on the coast of Norway.

Outputs from the program are time series of vessel motions and the
resulting loads in fenders and hawsers. Input parameters are
environmental parameters describing wind, waves, current and swell, data
for the vessel, the quay layout and the characteristics of hawsers and
fenders.

The simplification in the model gives a program that is efficient, fast
and well suited for parameter studies of fender - hawser interactions,
different pre-tension levels, environment conditions, etc. Typical time
requirement on a VAX 11/785 computer is 5 minutes of CPU-time for a
simulation of 3 hours duration.

2. SIMPLIFIED MATHEMATICAL MODEL.

In many cases the response simulation can be simplified by the following
assumptions:

* Wave frequency responses are modelled as linear responses to the waves.
  These are specified in the form of transfer functions for the 6 degrees
  of freedom. The linearized stiffness of the mooring system may be
  included in the wave frequency motion transfer functions.

* Low frequency motions in surge, sway and yaw as affected by low
  frequency wind- and wave drift forces, current and mooring system are
  modelled by a set of 3 coupled nonlinear second-order differential
  equations which are solved in the time domain.

* The wave frequency and the low frequency responses can be superimposed
  to represent the total motion of the vessel.

This simplification is straightforward for many types of moored and
combined mooring /thruster positioned offshore structures. It can,
however, also be used to study berth moorings for a wide range of cases.

298

E. Bratteland (ed.), Advances in Berthing and Mooring of Ships and Offshore Structures, 298–303.
© 1988 by Kluwer Academic Publishers.

3. SIMULATION MODELS.

The various models comprising the simulation program are described in the following:

3.1 Vessel data

The vessel is described using mass and damping parameters for the three degrees of freedom of motions in the horizontal plane. Additional data are force coefficients describing the forces acting from waves, wind and current. Force coefficients are described as functions of attack angle. Motions due to waves are described by 1.order transfer functions.

3.2. Wind

Davenport or Harris spectras may be specified as input, or constant wind speed may be specified. Wind direction is assumed constant.

3.3. Waves

Waves are described as standard frequency spectra, e.g. Pierson-Moscowitz or Jonswap spectra or a numerically defined input spectrum may be used.

3.4. Swell

A numerically defined input spectrum is used for swell.

3.5. Current

Current is assumed constant during one simulation period, and a depth current profile should be modelled in the current force coefficients describing the vessel.

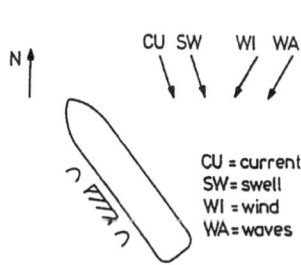

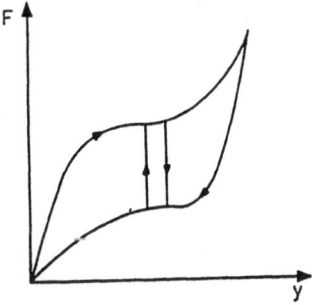

FIGURE 1. Environmental models.

FIGURE 2. Force-compression characteristics for a fender. Upper curve is for compression, and lower curve is for relief

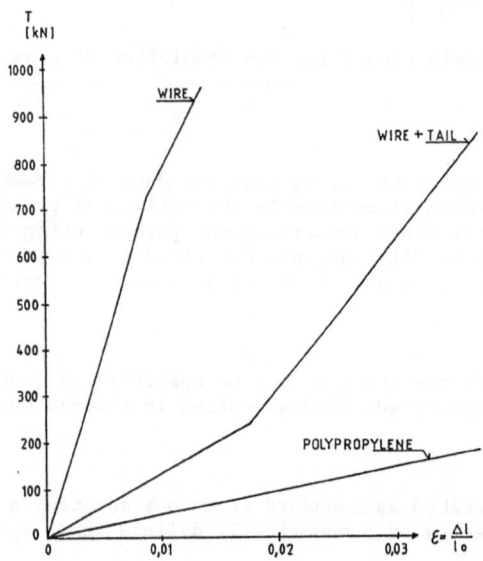

FIGURE 3. Typical force elongation characteristics for hawsers.

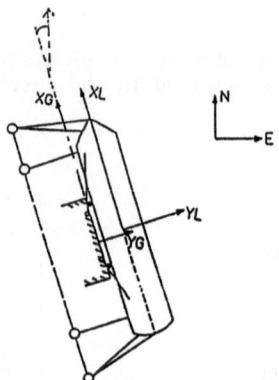

FIGURE 4. Coordinate systems used in the program. XG,YG is global system
and XL,YL is local vessel fixed system.

3.6. Fenders

Fenders are described by the fender positions on the quay. The relation between reaction force and fender compression for each fender is described by numerically defined curves, one for compression and one for reilef (fig. 2). This includes a simplified modelling of the fender hysteresis due to the energy absorption in the fender.

The component of the fender reaction force acting in the alongship direction is described by a friction coefficient, and in addition an elastic stiffness characteristic describes the static friction force.

3.7. Hawsers

The hawsers are described by their respective terminal point coordinates on the quay and on the vessel. The restoring force from each hawser is computed from a numerically defined force-elongation characteristic.

4. SIMULATION METHOD

Each parameter combination specified in the input may be simulated for a chosen period of time. The statistical properties of the output time series are analysed using post prosessing programs.

The simulation is performed by:

1. Computation of static equilibrium position disregarding environmental loads. Hawser pretensions are kept as specified in the input.

2. Computation of static equilibrium position including environmental loads. Static hawser tensions are computed using force displacement characteristics.

Further computations of vessel response separate the total motions in high frequency (wave frequency) and low frequency motions.

3. Computation of vessel motions due to waves and swell. Time series of motions for 6 degrees of freedom is computed from input spectra and transferfunctions. These motions are denoted high frequency motions. Note that the influence from fender and hawser system stiffness on the high frequency motions are not accounted for, unless the stiffness effect has been accounted for in the computation of the transfer functions.

4. The time domain simulation is performed for the surge, sway and yaw motions. Forces from wind and current, slowly varying wave drift forces, fender and hawser forces are used as input to a standard integration method. The results are time series denoted low frequency motions.

5. Finally the total motions are computed as the sum of high frequency and low frequency motions described above, and the corresponding total fender and hawser forces are computed.

5. ANALYSES OF STIFF FENDER SYSTEMS

As outlined above, the wave frequency responses are basically assumed to
be linear. The influence of the fender system on the wave frequency
responses can be investigated if:

a) The natural periods of motion in sway (or yaw) is short, and

b) The wave frequency responses are dominant compared with the low
   frequency responses.

These analyses are carried out in regular waves by including the wave
frequency excitation, added mass and damping, in the low frequency
simulation model.

Thus a full nonlinear simulation of the response to regular waves is
obtained.

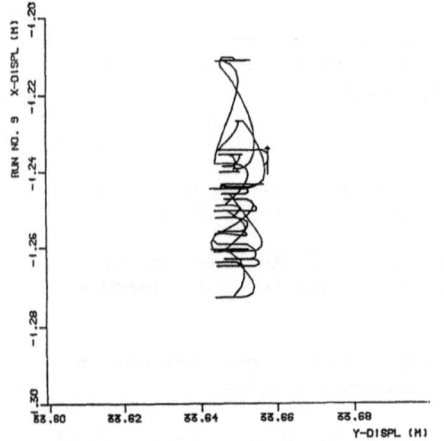 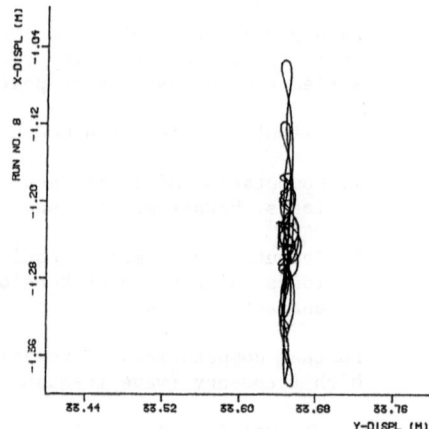

a) High friction coefficient (0.6)   b) Low friction coefficient (0.1)

FIGURE 5. Plots of the horizontal motion for tanker with environmental
forces acting in the longship direction.

TABLE 1. Example of result table of statistics for a simulation. Total
simulation time is typically 3 hours.

	AVERAGE	STANDARD DEVIATION	AVERAGE MINIMA EACH SEQ.	AVERAGE MAXIMA EACH SEQ.	TOTAL MINIMUM	TOTAL MAXIMUM
LF-SURGE (m)	0.38	0.05	0.21	0.54	0.18	0.57
LF-SWAY (m)	25.54	0.21	25.13	26.17	25.11	26.27
LF-YAW (deg)	0.00	0.00	0.00	0.00	0.00	0.00
TOT SURGE (m)	0.38	0.10	0.04	0.71	-0.01	0.75
TOT SWAY (m)	25.54	0.21	25.08	26.17	25.06	26.26
TOT YAW (deg)	0.00	0.00	-0.19	0.22	-0.20	0.23
FFEND 1 (kN)	140.0	540.7	0.0	5156.5	0.0	6089.3
FFEND 2 (kN)	52.9	234.5	0.0	2699.1	0.0	3234.1
FFEND 3 (kN)	46.0	215.6	0.0	2436.5	0.0	2859.9
FFEND 4 (kN)	111.5	474.3	0.0	4853.5	0.0	5424.2
TROPE 1 (kN)	427.6	145.6	77.4	1032.0	39.2	1119.8
TROPE 2 (kN)	503.6	189.6	114.3	1292.3	66.7	1438.1
TROPE 3 (kN)	351.9	54.5	176.1	599.2	160.9	625.8
TROPE 4 (kN)	293.9	49.0	129.7	487.8	99.0	561.7
TROPE 5 (kN)	534.2	198.3	140.2	1290.1	107.9	1565.3
TROPE 6 (kN)	471.0	149.7	125.0	1049.7	83.9	1199.3

FFEND N = Force from fender no. N
TROPE N = Tension in rope no. N

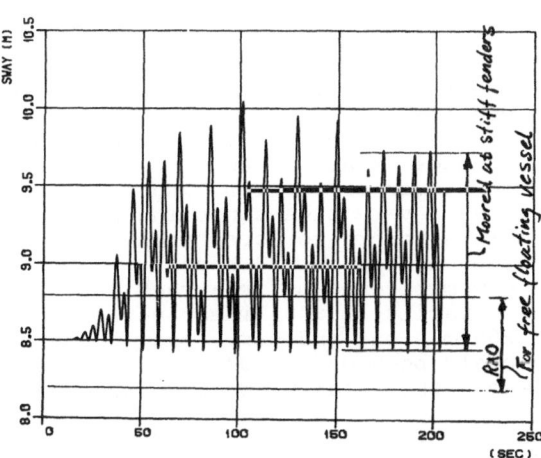

FIGURE 6. Example of
sway response
magnification and
nonlinearity due to a
stiff fender system.

RAPID IN FIELD FERRYBERTH CONSTRUCTION

ELLEN HESS THAYSEN

DANISH STATE RAILWAYS

1. INTRODUCTION
   As we all know, there are many things to be considered when
constructing a ferry berth, such as type of ferry, navigation
conditions, water levels, wawe and ice conditions, etc.
   This article discusses a new problem which had to be solved
lately in connection with the establishment of two ferry berths
for the Danish State Railways, wiz. the fact that there was
only very little time for building the ferry berth at the place
of construction.
   The two construction jobs to be discussed were the building
of a railway goods ferry berth in Copenhagen in 1986 and the
building of a car ferry berth at Elsinore in 1987.In connection
with the latter berth we also describe the problems that arose
because the berth was located outside the harbour jetties.

2. NEW RAILWAY GOODS BERTH IN COPENHAGEN
2.1 Background
   As part of the rationalization of railway goods traffic
between Scandinavia and the Continent via Denmark it was
decided to establish a goods service between Helsingborg in
Sweden and Copenhagen in Denmark.
   Railway goods waggons had previously been ferried across the
Sound between Elsinore and Helsingborg by small shuttle ferries
of a maximum length of 84 metres and with only one railway
track. The ferries also had to carry motor vehicles of every
description and passengers, mainly shoppers. The ferries
departed every 15 minutes.
   As the ferries could only carry a very limited number of
goods waggons across, goods trains had to be disconnected,
taken across by a large number of ferries, and reconnected on
the other side. This was a very time-consuming procedure. On
the direct service from Helsingborg to Copenhagen, the carriage
time to Germany was reduced by approx. 24 hours.
   The new goods service comprises two five-track goods ferries
with a goods waggon capacity of 800 linear metres each. The
ferries are just under 200 m. long and carry rail goods only.
   Agreements had previously been made about the implementation
of this service between the Swedish, West German and Danish
railways.
   Various circumstances - both of a political and an environmental nature - made it impossible to start the construction
work -in Copenhagen at the appointed time. Negotiations made
slow progress and we finally had to realize that the time
available for construction at the site was limited to five

E. Bratteland (ed.), Advances in Berthing and Mooring of Ships and Offshore Structures, 304–321.
© *1988 by Kluwer Academic Publishers.*

months if the service were to commence within the time stated and agreed on.

The normal building period originally scheduled for the berth was approx. 16-18 months. The berth was to be constructed in the conventional manner.

By working around the clock we hoped to reduce the construction period to 8 or 9 months.

As it was highly important for the railways to have the work completed in time, the project was reviewed in order to reduce construction time.

2.2 NEW DRAFT PROJECT

The berth in Copenhagen was to be established over an old ferry berth built in 1885 for much smaller ferries plying between Copenhagen and Malmö.

However, the new berth was to be moved further out into the dock basin than the old berth to provide sufficient space for railway tracks on shore.

This meant, among other things, that it was necessary to construct a long pier into the water as shown in figure 1. The existing pier only extended to the dotted line.

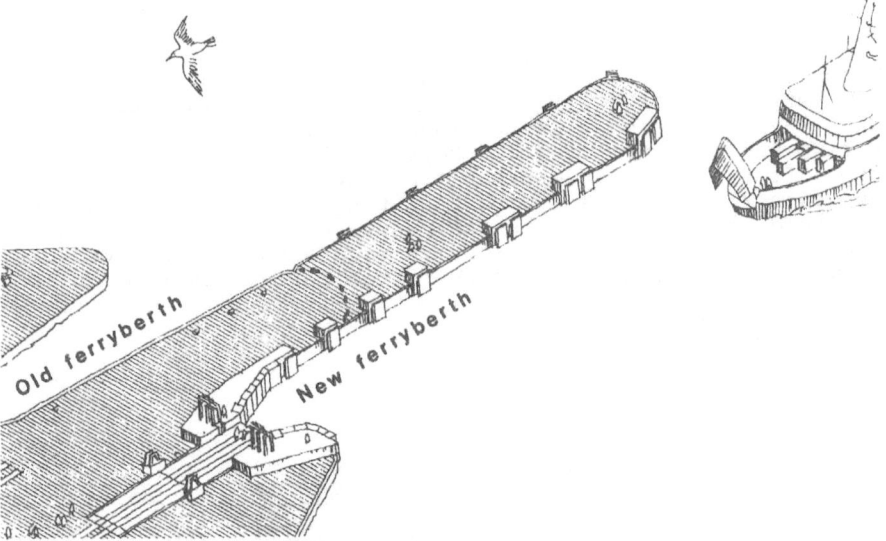

FIGURE 1. Original draft

The operation of driving down and anchoring steelsheet piling was one of the time-critical factors in the project. If this operation could be reduced and partly replaced by a more quickly completed operation which was independent of the driving operation, it might be possible to reduce construction time at the site to the five months that were available.

The key word was prefabrication, first of all of the long pier which included about half of the steel sheet piling. It should be mentioned that the required steel sheets had already been purchased because of the time of delivery for such sheets.

The project layout was therefore altered as shown in figure 2.

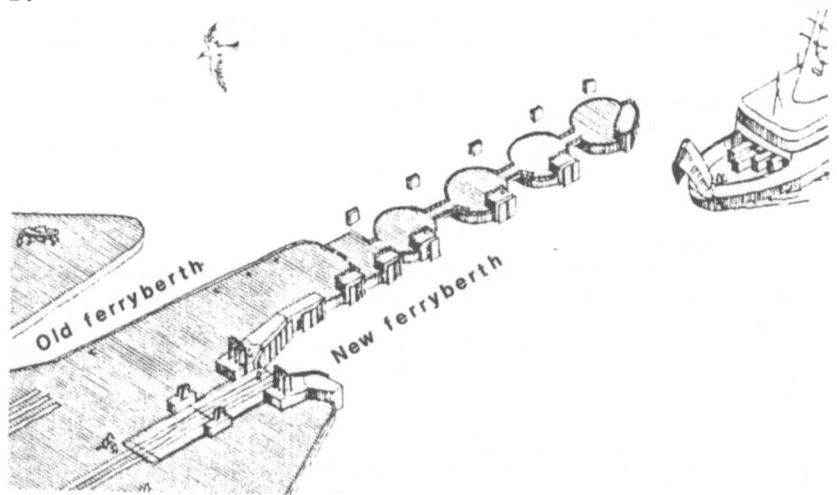

FIGURE 2. New draft

Here, the pier extension is made of four circular, prefabricated steel sheet cylinders, approx. 20 m in diameter and approx. 13 m in height. The cylinders were constructed around a top ring and a bottom ring and kept in place by internal strutting. After the cylinders had been lowered into position - by means of a floating crane with a lifting power of 1200 tons - a reinforced bottom was to be cast under water in the cylinders, and after they had been filled with sand also a reinforced lid. Bottom reinforcement had been fitted before positioning took place.

FIGURE 3. The building of the cylinders

The illustration in figure 3 shows a phase during the construction om the cylinders. Because of their shape the cylinders were called "Trunter", a Danish word for a special kind of
stump.

Prior to being positioned, the cylinders weighed approx. 300
t. Positioning took a total of 17 hours - there was about 500
m. from the place of manufacture to the ferry berth.

The illustration in figure 4 shows the procedure of positioning the cylinders.

FIGURE 4. Positioning of the cylinders

It should be mentioned that bottom conditions at the construction site made it very difficult to drive down sheet
piling since there were layers of flint and layers of small and
fairly uniform stones.

By altering the project and using "Trunter", a considerable
part of the difficult driving operation was eliminated.

Because of neighbouring buildings, driving would not have
been permitted in the period 19-7.

2.3 OTHER PREFABRICATION MEASURES

In connection with a ferry berth it is normal to prefabricate
as much as possible of the landing apron unit and also the fender assemblies. In the present case we evaluated the expediency
of further prefabrication and decided to implement the following activities:

1. Prefabrication of very large bottom protection slabs.
2. Prefabrication of bridges between "Trunter".
3. Prefabrication of concrete cassettes for quay walls.

As regards the bottom protection slabs we decided to make them so large that the available floating crane in the harbour would just be able to lift them. We called them Maxislabs.
The bottom protection slab pattern appears from Figure 5.

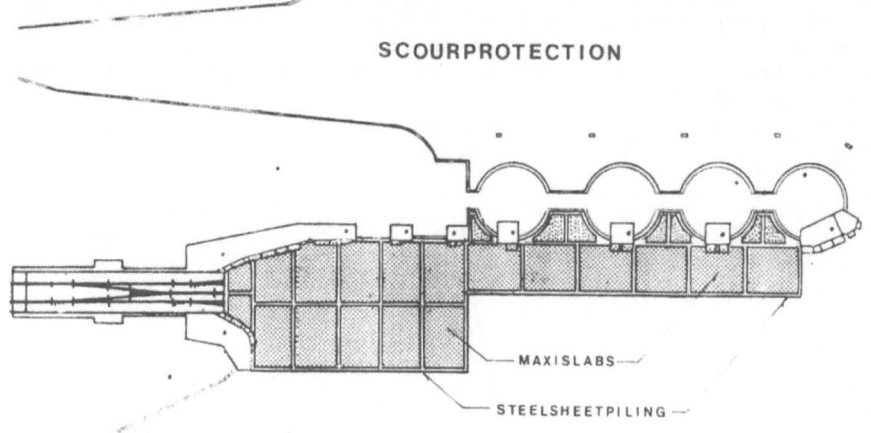

FIGURE 5. Slab pattern for bottom protection

The standard slabs covered an area og 160 m^2 and were made of reinforced concrete of a thickness of 0.3 m.
The reinforced slabs were cast on shore and on top of each other. Lifting anchors were embedded in the slabs.
Slabs to be placed against the "Trunter" were cast in shape to fit these.
The theoretical measurements were based on a spacing between the slabs of 0.6 m. The slabs were supported along two opposite sides by means of concrete piles placed horizontally on the bottom and adjusted to the correct height as shown in figure 6. The outside of the piles was recessed 0.3 m. from the slab edge. Under the rest of the slabs was water, eliminating the need for extensive levelling of the bottom.

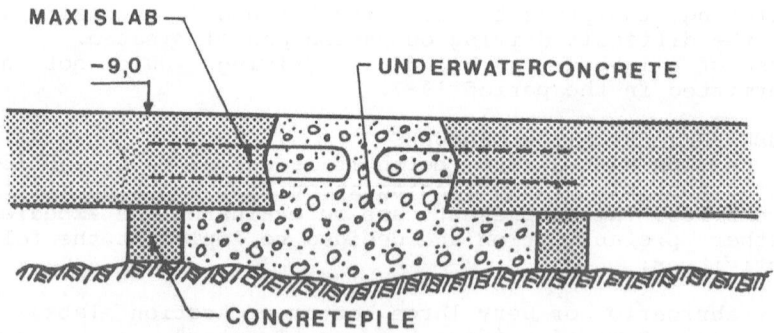

Figure 6. Slabsupport

The slabs were laid in series and the laying operation was quickly completed.

After the slabs had been laid, spaces between the slabs and between slabs and steel sheet piling (frame walls, "Trunter" and edgings) were filled with under-water concrete.

The bridges were made of inverted T-beams of prestressed concrete between which in situ concrete was cast. The outermost bridge, however, was prefabricated as an assembled unit and lowered in position by means of the floating crane.

The quay wall cassettes were reinforced concrete boxes cast with 0.3 m. thick walls. They were lowered in place by means of the floating crane, whereupon the inside of the boxes was filled with in situ concrete. The idea was to save time for erecting formwork at the site, but in this case the idea was probably not so good because the impact of the ferries was to be absorbed by the cassettes, and this led to the establishment of an in situ reinforcement inside the cassettes.

The altered project described above was a somewhat more expensive than the traditional solution, but working around the clock - combined with a shorter construction time requiring the presence of heavier equipment, such as dredgers, larger cranes, etc. at specific dates and times - turned out to be far more expensive.

When the work was performed, the normal working week was 40 hours, but the project required a working week of 168 hours divided between three shifts.

The work was carried out by the Danish contractors Monberg & Thorsen and perfect delivered within the stipulated time.

3. NEW OUTER FERRY BERTH AT ELSINORE
3.1 Background

The establishment of the previously mentioned railway goods service between Helsingborg and Copenhagen, which became operational on 3rd November 1986, released the small and not more than 84 m. long and 13.4 m. wide ferries plying between Elsinore and Helsingborg from all carriage of goods waggons.

This removal of goods waggons made the service more attractive especially to lorries, increasing the number of lorries during the following spring by over 20 % compared with previously approx. 5 %.

To further improve the service it had been decided already at the end of 1986 to put in two larger ferries. The ferries, which were no longer required on another service, were 105 m. long and 17.7 m. wide. Instead it was the intention to scrap the two oldest and smallest ferries in service.

Because of their width the ferries could not use the existing wedge berths in the harbour. To maintain the service during the construction period, and because there was no room for more ferry berths inside the harbour, it was decided to establish a new primitive corner ferry berth outside the harbour jetties.

3.2 Wind and Waves

The location outside the jetties was acceptable because there was suitable shelter from wind from most directions. From the N.E.-S.E. angle, however, the fetches - though of limited extent - were so long that a strong wind might create waves which would necessitate a suspension of the service. It should be mentioned that during hard frost, ice pack can occur at the construction site and also necessitate suspension of the service.

It was a condition that - apart from periods with ice pack - the berth could be used at least 95% of the time.

The significant wave heights from S.E. (wind from this direction generates the highest waves) were expected to be as stated in the wave statistics shown in figure 7 and 8 which are made partly on an annual basis and partly for the period June-September, both months inclusive (summer service period).

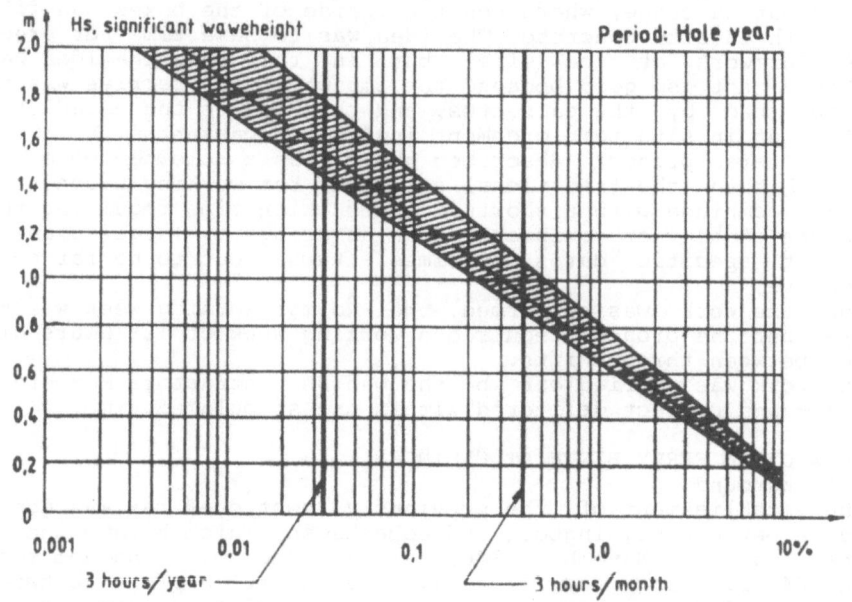

FIGURE 7. Wawe Statistics, annual basis

Because of the mentioned limited fetches, the wind only needs approx. 1-2 hours to generate fully developed waves.

There is therefore a very close correlation between wind and wave conditions, which makes it possible to calculate wave conditions directly on the basis of wind statistics.

The current at the construction site will be in the ferry's direction when it is berthed, and only a northerly current will be of any significance.

The current statistic is as follows:

Frequency %	Current Velocity m/s
5	0.35
1	0.50
0.1	0.65
0.01	0.85

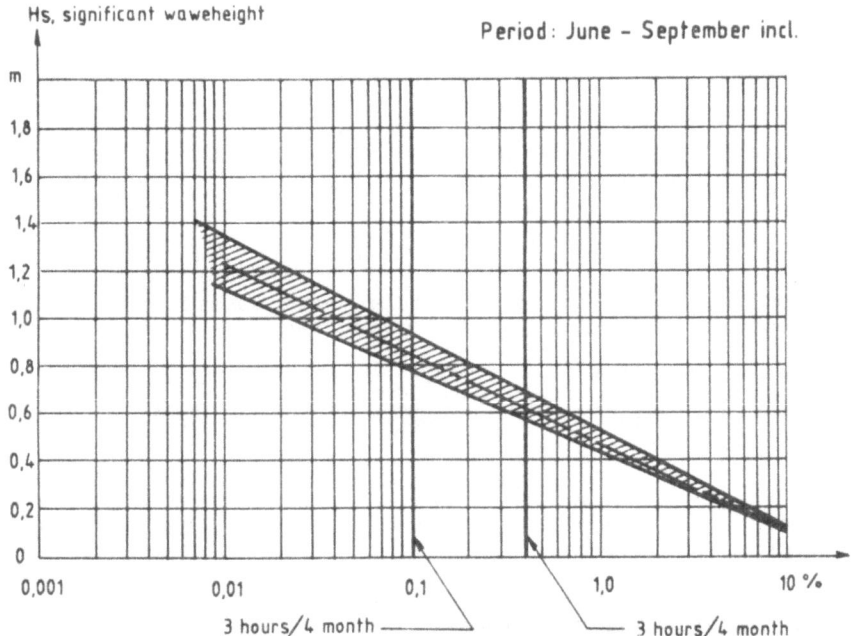

FIGURE 8. Wawe Statistics, summer service period

The Danish State Railways had stipulated - because of very short construction time due to the fact that a final decision on the project could not be made until the beginning of December 1986 and the berth was to be put into service on 1st June 1987 - partly that there was to be built a linkspan, i.e. an apron supported at the outermost end by boyant tanks, and partly that there was to be built a landing pier consisting of a large, moored pontoon.

However, the technicians seriously doubted the advisability of the pontoon solution, not least because the pontoon was to support the passenger gangway. Ferry and pontoon could not be expected to move together in reaction to the waves - on the contrary - and it would be extremely inadvisable to build a passenger gangway which was exposed to movements of that magnitude.

The technicians therefore forced through the making of physical model tests in order to ascertain the movements of the various components individually and in relation to each other under various wave conditions and various mooring methods. The ferry was to be fastened by moorings to the linkspan and have a breast fast farthest out.

The model tests were made by the Danish Hydraulic Institute.

The tests showed that, in general, a solution involving a pontoon as mooring side would neither serve the ferry nor the traffic to and from the ferry in a satisfactory way. It was necessary to find a solution involving a fixed mooring side. A dolphin solution was chosen as the quickest and probably also the least expensive.

An evaluation og the linkspan - dolphins solution was based on the criterion for acceptable movements contained in the Nordic report "Skibsbevægelser i havne" (Ship Movements in Harbours).

Type of ship	Sway (m)	Heave (m)	Yaw (°)	Pitch (°)	Roll (°)
Ferry (L_{oa}=100-150 m)	0.8	1.0	1.0	1.0	2.0

Based om the test, these criteria were considered to have been exceeded as stated in the following table.

Movement	Criteria	NE (m/s)	E (m/s)	SE (m/s)	S (m/s)
Roll	2°	> 25	> 25	9-10	13-14
Pitch	1°	> 25	> 25	9-10	12-13
Yaw	1°	> 25	> 25	8-9	9-10
Sway, stern	0.8 m	> 25	> 25	13-14	13-14
Sway, stem	0.8 m	> 25	> 25	10-11.	10-11
Heave, stern	1.0 m	> 25	> 25	17-18	15-16

These criteria do not apply to the passenger gangway, for which the criteria are much lower.

On an annual and seasonal basis, respectively, the above exceeding of criteria is expected to occur at the following frequency:

Annual basis	Roll	Pitch	Yaw	Sway stern	Sway stem	Heave stern
hours/year	101	92	175	24	88	11
%	1.15	1.05	2.00	0.27	1.00	0.13
Seasonal basis	Roll	Pitch	Yaw	Sway stern	Sway stem	Heave stern
hours/season	7	8	17	3	9	1
%	0.24	0.27	0.58	0.10	0.31	0.03

The Danish Hydraulic Institute concludes:

In general, the test shows that the most important factor regarding the operation and berthing situation is the moorings, i.e. the mooring system is subject to very great stresses. Based on the primary mooring system used, the load is expected to exeed 60 % of the ultimate strength (100 % ≈ 33 t.) during 5 % of the time, i.e. approx. 440 hours a year.

The test also shows that the frequently occurring great transverse movements at the stem of the ferry can be reduced considerably by supplementing the breast fast at the forepart with an additional breast fast. As an alternative or supplement to this, the bow propeller of the ferry can be used, as it will press the ferry against the dolphins' fenders and reduce movements as well as mooring stresses.

Further, navigation simulation tests were carried out at Skibsteknisk Laboratorium (Danish Maritime Institute) in order to determine at which wind directions and wind forces it would be impossible to enter the berth and whether an improvement of the ferry's manoeuvrability was required.

The simulation tests revealed that the ferries' manoeuvrability had to be improved by means of a new bow propeller of 1500 HP, a compass thruster abaft of 1000 HP and a spade rudder.

It now became possible to put in at the linkspan berth at wind directions critical to navigation at wind forces up to 15 m/s.

It should be mentioned that it is the Danish State Railways' experience that the navigation simulation results are slightly on the safe side.

3.3 Structures

Building of the berth at the place of construction was commenced about 1st April 1987 and the berth was put into service on 5th June 1987. The work was carried out by the Danish Contractors Christiani & Nielsen.

The structures had to be designed with due regard for an extremely short building period at a construction site at a rather exposed location.

The berth's landing apron assembly was designed as a linkspan, the mooring side as three fendered dolphins. Further, there was a passenger gangway supported by the centre dolphin an farthest out one mooring dolphin. The passenger gangway was connected to the existing landing system at the Danish State Railways' ferry berths inside the harbour. Figure 9 shows the project in perspective.

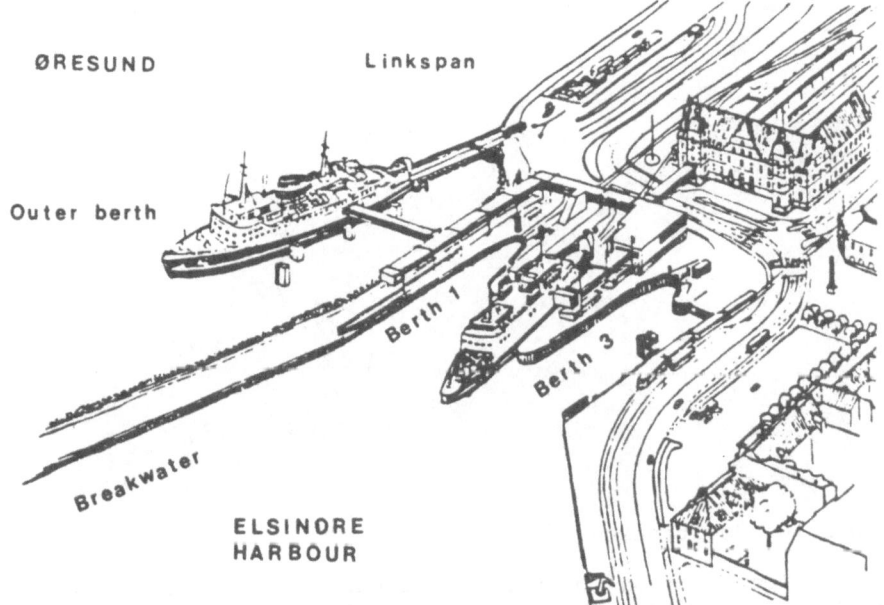

FIGURE 9. New outer berth at Elsinore

The linkspan was prefabricated to such an extent that erection at the site could be carried out in less than one week on an abutment built in advance. The land end of the linkspan was fendered by a 600 kNm fender.

The dolphins were made of steel pipes of a diameter of 3.2 m. At the waterline the diameter was reduced to approx. 1.5 m. to reduce the ice pressure as much as possible. Dimensions were based on ice pressure rather than on reactive forces from the fenders. Between the thick and the thin pipe a 10 m. high conical piece was inserted.

The bottom consisted of postglacial sand with streaks of mud. The pipes were vibrated into the bottom by means of Europe's largest vibrator and the operation was easily completed. There was one surprise, however: All pipes stopped at the same height.

It turned out that the conical shape, and maybe also the welded reinforcements at the transition from conical shape to cylindrical shape, was the cause of this. When the pipes were vibrated, the water inside the cone acted as a solid body. After the water had been pumped out of the cone, the pipes could be vibrated to the foreseen depth varying between 23 m. and 28 m. under sealevel.

It would have been possible to vibrate the piles down in less than two days.

Figure 10 shows the work with positioning a pile.

FIGURE 10. Positioning a pile

The upper section of the piles was filled with concrete on which was placed a specially shaped pile top carrying fenders designed to absorb the impact of the ship. The fender units

were designed for an energy absorption of 1100 kNm, a figure
which had been derived from the results of the navigation
simulation tests.
Figure 11 shows the fender element.

FIGURE 11. Fender element

4. SHORT CONSTRUCTION TIME

It should be mentioned that it is the Danish State Railways'
experience that reducing the construction period to such an
extent that it becomes necessary to work around the clock involves considerable additional expenditure - estimated at 25 -
35 %.

If construction time is short, an adaptation of structures
may be relevant. This additional expenditure should be compa-
rared with the value of time saved.

For financial reasons it is advisable always to try to achieve a construction time which is reasonable in relation to the
extent of the work and which makes it possible, in general, to
execute the work during normal working hours.

5. MOORING AND FENDER SYSTEMS, DANISH STATE RAILWAYS

5.1 Mooring System The Danish State Railways run a great number of ferry services. All services have to meet strict requirements conerning minimized ferry preparation time ashore, including time for entering and mooring and time for departure
from the berth. Fuother, it is our aim that arrivals and departures arc to take place with as little peronnel as possible
stationed ashore.
Until new large car- and passenger carrying ferries were put
in service between Kalundborg and Aarhus at the end of 1985,

all landing aprons were designed to rest on a bracket at the ferry's bow or stern, respectively. To position the ferry correctly in relation to the apron, two positioning fenders where provided which the ferry's rubbing piece should rest against and which could ensure, that there would always be approx. 3 cm air space between ferry and outer apron end, so that horizontal, longitudinal stresses from the ferry would not be transferred to the apron. The apron is not designed for such stresses.

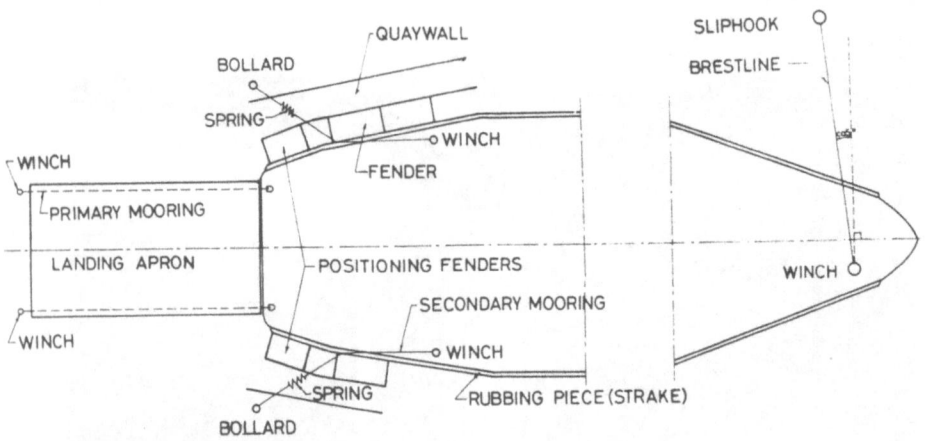

FIGURE 12. Normal mooring system

With reference to Figure 12, the mooring procedure is as follows:

1. Under its own power, the ferry positions itself close to apron unit and fender system.

2. Secondary moorings are attached to mooring hooks at both sides by the ferry's crew and the winches on the ferry are tightened until the ferry rests against the positioning fenders.

3. At the same time, from the outer end of the ferry, a line tied to a mooring rope is thrown ashore. A man on shore positions it on the sliphook, whereupon a winch on the ferry tightens the mooring until the ferry rests against the mooring side.

4. The landing apron is lowered until it rests on a bracket on the ferry. The apron is operated from the ferry.

5. At both sides - built into the apron - moorings along the apron are fastened to winches located in pits ashore. Towards the ferry these moorings end in a thimble. The ferry's crew places the thimbles around mooring irons recessed below the top of the deck. Then the winches are tightened, which can be done from the ferry.

6. If the ferry is of the passenger-carrying type the passenger gangway can now be positioned. Safety devices prevent the gangway from being positioned until steps 1-5 have been fully completed.

The procedure is reversed when the ferry departs. No crew will be required on shore as the sliphook' can be released from the ferry.

5.2 Types of ferry berth
The Danish State Railways use two principal ferry berth designs: wedge berths and corner berths.

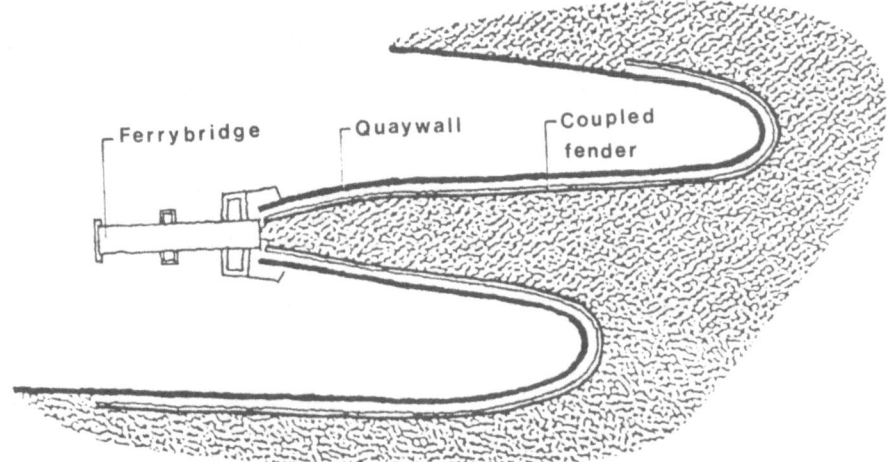

FIGURE 13. Wedge berth

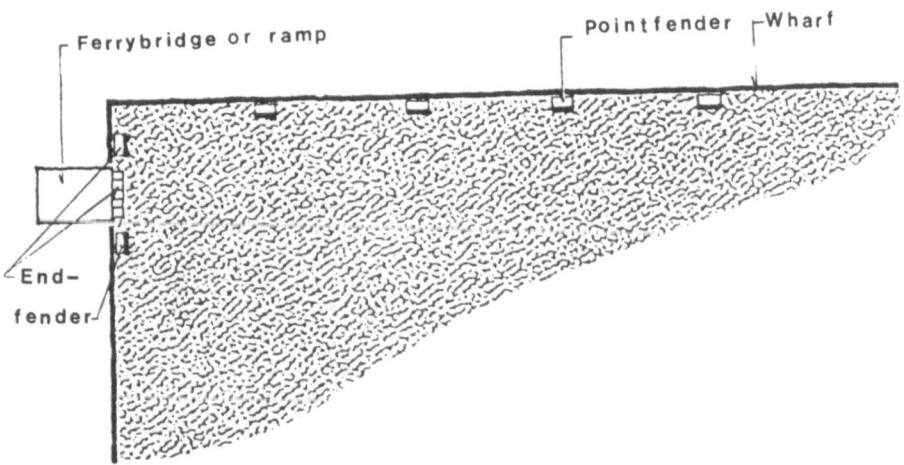

FIGURE 14. Corner berth

318

A wedge berth is shaped to ensure that the ferry's rubbing piece will be "wedged" into the innermost section of the berth preventing the ferry from moving sideways when berthed.
Two piers enclose the wedge berth, and the long pier also provides a mooring side for the ferry, see Figure 13.
A wedge berth is entered by sailing the ferry into the wedge-shaped mounth, whereupon the ferry will automatically be guided to the apron and the positioning fenders.

A corner berth is a ferry berth with only one pier, which provides the mooring side for the ferry, see Figure 14.
A corner berth is entered by sailing the ferry along and as close to the mooring side as possible, whereupon the ferry is stopped a few metres from apron front edge or land ramp. The ferry is then moved sideways towards the mooring side point fenders, inter alia by means of transverse propellers and sometimes also by mooring lines fastened ashore. It takes longer to enter a corner berth than a wedge berth under adverse wind conditions.

5.3 Dimensioning

The Danish State Railways' fender systems are normally dimensioned on the basis of impact ratings ascertained during simulated navigation in adverse weather. Various fender system cha-

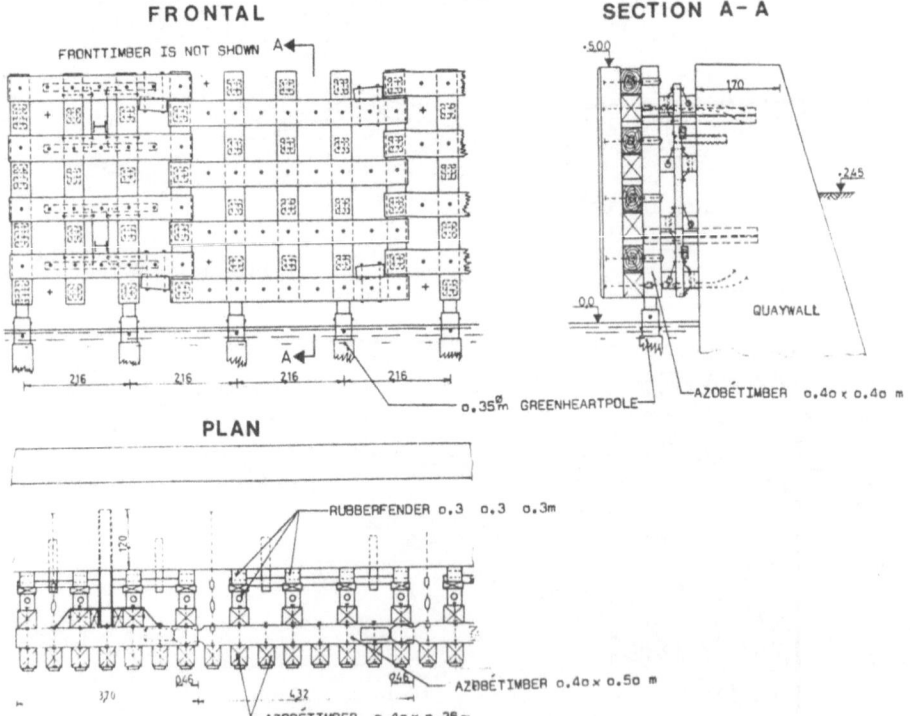

FIGURE 15. Coupled fenders of azobé timber

racteristics can be entered, whereupon the simulator outputs the relevant impact force.

Preliminary estimates of fender system energy absorption made by the Danish State Railways use the following dimensioning basis for wedge and corner berths, respectively:

In wedge berths the ferry sails at a rate of 3 m/sec. (approx 6 knots) when passing the berth pier heads, and the speed is gradually reduced to 1 m/sec. (approx. 2 knots) at the inmost end of the berth. The ferry's maximum angle to the berth centre line is reduced from 15° at the pier heads to 10° at the inmost end. The positioning fendes must be able to stand a direct impact at 0.3 m/sec.-0.5 m/sec. depending on the circumstances.

In corner berths the side fenders are dimensioned for the impact of a ferry moving sideways against the fenders at 0.3 m/sec.-0.5 m/sec. at right angles to the berth axis.

When energies are calculated, allowance must be made for ferry speed along and at right angles to the fender system, ferry rotation, ferry impact point, hydrodynamic mass and height of impact point.

An abundance of articles has been written about the calculation of fender systems. The Danish State Railways base their calculations on Jesper Larsen: Ship Impact in Ferry Berths, IABSE Reports, Volume 42, 1983.

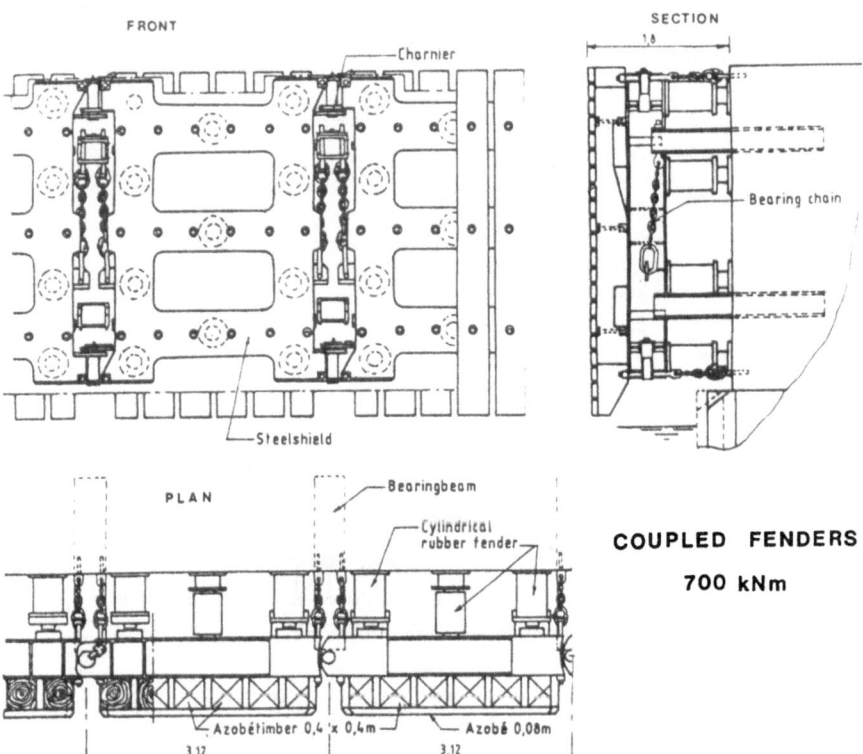

FIGURE 16. Fenders used at the main ferry service in Denmark

Since, depending on loading conditions and water level, the ferry rubbing pieces may hit the movable fender system both high and low, the fender system front will slope unless some form of parallel connection of the fender system is provided to ensure that the front will remain vertical irrespective of impact point.

Parallel connection can - if required - be effected by means of poles rigidly connected with the fender system, or by providing torsion rigid panels.

Figure 15 shows a continuous fender system with non-parallel connection for ferries with a displacement of approx. 4.000 - 5.000 t.

5.4 Fender Systems Used by the Danish State Railways

Ferries are normally provided with rubbing pieces located opposite the ferry main deck to ensure that impact forces are absorbed where the ferry structure is of sufficient strength.

Since, depending on loading conditions and water level, the ferry rubbing pieces may hit the movable fender system both high and low, the fender system front will slope unless some form of parallel connection of the fender system is provided to ensure that the front will remain vertical irrespective of impact point.

Parallel connection can - if required - be effected by means of poles rigidly connected with the fender system, or by providing torsion rigid panels.

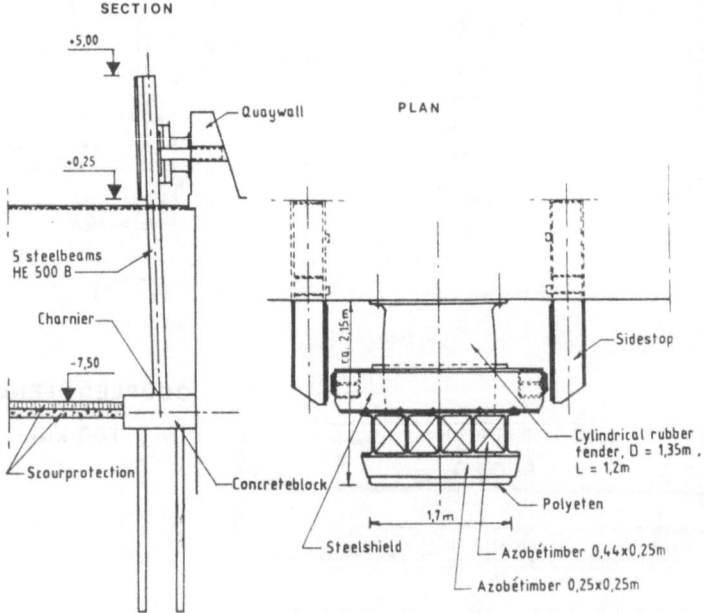

POINTFENDER 600 kNm

FIGURE 17. Parallel Connection Fender at Car Ferry Berth

Figure 16 shows a continuous fender system with non-parallel connection for ferries with a displacement og approx. 12.000 t. The fender system consists of steel panels covered by azobé timber and fendered by cylindrical rubber fenders.

Figure 17 and 18 show two types of point fender with parallel connection which are used in corner berths for ferries with a displacement of up to 15.000 t.

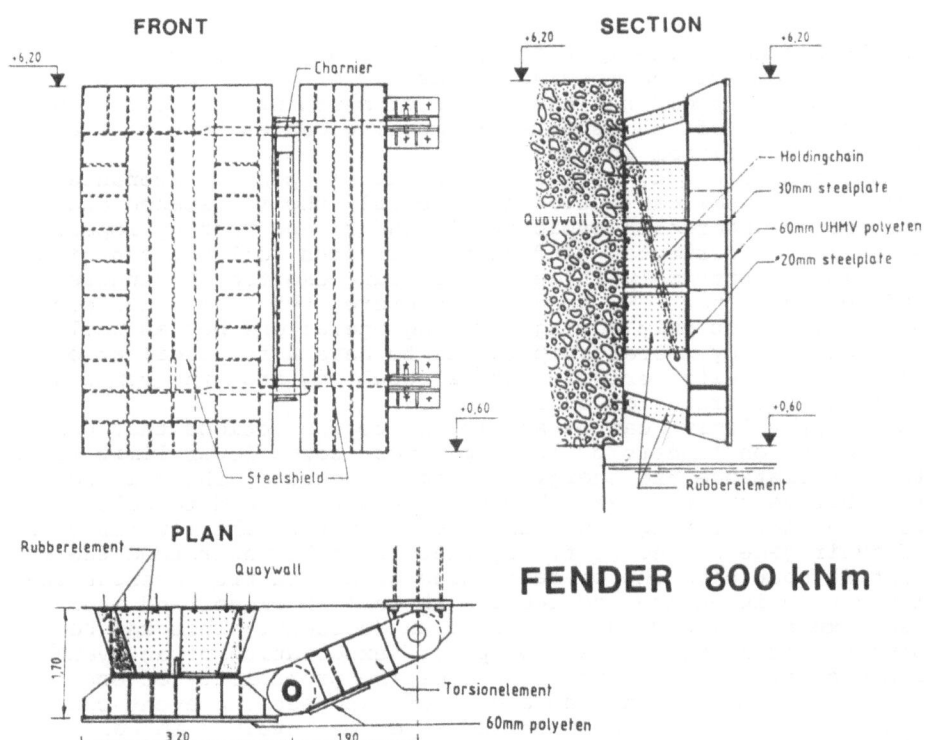

FIGURE 18. Parallel Connection Fender at railway goods Ferry Berth

THE PROBABILISTIC APPROACH IN THE SELECTION OF ROPES AND FENDERS

F. Vasco Costa

The knowledge of the statistical distributions of actions exerted on fenders and mooring cables and of their resistances
are essential tools for the design of berthing structures and
the selection of the fenders and mooring cables that will be
the most economical in the long run.
Damage accumulation, stress concentration, poor maintenance
and other causes can reduce considerably the resistance and
the energy absorption capacity of fenders and mooring ropes.
Wave trains, currents, wind gusts and human failures can give
occasion to extremely large forces being exerted on fenders
and mooring ropes. No wonder, therefore, that accidents occur
quite often not only during berthing operations but as well
when an already moored ship starts to have a few oscillations
in resonance with periodic forces resulting from the combination of wave trains or other causes.
As no upper limit can be attributed to the actions that can
be exerted on fenders or mooring ropes, nor a lower limit to
their resistance and energy absorption capacity, there is no
alternative but to design berthing structures and to select
berthing and mooring equipment having in view the minimization
of their generalized cost. By that is to be understood their
initial cost plus the present value of all future expenses related to maintenance and repairs as well as direct and indi-
rect expenses incurred in case of an accident. This can be
done by resorting to the concept of "expectation of an event"
which corresponds to the product of the probability of occurrence of the event and the amount of expenses the event can
cause. The larger the amount of expenses involved in case of
the berthing structure or its equipment being damaged, the
greater the convenience of increasing their strength, having
in view the reduction of the probability of such occurrances.

The availability of alternative berths and of spare fenders
and ropes has also to be taken into consideration when evaluating the expectation of the distinct modes and degrees of damage.

References: Vasco Costa, F. "Optimization of Structures", VIII
Congress of the Int.Ass. for Bridge and Structural Engineering,
New York 1968.
Vasco Costa, F."Reliability of Partly Damaged Structures" 1st
Working Conf.on Reliability and Optimization of Structural Sys-
tems - Aalborg, Denmark, May 1987.

E. Bratteland (ed.), Advances in Berthing and Mooring of Ships and Offshore Structures, 322.
© 1988 by Kluwer Academic Publishers.

DISTRIBUTION OF BERTHING SPEED AND ECCENTRICITY FACTOR OF
VLCC INVESTIGATED IN JAPAN

S.Ueda
Chief of the Offshore Structures Laboratory, Structures
Division, Port and Harbour Research Institute, Ministry
of Transport, Japan

1. INTRODUCTION
Berthing speed and the eccentricity factor were measured at
several deep water oil terminals as items in the investigation of berthing energy of VLCC. Data of berthing speed were
obtained at the Keiyo Seaberth, the Kashima Seaberth, Isewan
Seaberth and the Nisseki Kiire Seaberth by use of berthing
speed meters installed at those berths. For the Keiyo Seaberth, data were obtained in 1971, 1972 and 1978. For the
Kashima Seaberth data were obtained in 1972 and 1973. For the
Isewan Seaberth and the Nisseki Kiire Seaberth, data were obtained in 1978. Data of eccentricity factors were obtained at
the first two seaberths in 1971 and 1972. All of these seaberths are available for more than 200,000dwt oil tankers and
the structure of each berth is steel pipe pile dolphin type.
Therefore those berths can be regarded as being of the same
kind.

2. BERTHING SPEED
2.1 Frequency Distribution of Berthing Speed
The frequency distribution of berthing speed is obtained as
shown in Fig. 1 with regarding to all data obtained. Data at
the first impact and the second impact are secured. The first
impact means ship's first touch to the berth at either bow or
stern, and the second impact means the successive touch after
the first one. Berthing speeds both in the first and the
second impact are larger than those of other successive impact.
The maximum berthing speed observed in the data is 13cm/s, the
mean, μ, is 4.41cm/s and the standard deviation, σ, is 2.08 cm²/
s². As the number of data is 788, the maximum berthing speed
of 13cm/s corresponds to 99.61% of the cumulative probability.

2.2 Relation of Berthing speed at the First and the second
impact
Fig. 2 shows the frequency distribution of berthing speed at
the first impact and the second impact. The distributions are
quite similar. Furthermore, Figure 3 shows the relation of
berthing speed at the first impact and at the second impact
obtained at the Keiyo Seaberth. Generally, it seems that berthing speed at the first impact is somewhat larger than for the
second impact. However, sometimes berthing speed at the second impact is larger than for the first impact. Consequently,
we can recognize that there is not any significant difference
of berthing speed for the first and the second impact.

323

E. Bratteland (ed.), Advances in Berthing and Mooring of Ships and Offshore Structures, 323–327.
© 1988 by Kluwer Academic Publishers.

2.3 Comparison of Berthing Speed obtained at Several Seaberths
Figure 4-a,b,c show the frequency distribution of berthing
speed obtained at the Keiyo Seaberth, the Nisseki Kiire Seaberth and the Isewan Seaberth, respectively. Frequency distribution of berthing speed obtained at the Keiyo Seaberth and the
Nisseki Kiire Seaberth are fairly similar, while that for the
Isewan Seaberth differs to the other two in the mean and the
dispersion.
2.4 Relation between Berthing Speed and Displacement
Figure 5 shows the relation between berthing speed and displacement obtained at the Keiyo Seaberth and the Kashima Seaberth.
With regard to large tankers, it could be said that large berthing speed may occur for any displacement.
2.5 Expected Values of Berthing Speed
Table 1 lists the 1/1,000 expected values of berthing speed
calculated by using the Weibull distribution function comparing
with the design berthing speed. The 1/1,000 expected value of
berthing speed is less than 15cm/s. For seaberths receiving
about fifty tankers per year, this expected value would probably occur once in every ten years.

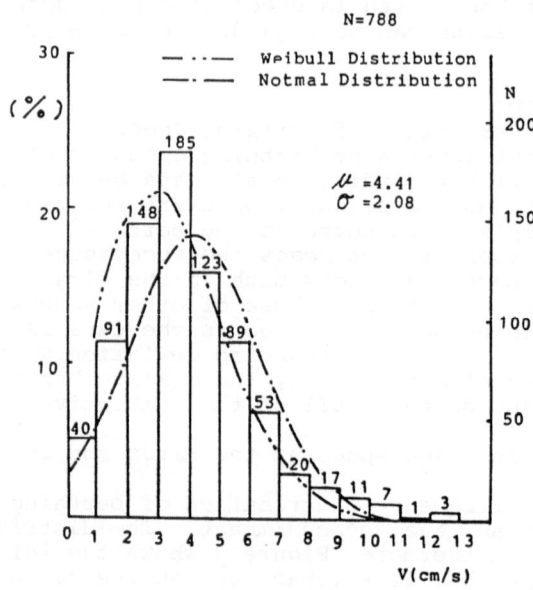

Figure 1   Frequency Distribution of Berthing Speed

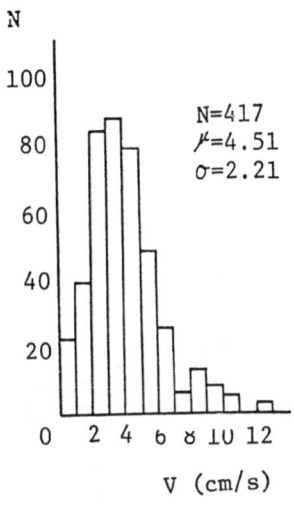 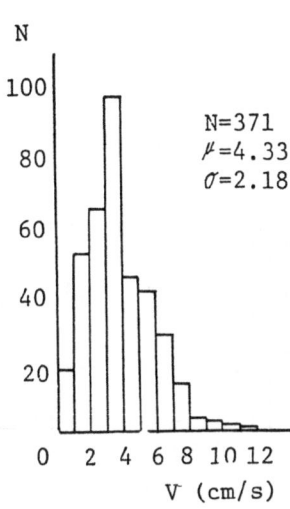

a) First Impact   b) Second Impact

Figure 2 Frequency Distribution of Berthing Speed

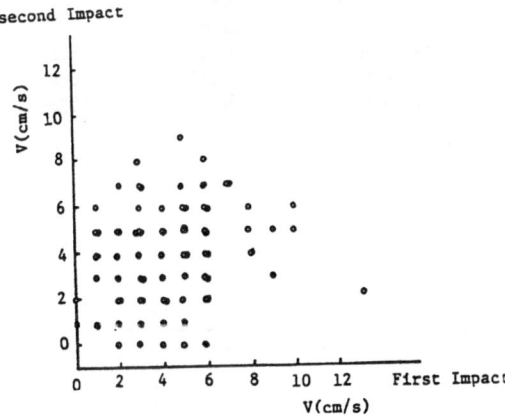

Figure 3 Relation of Berthing Speed between
First Impact and Second Impact

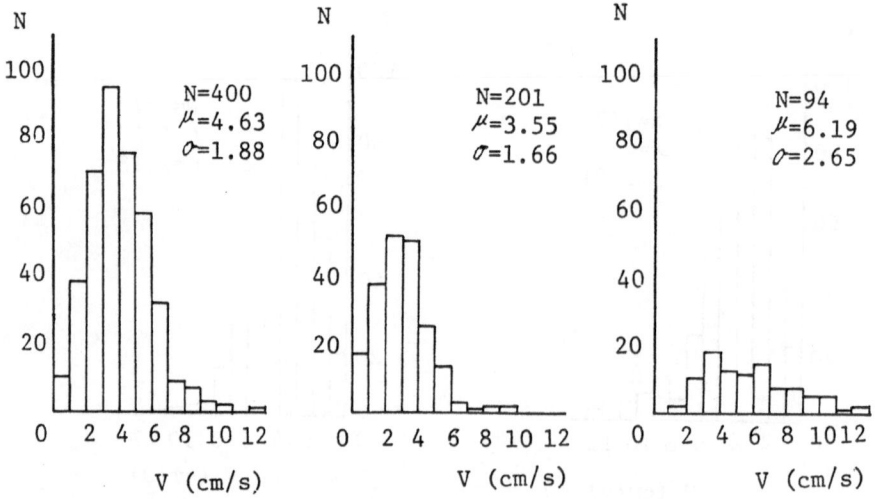

a) Keiyo Seaberth   b) Nisseki Kiire   c) Isewan Seaberth
      Figure 4 Frequency Distribution of Berthing Speed

Table 1 Expected Value of Berthing Speed

	1/1,000 Expected	Max.Observed Berthing Speed	Design Berthing speed
All of Data	14.5cm/s	13cm/s	
First Impact	15.5cm/s	13cm/s	
Second Impact	14.1cm/s	12cm/s	
Keiyo Seaberth	14.5cm/s	13cm/s	15cm/s
Nisseki Kiire	13.4cm/s	10cm/s	20cm/s
Isewan	20.2cm/s	13cm/s	15cm/s

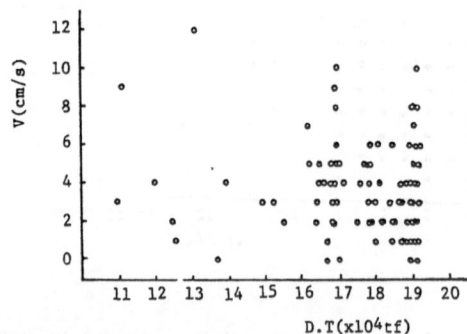

Figure 5 Relation between Berthing Speed and Displacement
       of Tankers

3. ECCENTRICITY FACTOR

The eccentricity factor is given by formula (1).

$$C_e = \frac{1}{1 + \left(\frac{l}{r}\right)^2} \qquad (1)$$

where, C_e:eccentricity factor, l:length of eccentricity, r:radius of gyration.

The eccentricity factors were calculated based on data obtained at the Keiyo Seaberth and the Kashima Seaberth. Figure 6 shows the distribution of eccentricity factors. They are nearly uniformly distributed in the range from 0.5 to 0.7. Consequently, it is recommended to use 0.7 as the eccentricity factor in the design work, for cases where there is no given data for the eccentricity factors.

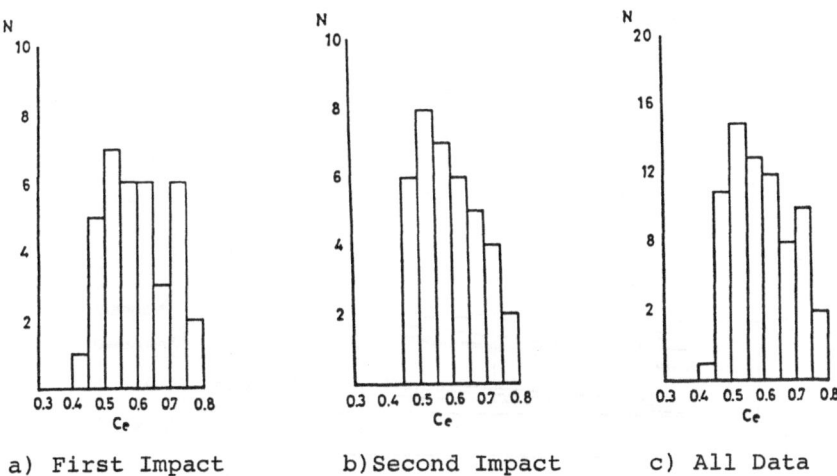

a) First Impact   b) Second Impact   c) All Data

Figure 6   Frequency Distribution of Eccenticity Factor

4. CONCLUSION

The 1/1,000 expected value of berthing speed of VLCC is about 15cm/s. The eccentricity factor C_e is determined considering the berthing point, the center of gravity and the radius of gyration of ship. If the values of these factors are not given in the design condition, it is recommended to use 0.7 as the eccentricity factor.

REFERENCE
1.   Ueda,S.: Study on Berthing Impact Force of Very Large Crude Oil Carries, Report of the Port and Harbour Research Institute, Vol. 20, No.2, June, 1981.

THE VIRTUAL MASS COEFFICIENT OF BERTHING SHIP

S.Ueda
Chief of the Offshore Structures Laboratory, Structures
Division, Port and Harbour Research Institute, Ministry of
Transport, Japan.

1. INTRODUCTION

The virtual mass of berthing ship is defined as the sum of the
mass and the added mass of berthing ship. There have been
many studies done by many researchers. Major results are
summarized in Table 1 and drawn in Figure 1.

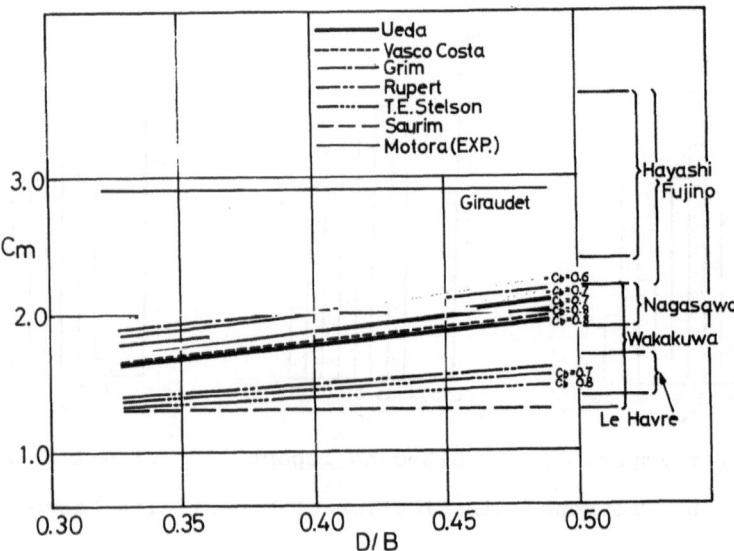

Figure 1. Summary of the Results of Virtual Mass
Coefficient.

Currently in Japan, the virtual mass of berthing ship is calculated according to the formula based on the experiment by
T.E. Stelson. However, comparing with above mentioned studies, it seems that the added mass calculated by this formula
is smaller than those by the other formulae. The reason for
such underestimation is that although Stelson carried out the
experiment only for the condition of L/d = 1 against various
condition of B/d, and also only for the condition of B/d=1

328

E. Bratteland (ed.), Advances in Berthing and Mooring of Ships and Offshore Structures, 328–337.
© *1988 by Kluwer Academic Publishers.*

against various conditions of L/d, the results of Stelson's experiment have been expanded and used for the condition of L/d = 8 against the condition of B/d $\lesseqgtr$ 1. Where L, B, and d correspond to ship length, width and draft, respectively. It must be said that if we calculate the added mass of berthing ship with B/d = 2 according to the Stelson's experiment, the added mass should be 1.46 to 1.60 times the mass of a cylinder of diameter d. As the result of review of berthing ship, model experiment was carried out in the Port and Harbour Research Institute, Ministry of Transport, Japan. A revised formula for the virtual mass of berthing ship was proposed and the result were examined comparing with formerly achieved results.

Table 1 Proposed Formulae and Values of
Virtual Mass of Berthing Ship

Name	Formula	range of value	
Grim	$C_M = 1.3 + 1.8 \ d/B$		
Vasco Costa	$C_M = 1.0 + 2.0 \ (d/B)$		
Saurin	$C_M = 1.3$	1.3	
Rupert	$C_M = 0.9 + 1.5 \ (d/B)$		
Giruadet	$C_M = 1.2 + 0.12 \ d/(H - d)$ where, $H \geq 1.07 \ d$		
Blok and Dekker		3.8	
Stelson	$C_M = 1.0 + 0.25 \ d^2 \ L \ w_0/DT$		
Observation in L'Havre		1.4	1.7
Ueda	$C_M = 1.0 + 0.5 \ C_B \ (d/B)$		
Motora		1.6	2.6
Nagasawa		1.9	2.1
Hayashi		2.4	3.6
Fujino		2.1	3.6
Wakakuwa		1.3	2.2

2. MODEL EXPERIMENT ON THE VIRTUAL MASS OF BERTHING SHIP
2.1 Outline of Model Experiment
Model experiment on the virtual mass of berthing ship was
carried out in the model basin belonging to the Port and
Harbour Research Institute by using a model ship of 500.000dwt
scaled into 1/50. The ratio of water depth to draft was set
in the range from 1.10 to 2.50. Berthing speed, berthing
angle ana berthing impact force were measured with deflection
meters and load cells installed on the H shaped steel beam on
the side wall of wave basin. The distance from the side wall
to the face line can be varied as 20cm, 130cm and 200cm by use
of different length H shaped steel bars.

2.2 Procedures of Experiment and Experiment Cases.
Model ship is pulled by wire ropes, driven by winch devices,
through the girder. The winch devices are stopped when the
model ship comes to a point 1 cm before colliding against the
model fender and make the model ship freely touch the fenders.
The depth of water was constant at 65 cm. The ratio of water
depth to draft, H/d, was varied as 1.10, 1.25, 1.50, 2.00 and
2.50. Three types of berth construction were used: gravity
type, piled pier type and dolphin type. These types are denoted as A, B, and C in this paper. Two kinds of springs of
which spring constant are 12.8kgf/cm (33tf/cm in Prototype)
and 25.0kg/cm (62.5tf/cm in prototype) are used as model
fenders. Berthing speed of model ship is set in three level
as 0.7cm/s (5.0cm/s in prototype), 1.4cm/s (10cm/s in prototype) and 2.2cm/s (15.6cm/s in prototype).

2.3 Results of Experiment
The virtual mass coefficient is calculated by formulae (1) and
(2).

$$P = c \sqrt{K_d C_e M_d} \cdot V \qquad (1)$$

$$c = \sqrt{C_M} = \sqrt{\frac{M_2}{M_d}} \qquad (2)$$

Where, P:berthing impact force, K_d:spring constant of model
fenders. C_e:eccentricity factor, M_d:displacement mass of berthing ship, M_a:Virtual mass coefficient of berthing ship,
V:berthing speed, C_M:virtual mass coefficient, c:square root
of virtual mass coefficient.
Now, the results of the model experiments are arranged orderly
in accordance with formula (1). Berthing speed is taken as
the abscissa and P/$K_d C_e M_d$ is taken as the ordinate. The
relation between these factors can be represented as shown in
Fig. 2-a,b,c. From these figures, it is known that berthing
speed is proportional to P/$K_d C_e M_d$ and its gradient is c.
The average value of gradient c was obtained for each case of
experiment and those values are summarized in Table 2. The
numbers in parentheses are the standard deviation. The deviation of data of each experimental case is, however, small. Fig.

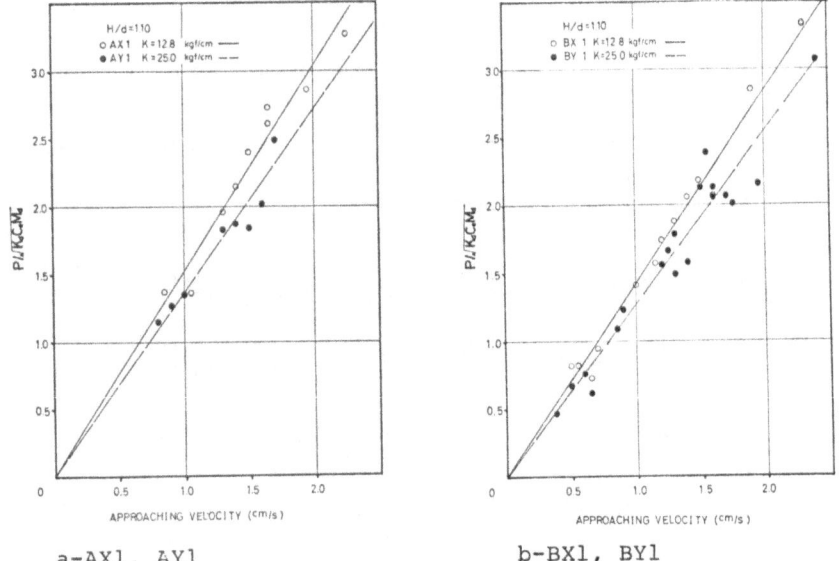

a-AX1, AY1   b-BX1, BY1

Fig.2. Relation between Berthing Speed and Berthing Impact Force

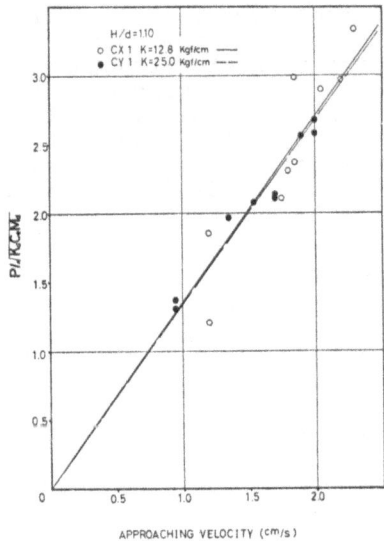

Figure-2   Relation between Berthing
Speed and Berthing Impact Force

332

Table 2 Average Values of Square Root of
Virtual Mass Coefficient

Structure	H/d Stiffness	1.10	1.25	1.50	2.00	2.50
A	12.8kgf/cm	1.52 (0.10)	1.41 (0.13)	1.36 (0.08)	1.17 (0.10)	1.07 (0.07)
A	25.0kgf/cm	1.36 (0.07)	1.31 (0.06)	1.32 (0.07)	1.18 (0.14)	1.08 (0.10)
B	12.8kgf/cm	1.42 (0.10)	1.37 (0.13)	1.37 (0.11)	1.26 (0.11)	1.16 (0.14)
B	25.0kgf/cm	1.27 (0.12)	1.27 (0.12)	1.30 (0.10)	1.15 (0.09)	1.13 (0.07)
C	12.8kgf/cm	1.36 (0.17)	1.42 (0.08)	1.35 (0.09)	1.15 (0.05)	1.05 (0.03)
C	25.0kgf/cm	1.35 (0.07)	1.39 (0.11)	1.28 (0.07)	1.13 (0.12)	1.10 (0.10)

3 shows the relation between square root of virtual mass coefficient, c, and the ratio of water depth to draft, H/d. It is known that the smaller the ratio of water depth to draft, H/d, becomes, the larger the virtual mass coefficient becomes. Fig. 4 shows the relation between square root of virtual mass coefficient, c, and distance between ship hull and face line. In other words, the figure shows the difference of virtual mass against different type of berth structures. The influence of structural type on the virtual mass coefficient is shown when the ratio of water depth to draft is small. Generally there is, however, no significant difference caused by structural type. Figure 5 shows the relation between the virtual mass coefficient and the ratio of draft to width, d/B. It is known that both factors are linear system relations.

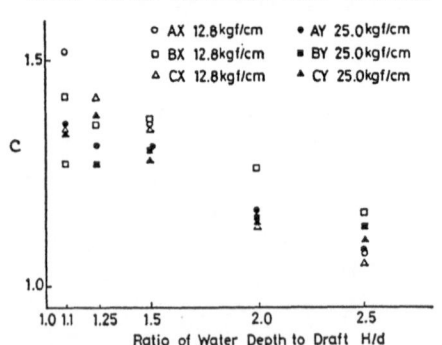

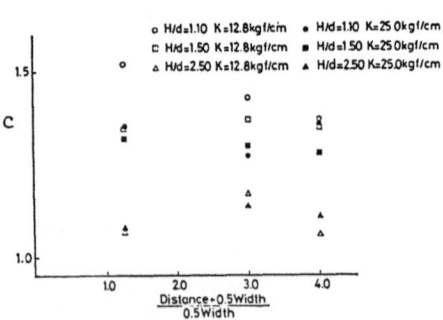

Fig.3. Relation between Square Root of Virtual mass Coefficient and Ratio of Water Depth to Draft

Fig.4. Relation between Square Root of Virtual mass Coefficient and Distance to ship hull

Summarizing the results of the experiment mentioned above, it is known that the square root of virtual mass coefficient, c, is in the range of 1.27 to 1.57 for the ratio of water depth to draft, H/d= 1.10. Therefore, it seems that the virtual mass coefficient C_M is in the range from 1.61 to 2.31.

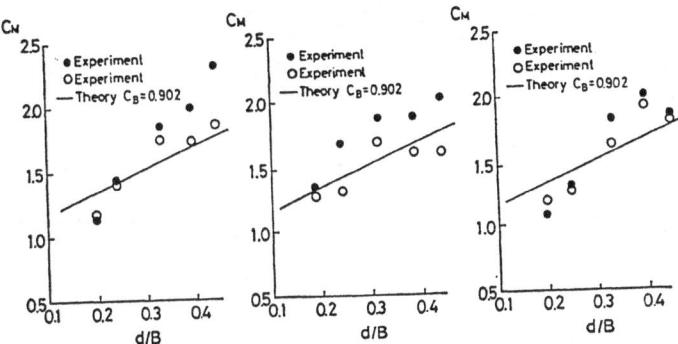

Figure 5. Relation between Square Root of Virtual Mass Coefficient and Ratio of Draft to Ship Width

3. REVISED FORMULA PROPOSED FOR CALCULATION OF THE VIRTUAL MASS

The author would like to propose a simple formula for calculation of virtual mass of berthing ship according to such experiments carried out by both T.E. Stelson and authors. Fig.6 shows modelled pattern wherein the difference between the vibration of a column and a plate in the Stelson's experiment and a berthing ship are visualized. In other words, in the Stelson's experiment the column and plate were completely immersed in the water, however, in the case of a berthing ship, the ship is just half immersed. Since the results of Stelson's experiment are hydrodynamically reasonable, his results can be applied as they are.

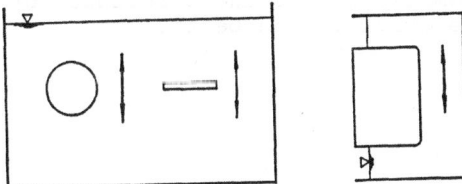

Figure 6. Modelled Pattern Wherein the Difference between the Vibration of a Column and a Plate and a Berthing Ship.

When an arbitrary cross-section of berthing ship with the surface of still water used as a symmetrical axis is assumed, it is possible to use a half of added mass for this cross-section as the added mass for berhing ship.
Consequently, the added mass of berthing ship can be approximately given by the following formylar (3).

$$M_a = \frac{1}{2} \times \frac{\pi}{4g}(2d)^2 Lw_0 = \frac{\pi}{2g} d^2 Lw_0 \qquad (3)$$

where, d:draft of berthing ship, L:ship length, w_0:unit volume weight of sea water, g:acceleration of gravity.
The virtual mass coefficient can be represented by formula (4).

$$C_M = \frac{M_d + M_a}{M_d} = 1 + \frac{\pi}{2C_B} \times \frac{d}{B} \qquad (4)$$

where, C_B is the block coefficient and is represented by formula (5)

$$C_B = \frac{M_d g}{LBdw_0} \qquad (5)$$

Formula (4) is substantially in the same form as that of formula (6) which was proposed by Vasco Costa.

$$C_M = 1 + \frac{2d}{B} \qquad (6)$$

Formula (4) gives the same value of added mass coefficient as formula (6) when $C_B = 0.785$. Vasco Costa corrected Grim's experimental formula by comparing it with the results of experiments carried out by Saurin. Although these formula were obtained in different ways, they give almost the same values.
Table 3 lists the virtual mass coefficient calculated by formula (4) comparing with the values obtained by the experiments. Both the virtual mass coefficient calculated by formula (4), and results of field investigation are plotted in Fig.7. The results of model experiments does not perfectly conform to the virtual mass coefficient calculated by formula (4).
The gradient of linear relation between the virtual mass coefficient and the ratio of draft to width, d/B obtained from the experiment is seemingly larger than that obtained from formula (4). And this is remarkable for the structures of type A which is equivalent to the gravity type berth.

Table 3 Virtual Mass Coefficient Obtained by
Experiment and Calculation

Case	H/d	d/B	Exp	Cal
AX1	1.10	0.442	2.31	1.77
AX2	1.25	0.396	1.99	1.69
AX3	1.50	0.333	1.85	1.58
AX4	2.00	0.243	1.37	1.42
AX5	2.50	0.199	1.14	1.35
AY1	1.10	0.442	1.85	1.77
AY2	1.25	0.396	1.72	1.69
AY3	1.50	0.333	1.74	1.58
AY4	2.00	0.243	1.39	1.42
AY5	2.50	0.199	1.17	1.35
BX1	1.10	0.440	2.02	1.77
BX2	1.25	0.386	1.88	1.67
BX3	1.50	0,316	1.88	1.55
BX4	2.00	0.242	1.69	1.42
BX5	2.50	0.190	1.35	1.33
BY1	1.10	0.440	1.61	1.77
BY2	1.25	0.386	1.61	1.67
BY3	1.50	0.316	1.69	1.55
BY4	2.00	0.242	1.32	1.42
BY5	2.50	0.190	1.28	1.33
CX1	1.10	0.447	1.85	1.78
CX2	1.25	0.394	2.02	1.69
CX3	1.50	0.329	1.83	1.58
CX4	2.00	0.247	1.32	1.43
CX5	2.50	0.198	1.10	1.34
CY1	1.10	0.447	1.82	1.78
CY2	1.25	0.394	1.93	1.69
CY3	1.50	0.329	1.64	1.58
CY4	2.00	0.247	1.28	1.43
CY5	2.50	0.198	1.21	1.34

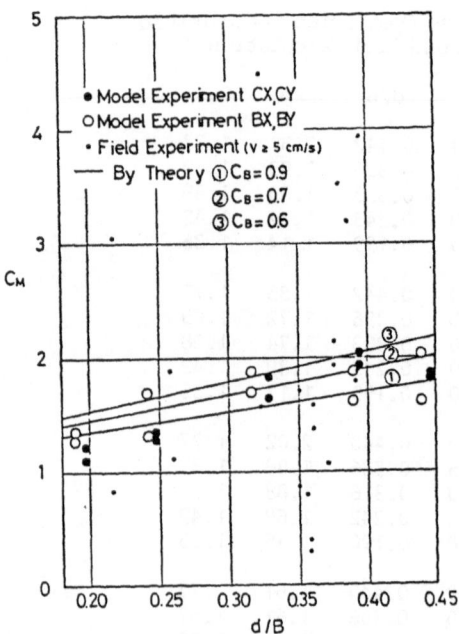

Figure 7. Virtual Mass Coefficient obtained by
Calculation and Investigation

3.1 Influence of Structural Type to the Virtual Mass Coefficient.

As for the piers or dolphins, the virtual mass coefficient calculated by formula (4) agrees well with the results of model experiments. However, for the gravity type berth the virtual mass coefficient for 10,000dwt cargo ship (144m in length, 19,4m in width, 8.2m in draft fully laden and 0.59 in block coefficient) calculated by formula (4) is C_M= 2.1 against that obtained from experiment, C_M=2.3. But, in the design work, this difference is considered not so significant.

4. CONCLUSION

The virtual mass coefficient is proposed given by formula (4) for the design of mooring facilities.

$$C_M = \frac{M_d + M_a}{M_d} = 1 + \frac{\pi}{2C_B} \times \frac{d}{B} \qquad (4)$$

REFERENCES:

1.Ueda,S.:Study on Berthing Impact Force of Very Large Crud Oil Carries, Report of the Port and Harbour Research Institute, Vol.20, No.2, June,1981.

2.Japanese National Section of PIANC: Design of Fender Systems, March,1980.

3.PIANC:Report of the International Commission for Improving the Design of Fender Systems 1984.

4.Brolsma,J.U.et al: On Fender Design and Berthing Velocities, Proc. of PIANC, Leningrad, 1977, Sect 2, Subject 4, pp.87-100.

5.Hayashi,T. and M.Shirai:Study on the Berthing Impact Force, 8, 9, 19th Conference on Coastal Engineering in Japan, 1961, 1962, 1972, pp.101-106, 112-115, 427-431.

6.Vasco Costa,F.:The berthing Ship, The Dock & Harbour Authority, Vol.45,Nos.523,524,525,May,June,July,PP.22-26,49-52,90-94.

7.Rupert,D.:Zir Bemessung und Konstruction von Fendern un dalben, Meittelungen des Franzius-Institute fur Wasserbau und Kusten Ingenieurs wesen der Technishe Universitate, Hanover, Heft 44.

8.Grim,O.: Das Schiff und des Dalben, Schiff und Haven, 7, p.535.

9.Giraudet,P.: Researches Experimentale sur l'energie d'accostage des navires, Annaless des ponts et chaussees No.2 de mars-arvil, 1966.

10.Blok,J.J. and J.N.Dekker: On Hydrodynamics aspects of Ship Collision with rigid or non-rigid Structures, 11th OTC,1979, #3664, pp.2683-2698.

11.Stelson,T.E. and F.T.Mavis:Virtual Mass and Oscileration in Fluid, Proc. of ASCE, Vol.81, No. 670,EM, 1955, pp.670=1-670=9.

12.Motora,S.:On the Measurement of Added Mass and Added Moment of Inertia for Ship Motions, Journal of the Society of Naval Architects of Japan, Nos.105, 106, 1959, 1959, pp.83-92, 59-68.

13.Nagasawa,J.: Hull Strength on the Ship's Contact with the Quay wall, Journal of the Society of Naval Architects of Japan, Nos. 106, 107, 1959, 1960, PP.189-194, 221-228.

14.Fujino,M.:Studies on Manoeuvrability of Ship in Restricted Waters, Journal of the Society of Naval Architects of Japan, No.124, 1978, pp.51-72.

15.Wakakuwa,T.:Estimation of the Magnitude of Lateral Loads on Piled Jetties under Impact of Berthing Vessels, Report of the Ship Research Institute, Vol.1, No.1, pp.1-32.

SHIPS AND BERTH STRUCTURES INTERACTIONS

F.F.M. VELOSO GOMES

Ph.D., M.Sc., Civil Eng.
Associate Professor at the Laboratorio Hidraulica
Faculdade de Engenharia da Universidade do Porto - PORTUGAL

1. INTRODUCTION
The advanced prediction techniques of ship behaviour at berthing
facilities need to include several hydrodynamic and   mechanical
aspects influenced by several local structures.

2. FEATURES CONCERNING BERTHING OF   SHIPS   RELATED   TO   BERTH
   STRUCTURES
Some of the important features concerning berthing of ships
related to berth structures, in a sense of safety, operational
limits and berthing design, are:

- environmental short waves in the vicinity of the berthing
  position as a result of interactions between the incoming wave
  climate, the local wind (external conditions), the   bottom
  configurations and the following harbour components:

  . breakwaters
  . terminal or berth lay-out
  . berth or quay structure.

  These aspects have to be considered during the manoeuvring
  operations and while the ship is moored (wave induced dynamic
  loading).

  - long ship motions induced by water resonance inside the
    harbour, particularly at the nodes. They depend on the
    geometry and dimensions of the basins, harbour entrance,
    water depth, reflections on quay walls and other boundaries,
    berthing position and on the overall mooring and berth arrange
    ments.

  - water cushion effect between the ship's hull and the berthing
    structure (open pile, vertical continuous quay).

  - interactions between excitations and mooring and fendering
    arrangements (conventional systems and new low recoiling
    fenders with tension moorings based on pier or quay winches).

  - stand-off actions on moored ships associated to the effect
    of currents running through pile-supported berthing structures
    (hull pressure changes, forces and moments).   This actions
    are dependent on the berth lay-out, transverse and longitudinal
    pile spancings, pile diameters, current flow, ship position,
    underkeel clearance).

  - interaction between a passing ship and a moored   ship   as a

338

E. Bratteland (ed.), Advances in Berthing and Mooring of Ships and Offshore Structures, 338–342.
© 1988 by Kluwer Academic Publishers.

result of a close distance between the berth and a waterway.

- motion induced hydrodynamic coefficients (added mass and damping coefficients). They depend on the approaching motion history, on the distance between the ship and the quay, on shallow water effects.

3. THREE USUAL CALCULATION METHODS

In traditional calculations of berthing impact, the approaching velocity of ship V and some few coefficients account for everything which is unknown but belived to be important to quantify the effective berthing energy E. The final value depends very much on the designers skill and judgement.

For instance, the "Design Standards" referred by the "Working Group on Fender System Design of the Japanese National Section of PIANC"(1980), only consider two cases of structural situations through a berthing manoeuvre coefficient C_E (eccentricity factor, falling in 0.5 to 0.8 range):

- "wharf with a number of fenders arranged"

$$E = 1/2 \ (WV^2/2g) \qquad W - \text{"virtual" weight of ship}$$

- "dolphine or wharf with fenders placed at a large spacing between"

$$E = 1/2 \ (WV^2/2g) \ C_E$$

And the report by the "Commission Internationale pour l'Amélio ration de la Conception des Systèmes de Défense", PIANC (1984) considers

$$E = E_C \ x \ C_M \ C_E \ C_C \ C_S$$

C_C - wharf configuration coefficient (or berth type factor) which lies in the range 0.8 (for a vertical face and a parallel berthing manoeuvre) to 1.0 (for an open pile berthing structure).

C_M - added mass coefficient

C_E - eccentricity coefficient

C_S - softness coefficient

A purely deterministic approach is nowadays unsatisfactory to a rational design of docks, harbours and mooring facilities. So, a far better understanding of the physics and mechanics involved in the dynamic behaviour of moored ships must be a priority.

The full-scale measurements of approaching velocity, eccentricity factors and impact energies from monitored berths are so closed dependent on local conditions that such data must be widely extended to be general accepted to support statistical methods in order to provide design probability curves and risk analysis.

It is well known that small scale hydraulic modelling is an important design tool. But as a simplified representation of a complex real oscillating system the results and their interpretation must be considered with precautions.

Proper extensive prototype measurements would be very helpfull to adjust the testing conditions, the scaling techniques and to improve levels of predictions.

4. QUESTIONS ABOUT MATHEMATICAL MODELS

Of course we have recent advances on analytical expressions and on numerical models, which are constantly updating and work is hard in progress for further development.

But two vital questions arise about the actual applicability of such models:

- Are the available mathematical models limited, as a useful prediction technique, to the early stages in the design of a berthing/mooring system (including berth location, ship motions, fenders, mooring lines)?

- If the most sophysticated mathematical models can successfully substitute the physical models, as some of them are commercially presented to the harbour authorities, we would like to learn more about the way they actually can compute the following problems and the simplifications introduced to them, specially at more exposed locations:

  - non-linear (and non-permanent) excitations:

    . low frequency waves and resonant harbour oscillations
    . wave overtopping of breakwaters
    . wave breaking
    . flow separation effects
    . coupling between several short waves deformations inside the harbour
    . drift forces on the ship's hull (second order wave induced forces and moments, wave grouping effects).

  - non-linear relations between fender excitation and compression, mooring lines loads and elongation, under cyclic loading. Multiple possibilities of mooring arrangements, including forced fendering. Interactions.

  - determination of the elements of the motion induced added mass matrix, considering:

    . real hull shapes
    . approaching motion including the rate of change of the berthing velocity and the distance from the ship to the quay
    . underkeel clearance
    . fender and berthing structure characteristics

  - determination of the elements of the motion induced damping matrix associated to the wave system generated by the motion of the vessel, the vortice generation and the friction forces on the ship's hull (considering viscous effects).

To give confidence and reliability it is important to know which prototype situations or hydraulic model comparations have been considered. This is necessary to verify the accuracy and validity of such mathematical models in order to quantify the complex real excitations and induced interactions.Computer time and costs will provide a better insight.

5. SIMPLIFIED ASSUMPTIONS

In fact, the set of differential equations of motion of a moored vessel with non-linear and frequency dependent hydro-dynamic coefficients are being approached by analytical methods which consider simplified assumptions. The question is how far are those assumptions unimportant to achieve realistic results since they consider one of the options within the following items:

- some or all the six fundamental forms of ship motions (they are not equally important for different ship types and cargo operations)

- free ship or free sailing ship or moored ship

- simplified hull shape or real hull shape

- regular wave excitation (sinusoidal) or irregular waves

- one wave direction or several wave directions

- inclusion or non-inclusion of wind and current excitation forces

- inclusion or non-inclusion of drift forces (at least for beam and head waves)

- without considering or considering underkeel clearance (at least with some of the motions)

- consideration of all the hydrodynamic coefficients or only some of them

- uncoupled motions or coupled motions (coupled hydrodynamic coefficients)

- linearized assumption of small amplitude of motion with constant hydrodynamic coefficients for a given form (independent of the motion amplitude and time) or frequency dependent hydrodynamic coefficients

- considering the quay influence or without considering such influence

- linear or non-linear behaviour of fenders

- linear or non-linear behaviour of mooring lines.

6. COMPLEMENTARY PREVISION TECNHIQUES

As a conclusion and as far as a recent literature survey can evidence, it is more realistic and sound to defend that it will be still necessary to run model tests at least to improve the assessment of the matrix hydrodynamic coefficients for real hull shapes and local conditions.

So, numerical models and physical models for berthing and mooring studies can be regarded as being complementary prevision techniques. And for the design of simple situations (sheltered berths, medium size ships, conventional berth and mooring arrangements, local knowledge on the behaviour of ships and structures) it is enough to use energy probability curves including the effect of all the stated berthing factors.

REFERENCES

Bruun, P.
Mooring and Fendering Rational Principles in Design. The Dock and Harbour Authority, February. Volume 64 Number 758 (1984)

DHL
Dynamic Behaviour of Moored Ships in Harbours. The Dock and Harbour Authority. February 1984

P.I.A.N.C.
(Japanese National Section). Design of Fender Systems.
Japan 1980

P.I.A.N.C.
Rapport de la Commission Internationale pour l'Amelioration de la Conception des Systemes de Defense. Supplement au Bulletin 45. Bruxelles 1984

Rita, M.A.B.M.
On the Behaviour of Moored Ships in Harbours - Theory,Practice and Model Tests. Dissertation. Laboratorio Nacional de Engenha ria Civil. Lisboa 1984

Vasco Costa, F.
Moored Ships as Oscillating Systems. The Dock and Harbour Authority, February and June 1983.

EFFECT OF FLUCTUATION OF WIND ON SHIP MOTIONS

S. Ueda
Chief of the Offshore Structures Laboratory, Structures
Division, Port and Harbour Research Institute, Ministry
of Transport, Japan.

1. INTRODUCTION
This paper describes the effect of fluctuation of wind on
ship motions examined by means of the numerical simulation
method. Ship sizes 258,000 dwt and 237,000 dwt oil tankers.
Ships are moored to fixed deep water terminals with several
synthetic ropes. Gusty wind is generated according to the
frequency spectrum proposed by Davenport. The results of computation show that long period large oscillations caused by
the fluctuation of wind could occur.

2. CONDITIONS OF COMPUTATION
2.1 Ships
Ships to be the object of this study are 258,000dwt,and 237,C00
dwt oil tankers. Ships are moored to dolphins with mooring
ropes and rubber fenders. The properties of ships are listed
in Table 1.

2.2 Mooring Systems
Ships are moored to dolphins with fenders and mooring ropes
made of nylon as shown in Figure 1. As described later, surging motion of the 237,000 dwt tanker is larger than for the
258,000 dwt tanker. The arrangement of mooring ropes are
revised so that the spring constant of ropes becomes smaller.

2.3 Wind and Wave Conditions
Figure 2 shows the frequency spectrum of gusty wind generated
by using the frequency spectrum proposed by Davenport. As
shown in the figure, the peak frequency becomes smaller with
the incremental decrease of wind speed. The dominant period
of gusty wind is longer than 50s. Therefore, we must pay
attention for surging motion against wind. In this case
study, wind direction was set at zero, that means wind blowing
from head to stern. The significant wave height is 1.5m and
the significant wave period is 10s. In this computation, wave
drift forces are not considered.

E. Bratteland (ed.), Advances in Berthing and Mooring of Ships and Offshore Structures, 343–346.
© 1988 by Kluwer Academic Publishers.

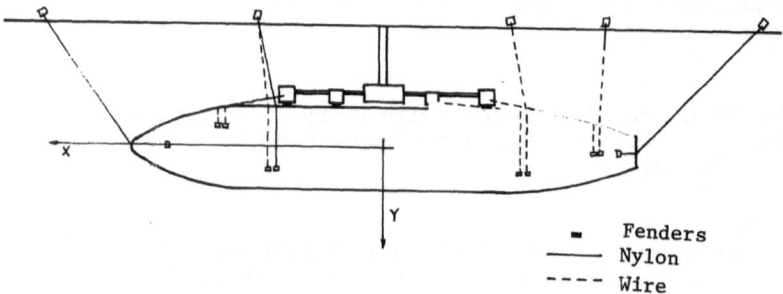

a) 258,000dwt Tanker

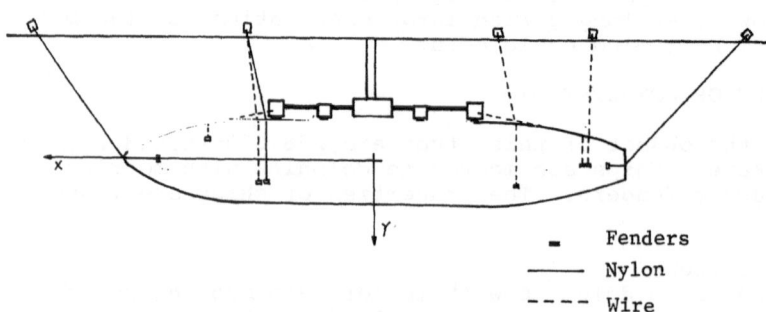

b) 237,000dwt Tanker
Figure 1 Mooring of Tanker

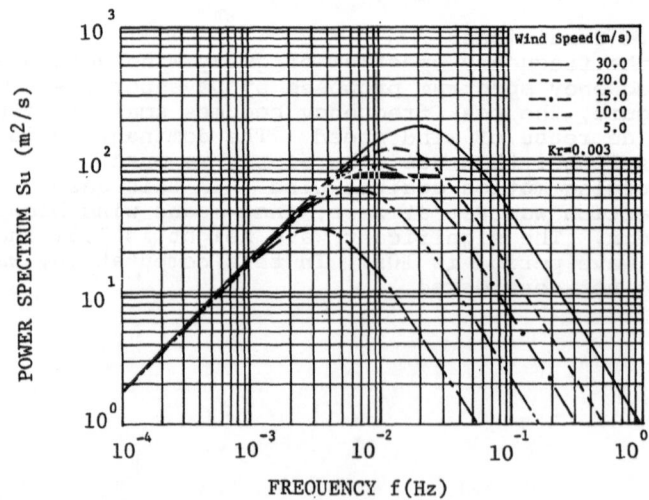

Figure 2 Frequency Spectrum of Gusty Wind by Davenport Spectrum

Table 1.   Properties of ship.

Items	Ship A	Ship B
D W T	258,000tf	237,000tf
L O A	321m	321m
L P P	310m	304m
B	58m	52m
D	29.5m	25.7m
d(Full)	19.5m	19.9m
d(Ballasted)	9.9m	7.34m
DT(FUll)	271,000tf	271,000tf
DT(Ballasted)	136,400tf	91,000tf

3.   RESULTS OF COMPUTATION

Table 2 lists the results of computation. Here, only 'surging and swaying motion are discussed. Examining Case No.1 to 10,it seems that there is no difference between the results with and without wave action. Therefore, it may be said that the effect of waves on these motions is not significant. Comparing the results of Case No.6 to 10 with the results of Case No.16 to 20, motions of the 237,000dwt tankers are larger than those of the 258,000 dwt tankers. As the virtual spring constant of longitudinal mooring system for 237,000 dwt tankers is calculated to about 38.5tf/m, the virtual natural period becomes close to the dominant period of gusty wind.

Table 2 Results of Computation.

NO.	DWT x10³	Wind Dir	V(m/s)	Wave Dir	$H_{1/3}$	$T_{1/3}$	Surge (cm) Max	Min	Ave	Sway (cm) Max	Min	Ave
1	258	0	5	-	-	-	-5.6	-20.9	-13.1	-0.0	-0.1	-0.1
2	258	0	10	-	-	-	-2.36	-80.2	-52.2	-0.1	-0.4	-0.3
3	258	0	15	-	-	-	-53.1	-177.4	-117.0	-0.3	-0.9	-0.6
4	258	0	20	-	-	-	-90.2	-316.3	-206.0	-0.5	-1.8	-1.1
5	258	0	30	-	-	-	316.7	-874.9	-299.7	25.7	-8.3	4.0
6	258	0	5	0	1.5	10	3.5	-35.1	-13.1	0.0	-0.2	-0.1
7	258	0	10	0	1.5	10	-16.4	-91.6	-52.1	-0.1	-0.4	-0.3
8	258	0	15	0	1.5	10	-47.1	-185.7	-116.9	-0.2	-1.0	-0.6
9	258	0	20	0	1.5	10	-79.4	-325.0	-206.0	-0.4	-1.8	-1.1
10	258	0	30	0	1.5	10	318.2	-871.0	-300.6	25.9	-8.5	4.0
11	237	0	5	-	-	-	-10.0	-32.8	-21.6	-0.1	-0.2	-0.1
12	237	0	10	-	-	-	-40.6	-129.5	-85.5	-0.2	-0.6	-0.4
13	237	0	15	-	-	-	-88.5	-292.1	-187.4	-0.4	-1.5	-0.9
14	237	0	20	-	-	-	-101.1	-504.7	-307.1	-0.5	-3.0	-1.6
15	237	0	30	-	-	-	25.6	-915.1	-487.0	16.2	-9.9	-1.7
16	237	0	5	0	1.5	10	4.0	-46.0	-21.6	0.0	-0.2	-0.1
17	237	0	10	0	1.5	10	-32.8	-143.2	-85.5	-0.1	-0.7	-0.4
18	237	0	15	0	1.5	10	-82.6	-301.5	-187.2	-0.4	-1.6	-0.9
19	237	0	20	0	1.5	10	-96.8	-508.4	-307.1	-0.4	-3.1	-1.6
20	237	0	30	0	1.5	10	21.4	-919.7	-490.5	16.1	-9.7	-1.7

4. CONCLUSION
As shown by the computation, attention must be paid to surge motions induced by gusty winds.

REFERENCES
1. Ueda, S.: Analytical Method of Ship Motions Moored to Quay Walls and the Applications, Note of the Port Harbour Research Institute, No. 504, 371p.
2. Ueda, S. et al: Numerical Simulation Method of Ship Motions Moored to Quay Wall and Same Characteristics of the Motions, Coastal Engineering in Japan, Vol. 27, 1984, pp. 293-301.
3. Davenport, A.G.: Gust Loading Factors, Journal of the Proc. of ASCE, ST3, June, 1967, pp. 11-34.

CONSIDERATION ON SAFETY OF FENDERING SYSTEM

S.UEDA
Chief of the Offshore Structures Laboratory, Structures
Division, Port and Harbour Research Institute, Ministry
of Transport, Japan.

1. INTRODUCTION

It is difficult to determine uniformly the safety factor of
fendering system against all the type of fenders, however, it
is important. Here, discussion is made on allowable load and
allowable stress of infra-structures. PIANC Fender Commission
issued a report for improving design of fendering systems in
1984. In that report, description on the safety of fendering
system is made for two types of fenders. The author will
present some supplementary discussion to that report.

2. CONSIDERATION ON RESERVE ENERGY

We discuss the safety of fendering system for both exsisting
breasting dolphin with buckling type fenders and the imaginary
ones with pneumatic fenders. The dolphin consists of steel
pipe piles of 1500mm diameter with yield stress 2.700kgf/cm^2.

2.1 Case Study for the Dolphin with Buckling Type Fenders.
The dolphin is designed so the stress induced in the steel
pipe piles is less than the allowable stress. Currently in
Japan, as the load from berthing ship is treated as extraordinary load, the allowable stress is 1.5 times that of ordinary load. Figure 1 shows the relation between both reaction
force and energy absorption of fenders and dolphins against
deflection of each. If we use the same steel pipe piles when
we design the breasting dolphin so that the stress of steel
pipe piles is $2/3$ or $0.82\ \sigma_y$ the number of steel pipe piles and
the lateral spring constant of dolphin become nine and 21.3tf/
cm, respectively. However, changes in the laterat spring
constant of dolphin make considerations complicated. Therefore,
we suppose that the lateral spring constant of dolphins does
not change for any conditions of allowable stress by djusting
the wall thickness of steel pipe piles. According to Figure 1,
calculation of energy absorption was made and the results of
energy absorption of fenders and dolphins are listed in Table 1
against both conditions that allowable stress is $2/3$ or $0.82\ \sigma_y$

347

E. Bratteland (ed.), Advances in Berthing and Mooring of Ships and Offshore Structures, 347–351.

Where,

E_1 = total energy absorption by mooring system up to the allowable stress in the steel pipe piles

E_{F1}= energy absorption by fender unit only up to the allowable stress in the steel pipe piles

E_{D1}= energy absorption by dolphin only up to the allowable
stress in the steel pipe piles

E_2 = total energy absorption by mooring system up to yield
stress in the steel pipe piles

E_{F2}= energy absorption by fender unit only between allowable
stress and the yield stress in the steel pipe piles

E_{D2}= energy absorption by dolphin only between the allowable
stress to the yield stress in the steel pipe piles.

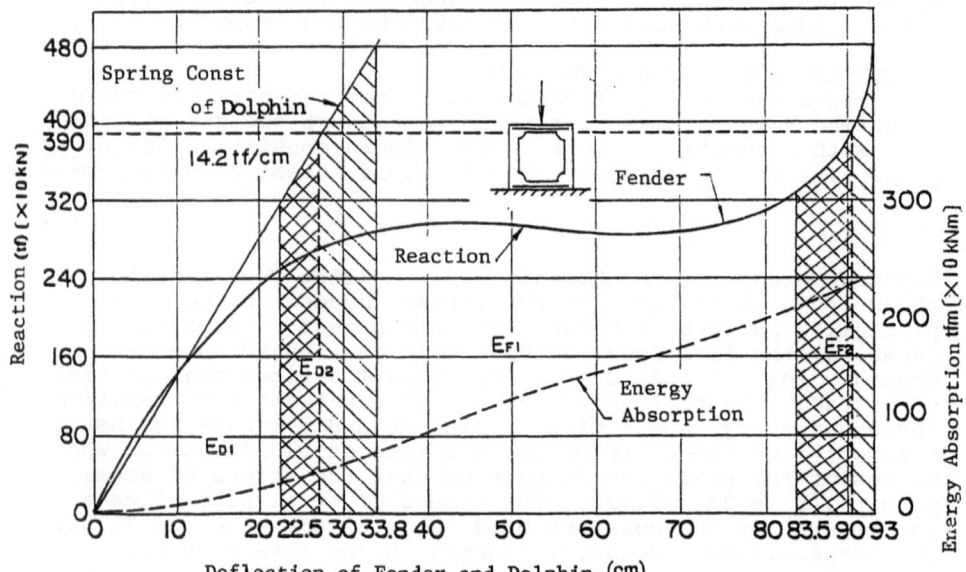

Figure 1.   Reaction Force and Energy Absorption for Dolphin
Pile and Fender Unit related to their respective
Deflection (Buckling Type)

Suppose the dolphin is designed on the condition that the allowable stress is $2/3\sigma_y$, then the reaction force will be 320tf
and 480tf for working stresses of $2/3\sigma_y$, and $1.0\sigma_y$, respectively. Supposing the dolphin is designed on the condition that
the allowable stress is $0.82\sigma_y$, reaction force will be 320tf
and 390tf for working stresses of $0.82\sigma_y$ and $1.0\sigma_y$. This means
that if we design the dolphin on the condition that the allowable stress is $0.82\sigma_y$, reserve energy of mooring system when
the working stress becomes the yield stress is only 17% of

energy absorption of mooring system when the working stress is $0.82\sigma_y$, while if we design the dolphin on the condition that the allowable stress is $2/3\sigma_y$, reserve energy of mooring system is 33 % of energy absorption of mooring system when the working stress is $2/3\sigma_y$. Therefore, reserve energy is not expected when we use the buckling type fenders if we take the allowable deflection as large as possible.

Table 1. Reserve Energy of Mooring System (tfm)
(by using Buckling Type Fender)

σ_{all}	E_{F1}	E_{D1}	E_1	E_{F2}	E_{D2}	E_2	E_2/E_1
$2/3\sigma_y$	210	36	246	38	45	329	1.33
$0.82\sigma_y$	210	36	246	15	17	278	1.13

2.2 Case Study for the Dolphin with Pneumatic Type Fenders. If we use the pneumatic type fenders, the relations between both reaction force and energy absorption of fenders and dolphins against each deflection are as shown in Fig. 2. And the reserve energies are calculated as listed in Table 2. Suppose the dolphin is designed on the condition that the allowable stress is $2/3\sigma_y$, reaction force will be 427tf and 641tf for working stress of $2/3\sigma_y$ and $1.0\sigma_y$. And suppose the dolphin is designed on the condition that the allowable stress is $0.82\sigma_y$, reaction force will be 427tf or 521tf for working stresses of $0.82\sigma_y$, or $1.0\sigma_y$, respectively. If we design the dolphin on the condition that the allowable stress is $0.82\sigma_y$, reserve energy of mooring system when the working stress becomes the yield stress is 26% of energy absorption of mooring system when the working stress is $0.82\sigma_y$, while if we design the dolphin on the condition that the allowable stress is $2/3\sigma_y$, reserve energy of mooring system is 68% of energy absorption of mooring system when the working stress is $2/3\sigma_y$.

Table 2. Reserve Energy of Mooring System (tfm)
by using Pneumatic Type Fender)

σ_{all}	E_{F1}	E_{D1}	E_1	E_{F2}	E_{D2}	E_2	E_2/E_1
$2/3\sigma_y$	210	64	274	104	81	459	1.68
$0.82\sigma_y$	210	64	274	41	31	346	1.26

350

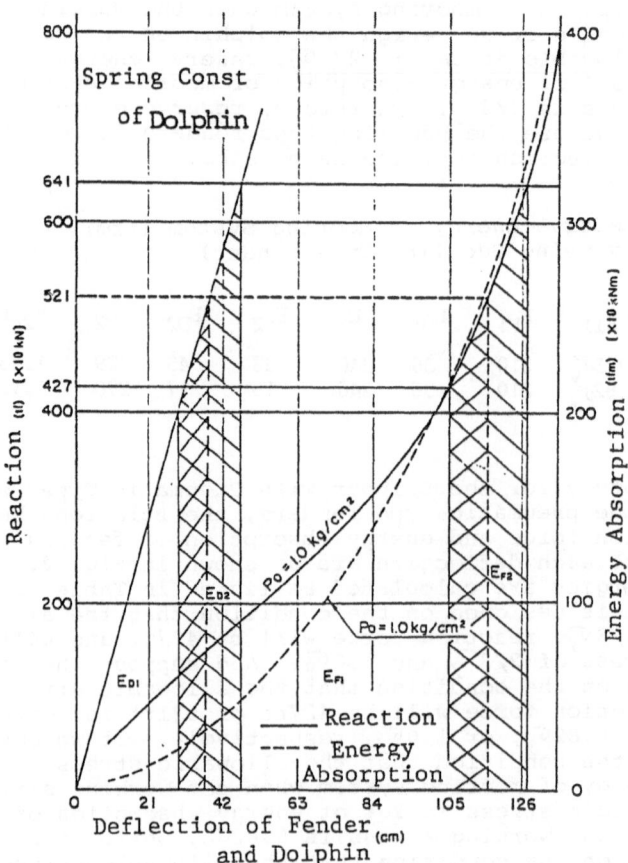

Figure 2. Reaction Force and Energy Absorption for Dolphin
Pile and Fender Unit related to their respective
Deflection (Pneumatic Type)

3. PROBABILITY OF EXCEEDANCE OF WORKING STRESS IN STEEL PILES
Table 3 lists the probability of exceedance of 50%, 60%, 70%,
80% 90% and 100 % of yield stress of steel pipe pile. The
dolphin is designed assuming the allowable stresses are 1.0$\bar{\sigma}_y$,
0.9$\bar{\sigma}_y$ and 2/3$\bar{\sigma}_y$ in the case of the buckling type fenders,
while the allowable stresses are 1.0$\bar{\sigma}_y$ and 2/3$\bar{\sigma}_y$ in the case
of the pneumatic type fenders. Reaction force both of
fenders and dolphin are 320tf, 427tf, 480tf, 427tf and 641tf,
respectively, when the working stress equals the allowable
stress. It seems that the probability of exceedance when using the buckling type fenders are rather larger than that when
using the pneumatic type fenders. Therefore, the allowable
stress of infra-structures must be determined considering the
fender type and the frequency of occurence of working stress.

It may be recommended to take care when using the buckling type fenders. Whether berthing impact force should be treated as ordinary load or extra-ordinary load should be discussed on the basis of energy absorption as long as the working stress is less than the yield stress.

Table 3. Probability Exceedance of 50%, 60%, 70%, 80%, 90% and 100% of Yield Stress of Steel Pipe Pile.

	Buckling Type fender			Pneumatic Type Fender	
	$\sigma_{all}=\sigma_y$	$\sigma_{all}=0.9\sigma_y$	$\sigma_{all}=2/3\sigma_y$	$\sigma_{all}=\sigma_y$	$\sigma_{all}=2/3\sigma_y$
$\sigma>0.5\sigma_y$	0.273	0.187	0.130	0.023	1.17×10^{-4}
$\sigma>0.6\sigma_y$	0.169	0.086	0.010	3.73×10^{-3}	4.11×10^{-6}
$\sigma>0.7\sigma_y$	0.112	8.36×10^{-3}	3.57×10^{-7}	5.06×10^{-4}	1.05×10^{-7}
$\sigma>0.8\sigma_y$	0.058	1.64×10^{-7}	4.62×10^{-8}	1.51×10^{-5}	2.82×10^{-9}
$\sigma>0.9\sigma_y$	0.027	6.29×10^{-8}	1.23×10^{-8}	1.75×10^{-6}	1.68×10^{-10}
$\sigma>1.0\sigma_y$	7.39×10^{-7}	2.90×10^{-8}	3.11×10^{-9}	1.64×10^{-7}	7.15×10^{-12}

REFERENCES
1. PIANC: Report of the International Commission for Improving the Design of Fender Systems, 1984.
2. Ueda, S. and E. Ooi: On the design of Fending Systems for Mooring Facilities in Port, Note of the Port and Harbour Research Institute, No. 596, 1987.

LONG WAVES AND MOORED SHIP BEHAVIOUR AT THE PORT OF SINES - A
RESEARCH PROJECT

MANUEL A. B. MARCOS RITA

Civil Engineer, Research Officer of the Laboratôrio Nacional de
Engenharia Civil (LNEC), Lisbon, Portugal

SUMMARY
    A 5 years research project to study the long waves and the
moored ship behaviour at the port of Sines (Portugal) is
briefly presented. The project is included in a more general
research project on the wave climatology of the portuguese
coast - the PO-WAVES project - presented in May 1987 to the
NATO Scientific Affairs Division, to be partly funded by the
Science for Stability Programme. The project is expected to
start in October 1987.

1. INTRODUCTION
    Recent studies for the design of harbours clearly pointed out
that the knowledge of the long wave climate near the harbour
entrance is essential for proper harbour design.
    In fact, long period waves, rather than short period waves,
can easily penetrate the harbour basins and cause harbour
resonance which in turn induces severe motions on moored ships.
Bound long waves, in particular, through the slow drift forces
they exert on moored ships, are an important causing agent of
moored ship problems, even in the absence of harbour resonance.
    A method to establish the long wave climate at the entrance
of a harbour is then essential to proper harbour design. And to
do that, prototype measurements of both short period and long
period waves must be carried out. In addition, moored ship
behaviour under the effect of the long waves must be analised.
    Results will allow harbour engineers, in absence of long wave
measurements at the harbour site, to lay out a first
approximation long wave climate for better harbour design.
    In addition, it will contribute to anticipate the occurrence
of hazardous moored ship situations, that is, it will allow a
better control of harbour downtime which has a strong influence
in the economics of the harbours and, consequently, of the
country.

2. BACKGROUND
    Apart from the usual recognised causes (landslides,
meteorological fluctuations and tsunamis) which are responsible
for the generation of free long waves, another cause of the
occurrence of long period waves has been detected some years
ago, MUNK (1962), but only recently considered in harbour
problems: grouping of natural wave trains, BOWERS (1977),
GRAVESEN et al. (1978), BOWERS (1980).
    In fact, storm waves propagate in groups of varying lenghts
and heights that give rise to long period oscillations of the

352

E. Bratteland (ed.), Advances in Berthing and Mooring of Ships and Offshore Structures, 352–357.
© 1988 by Kluwer Academic Publishers.

water level. These long waves are normally called "bound long waves", "surf-beats" or "second-order long waves".

In recent years the grouping of natural waves has gained an increasing interest and several attempts have been made to characterize the wave groups, NOLTE et al. (1972), RYE (1974), GODA (1976), FUNKE et al. (1979), and to relate them with the wind wave spectral information, CARVALHO (1981), RYE (1983).

More recently, the mathematical relationship between wave grouping and bound long waves has been derived, HANSEN (1978), DEAN et al. (1981), SAND (1981), SAND (1982) (see list of references at the end of the text).

According to SAND (1981), for instance, the bound long waves are coupled to the incoming short wave system through the theoretical relation:

$$H_{s1} = C H_s^2$$

in which H_{s1} is the long period significant wave height, H_s is the short period significant wave height and C is a factor largely influenced by the shape of the directional spectrum of the short waves, that is, it is a function of the directional spreading of the waves.

In addition to this, bound long waves travel with the group velocity, that is, they follow the wave groups, and consequently the long wave periods are equal to those of the groups.

These periods may be near the periods of oscillation of moored ships due to the spring effect of the moorings. In fact, since second-order waves consist of a broad range of low frequencies which covers the typical frequencies of oscillation of large moored ships, second-order long waves are more effective than first-order waves in causing ship motion, BOWERS (1975), JOHNSON et al. (1982), PINKSTER (1982), MYNETT et al. (1985), O'BRIEN (1985).

It is generally believed that the relations between long waves and short waves, as the one by SAND (1981), indeed describe long waves rather well. However, proper verifications in nature have not been carried out yet.

In 1984, the Delft Hydraulics Laboratory (DHL) and the LNEC, using a combination of results from limited measurements carried out in the port of Sines (Portugal) and from a DHL physical model, tried to establish an approach to the occurrence of long waves in Sines based on the work of SAND (1981) and SAND (1982).

From the conclusions of that study, VIS et al. (1985) (presented to the International Conference on Numerical and Hydraulic Modelling of Ports and Harbours, Birmingham, April 1985) it is remarked that to study the problem in more detail and with a higher accuracy it is necessary to carry out much more measurements with more sophisticated equipment that the one used before, so that a better understanding of the problem of long waves can be achieved, in terms of its relation to the short waves, the response of the harbour basin, and its influence on moored ship behaviour.

354

3.OBJECTIVES

Following the results of VIS et al. (1985) a research project on the relationship between long waves and short waves and on the influence of long waves in moored ship behaviour is proposed.

For that, a field measuring campaign is proposed in which the following phenomena would be simultaneously measured:

- The short waves, including the directional spreading, off the harbour.
- The short waves, including the directional spreading, near the harbour entrance.
- The long waves near the harbour entrance (if possible in the same measuring point of the short waves).
- The short waves and the long waves, including the directional spreading, in shallow water.
- The long waves inside the harbour.
- The motions of a moored ship (surge, sway and yaw).
- The forces in the mooring lines of that ship.
- The forces in the fenders (eventually).

The most appropriate location to carry out these measurements is the port of Sines, where mooring force gauge systems are already available in three berths (which in Portugal exist only in that harbour) and one quick tide gauge is installed at the harbour limits. The rest of the equipment will have to be installed.

In order to have enough statistical data, measurements should be carried out at least during four winter seasons (totalling appr. 24 months).

With these data the following objectives are expected to be accomplished:

1 - To devise a method of establishing a first approximation long wave climate at the entrance of a harbour, for design purposes.
2 - To determine the correlation between the moored ship behaviour and the combined action of short and long waves in order to estimate berth downtime.
3 - To validate the results of wave disturbance tests (physical model), as far as the ship motions and mooring forces are concerned, by comparison to field measurements.
4 - To validate the results of a mathematical model of moored ship behaviour (to be developed).

4.WAVE AND SHIP MEASUREMENTS

It is proposed to measure:

- The short waves, including the directional spread (1) near the harbour entrance and (2) outside the harbour at a depth of 100 meters, with two WAVEC buoys.
- The long waves, including the directional spread (1) near the harbour entrance (close to the location where a WAVEC buoy will be installed) and (2) close to the shore at a water depth of 20 meters. These long waves will be measured with two SEA-DATA directional pressure sensors, mounted on the sea bed, coupled to a surface buoy equiped with a telemetric link to a receiving station located at the Centralized Command Room.

- The short waves at a water depth of 20 meters, close to the location where one of the directional pressure sensors will be installed, with a DATAWELL WAVERIDER buoy (already in operation, nearby). The directions of these short waves will be obtained from the directional pressure sensor.
- The long waves inside the harbour, with three SEA DATA pressure sensors each mounted onto the wall of a caisson (being part of berths 7/8 and 4/5), coupled to a telemetric system installed on top of the caisson. The continously transmitted signals from these three stations will also be received at the Centralized Command Room.
- The motions of (a) moored ship(s) at berths 4 and/or 5 using:
  - -2 laser rangefinders for surge motion detection, one for berth 4 and one for berth 5, aligned with the main horizontal axis of the ship;
  - -4 acoustic ranging systems, for sway and yaw motions, two on each berth, laying across the main horizontal axis of the ships.
- The mooring line forces of the ship(s) using the already existing special bollards at berths 4 and 5.

The signals of all the above mentioned ship motion sensors as well as the mooring line force signals will be collected in the Control Room of berths 4/5 by a data acquisition system and transmitted to the Centralized Command Room over a (specially to be installed) fibre optic cable.

In the Centralized Command Room a HP 300 series computer type will sample all signals (i.e. short and long wave heights and directions, ship motions and mooring line forces) and it will store them, in real time, on a high capacity magnetic tape cartridge system.

At LNEC, a similar computer system (HP 300 series type) equipped with a tape cartridge reading facility, printer, plotter, etc., will enable pre-processing of the recorded data, after which the signals can be sent, on line, to the mainframe computer of LNEC (VAX 8700 system), where they can be further processed.

The measured data will be compared with physical and mathematical models in order to verify the validity of these models.

To this end, a physical model of the port of Sines will be constructed at the LNEC and an integral software package will be developed to simulate wave propagation and resulting ship motions.

The main elements of that software package will be:
- A wave propagation model.
- A model to compute the forces on a ship caused by the local wave action.
- A model which simulates the behaviour of the moored ship under this wave action and other external forces like wind.

5. PARTICIPANTS

For assistance, instalation and maintenance of the wave measuring equipment, the Instituto Hidrográfico (IH) is being

asked to participate in this project. Its role will also
include the exploitation of the equipment by teams of experts
which will be in charge of the long wave data recording,
reading and decoding steps.

On the other hand, APS - the Sines Port Authority - will be
in charge of some supporting tasks.

The role of the LNEC will include the responsability of the
project (both technical and managerial) and the analysis of all
measured data.

Prior to the begining of the measurements, however, the LNEC
will also be responsible for the selection of the land based
equipment to be bought, whereas IH will be responsible for the
procurement of the sea equipment of this project.

The project will be consulted by a technical adviser
(Ferdinand Vis of Delft Hydraulics Laboratory) and by an
international board of experts.

6. BUDGET
For the 5 years project term the overall budget is as
follows:

NATO contribution	: 26,000 x 10^3 BF	(675,000 US$)
National contribution	: 47,000 x 10^3 BF	(1,221,000 US$)
Total	: 73,000 x 10^3 BF	(1,896,000 US$)

REFERENCES

BOWERS, E. C. (1975) - "Long period oscillations of moored
ships subject to short wave seas". Transactions of the Royal
Institution of Naval Architects, Vol. 117, 1975.
BOWERS, E. C. (1977) - "Harbour resonance due to set-down
beneath wave groups". Journal of Fluid Mechanics, Vol. 79, N.1,
1977.
BOWERS, E. C. (1980) - "Long period disturbances due to wave
groups". Proceedings, Seventeenth Coastal Engineering
Conference, Sidney, Vol.I, March 1980.
CARVALHO, M. M. (1981) - "Modelação estocástica da agitação
maritima". Notas de curso, Curso de Pós-Graduação em
Oceanografia, COPPE, Universidade do Rio de Janeiro, Rio de
Janeiro, 1981.
DEAN, R. G. and J. N. SHARMA (1981) - "Simulation of wave
systems due to nonlinear directional spectra". Proceedings,
International Symposium on Hydrodynamics in Ocean Engineering,
The Norwegian Institute of Thechnology, 1981.
FUNKE, E. R. and E. P. MANSARD (1979) - "On the synthesis of
realistic sea states in a laboratory flume". Thechnical report
LTR-HY-66, National Research Council, Ottawa, 1979.
GODA, Y. (1976) - "On wave groups". Proceedings, First
International Conference on the Behaviour of Off-shore
Structures (BOSS'76, Trondheim), Vol.1, August 1976. Published

by the Norwegian Institute of Thechnology.

GRAVESEN, H. ; JENSEN, O. J. and T. SORENSEN (1978) - "Harbour resonance generated by storm waves". Proceedings, Seventh International Harbour Congress, Antwerp, May 1978.

HANSEN N. E. O. (1978) - "Long period waves in natural wave trains". Progress report N.46, Institute of Hydrodynamics and Hydraulic Engineering, Technical University of Denmark, August 1978.

JOHNSON, R. R. ; MANSARD, E. P. and B. D. PRATTE (1982) - "Moored ship motions in irregular waves". Proceedings, Eighteenth Coastal Engineering Conference, Cape Town, November 1982.

MUNK, W. H. (1962) - "Long ocean waves". Chapter 18 of Vol.I of "The Sea", Wiley Interscience Publishers, New York, 1962.

MYNETT, A. E. ; KEUNING, P. J. and F. C. VIS (1985) - "The dynamic behaviour of moored vessels inside a harbour configuration". International Conference on Numerical and Hydraulic Modelling of Ports and Harbours, Birmingham, BHRA, April 1985.

NOLTE, K. G. and F. H. HSU (1972) - "Statistics of ocean wave groups". Paper OTC-1688, Fourth Offshore Technology Conference, Houston, Texas, April 1972.

O'BRIEN, W. T. (1985) - "The slow drift oscillations of moored vessels: field measurements, model calibration and parametric analyses". International Conference on Numerical and Hydraulic Modelling of Ports and Harbours, Birmingham, BHRA, April 1985.

PINKSTER, J. A. (1982) - "Set-down and grouping in irregular waves". Proceedings, Eighteenth Coastal Engineering Conference, Cape Town, November 1982.

RYE, H. R. (1974) - "Wave group formation among storm waves". Proceedings, Fourteenth Coastal Engineering Conference, Copenhagen, Vol.I, June 1974.

RYE, H. R. (1983) - "Ocean wave groups - Their relation to spectral information and their reproduction by means of laboratory or numerical simulation". Proceedings, International Conference on Coast and Port Engineering in Developing Countries, Sri Lanka, Vol.2, March 1983.

SAND, S. E. (1981) - "Long waves in directional seas". Manuscript submitted to the Journal of Coastal Engineering, Elsevier Scientific Publishing Company, Amsterdam, July 1981.

SAND, S. E. (1982) - "Wave grouping described by bounded long waves". Ocean Engineering, Vol.9, N.6, 1982.

OBSERVATION OF SHIP MOTIONS IN ROUGH WEATHER CONDITIONS

S.Ueda
Chief of the Offshore Structures Laboratory, Structures
Division, Port and Harbour Research Institute, Ministry
of Transport, Japan

1. INTRODUCTION
This paper describes the results of field observation of ship
motion in rough weather conditions. Ships investigated are
general cargo ship of 2,000dwt, 2,500dwt, 6,800dwt and 48,000
dwt. Some investigations were carried out in conditions with
wind speed of more than 20m/s. Some differences on the characteristics of motions have been observed for the different
ships.

2. OBSERVATION IN PORT OF YOKKAICHI
Observations have been carried out twice at the Mie Ship
Building Yard in the Port of Yokkaichi. Ships observed were
6,800 dwt Ro-Ro ship under equipment. Figure 1 shows the plan
view of the observation site and location of the wave recorder
and the wind meter. Table 1 lists the wave and wind conditions
when the observations were carried out. Wind direction is measured clockwise from bow. The first observation was carried
out when the typhoon was approaching   the district in 1980,
and the second observation was carried out during winter when
a depression affected the district in 1981. The maximum mean
wind speeds were 28m/s and 19m/s, respectively. The wind direction for the first observation was from the seaward while
that for the second observation was toward the sea. Significant wave heights in the first observation were in the range
from 0.49m to 0.75m, however, wave height in the second observation was almost nil and supposedly less than 10cm.

Table 1 Wind and Wave Conditions during Observation
(Port of Yokkaichi)

| | Y-80-1 | | | | Y-80-2 | |
| | Wave | | Wind | | Wind | |
No.	$H_{1/3}$ (m)	$T_{1/3}$ (s)	V. (m/s)	Dir. deg.	V (m/s)	Dir. deg.
1	0.59	5.1	26	107	14	-118
2	0.65	5.0	26	107	17	-109
3	0.75	4.7	28	121	19	-104
4	0.49	5.1	21	112	16	-102
5	0.53	5.1	20	107	9	-55
6	0.51	5.0	20	130	8	-42

E. Bratteland (ed.), Advances in Berthing and Mooring of Ships and Offshore Structures, 358–367.
© 1988 by Kluwer Academic Publishers.

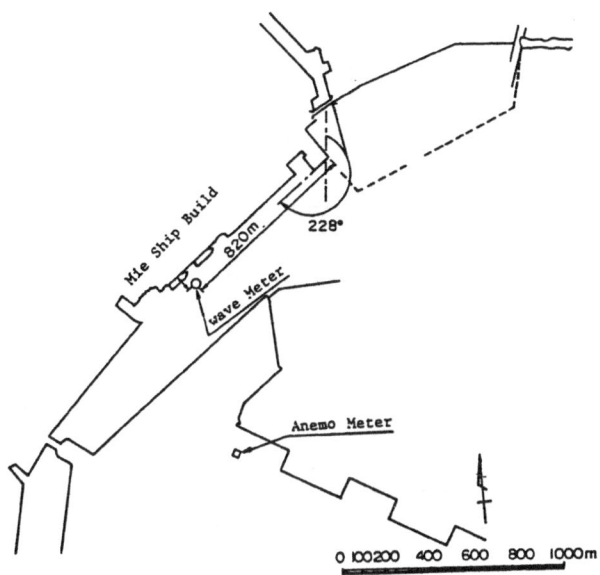

Figure 1 **Plan** View of Observation Site   (Port of Yokkaichi)

Figure2shows arrangement of instruments such as displacement
meter, camera and clinometer. Scales were placed on the hull
in order to measure surge and heave motions by means of camera.
Ship was moored by thirteen synthetic ropes and tire fenders
were used. The tire fenders were made by inserting several
rods into the inner hole of the tires. Wind directions att-
acking the ship are shown in Figure 2.

Figure 3 shows surge and heave motion obtained during the
first observation. Surge motions observed in the first observation were in the range from 0.19m to 0.29m and heave motions
were in the range from 9cm to 12cm. Surge motion is not nec-
cesarily proportional to the wave height. Figure 4 shows surge, sway and heave motions obtained in the second observation.
Surge motions observed in the second observation are in the
range from 0.52m to 1.08m while the wave height is less than
10cm and the motion increased and became long period motion
in cases No 3 to 6 when wind speed decreased.

360

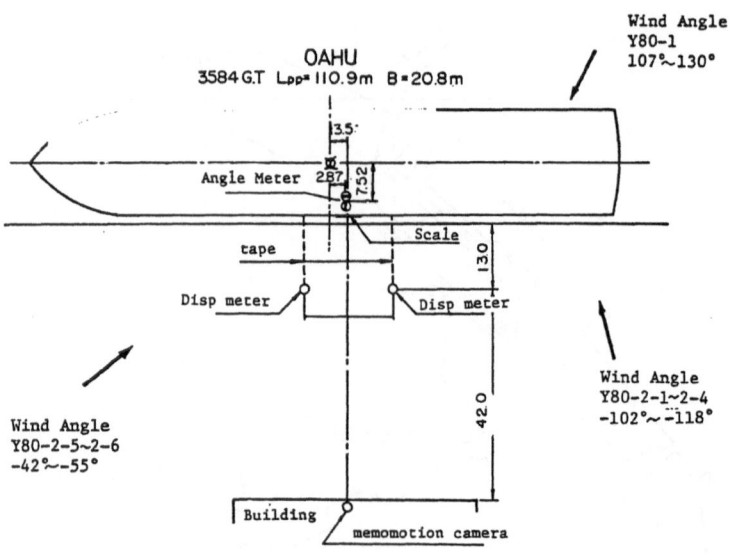

Figure 2   Arrangement of Instrument (Port of Yokkaichi)

Sway motions are in the range 9 to 55cm, however, the motion
increased, and became long period motion in the cases No.3 to
6 as well as surge motions. Heave motions were in the range
2to4cm. No time histories are presented, however, roll, pitch
and yaw motions were less than 0.32 degrees. The reason for
increase of the surge and sway motions in the cases No.3 to 6
is thought to be change of the wind directions. As shown in
Figure 2, wind blew from stern to bow toward the quay wall in
cases No.1 and 2, then, wind direction changed to blow from
bow to stern but still toward the quay wall. It seemed that
in cases No.1 and 2 ship drifted toward the bow and the bow
lines were slackened while stern lines were taught. But,
after the wind direction changed, the ship began to drift
toward stern tightening the bow lines. It is believed that
this caused the large and long period motions.

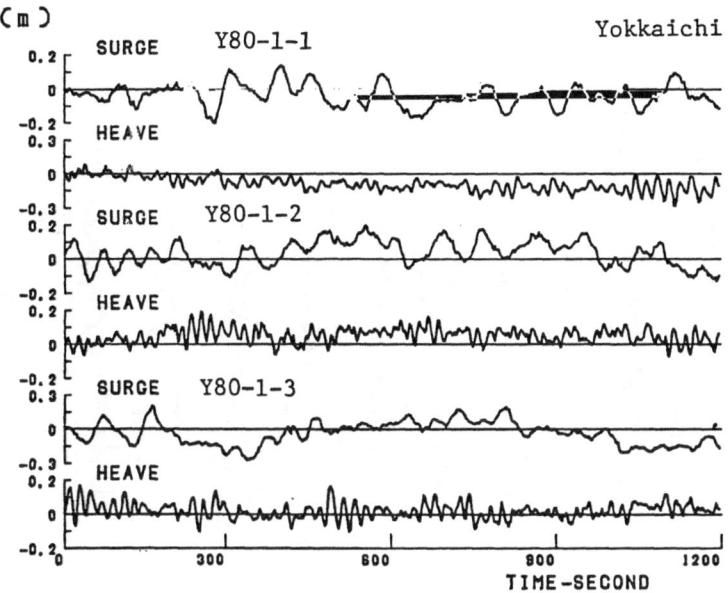

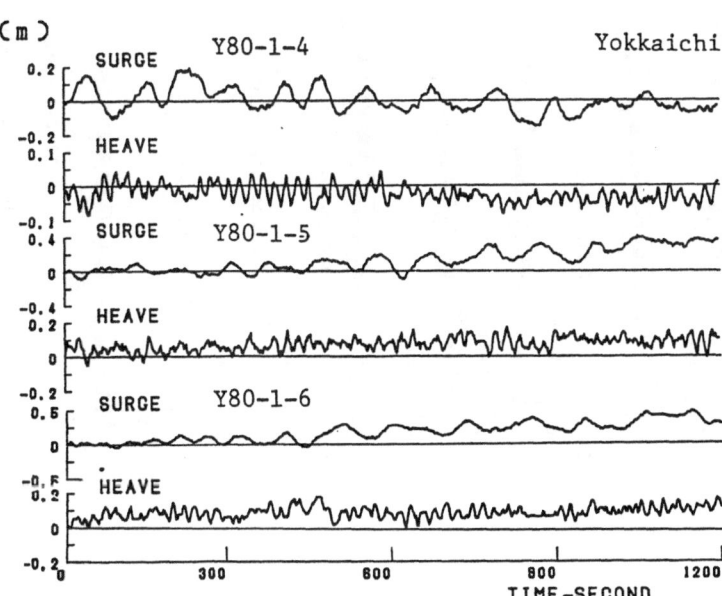

Figure 3   Time Histories of Ship Motions (Y-80-1)

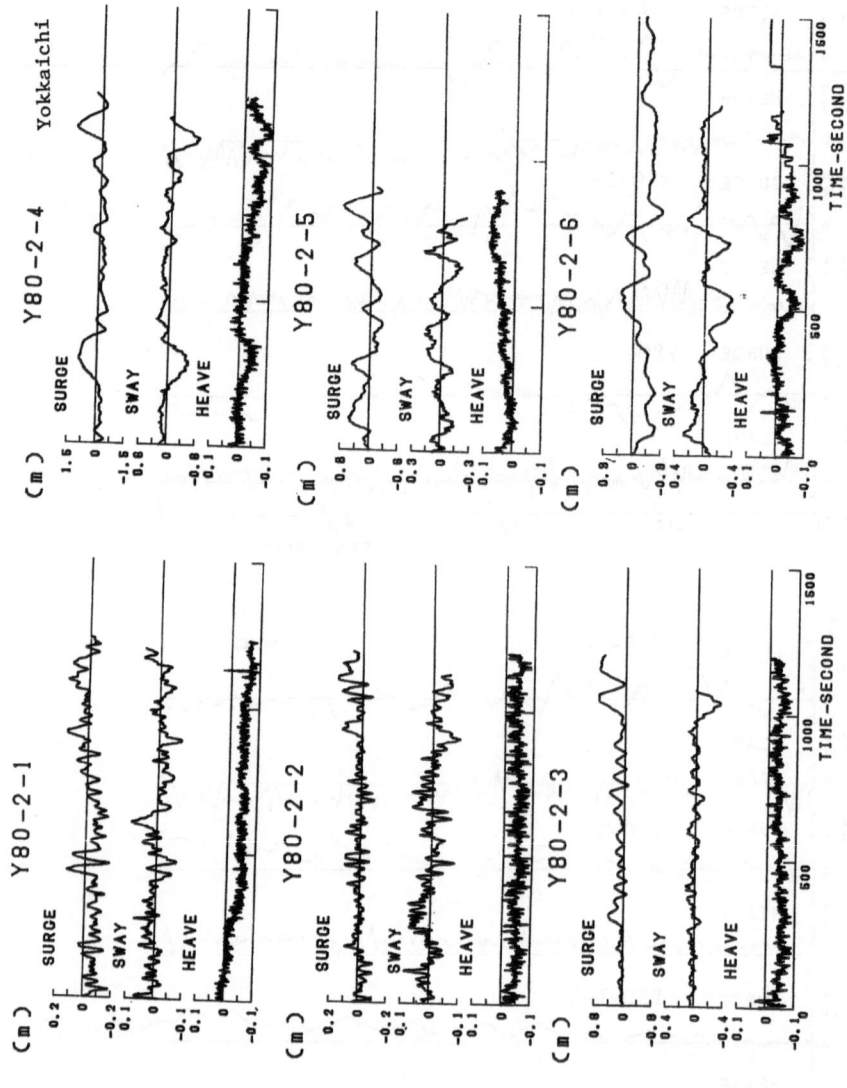

Figure 4 Time Histories of Ship Motions (Y-80-2)

3. OBSERVATION IN PORT OF NIIGATA

Observations have been carried out three times at the Rinko Pier in the Port of Niigata. Ships observed were 2,000dwt, 2.500dwt and 48,000dwt cargo ships. Figure 5 shows the plan view of the observation site and the location of the wave recorder and the wind meter. Table 2 lists the wave and wind conditions when the observations were carried out, where the wind direction is measured clockwise from the bow. Those observations were carried out during January to March, 1982. Figure 6 shows arrangement of instruments such as displacement meter, camera and clinometer. In this observation PSS (position sensor system) target was set on the hull. Ship was moored by synthetic and wire ropes. At the berth where No.1 and No.2 of observations were done, both tire fenders and floating type pneumatic fenders were used. Floating fenders were used for the larger ship. Tire fenders were made of two pieces of tires with 1 m in diameter and 0.24m in width, sandwiched between two steel plates. For the No.1 case pneumatic fenders were used, while for the No.2 case tire fender were used. Regarding the No.3 case, fixed type pneumatic fenders were used. Wind directions are given in Figure 7.

Table 2 Wind and Wave Conditions at Observation
(Port of Niigata)

| | N-82-1 | | | | Y-82-2 | | | | N-82-3 | | | |
| | Wave | | Wind | | Wave | | Wind | | Wave | | Wind | |
No.	$H_{1/3}$ (m)	$T_{1/3}$ (s)	V (m/s)	Dir. deg.	$H_{1/3}$ (m)	$T_{1/3}$ (s)	V (m/s)	Dir. deg.	$H_{1/3}$ (m)	$T_{1/3}$ (s)	V (m/s)	Dir. deg.
1	0.19	3.5	11	82	0.32	3.1	9	23	0.54	2.9	5	129
2	0.21	3.5	9	77	0.32	3.4	9	19	0.52	2.7	5	127
3	0.22	3.1	9	73	0.31	2.7	8	19	0.51	2.7	5	112
4	0.20	3.3	9	73	0.25	2.8	9	19	0.54	3.0	7	118
5	0.25	2.7	11	55	0.30	3.1	10	32	0.57	3.1	5	136
6	0.25	2.4	10	49	0.31	3.3	9	23	0.57	3.3	6	127

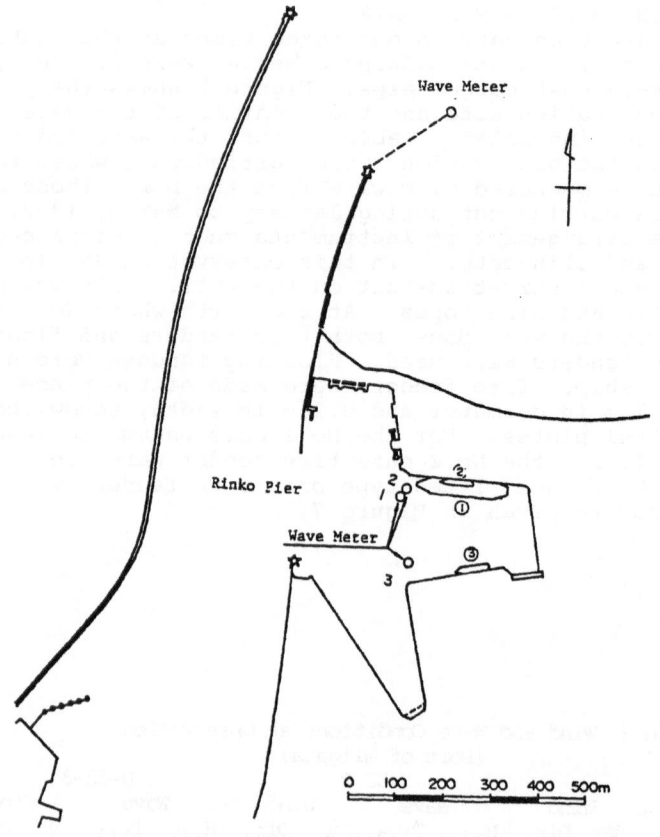

Figure 5   Plan View of Observation Site (Port of Niigata)

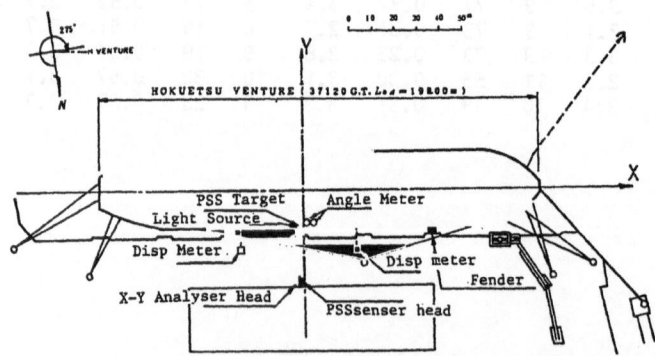

Figure 6   Arrangement of Instrument   (Port of Niigata)

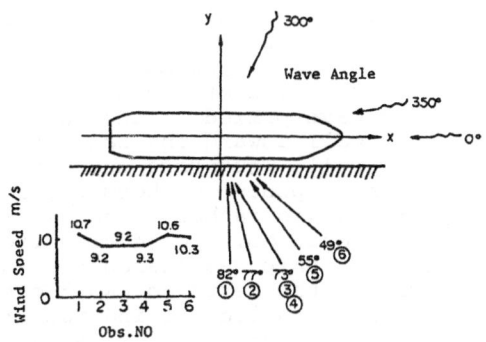

a) N-82-1

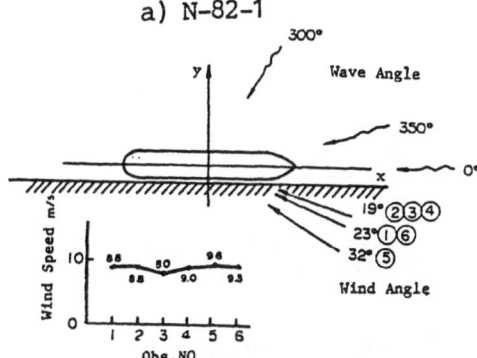

b) N-82-2

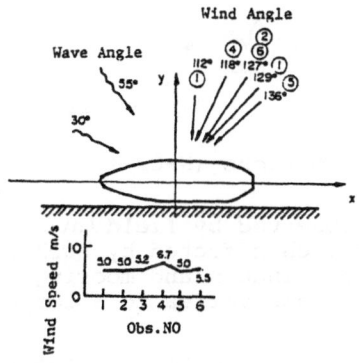

c) N-82-3

Figure 7 Wind Conditions (Port of Niigata)

366

Figure 8 to 10 show time histories of ship motions. Difference of ship motions can be found between No.1,2andNo.3. In the case of No.1and2 long period surge, sway and yaw motions were observed, while in the case No.3, the period of those motions are shorter and nearly equal to the wave period.In the case No. 1 and No.2, wind blew from the quay wall toward the sea, then the ship drifted away from the quay wall. Therefore,the ship was held in place only by the mooring ropes. Furthermore, in the case No.2, as the fender stiffness was rather heigh, suh-harmonic sway motion occured. In the case No.3, wind blew from the sea toward the quay wall, then the ship was pushed towards the quay wall. As mentioned above, at the berth for No.3 case, the fixed type pneumatic fender were installed, giving rather symmetric load-deflection characteristics resulting in that the subharmonic motion did not occur.

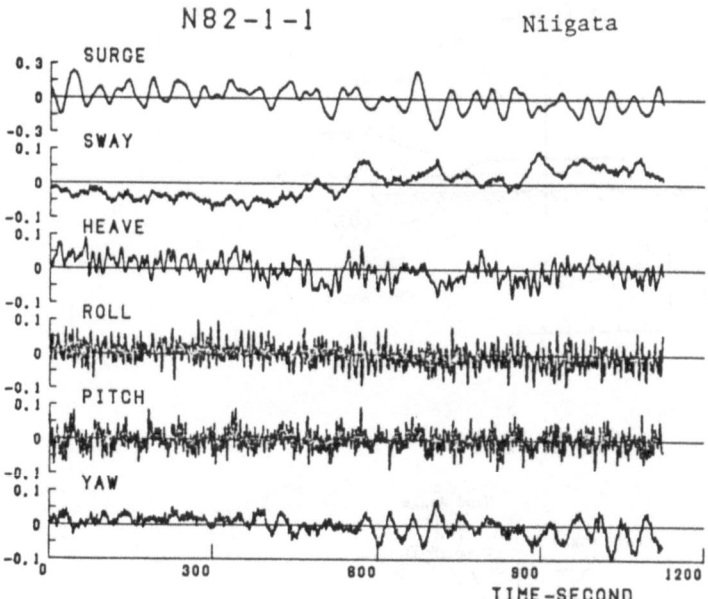

Figure 8 Time Histories of Ship Motions (N-82-1)

CONCLUSION
Interesting phenomena have been observed by field investigations. Ship motions are complicated and much affected by such factors as wind, waves, characteristics of fenders and mooring ropes. Combining those factors, ship motions varies to a large extent.

REFERENCES
1.Ueda, S.:   Analytical Method of Ship Motions Moored to Quay Walls and the Applications, Note of the Port and Harbour Research Institute, No.504, 371p.
2.Ueda. S. and et al: Observation of Ship Motions Moored to Quay Wall in Storm, 28th Conference on Coastal Engineering in Japan, 1981, pp.431-435.

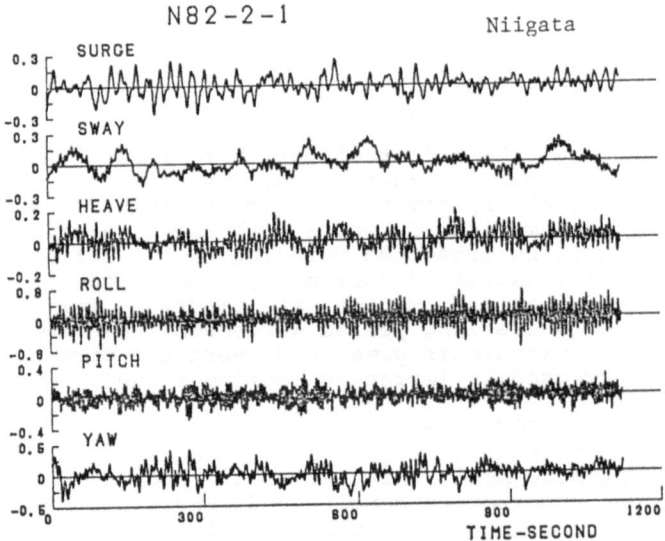

Figure 9. Time Histories of Ship Motions (N-82-2)

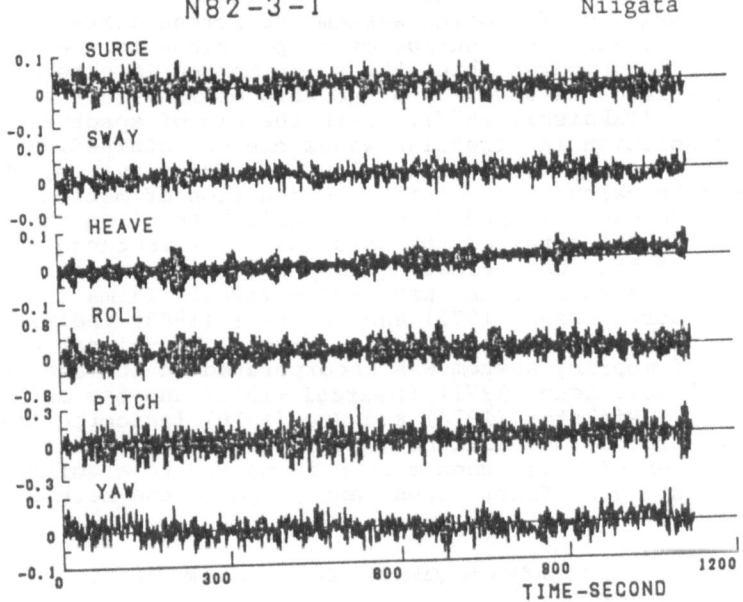

Figure 10. Time Histories of Ship Motions (N-82-3)

EFFECT OF CHARACTERISTICS OF MOORING SYSTEMS AND EXTERNAL
FORCES TO MOTIONS OF MOORED SHIPS

S.Ueda
Chief of the Offshore Structures Laboratory, Structures
Division, Port and Harbour Research Institute, Ministry
of Transport, Japan.

1. INTRODUCTION
This paper describes the effect of the characteristics of mooring systems and external forces to motions of moored ships in
accordance with model experiment and computations. Here, the
differences of ship motions in waves with and without the action from wind will be presented relating to two different load-
deflection characteristics of mooring systems.Furthermore, it
will be indicated that ship motions are important for ports
located at places where long period waves sometimes come into
the basin, and an example is presented where the port operating rate was increased by improving mooring system.

2. NUMERICAL SIMULATION METHOD
For a long time, ship motions has been estimated by means of
model tests. Although model tests still are a useful measure
because they can incorporate complicated conditions of geograp-
hics and non-linear effects between the ship and the waves, it
becomes rather difficult to perform such tests in a limited
wave basin when a ship or a floating body become large. Furthermore it takes a long time to run many experimental tests,
and the costs involved would also be high.
In this respect, a numerical simulation method is thought to
be useful because it can incorporate non-linear load deflection characteristic of mooring system and irregularity of wind
and wave loads, and yield output of ship motions in the time
domain. For an unmoored ship, the Strip Method is extensively
used to calculate the wave loads and ship motions in the frequency domain (Takaishi, 1977). With the aid of spectral analysis, the solution in irregular waves can be obtained. Furthermore, for a ship moored with a linear mooring system, ship
motions can be calculated solving the equation of motions by
use of wave forces obtained from the Strip Method. On the other
hand, Ijima (1975) obtained the solution for a rectangular sectional three dimensional floating body which is moored with a
linear mooring system in the proceeding waves. Ijima (1972),
Ito (1972), Oortmerssen (1975) and Sawaragi (1983) individually
obtained a solution for the standing waves. Non-linear characteristic of mooring system was incorporated in the analysis
by Russel (1958), Lean (1971), Sawaragi (1983) and the authors
(1983, 1985). Pinkster (1974) and Hsu (1970) indicated that
slow drift oscillations will be induced according to the second
order pressures of waves when a ship is moored to a soft mooring
system consisting of fiber ropes, and proposed the method of

368

E. Bratteland (ed.), Advances in Berthing and Mooring of Ships and Offshore Structures, 368–379.

computations individually. In those methods mentioned above, ship motions are calculated under the condition that a ship is subjected to waves only, and the effect of wind is not incorporated. Accordingly, in this paper, ship motions are calculated in the time domain under conditions that a ship is moored to a non-linear mooring system and is subjected to both irregular waves and gusty wind by using the numerical simulation method.

In the procedure of numerical simulation method, first of all, computational conditions such as ship size, draughts water depth, the waves and the wind, and the characteristics of mooring system shall be determined. The radiation forces are calculated as the hydrodynamic forces due to waves generated by ship motions of unit amplitude. And these forces are to be treated as the added masses and the damping coefficients in the equation of motions. The wave forces are calculated by integrating the wave pressures acting on about twenty Lewis form approximated cross sections of the ship hull.Regarding surge motion, the wave force is calculated separately treating it as a Froude Kriloff Force. The wave forces for irregular waves are calculated by superpositioning of components of regular waves by taking into account the phase difference. Any frequency spectra such as those proposed by Brectschneider - Mituyasu, JONSWAP, Ochi-Hubble or any other optional ones can be used to generate the irregular waves. The directional spreading of spectra can also be considered in the calculation of wave forces, although this takes large computational time. The wind force acting on the ship hull is calculated by use of Hughes's experimental formula. The drag coefficients for wind force should be appropriately determined according to the results of wind tunnel tests (Tuji,1972). The frequency spectra of wind proposed by Davenport and Hino can be used to generate the gusty wind. Furthermore, the current force and the wave drift force can be incorporated in this numerical simulation method. Generally, as the load-deflection characteristics of fenders and mooring ropes are non-linear and some fenders exhibit large hysterisis, this non-linearity is incorporated in the numerical simulation method. The solution will be obtained in the time domain by numerically integrating the equation of motions by use of Wilson θ method.

3. HYDRAULIC MODEL TESTS

In order not only to investigate the characteristics of ship motions but also to evaluate the results of numerical simulation, hydraulic model tests were carried out in a wave basin belonging to the Port and Harbour Research Institute. The basin is 25 m in length, 15 m in width and 1 m in depth. Wave generators and wind fans (movable type) were set in the wave basin. Tests could be carried out under conditions of regular waves, irregular waves, steady wind and gusty wind. The model ship was a 10,000 DWT cargo ship scaled into 1/30 to the size of 440 cm in length, 61 cm in width and 41 cm in depth. The draughts of the model ship are 27.7cm, 22.6cm and 14.4 cm in fully laden, half loaded and in ballasted conditions, respectively. The model ship is moored to a model quay wall with six mooring ropes and two fenders.

As the quay structure is a vertical wall, the model ship moves
under the action of both incident and reflected waves.
Among three types of model fenders the type No.1 and the No. 3
model fenders were mainly used. Load-deflection characteristics of model fenders are shown in Figure 1. The load-deflection characteristic of type No.1 model fender exhibits the
steady reaction force in loading against the deflection in the
range from about 15 to 35% of its height and also exhibits
large hysteresis in unloading. The No. 3 type model fender,
exhibits a hyperbolic load-deflection characteristic and small
hysteresis. Model mooring ropes are made by twisting a couple
of pieces taken from nylon stockings. The load-deflection
characteristic of mooring ropes is shown in Figure 2. Both
model fenders and model mooring ropes exhibit quite similar
load-deflection characteristics to those of prototypes.
Directions of waves are 30, 60, 90 and 180 degrees, while wind
directions are 90, 120, 180, 240, and 270 degrees relative to
the longitudinal axis of the ship. Where 90 degrees is perpendicular to the face line of the quay wall and a ship is
subjected to waves from the seaside. Several combinations of
wave and wind directions are used in the hydraulic model test.
Significant wave periods in the hydraulic model test are 0.73,
1.10, 1.46, 1.83, 2.19, and 2.56s, which are equivalent to 4,
6, 8, 10, 12, and 14s in the prototype, respectively. The significant wave height in the model test is 1.67 cm which is
equivalent to 0.5 m in prototype. Wind speeds in the hydraulic model tests are 0, 3.65, 4.56, 5.48, and 6.39m/s, which
are equivalent to 0, 20, 25, 30 and 35m/s in prototype, respectively. In several of the hydraulic model tests, model wind
screens of 20, 40, and 60 cm height were introduced at location
1.2 and 2.0m from the quay line, which are equivalent to 6, 12
and 18m height and 36 and 60 m away from the quay line in the
prototype.

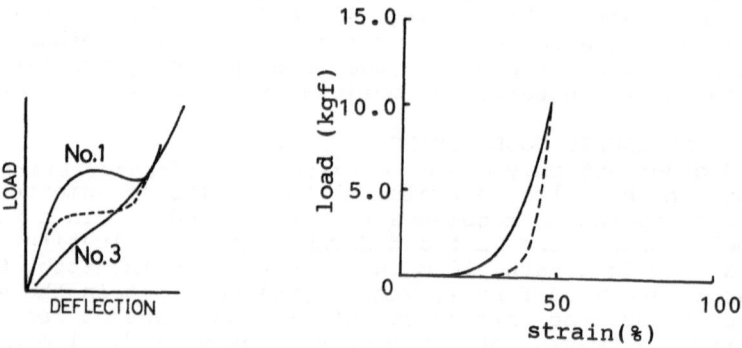

Figure 1 Load Deflection   Figure 2 Load Deflection
Characteristics of Fenders   Characteristics of Ropes

4. CHARACTERISTICS OF SHIP MOTIONS SUBJECTED TO WIND AND WAVES, AND COMPARISON OF COMPUTATION AND MODEL TESTS

In this chapter, the characteristics of ship motions are described. Several interesting phenomena are observed especially in swaying motions according to the effects of wind, waves and characteristics of fenders. Firstly, the characteristics of swaying motions are described, next comparisons are made between results of hydraulic model tests and numerical simulation.

4.1 Effect of Wind

Figures 3-a,b show the typical time histories of swaying motions of the model ship obtained from the hydraulic model test and the numerical simulation respectively, to show the effect of wind. The model ship is in ballasted condition and is moored to the model quay wall with the type No.1 fenders and six mooring ropes. Here, both the wave and the wind directions are 90 degrees. These series of hydraulic model tests are denoted ABQH in this paper. The uppermost diagram is the time history of swaying motion for regular waves without any wind. From the second to the last diagrams the time histories are for swaying motions in irregular waves with and without action of wind. The wave height is 1.67cm, the significant wave period is 2.19s, and wind speeds are 0, 3.65, 4.56, 5.48, and 6.39m/s, respectively. In this figure, the zero line corresponds to the location of the face line of the fenders, whereas positive values correspond to the offshore motions and the negative values correspond to the onshore motions. Characters R and IR indicate that waves are regular and irregular respectively. A character of CIR indicates that the waves is irregular and the wind is steady. And digits of 20, 25, 30 and 35 at the end indicate that prototype wind speeds are 20, 25, 30 and 35 m/s, respectively.

When the wind speed is 0m/s, the offshore swaying motion both under regular and irregular waves becomes large and the period of swaying motion becomes three to four times that of the wave period. This motion is the so-called subharmonic motion caused by the asymmetry of the load-deflection characteristics of the mooring system consisting of fenders and mooring ropes and for the conditions when the steady wind force is not so strong. With increase of the wind speed from the seaside, compression of the fenders increase according to the increase of the steady wind force. Then the neutral position of the swaying motion moves closer to the fender side. However, the amplitudes of swaying motion decrease as the wind speed increases, because the asymmetry of the load-deflection characteristics becomes reduced as the wind speed increases. In these hydraulic model tests, the fenders have been designed so that the deflection of it becomes about 10% of its height when a model ship is subjected to wind of 6.39m/s average speed. The asymmetry of the mooring system previously mentioned, therefore becomes relatively smaller according to increase of fender compression corresponding to the increase of wind speed.

372

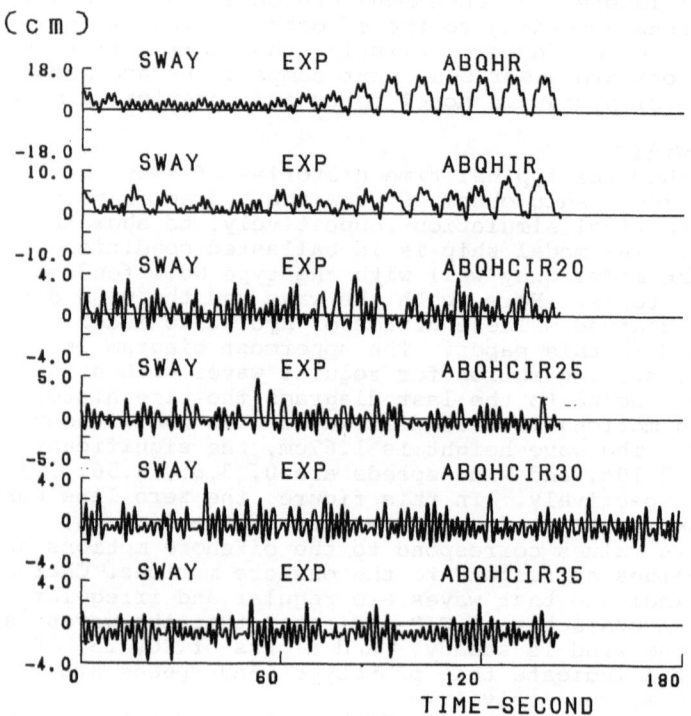

Figure 3-a Time Histories of Swaying (Model Test)

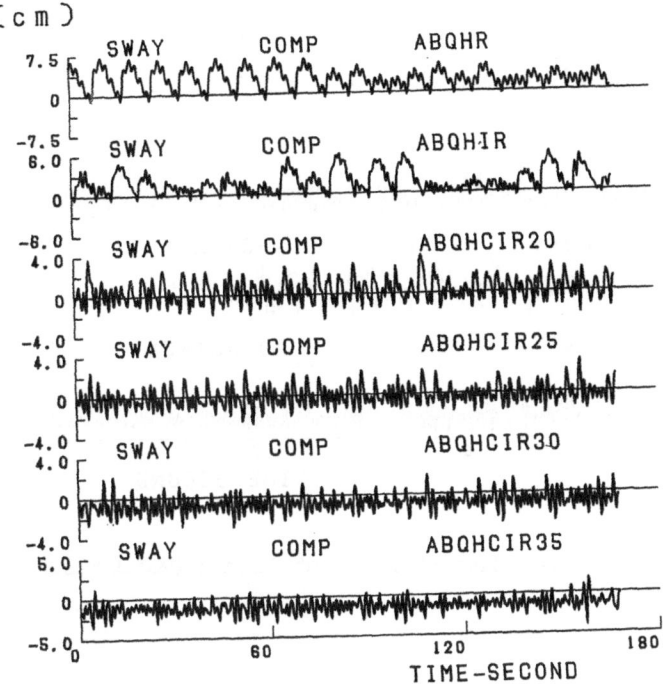

Figure 3-b Time Histories of Swaying (Computation)

Time histories of swaying motion of the model ship in the
numerical simulation are very similar to those in the hydraulic
model tests. Although the maximum amplitude of swaying motion
differs somewhat, the characteristics of swaying motion are
very similar for the hydraulic model tests and the numerical
simulation. It can be said that the computation simulates the
ship very well, and even the subharmonic motion is well reproduced by the numerical simulation. In order to examine in
further detail, the frequency spectra are shown later on.
4.2 Effect of Characteristics of Mooring System
Fig.4-a shows time histories of swaying motions of the model
ship, subjected to waves and wind from the 90 degrees direction. The significant wave height and the significant wave
period of irregular waves are 1.67cm and 1.83s, respectively.
The upper two diagrams correspond to cases with the type No.1
model fenders while the lower two diagrams are for cases with
the type No.3 model fenders. The series of hydraulic model
test using the type No.3 model fenders are denoted ABQA in this
paper. For each pair of diagrams, the upper one is the case
without wind action and the lower one is the case with steady
wind action of 5.48m/s speed (the digit of 30 at the end indicate the prototype wind speed being 30m/s).

374

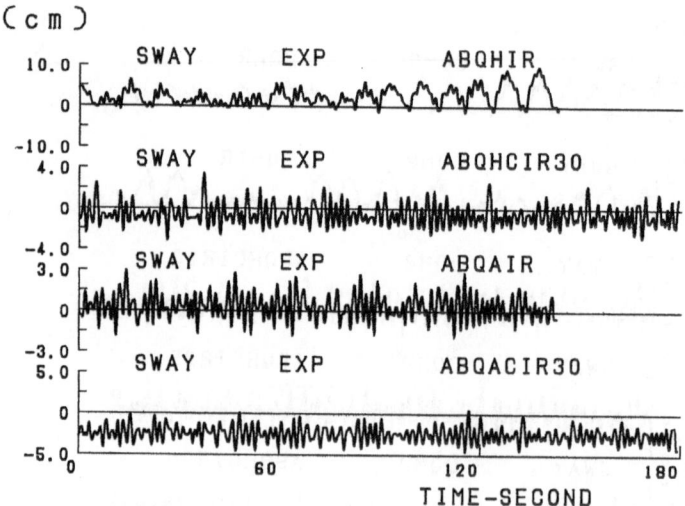

Figure 4-a   Time Histories of Swaying (Model Test)

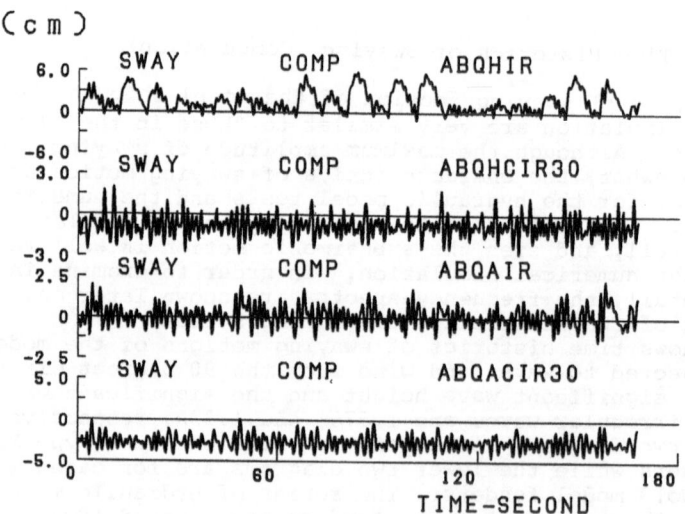

Figure 4-b   Time Histories of Swaying (Computation)

When the wind speed is 0m/s, the difference in time histories
of swaying motions between cases using the type No.1 and the
type No.3 model fenders can be easily seen. This phenomenon
depends on the asymmetry of the load-deflection characteristics
of the mooring system. Because the asymmetry of the load-deflection characteristics of the mooring system with the type No.1
model fenders is strong, large offshore swaying motions of long
period are observed. For the type No.3 model fenders, the asymmetry of the load-deflection characteristics of the mooring
system is relatively small and therefore the swaying motion
period is not so large.
Figure 4-b shows time histories of swaying motions of the model
ship in the numerical simulation method corresponding to those
of the hydraulic model tests. Good agreement between the results
of the hydraulic model tests and the numerical simulation is
easily observed.
Figure 5 shows frequency spectrums of swaying motions for the
hydraulic model tests and the numerical simulation, for cases
of ABQHIR (type No.1 fenders, with irregular wave and without
wind), ABQAIR (type No.3 fenders, with irregular waves and without wind), ABQHCIR 30 (type No.1 fenders, with irregular waves
and steady wind) and ABQACIR 30 (type No.3 fenders, with irregular waves and steady wind). Where the significant wave
height, the significant wave period and the wind speed are
1.67 cm, 1.83s and 5.48m/s, respectively. For the frequency
spectrum of swaying motion of ABQHIR, power peaks are observed
at 10s ans 1.8s and the power intensity at 10s is larger than
that for 1.83s. However, for the frequency spectrum of swaying
motion of ABQAIR, the power intensity at 10s is smaller than
not only that of 1.83s but also that of ABQHIR. Frequency spectrums both of ABQHIR30 and ABQACIR30 show almost the same char-
ateristics. Good agreements of frequency spectrums were obtained
between the hydraulic model test and the numerical simulation.

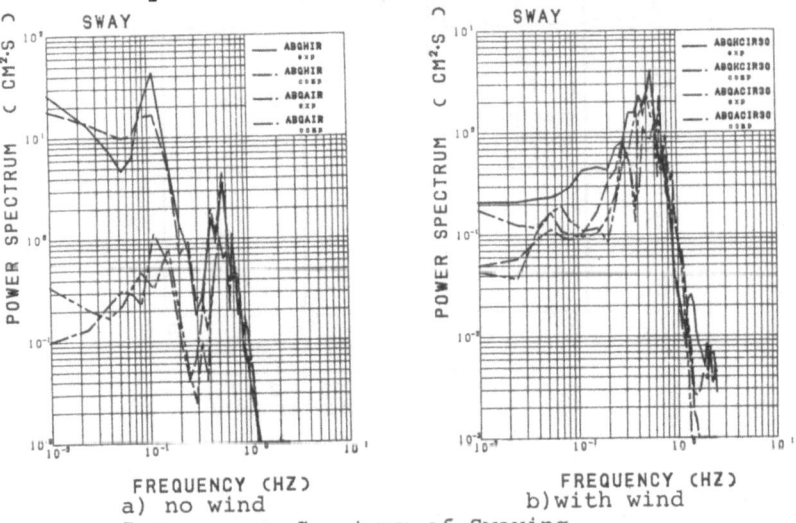

a) no wind   b) with wind
Figure 5 Frequency Spectrum of Swaying

376

4.3 Effect of Waves

Figure 6 shows the maximum amplitudes of offshore and onshore swaying motions, for ABQHIR30 and ABQACIR30 comparing the results of the hydraulic model tests and the numerical simulation method. The open and the closed symbols indicate the maximum (denoted Max) and minimum (denoted Min) motion of the numerical simulation (denoted as comp) and the hydraulic model tests(denoted as exp), respectively. Here the conditions of waves and wind are the same as those in Figure 5. Generally, the amplitude of ship motion increases when the wave period becomes longer. Good agreements were obtained between the results of the hydraulic model tests and the numerical simulation. Data obtained for condition when the significant wave height is 0 m and the gusty wind is 5.48m/s mean speed are plotted. The model ship is in this case subjected to wind and waves from the same directions as mentioned above. The amplitude of swaying motion is considerably smaller when the wave height is 0m. For this case study, it can thus be said that wind gustiness does not affect ship motions as much as waves.

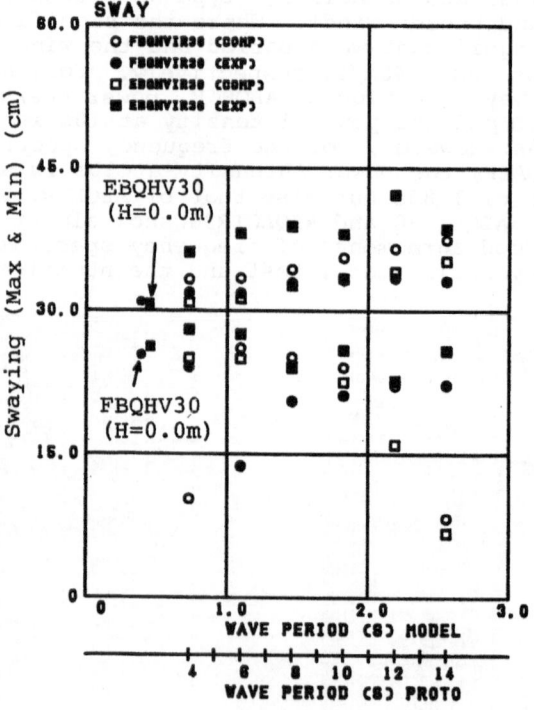

a) Swaying

Figure 6 Amplitude of Swaying

5. EXAMPLE OF INCREASE OF PORT OPERATION RATE BY IMPROVING MOORING SYSTEM

This author will emphasize the importance of fender characteristics for port planning, when long period waves exist, and will present an example of improved performance at the C Port in Japan.

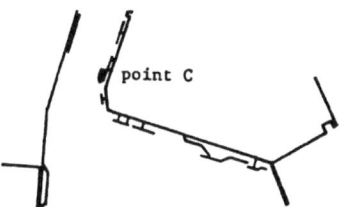

Figure 7   Layout of C Port

Figure 7 shows the layout of facilities at C Port.In this port several berths are available for oil tankers in the range from 3,000DWT to 150,000DWT. The object of this study is the berth located at point C in Figure 7. The berth is constructed of steel pipe piles, and is of the dolphin type for 3,000DWT tankers. There are two breasting dolphins and two V shaped rubber fenders with height 400mm and with spacing 8 m installed for each dolphin. These fenders were designed to absorb the berthing energy of the ship. However, the kinetic energy of the ship moving in waves could possibly become larger than that of berthing. Accordingly, the load-deflection characteristics become rather asymmetric so that subharmonic motions hindering loading and unloading operations could occur during October to March. Possible improvement measure is thought to replace the fenders with a type that exhibits load-deflection characteristics comparable to type No.3 (pneumatic type). Several cases of numerical simulations were carried out and good results were obtained for those cases using the pneumatic type fenders with height 1,000 or 1,200 mm. Figure 8 shows sway motions comparing results of cases using present fenders (buckling type) and replaced fenders (pneumatic type). There is a clear difference between the results for the two cases, showing that ship motion in the case of replaced fenders (pneumatic type) are smaller than for present fenders (buckling type). Furthermore offshore motion which means ship motion away from the quay wall is reduced considerably in case of the replaced fenders (pneumatic type). Figure 9 shows the fender deformations and the reaction forces for both cases. It is noticed by the Author that the port operating rate has been improved after replacing the fenders.

378

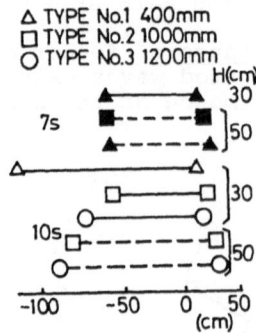

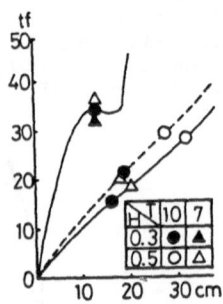

Figure 8 Swaying Motion   Figure 9 Fender Deflection

6. CONCLUSIONS

A numerical simulation method is developed to compute motions and mooring forces of moored ship in rough weather. The results of the numerical simulation method are in good agreement with that of hydraulic model tests carried out in the wave basin belonging to the Port and Harbour Research Institute. Moreover, the relationship between ship motion and characteristics of waves, wind and the mooring system are clarified through the hydraulic model tests and the numerical simulation. The conclusions of this paper are as follows.

1) When the asymmetry of the mooring system is large and a ship is subjected to waves and the steady force is not so strong, the subharmonic swaying motion occurs and the amplitudes become large.

2) When the wind speed increases, i.e., the steady force increases, the asymmetry of the mooring system becomes relatively small, and the amplitude of the swaying motion decreases.

3) Characteristics of ship motions achieved from both hydraulic model tests and numerical simulation method are very similar as shown in time histories and frequency spectrum of ship motions, and even the subharmonic swaying motion was well reproduced by the numerical simulation.

4) Waves affect the amplitude of ship motions more than wind does in this case study.

5) When a ship is moored in an asymmetric mooring system and also subjected to waves and a small steady force, subharmonic swaying motion with large amplitude will occur. Replacing fenders, making the mooring system more symmetric could be an effective measure to reduce the swaying motion.

REFERENCES

1.Hsu,F.A. and Blenkarn, F.A.[1970]:Analysis of Peak Mooring Forces by Slow Vessel Drift Oscillations in Random Seas, 2nd Offshore Technology Conference #1159,pp.I-135-I-146.

2.Ijima,T. et al[1972]:Scattering of Surface Waves and the Motions of a Rectangular Body by Waves in Finite Water Depth. Trans. of Japan Society of Civil Engineers, No.202, pp.33-48.

3.Ijima,T. et al[1975]:On the Motions of Elliptical or Rectangular Floating Body in Waves of Finite Water Depth, Trans. of Japan Society of Civil Engineers, No. 244,pp.91-105.

4.Ito,Y. and Chiba, S.[.1972]:An Approximate Theory of Floating Breakwater, Report of the Port and Harbour Research Insititute, Vol.11, No.2, pp.141-166.

5.Lean,G.H.[1971]:Subharmonic Motions of Moored Ships Subjected to Wave Action, Trans. Royal Institute of Naval Architects, London, No.113, pp.387-399.

6.Oortmerssen,G.Von [1975]:The Motions of a Ship in Shallow Water,Paper presented to the meeting of Society of Naval Architects and Marine Engineers.

7.Pinkstar,J.A.[1974]:Low Frequency Phenomena Associated with Vessels Moored at Sea, Society of Petroleum Engineers of Aerican Institute of Mechanical Engineers, SPE, 4834.

8.Russel,R.C.H.[1958]:A Study of the Movement of Moored Ships Subjected to Wave Action, Proceedings of Institution of Civil Engineering. Vol.12, pp.379-398.

9.Sawaragi,T. et al[1983]:New Mooring Systems to Reduce Ship Motion and Berthing Energy,Proc. of 30th Japanese Conference on Coastal Engineering, pp.460-464.

10.Takaishi,Y. and Kuroi, M.[1977]: Practical Calculation Method of Ship motions in Waves, 2nd Symposium on Sea Keeping, pp.109-133.

11. Ueda, S. and S. Shiraishi[1983]: Method and its Evaluation for Computation of Moored Ship's Motions, Report of the Port and Harbour Research Institute, Vol. 22, No. 4, pp. 181-218.

12. Ueda, S.[1985]: Analytical Method of Ship Motions Moored to Quay Walls and the Applications, Technical Note of the Port and Harbour Research Institute, No. 504, 371p, 1984.1.

13. Ueda, S.[1986]: Analytical Treatment of Ship's Mooring Problems in Rough Weather, Proc. of 5th OMAE, Tokyo, 1986, pp.522-528

14. Ueda, S[1987]: Analytical Treatment of Ship Motions Moored to Quay walls and Application, Proc. of 5th Coastal Zone, Seattle, 1987, pp.1545-1559.

ASYMPTOTIC AND VARIATIONAL METHODS APPLIED TO THE EVALUATION OF
WAVE DRIFT FORCES

C.P.PESCE
J.A.P.ARANHA
Ship and Ocean Research Division
IPT-, São Paulo,Brazil

1.INTRODUCTION

One of the most important tasks in the prediction of moored
system dynamics is the satisfactory evaluation of slow-drift
wave forces acting upon the body in irregular waves.

By satisfactory we mean the use of a precise and relatively
simple computational scheme.

Under the assumption of small amplitude waves, slow-drift wave forces are usually separated into two distinct components:
$F_1(t)$, associated with the quadratic interaction of the first
order (linear) potential and $F_2(t)$, related to the low-frequency second-order potential.

For $F_1(t)$ either a near-field method which integrates the second-order pressure over body's surface (see, for instance Faltinsen & Lockens (1980), Pinkster (1979)), or a far-field integration procedure, based on the knowledge of the total linear
scattered wave at infinity (see Ogilvie (1983) or Mei (1983),
can be sucessfully used.

The computation of $F_2(t)$ needs special considerations. The
low frequency second-order potential $\Phi(x,y,z,t)$ is excited by
quadratic interactions of the first-order potential at free and
body's surfaces. On one hand $\Phi_2(x,y,z,t)$ can be viewed as an
amplitude modulation of the steady second-order potential in
harmonic waves, calculated at the mean frequency ω_0 (see Aranha
& Pesce (1986)).On the other hand the quadratic interactions
can be calculated at two near frequencies. Despite the approach
used $F_2(t)$ can be precisely determined without computing
$\Phi_2(x,y,z,t)$. This can be accomplished by applying Green's theo
rem to ϕ_2 and $\phi_o{}^{(j)}$, being $\phi_o{}^{(j)}$ the jth radiation potential
at zero (or $\Delta\omega$) frequency:- see Aranha & Pesce (1986), for the
two dimensional case and Ligthill (1979) or Benschop et alii
(1987) for the general three-dimensional case.

Thus, once the total linear potential is known the wave drift
forces can be evaluated. But even the linear problem solution,
although well established can be somewhat time consuming.

The purpose of the present work is to present some techniques
that can improve computational efficiency, leading to the wave-
drift coefficients.

Section 2 presents an overview on the formulation of slow-
drift forces in irregular waves and an asymptotic expansion on
the small bandwidth. Sections 3 and 4 restrict attention to
the two dimensional case, presenting a variational method for
the problem and analysing some typical geometries.

E. Bratteland (ed.), Advances in Berthing and Mooring of Ships and Offshore Structures, 380–395.
© 1988 by Kluwer Academic Publishers.

2. WAVE DRIFT FORCES IN IRREGULAR WAVES

Consider a body floating in waves where x,y,z is the right hand coordinate system with x,y the undisturbed free-sufrace plane, z pointing upwards (fig.1).

Let the potential associated with a plane irregular wave train, of narrow band spectrum, propagating in finite depth water, be given by

$$\tilde{\Phi}_I(x,y,z,t) = \frac{g\bar{A}}{2\omega_0} a(x,y,z,t) \frac{\cosh K_0(z+h)}{\cosh K_0 h} e^{i(K_0\cos\theta x + K_0\sin\theta y - \omega_0 t)} + (*) \qquad (2.1)$$

where (*) means complex conjugate, $\bar{A}$ = average wave amplitude, ω_0 = central frequency, $\frac{\omega_0^2}{g} = K_0 \tanh K_0 h$, h = depth, $a(x,y,z,t)$ = =nondimensional amplitude modulation. If $\Delta\omega$, the nondimensional bandwidth, is small them $(\frac{\partial}{\partial x}, \frac{\partial}{\partial y}, \frac{\partial}{\partial z}; \frac{\partial}{\partial t}) a(x,y,z,t) \sim 0(\Delta\omega) << 1$, and the linear diffracted wave $\tilde{\Phi}_1(x,y,z,t)$ can be written as:

$$\tilde{\Phi}_1(x,y,z,t) \simeq a(x,y,z,t) \, \Phi_1(x,y,z,t) \, (1+0(\Delta\omega)) \qquad (2.2)a$$

with

$$\Phi_1(x,y,z,t) = \frac{1}{2} \frac{g\bar{A}}{\omega_0} \phi_1(x,y,z) e^{-i\omega_0 t} + (*) \qquad (2.2)b$$

where $\phi_1(x,y,z)$ is the nondimensional linear diffracted potential at frequency ω_0. The leading order component of the slow drift force associated to the quadratic interaction of the linear potential can then be written as:

$$\mathbf{F}_1(t) = \rho g \bar{A} D^2 \delta \, \mathbf{Q}_1(\omega_0) |a(t)|^2 \quad (1+0(\Delta\omega)) \qquad (2.3)$$

where ρ = water density; D = typical body length scale, $\delta = \bar{A}/D$ - small amplitude parameter; $a(t) = a(o,o,t)$, and $\mathbf{Q}_1(\omega_0)$ is the second-order steady force coefficient for a harmonic wave with frequency ω_0. Expression (2.3) was suggested by Newman (1979), for the two-dimensional case, whith the sway mean drift force, derived by Maruo (1960), and given by

$$Q_1^{(2)}(\omega_0) = \frac{1}{2} |R|^2 (1+\frac{2K_0 h}{\sinh 2K_0 h}) \qquad (2.4)$$

where R is the reflection coefficient and (2) designates sway motion.

For the three-dimensional case Newman (1967) extended Maruo's result, relating the surge, sway and yaw drift force coefficient to the total nondimensional scattered wave amplitude $A(\theta)$ at infinity. From momentum conservation, applying divergence theorem and taking the average value over one cycle the drift force can be expressed as a function of $\phi_1(r,\theta)$ as $r \to \infty$. Then from the asymptotic form for $\phi_1(r,\theta)$ as $r \to \infty$, and applying the stationary phase method it follows:

$$Q_1^{(1)}(\omega_0) = -\frac{1}{K_0 D} \frac{C_g}{C} \left\{ \frac{1}{\pi} \int_0^{2\pi} \cos\theta |A(\theta)|^2 d\theta + 2 R_e A(o) \right\} ; \text{ surge} \qquad (2.5)a$$

$$Q_1^{(2)}(\omega_c) = \frac{1}{K_o D} \frac{C_g}{C} \frac{1}{\pi} \int_0^{2\pi} \sin\theta \ |A(\theta)|^2 d\theta \quad ; \text{ sway} \qquad (2.5)\,b$$

$$Q_1^{(6)}(\omega_0) = \frac{1}{K_o^2 D^2} \frac{C_g}{C} \text{ Im} \left\{ \frac{dA^*}{d\theta} \Big|_{\theta=o} + \frac{1}{\pi} \int_0^{2\pi} A(\theta) d \frac{A^*(\theta)}{d\theta} d\theta \right\}$$

$$\text{; yaw} \qquad (2.5)\,c$$

where C_g and C are respectively group and phase celerity.

Now let $\Phi_2(x,y,z,t)$ be the second-order potential excited by the first order potential at ω_0

. $\nabla^2 \Phi_2 = 0$; in the fluid domain $\qquad\qquad (2.6)\,a$

. $\dfrac{\partial \Phi_2}{\partial n}\Big|_{z=-h} = 0$ on the bottom, $\qquad\qquad (2.6)\,b$

$$. \frac{\partial^2 \Phi_2}{\partial t^2} + g \frac{\partial \Phi_2}{\partial z}\Big|_F = \left\{ -\frac{\partial}{\partial t}(\nabla\Phi_1)^2 + \frac{1}{g}\frac{\partial\Phi_1}{\partial t}\frac{\partial}{\partial z}\left(\frac{\partial^2\Phi_1}{\partial t^2} + g\frac{\partial\Phi_1}{\partial z}\right) \right\}_F =$$

$$= L_2\ (x,y,t)\Big|_F \qquad ; \text{ at } z = 0 \qquad\qquad (2.6)\,c$$

$$(*)$$

. $\nabla\Phi_2.n\Big|_{SB} = B_2\ (x,y,z,t)\Big|_{SB}$ at body's surface $\qquad (2.6)\,d$

. radiation condition at $r = \sqrt{x^2 + y^2} \to \infty$

$L_2(x,y,t)\Big|_F$ and $B_2(x,y,z,t)$ are quadratic operators, depending on the total linear potential and body's motions. As $\Phi_1(x,y,z,t)$ is harmonic in ω_0, $L_2(x,y,t)\Big|_F$ and $B_2(x,y,z,t)\Big|_{SB}$ can be separated into steady components $L_{20}(x,y)\Big|_F$ and $B_{20}(x,y,z)\Big|_{SB}$, and $2\omega_0$ pulsating parcels $L_{22}(x,y,z,t)\Big|_F$ and $B_{22}(x,y,z,t)\Big|_{SB}$. Since we are mainly concerned with the low frequency potential $\Phi_2(x,y,z,t)$, only $L_{20}(x,y)$ and $B_2(x,y,z)$ will be considered from now on. Taking $\phi_{20}(x,y,z)$ as the dimensionless steady component of $\Phi_2(x,y,z,t)$ at frequency ω_0 such that

. $\nabla^2\phi_{20}=0$ in the fluid domain $\qquad\qquad (2.7)\,a$

. $\dfrac{\partial\phi_{20}}{\partial n}\Big|_{z=-h} = 0$ on the bottom $\qquad\qquad (2.7)\,b$

. $\dfrac{\partial\phi_{20}}{\partial z}\Big|_F = L_{20}(x,y) = -2\omega\text{Im}(\phi_1^* \bar{\nabla}^2\phi_1) ; \bar{\nabla}^2 = \dfrac{\partial^2}{\partial x^2} + \dfrac{\partial^2}{\partial y^2} \qquad (2.7)\,c$

---

* for conciseness sake $B_2(x,y,z,t)$ will be ommited (see Ogilvie,1983)

$$. \nabla\phi_{20} \left. \vec{n} \right|_{SB} = B_{20}(x,y,z) \quad \text{at body's surface} \qquad (2.7)d$$

$$. \nabla\phi_{20} \to 0 \quad \text{as } r \to \infty \qquad (2.7)e$$

the second-order potential "drift coefficient" can be defined as:

$$Q_2(\omega_0) = -\omega \iint\limits_{SB} \phi_{20}(x,y,z) \, n \, dS_B \qquad (2.8)$$

In the expressions above $\omega = \omega_0/\omega_a$; $K = K_0 \cdot \dfrac{\omega a^2}{g}$; $\omega a = \sqrt{\dfrac{D}{g}}$.

Since the amplitude modulation $a(x,y,z,t)$ is "nearly constant", the exciting terms for the second-order low-frequency potential $\tilde{\Phi}_2(x,y,z,t)$ are given by

$$\tilde{L}_{20}(x,y,t) = |a(x,y,o,t)|^2 \, L_{20}(x,y) \quad (1+0(\Delta\omega)) \qquad (2.9)a$$

and

$$\tilde{B}_{20}(x,y,z,t) = |a(t)|^2 B_{20}(x,y) \quad \text{at body's surface yielding}$$

$$\tilde{\Phi}_2(x,y,z,t) = |a(x,y,o,t)|^2 \, \frac{g\bar{A}}{\omega_0} \phi_{20}(x,y,z) \quad (1+0(\Delta\omega)) \qquad (2.10)$$

and from Bernoulli equation, up to secon-order, it follows

$$\mathbf{F}_2(t) = g \, \bar{A}D^2 \, Q_2(\omega_0)\frac{1}{\omega} \cdot \frac{d}{dt} |a(t)|^2 (1+0(\Delta\omega))$$

$$(2.11)$$

Notice that $\dfrac{\partial\phi_{20}}{\partial t} = 0$ and the mean drift-force component $\bar{F}_2(t) = 0$. Or in other words the low-frequency force component associated with the second-order potential has a strictly oscillatory character.

Now if Green's theorem is applied to $\phi_{20}(x,y,z)$ and to $\phi_0^{(j)}$ - the nondimensional radiated potential at $\omega \to o$, due to a unit amplitude motion in the jth mode (j=1,2,6), $-Q_2^{(j)}(\omega_0)$ can be written as (see Lighthill (1979) or Benschop et alii (1987), or Aranha & Pesce (1986) for the two-dimensional case) (*)

$$Q_2^{(j)}(\omega_0) = \iint\limits_{S} \phi_0^{(j)} (\nabla\phi_{20} \cdot \mathbf{n}) \, dS + Q_\infty^{(j)}(\omega_0) \qquad (2.12)$$

where $S=S_B \cup F$, being S_B the body's surface and F the free surface. Expression (2.12) allows one to calculate $Q_2(\omega_0)$ without computing $\phi_{20}(x,y,z)$ since $(\nabla\phi_{20} \cdot n)$ is given by $L_{20}(x,y)$ on F and $B_{20}(x,y,z)$ on S_B. In some sense (2.12) can be viewed as an extension of Haskind relation for the steady second-order problem.

It should be notice that if diffraction and radiation were non-existent $L_{20}(x,y) \sim 0$ (1)-it would be given only by quadratic terms of the incident wave potential. As, for infinite depth, $L_{20}(x,y) = 0$, $Q_2(\omega_0)$ is expected to be null in these situations.

*. $Q_\infty^{(j)}(\omega_0)$ is the far field contribution associated with mass conservation; see Aranha & Pesce (1986)

This reasoning certainly suggests that the effect of the second-order potential could be disregarded in the general case where diffraction is small. (see for instance Faltinsen & Løckens (1979)).

This argument, however, is only partially correct.

For simplicity consider the two-dimensional case. Then $Q_1^{(2)}(\omega_0) = \frac{1}{2}|R|^2$ and as $\frac{1}{\omega}\frac{d}{dt}|a(t)|^2 \sim 0 (\Delta\omega)$ the ratio $F_1(t)/F_2(t)$ can be gauged by the parameter $r_1(\omega_0)$ where

$$r_1(\omega_0) = \frac{|R|^2}{2\Delta\omega|Q_2^{(2)}(\omega_0)|} \qquad (2-13)$$

For a surface piercing body $|R| \sim 0(1)$ - and as $Q_2^{(2)}(\omega_0) \sim \sim 0 (|R|)$ -, $r_1(\omega_0) \sim 0(1/\Delta\omega) >> 1$: the relative effect of the second-order potential is small and it can be neglected, a result in accordance with Faltinsen & Lockens (1980). For a submerged body the same conclusion can be false. A submerged circle, e.g., has $R = 0$, irrespective the frequency and submergence (see Ogilvie (1963), and then $r_1(\omega_0) = 0$: the effect of the second-order potential must be considered in this last case. For an arbitrary submerged body $|R| << 1$ if S, the submergence, is large, leading to $r_1(\omega_0) \sim 0 (|R|/\Delta\omega)$, a result that will be explored in section 4.

3. THE TWO DIMENSIONAL CASE. A VARIATIONAL APPROXIMATION

In the preceding sections it was shown that only the linear problem must be solved in order to compute the drift forces.

The attention is now turned to some global first-order quantities as added mass, radiation damping, exciting forces,since then the reflection (and transmission) coefficient can be calculated as well as the boundary conditions $L_{2_0}(y)$ and $B_{2_0}(y,z)$.

Neverthless, the existing methods to determine such global parameters do not take into account an important fact: these quantities are weakly dependent on the local features of the flow field. On the contrary, the wide spread tendency seems to be to reconstruct the local flow field and to obtain these parameters by integration. A question is naturally risen:why not to access them directly?

It will be shown that a variational approach is well suited, as in several branches of applied physics. Miles (1967), for example, using a variational method with only one trial function, have computed, with great accuracy, the reflection and transmission coefficients for a shelf. Following this reasoning, Aranha & Pesce (1987) have developed a variational method for the full two-dimensional linear problem.

In this method the quantities mentioned above, such as added mass,radiation damping, reflection coefficient, are obtained as stationary values of some functionals. The results are rewarding with typical four trial functions the two dimensional hydro dynamic coefficients were computed for several usual cross-sectional geometries, yielding results that agree, in the whole range of frequencies, with values determined by classical means. By using the variational method the amount of numerical work is minimum: it corresponds to construct and invert typically a 4x4 real symmetric matrix, compared to a usual 25x25 complex and full matrix handled in Green's functions method, or, to a 70x70 real and symmetric matrix in the Hybrid Element Method.

The application of the variational method to the present problem will be done in the next section, after a brief outline.

3.1.An overview on the variational approach

As it is well known a definitive feature of a variational method is to give an order δ^2 approximation to a stationary value λ of a functional $F(\phi)$, if an order δ approximation is used for ϕ , being ϕ the solution.

In the present context, if the global quantities (added mass, radiation damping, exciting forces, reflection and transmission coefficients) can be related to stationary values of well defined functionals, this means that a relatively crude approximation for the linear radiation and diffraction potentials well provide a very good result for these coefficients. Such crude approximations should, however, resemble the fundamental characteristics of the flow field, as for example, simulating the boundary conditions in a loose but proper manner. Judicious choices of these functions are necessary, in order to extract correct results.

For constructing the proper functional it is necessary to take a look at the weak formulation for the two-dimensional radiation (and diffraction) problems.

The basic ideas are as follows - see Aranha & Pesce (1987), for details.

Referring to figure 2, suppose the flow region divided into two sub-regions, one close to (inner) and another far from (outer) the cross-section. To the outer solution one associates a Fourier series (*)solution, with undetermined coefficients.(**) To the inner region (5) an unknown solution is attributed, satisfying the pertinent boundary conditions.

Both solutions (inner and outer) must be made compatible along the common - vertical boundary (y=b) - by equating the potential functions and their y-derivatives.

In the inner region S, the weak equation for the problem can then be constructed by supposing that ϕ has finite energy($(\nabla\phi)^2$ is Lebesgue integrable), and by considering the continuity equation. Such an weak equation can be obtained by applying the divergence theorem to the inner region, for each mode j.

Defining j = 1,2,3,4 corresponding to surge, sway, heave and roll (unit velocity) nondimensional potentials and j = 7,8, to the even and odd parts of the nondimensional diffracted potential ϕ_D (y,z) such that (see Aranha & Pesce(1987))

$$\phi_D(y,z) = \phi_7(y,z) + \text{sign } y. \quad \phi_8(y,z) \qquad (3.1)$$

the weak equations are:

$$G(\phi^{(j)},\psi) = iK_0(A\circ_j - \frac{\delta j}{2}) L_0(\psi) + (1-|\delta_j|) V_j(\psi) \quad (3-2)$$

(*) or integral, depending upon the depth.
(**) from here on, unless explicity indicated, all variables are dimensionless where: $\sqrt{B/2g}$ is the time-scale and B/2 the lenght-scale.

valid for all Ψ such that $\iint_S (\nabla \Psi)^2 \, dS < \infty$.

In equation (3.2) $G(.;.)$ and $V_j(.)$ are functionals defined as

$$G(\phi, \Psi) = \iint_S \nabla \phi \cdot \nabla \Psi \, dS - \omega_0 \int_F \phi(y,0) \, \Psi(y,0) \, dy +$$

(3-3)

$$+ \sum_{n=1}^{\infty} Kn \, Ln(\phi) \, Ln(\Psi)$$

$$V_j(\phi) = \int_{\partial B} \Psi v_j \, d\partial B \tag{3.4}$$

where the nondimensional parameters are defined as

ω : non dimensional wave frequency

K_n : non dimensional wave numbers: $\omega_n = - K_n \tan Knh$

$A_0 j$: non dimensional radiated (diffracted) wave amplitude in the jth mode

$$v_j = n_j \; ; \; \text{for} \quad j = 1,2,3,4 \quad \text{radiation modes}$$
$$0 \; ; \; \text{for} \quad j = 7,8 \quad \text{(diffraction-even and odd parts)}$$

$$\delta_j = \begin{array}{l} 0 \quad\quad ; \; \text{for } j = 1, \, 2,3,4 \\ (-1)^{j+1} \; ; \; \text{for } j = 7,8 \end{array}$$

$$L_n(\Psi) = \int_{-h}^{0} \Psi(\bar{b},z) \, f_n(z) \, dz \; ; \; n = 0,1,2 \tag{3.5}a$$

being

$$f_\theta(z) = F_0 \cosh K_0(z+h), \; F_0 = (\frac{1}{h} \frac{4K_0 h}{2K_0 h + \sinh 2K_0 h})^{1/2} \tag{3.5}b$$

the propagating Fourier component and

$$f_n(z) = F_n \cos K_n(z+h); \; F_n = (\frac{4K_n h}{2K_n h - \sin 2K_n h})^{1/2} \tag{3.5}c$$

the evanescent ones.

By isolating the propagating component the potential $\phi^{(j)}$ can be written as (see Aranha & Pesce (1987))

$$\phi^{(j)}(y,z) = (A_{oj} + \frac{\delta j}{2})p^{\pm}(y,z) + (1-|\delta_j|) \, \phi_{Rj}(y,z) + (*) \tag{3.6}$$

where

$$p^{\pm}(y,z) = q^{\pm}(y,z) + \begin{cases} \phi_{R7}(y,z): \text{for even potential functions} \\ \phi_{R8}(y,z): \text{for odd potential functions} \end{cases}$$

(3.7)

with

$$q^{\pm} (y,z) = f_o (z) . \begin{cases} 1 \\ y/b \end{cases} \qquad (3.8)$$

and the suffix R denoting the part of $\phi^{(j)}$ that is free of radiated wave $L_o(\phi_R) = 0)$, satisfying

$$G(\phi^{(j)}, \Psi_R) = V_j(\Psi_R) \quad ; \quad L_o(\Psi_R) = 0 \qquad (3.9)$$

From (3.8) it follows that the parameters $\Lambda_{j\ell}$ defined as

$$\Lambda_j\ell = G(\phi_R^{(j)}; \phi_R^{(\ell)}) \qquad (3.10)$$

can be written as

$$\Lambda_j\ell = F_j (\phi_R^{(j)}; \phi_R^{(\ell)}) = \frac{V_j(\phi_R)^{(\ell)} V_\ell(\phi_R^{(j)})}{G(\phi_R^{(j)}; \phi_R^{(\ell)})} \qquad (3.11)$$

where $F_j (.;.)$ is a functional defined in $\overset{o}{W}_2^{(1)}(S) \times \overset{o}{W}_2^{(1)}(S)$ and such that

$$F_{j\ell} (\phi_R, \Psi_R) = \frac{V_j(\phi_R) V_\ell(\Psi_R)}{G(\phi_R, \Psi_R)} \qquad (3.12)$$

where

$$\overset{o}{W}_2^{(1)}(S) = \left\{ \Psi_R: \Psi_R \in W_2^{(1)}(S); \quad L_o(\Psi_R) = 0 \right\}$$

It can be shown (see Aranha & Pesce (1987)) that $F_j\ell(.;.)$ is stationary in the "point" $(\phi_R^{(j)}; \phi_R^{(\ell)})$. More than that:

if $F_j(\phi_R) = F_{jj}(\phi_R, \phi_R)$, defined in $\overset{o}{W}^{(1)}(S)$, among all ϕ_R, the one that makes $F_j(.)$ stationary is the j-solution of the radiation and diffraction problems.

Then, from the definitions of added mass radiation damping, exciting forces, etc., using (3-6)-(3.11) it can be shown that all first-order quantities can be expressed by the following identities (see Aranha & Pesce (1987) $(\Delta_{j\ell} = 1+(-1)^{j+\ell})$

added mass coefficient

$$a_{j\ell}(\omega_o) = \Delta_{j\ell} \left\{ \frac{V_j(P\pm) \ V_\ell(P\pm)}{K_o^2 + (G(p\pm, p\pm))^2} \ G(p^\pm, p^\pm) + \Lambda_{j\ell} \right\} \qquad (3.13)$$

radiation damping coefficient

$$b_{j\ell}(\omega_o) = \Delta_{j\ell} \ \omega_o K_o \ \frac{V_j(P^\pm) \ V_\ell(P^\pm)}{K_o^2 + (G(p^\pm, p^\pm))^2} \qquad (3.14)$$

exciting forces

$$E_\ell (\omega_0) = (-1) \quad iK_0 \frac{V_\ell (p^\pm)}{G(\pm, \pm) - iK_0} \tag{3.15}$$

wave __amplitude coefficients

$$A_{oj} = -\frac{\delta_j}{2} \frac{G(p^\mp, p^\mp) + iK_0}{G(p^\pm, p^\pm - iK_0} + (1-|\delta_j|) \frac{V_j (p^\pm)}{G(p^\pm, p^\pm) - iK_0} \tag{3.16}$$

where $\pm$ stand for even or odd potential functions and

$$V_j (p^\pm) = V_j (q^\pm) + \begin{cases} \Lambda_{j7} \\ \Lambda_{j8} \end{cases} \tag{3.17}$$

$$G(p^\pm, p^\pm) = G(q^\pm, q^\pm) + \begin{cases} \Lambda_{55} \\ \Lambda_{66} \end{cases} \tag{3.18}$$

On the other hand, as $y \to \infty$ the jth potential can be written as

$$\phi^{(j)}(y,z) \approx \left\{ A_{oj} e^{iK_0 (|y| -b)} + \frac{\delta_j}{2} e^{-iK_0 (|y| -b)} \right\} f_0(z); \quad y \to \infty \tag{3.19}$$

and then the reflection and transmission coefficients follow as

$$R = R_D - \omega^2 (\sum_{j=2}^{4} A_{oj} \cdot \eta_j) e^{-2iK_0 \bar{b}} \tag{3.20}$$

$$T = T_D - \omega^2 (\sum_{j=2}^{4} A_{oj} \cdot \eta_j) e^{-2iK_0 \bar{b}} \tag{3.21}$$

where D designates the part associated to diffraction such that

$$R_D = (A_{07} - A_{08}) e^{-2iK_0 b} \tag{3.22}$$

$$T_D = (A_{07} + A_{08}) e^{-2iK_0 b} \tag{3.23}$$

and η_j is the complex amplitude of the jth mode of oscillation. Expression (3.20) used in (2.4) gives the mean drift force $Q_1^{(2)} (\omega_0)$.

It should be noticed the similarity of expression (3.11) to Rayleigh quotient in vibration problems.
More, if we define

$$\mathcal{L}_j (\phi_R) = \frac{1}{2} G(\phi_R ; \phi_R) - V_j (\phi_R)$$

as the Langrangean associated to the jth problem, free of propa gating waves, it is not difficult to see that $\mathcal{L}_j^* (\phi_R)$ indeed represents the excess of kinetic energy over the total (free surface plus body'surface) potential energy. The stationary condition for $\mathcal{L}_{ej}^{*}(\phi_R)$ leads to the weak equation (3.2).

The search for stationary conditions for (3.10) is then equivalent to solve (3.2) in a space spanned by a finite series of elementary trial functions $T_n(y,z)$. Once the series coefficients are determined, the parameters $\Lambda j\ell$ are promptly computed.

3.2. Variational approximation for $Q^{2(2)}(\omega_0)$

Let

$$G_o(\phi,\Psi) = \iint_{A\infty} \nabla\phi.\nabla\Psi dA \qquad (3.24)$$

$$V_{20}(\Psi) = \int_{\partial F} L_{20}(y) \ \Psi \ (y,o) \ dy + \int_{\partial B} B_{20}(y) \ \Psi \ (y,z)\big|_{\partial B} \ d\partial B \qquad (3.25)$$

Then $\phi_{20}(y,z)$, the steady second-order potential at ω_0, and $\phi^{(2)}(y,z)$, the sway-(j=2) radiation potential at zero frequency,are solutions of the following weak equations (see Aranha & Pesce (1987)

$$G_0(\phi_{20};\Psi) = V_{20}(\Psi); \quad \text{all } \Psi \qquad (3.26)$$

$$G \ (\phi_o^{(2)};\psi) = \ V_2(\Psi) \ ; \ \text{all } \Psi \qquad (3.27)$$

By definition

$$Q_2^{(2)} \ (\omega_0) = V_2(\phi_{20}) = V_{20}(\phi_o^{(2)}) \qquad (3.28)$$

This is an alternative form of equation (2.12), restricted to the two-dimensional case.

Let $\{\hat{\phi}_{20}(y,z); \hat{\phi}_o^{(2)}(y,z)\}$ be the solutions of (3-26)and (3.27) in a finite dimensional space $W_N(A)$ and $\{\Delta\phi_{20}(y,z); \Delta\phi_o^{(2)}(y,z)\}$ the difference between the exact solutions and the approximated ones. These functions are in a space orthogonal to $W_N(A)$ with respect to the "inner product" $G(.;.)$.

If $\{||\Delta\phi_{20}|| \ ; ||\Delta\phi_o^{(2)}||\}$ are their norm then the approximated value $Q_2^{(2)}(\omega_0) = V_{20}(\hat{\phi}_o^{(2)}) = G_o(\hat{\phi}_2;\hat{\phi}_o^{(2)})$

have an error smaller than $||\Delta\phi_{20}||.||\Delta\phi_o^{(2)}||$, a direct result from Schwarz inequality. Or in other words rough approximations for ϕ_{20} and $\phi_o^{(2)}$ produce a improved result for $Q_2^{(2)}(\omega_0)$.

Notice that if a ε approximation is used for the linear potential $\phi_1(y,z)$, in order to compute $L_{20}(y)$ and $\hat{B}_{20}(y,z)$, then $\hat{Q}_1^{(2)}(\omega_0)$ has an error of order ε^2 (see section 3.1) and $\hat{Q}_2^{(2)}(\omega_0)$ an error of order ε.

3.3. The choice of trial functions

If a trial function $\hat{\Phi}_R^{(j)}$ is properly chosen for the jth-problem in 3.1, equation (3.11) provides a variational approximation for the $\Lambda j\ell$ parameters.

A satisfactory trial function should resemble the fundamental features of the flow field, which are distinct for each mode, deserving thus specific considerations. Not only the macroscopic but also the local characteristics of the flow induced by the body's presence must be taken into account (as, for example the infinite velociy associated to the rotation of the fluid around a sharp corner).

A proper definition for $\hat{\Phi}_R^{(j)}$ is given by a finite series of elementary trial functions $T_n(y,z)$ composed, for example, by Green's functions and related high-order singularities such as dipoles and vortices, all placed in convenient positions, inside the body contour.

A pole (source or sink) can be used to represent the flow induced by the low frequency heave motion as well as dipoles imitates the flow associated with high frequency sway or heave motions. A vortex(*) point, on another hand, can be though to simulate the rolling motion of the body, or a rotating flow around a sharp corner. Figure 3 shows, for symmetric cross-sections, pairs (or systems) of elementary singularities that have been used to construct the set of elementary trial functions.

The free wave behaviour should also be recovered and a special elementary trial function $T_o^{\pm}(y,z)$ based on the $f_o(z)$ Fourier component can be successfully used.

$$T_o^{\pm}(y,z) = f_o(z) \cdot \begin{cases} \cos K_o y - \cos K_o b & ; \text{ even modes} \\ \sin K_o y - \frac{y}{b} \sin K_o b & ; \text{ odd modes} \end{cases} \tag{3.29}$$

4. NUMERICAL RESULTS FOR TWO-DIMENSIONAL BODIES

Rectangular and circular cross sections were chosen to ilustrate some meaningful results. Only the infinite depth case was considered.

For the rectangle two conditions were analysed: the surface piercing and the submerged cases. For the circle only the submerged case was exemplified.

The variational method was applied by choosing the following elementary trial functions that, besides $T_o^{\pm}(y,z)$ (see section 3.3), represent the near field flow around the body
a) rectangle:
   - even mode (heave): pole at 0, vortices in the vicinity of the corners and a pole at M for the surface piercing case.

   - odd modes (sway, roll) source-sinK pair at F, vortices in the vicinity of the corners and a vortex system centered at 0 (see figure 3).

_____

(*) Care must be taken in order to choose the correct system of vortices such that the center and the branch cut remain within the body, keeping finite the total energy

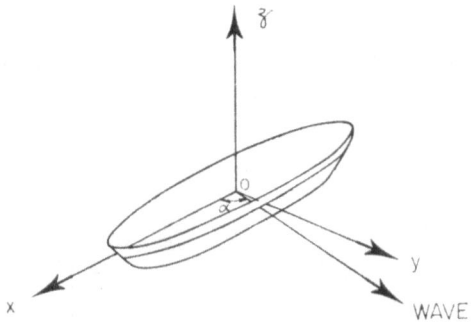

Figure 1. Coordinate system

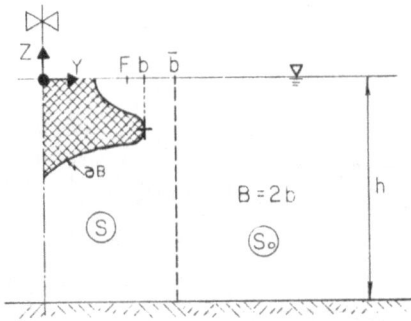

Figue 2. The geometry of the two-dimensional body and the finite
fluid region S.

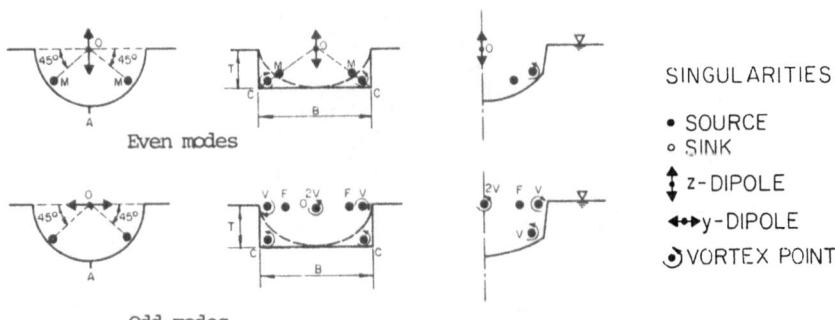

Figure 3. Two dimensional geometries and singularities to be used in
the variational approximation.

392

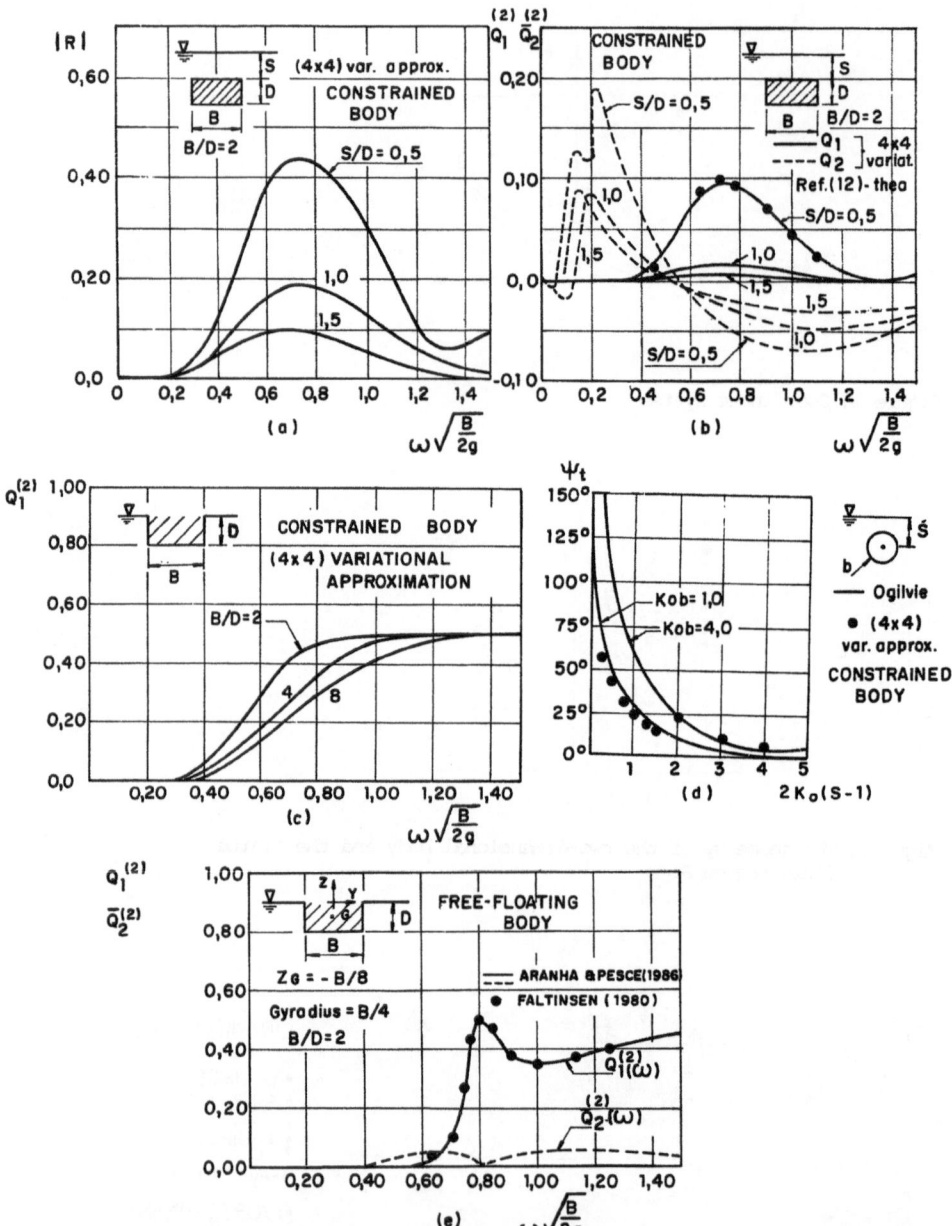

Figure 4. (a),(b) constrained submerged rectangle: reflection coefficient |R| and drift force coefficients $Q_1^{(2)}$, $\bar{Q}_2^{(2)}$ ($\Delta\omega = 0,2$) (c) drift coefficient for a surface piercing constrained rectangle (d) transmission coefficient phase Ψ_T for a constrained submerged circle (e) drift coefficients for a surface piercing, free-floating rectangle ($\Delta\omega$=0.2)

b) Circle:
- even mode (heave): z-dipole at 0 and poles (sources) pla
  ced at 0.9R, 45° degree aparted.

- odd modes (sway, roll): y-dipole at 0, and poles placed at
  0.9R, 45° degree aparted.

To compute $Q_2^{(2)}(\omega)$ for the submerged rectangle a source-sinK
pair, placed at F, was used as trial function for $\phi^{(2)}$.

Figure 4a and 4b correspond to a constrained submerged rec -
tangle, for three submergence values (S/D= 0.5,1.0,1.5).

Figure 4a presents the reflection coefficient $|R|$.

Figure 4b shows the wave drift force coefficients $Q_1^{(2)}(\omega)$ and
$\bar{Q}_2^{(2)}(\omega) = \Delta\omega . Q_2^{(2)}(\omega)$, where the dimensionless spectrum bandwidth was chosen to be $\Delta\omega = 0.2$. Comparisons made with Inoue &
Kyosuka (1985) results show a quite satisfactory agreement. No
tice that $r = |\bar{Q}_2^{(2)}|/Q_1^{(2)} >> 1$ if $\omega_p = \omega\sqrt{B/2 g} \leq 0.5$ or $\omega_p \geq 0.8$,
and $r \sim 0(1)$ otherwise. Also $\bar{Q}_2^{(2)}$ decreases faster than $Q_1^{(2)}$
with S. These observations indicate that the second-order po-
tential effect should be considered for submerged bodies.

Figure 4.c and 4.e correspond to a surface piercing rectangle.
Figure 4.c shows, for various B/D (beam/draft) ratio values, the
drift coefficient $Q_1^{(2)}$ for the constrained body. A (4x4) varia
tional approximation was used. Figure 4.e, extracted from ref.
(1), presents for a free-floating case, B/D = 2, the drift
coefficients $Q_1^{(2)}$ and $\bar{Q}_2^{(2)}$ ($\Delta\omega$= 0,2). To be noticed the difference between the constrained and the free-floating cases (B/D=
2).

Also to be mentioned the small relative contribution from the
second-order potential, represented by $\bar{Q}_2^{(2)}$, if compared to the
submerged case.

Figure 4.d presents the transmission coefficient phase Ψ_T as
function of submergence for two different frequencies. The results correspond to a (4x4) variational approximation and were
compared to Ogilvie (1963) curves. It should be mentioned that
the computed value for the reflection coefficient $|R|$ was always
very small (less than 0.01), a result in accordance with Dean
theoretical predictions (see also Ogilvie (1963)).

5.SOME CONCLUDING REMARKS

An asymptotic expansion in the small spectrum bandwidth, resulting in a modulate form representation for an irregular wave train, was used to separate the slow-drift force into two
different components $F_1(t)$ associated with the quadratic interactions of first-order potential and, $F_2(t)$ corresponding to
the slow-modulation of the steady second order potential at the
central frequency ω_0.

For a general three dimensional case the extended Maruo's far-
field method can be applied to the computation of $F_1(t)$ and
Ligthill relation allows us to evaluate $F_2(t)$ without computing
the second-order potential.

For the two dimensional case a variational method was outlined. This method takes into-account the full linear problem
and gives directly the main global hydrodynamic quantities, a-
mong them the drift force coefficients $Q_1^{(2)}$ an $Q_2^{(2)}$. For the
latter a particular form of Ligthill result, deduced indepen-

394

dently by Aranha & Pesce (1986), was recovered from the weak
formulation for the steady second-order potential.
A point should be touched concerning a special class of body
geometry: the slender body case.
Application of the asymptotic and variational methods quoted-
above in conjunction with the slender body theory   (see for ins
tance Newman & Sclavounos or Aranha & Martins (1987) or Aranha
& Pesce (1987)),can be viewed as a efficient manner to deal
with this important case.
Finally a important point should be mentioned. Although nume
rical results where presented only for the two-dimensional case,
it could be observed that the second-order potential contribution can become relevant in the evaluation of wave drift-forces
acting upon submerged bodies, as for example the pontoons of a
semi-submersible platform.

REFERENCES

1. Aranha, J.A.P. & Pesce, C.P.: Effect of the Second-order Po
   tential in the Slow-Drift Oscillation of a Floating Structure in Irregular Waves. Journal of Ship Research, vol. 30,nº
   2,June 1986, pp. 103-122.
2. Aranha, J.A.P. & Martins, C.A.: Diffraction of Sea Waves by
   a Slender Body. Part (2) Water of Arbitrary. Depth to be published in the Journal of Fluid Mechanics.
3. Aranha, J.A.P. & Pesce, C.P.: A Variational Method for the
   Two-dimensional Water Wave Radiation and Diffraction Problems.
   Submitted to the Journal of Fluid Mechanics, 1987.
4. Aranha, J.A.P. & Pesce, C.P.: Variational Approximation and
   Slender Body Theory Applied to the Motion Analysis of Floating Structures. 6th International Symposium on Offshore Engineering, Brazil Offshore'87. R.Janeiro, August 24-28, 1987.
5. Aranha, J.A.P. & Pesce, C.P: Slow Drift and Trapping of Waves
   on Submerged Structures. IUTAM Symposium on Nonlinear Water
   Waves, Tokio, August 24-28, 1987.
6. Benschop, A.; Hermans, A.J. & Huijmans, R.N.: Second-order
   Diffraction Forces on a Ship in Irregular Waves. Applied
   Ocean Research, 1987, vol. 9., nº 2, pp. 96-103.
7. Faltinsen, D.M. & Løckens, A.E.: Slow Drift Oscillations of
   a Ship in Irregular Waves. Modelling, Identification and
   Control, Vol. 1, nº 4, 1980,pp.125-213
8. Inoue, R. & Kyosuka, Y.: On the Nonlinear Wave Forces acting
   on Submerged Cylinders, Naval Architecture and Ocean Enginee
   ring, vol.(23), pp. 83-96.
9. Lighthill, M.J.: Waves and Hydrodynamic Loading. Proc.of Second Int. Conf. on the Behaviour of Offshore Structures,vol.
   1, pp. 1-40, 1979.
10. Maruo, H.: The Drift of a Body Floating on Waves. Journal of
    Ship Research. vol. 4, nº 3, Dec. 1960, pp. 1-10.
11. Newman, J.N.: The Drift Force and Moment on Ships in Waves.
    Journal of Ship Research, vol. 11, nº 1, Mar. 1967.
12. Newman, J.N.: Second-order, slowly-varying Forces on Vessels
    in Irregular Waves. In: International Symposium on the Dyna-
    mics of Marine Vehicles and Structures in Waves, London, Apr.
    1-5, 1974, London IME, 1975, pp. 182-186.
13. Newman, J.N.: The Theory of Ship Motion. Adv. Appl. Mech. 18,
    221-283,1978.

14. Ogilvie, T.F.: First-and Second-order Forces on a Cylinder Submerged under a Free Surface. Journal of Fluid Mechanics, 1963, pp. 451-972.
15. Ogilvie,T.T. : Second-order Hydrodynamics Effects on Ocean Platforms. International Workshop on Ship & Platform Motions- Berkeley, Oct. 1983,pp.205-263.
16. Pinkster, J.A.: Mean and Low Frequency Wave Drifting Forces on Floating Structures. Ocean Engineering, vol. 6, nº 6, 1979, pp-593-615.

This work has been supported in part by CNPq - Conselho Nacional de Desenvolvimento Científico e Tecnológico, proc. nº 304062-85, MCT-Ministério da Ciência e Tecnologia, and in part by FAPESP- Fundacão de Amparo à Pesquisa do Estado de S.Paulo.

Aknowledgements -

We are indebt to Dr. Antônio Carlos Fernandes for his valuable comments, and to Mrs. Nair de Toledo for typing the manuscript.

MOORING SYSTEMS OF VLCC, LPGC AND LNGC IN Japan

S.Ueda
Chief of the Offshore Structures Laboratory, Structures
Division, Port and Harbour Research Institute, Ministry
of Transport, Japan

1. INTRODUCTION
This paper gives a description of present state of mooring
systems of deep water oil terminals for VLCC, LPGC and LNGC in
Japan. Some new type of fenders and mooring equipments are
installed.

2. FACILITIES
2.1. Facilities Available for VLCC
Facilities available for VLCC larger than 100,000dwt oil tankers are listed in Table 1. There are 57 such facilities.
Thirty-two of them are available for oil tankers of more than
200,000dwt. Generally speaking the berth structure could be
classified as a fixed type berth or a floating type berth.
There are fourty-six fixed type berths, and among them twenty-
six are dolphin type berth and sixteen are pier type berth.
The floating type berths consist of twelve single point buoy
berth and three multi mooring buoy berth. Fenders installed
to the fixed type are highgrade rubber fenders such as H type,
Cellular type and Pneumatic type with height in the range of
1000 to 3000mm.
2.2. Facilities Available for LNGC
Facilities available for LNGC are listed in Table 2. There are
twelve berths operating, and some berths are under construction.
2.3. Facilities Available for LPGC
Facilities available for LPGC are listed in Table 3. There are
thirty-one berths operating.

REFERENCES
1. Horii,O., S.Ueda and T.Ichikawa: Investigation of Crude
Oil Tanker Berth, Note of the Port and Harbour Research
Institute, No. 201, 1975 126p.
2. Japanese Association for Maritime Information: Port in
Japan, 1985.

E. Bratteland (ed.), Advances in Berthing and Mooring of Ships and Offshore Structures, 396–400.
© 1988 by Kluwer Academic Publishers.

Table 1 Deepwater Terminal available for more than 100,000dwt VLCC

Name of Port	Name of Berth	Type of Berth	Available Ship Size dwt 10^3tf	Fenders Name Size	Number
Tomakomai	Idemitsu SB	D	280	C-3,000	3/each
Tomakomai	Hokkaido Kyobi SB	D	127	C-2,000	4
Muroran	Nihon Refinery SB	D	100	C-2,000	2/each
Shiogama	Touhoku Sekiyu SB	D	234	H-1,750	2/each
Niigata	Showa Sekiyu SB	SPM	130	−	
Kashima	Kashima Sekiyu SB	D	110	C-1,250	3/each
Kashima	Kashima Terminal SB	D	250	H-1,700	2/each
Chiba	Keiyo SB-1	D	258	C-2,000/2,250	2/6
Chiba	Keiyo SB-2	D	258	C-2,000/2,250	2/6
Chiba	Cosmo Sekiyu SB-1	D	100	C-1,600	8
Chiba	Cosmo Sekiyu SB-2	SPM	108	−	
Chiba	Fuji Sekiyu SB	D	120	H- 800	3/each
Kawasaki	Mitsubishi			SUC-1,600	2
Kawasaki	Mitubishi-Shouwa SB	SPM	265	−	
Kawasaki	Toa Nenryou SB-1	D	120	H-1,400	6/each
Kawasaki	Toa Nenryou SB-2	D	260	C-2,000	3/each
Yokohama	Ohgishima Sekiyu SB	SPM	248	−	
Yokohama	Asia Sekiyu			C-1,450	2/each
Yokohama	Nihon Refinery SB-1	P	150	C-2,000	2/each
Yokohama	Nihon Refinery SB-2	P	150	SUC-2,500/2,250	2/1
Shimizu	Toa Nenryou SB	D	250	C-2,000	2/each
Atsumi	Chubu Denryoku SB	SPM	210	−	
Nagoya	Idemitsu SB	D	100	C-1,600	2
Nagoya	Isewan SB-1	D	280	C-3,000	4
Nagoya	Isewan SB-2	D	280	C-3,000	4
Nagoya	Nihon Kogyo SB	D	100		
Yokkaichi	Cosmo Sekiyu SB	SPM	240	−	
Yokkaichi	Showa Yokkaichi SB-1	SPM	170	−	
Yokkaichi	Showa Yokkaichi SB-2	SPM	275	−	
Owase	Toho Sekiyu SB	MPM	210	−	
Owase	Toho Sekiyu SB	D	100	700,900	4+2
Wakayama	Toa Nenryou SB	D	236	C-2,000	2
-Shimotsu	Fuji Kosan SB	D	245	C-2,000	2/each
Sakai	General Sekiyu SB	P	230	1,100	6
-Senboku	Koa Sekiyu SB	D	150	750	9
Sakai	Cosmo Sekiyu SB-1	D	200	1,000	4
-Senboku	Cosmo Sekiyu SB-2	D	200	1,000	4
Sakai -Senboku	Marubeni	D	100	C-2,000	2
Himeji	Idemitsu SB	SPM	258	−	
Mizushima	Mitsubishi Sekiyu SB	D	240	H-1,700	2/each
Mizushima	Nihon Kogyo SB	P	240	C-2,250	2/each
Sakaide	Asia Kyouseki SB	D	195	C-1,600	16
Kagawa	Mitsubishi SB	P	125		
Kikuma	Taiyo Sekiyu SB	MPM	130	−	
Iwakuni	Koa Sekiyu SB	D	150	750	9/each
Tokuyama	Idemitsu SB-1	MPM	275	−	
-Kudamatsu	Idemitsu SB-2	P	150	C-1,600/800	7/4
Tokuyama	Nihon Refinery SB	P	178		

398

Ube	Seibu Sekiyu SB	SPM	250	–	
Mutsure	Nisshin Tanker	D	150	C-2,000	5
Kiire	Nihon Staging SB-1	D	150	H-1,700	2/each
Kiire	Nihon Staging SB-2	D	150	H-2,000	2/each
Kiire	Nihon Staging SB-3	D	450	H-2,000	6/each
Kiire	Nihon Staging SB-4	D	500	H-2,500	2/each
Kin	Nansei Sekiyu SB	SPM	270	–	
-Nakagusuku	Okinawa Terminal SB-1	D	500	Raykin K-60	2/each
	Okinawa Terminal SB-2	D	150	Raykin K-60	2/each
Kin	Okinawa Staging SB	D	500	H-2500	2/each
-Nakagusuku	Okinawa Staging SB	D	300	H-2500	2/each

Note: SB:Sea Berth, D:Dolphin, P:Pier,
SPM:Single Point Mooring, MPM:Multi Point Mooring

Table 2   Major DeepWater Terminal available LPGC

Name of Port	Name of Berth	Type of Berth	Available Ship Size dwt 10^3tf	Fenders Name	Size	Number
Aomori	Mitsui LPG	D	53	C-1,000		3/each
Shogama	Tohoku Sekiyu	D	60	H-1,700		3/each
Nanao	LPG Terminal	D	60	C-1,000/800		12/6
Kashima	Kashima Sekiyu	D	100	C-1,250		3/each
Keihin	Tokyo Gas	D	35	V- 600		2/each
Chiba	Idemitsu	D	80	300		48
Chiba	Fuji Sekiyu	P	120	C-2,000		6
Chiba	Tokyo Denryoku	D	60	C-2,000		3
Chiba	Kyokuto Sekiyu	D	100			
Chiba	Mrubeni	D	57	C-1,250/1,000		2/2
Kawasaki	Mitsui LPG	D	50			
Kawasaki	Kyoudo Sekiyu	D	35	V- 300		3/each
Kawasaki	General Sekiyu	D	46	C- 800		1x4
Kawasaki	Nihon Sekiyu Gas	D	90	H-1,400		4/each
Nagoya	Idemitsu	D	65	C-2,000		1/each
Kinuura	Shell Sekiyu	D	55	800		4/each
Oosaka	Toyo Futo	D	45	SA- 600		6/each
Sakai	Mitsui Toatsu	P	66	C-1,700		5
-Senboku	General Sekiyu	D	230	C-2,000		2/each
Sakai	Iwatani	D	150	V- 400		30
-Senboku	Maruzen	D	200	C-1,000		
Wakayama -Shimotsu	Sumitomo Kinzoku	D	55	C-1,250		20
Kobe	Mitsubishi	D	47	V- 800		2/each
Higashi -Harima	Kobe Seiko	D	70			
Mizushima	Nihon Kogyo	P	180	C-2,250		2/each
Tokuyama -Kudamatsu	Idemitsu	D	70	C-1,600		7
Namikata	Namikata Terminal	D	125	C-2,000		7
Namikata	Namikata Terminal	P	3	C-1,000		2/each
Ohita	Ohita Kyoubi	D		SUC-2,000		2
Ohita	Showa Denko	D	97			
Karatsu	LPG Terminal	D	57	H-1,700		2/each
				C-1,600		7
		D	60	C-1,250		2/each
Imari	Kyushu LPG	D	70	C-1,700/1,600		4/2

Note: D:Dolphin, P:Pier

Table 3 Major Deepwater Terminal available for LNGC

Name of Port	Name of Berth	Type of Berth	Available Ship Size $10^3 m^3$	Fenders Name	Size	Number
Chiba	Tokyo Denryoku	D	75	H-1,400		6/each
	Tokyo Gas			C-1,750/1,450		9/3
Chiba	Tokyo Denryoku	D	125	H-1,400		6/each
Kawasaki	Tokyo Denryoku	D	130	C-2,000/1,700		8/2
Yokohama	Tokyo Gas	D	75	H- 800		6/each
Ngoya	Chubu Denryoku	D	125	C-2,000		2/each
Sakai	Ohsaka Gas	D	75	H		2/each
-Senboku	Ohsaka Gas	D	125	H		5/each
Himeji	Kansai Denryoku	D	125	H,PI-1,700		10/each
Kanmon	Kitakyushu LNG	D	125	H-2,000		4/each
Niigata	Nihonkai LNG	D	125	UNIT-2,500		5/each
Futtsu	Tokyo Denryoku	D	130	SUC-2,000		10
Yokkaichi	Chubu Denryoku	D	125	ABF-P-2,250		1/each

Note: D:dolphin, P:pier

Effect of Mooring Lines on Ship Motions

Professor Michael Triantafyllou
MIT, Room 5-323
Cambridge, MA 02139, U.S.A.

It is commonly accepted that for open sea moorings the effect
of the mooring lines on the dynamics of a vessel is restricted
to the lower frequency range below the wave spectrum frequencies.
For high frequency (wave-induced) motions, vessel motions are
evaluated as if the mooring lines were absent. These motions
are then used as input to study the dynamic tension build-up
in the lines under the assumption again that the upper end of
the lines follows the motions of the vessel.
Recent work has shown (1) that for deeper water moorings (over
250 m in water depth), the mooring lines respond to imposed
motions through stretching rather than through changes in the
catenary. In fact, for sufficiently high frequencies, it is a
good approximation for the dynamic tension to use the product
of the component of the motion along the tangential direction
of the cable configuration p (t), times the elastic stiffness
of the cable, K_C, which for a uniform cable is

$$K_C = \frac{EA}{L} \tag{1}$$

where E is Young's modulus, A the cross sectional area and L
the cable length. The dynamic tension magnitude $\tilde{T}$ (t) is then

$$\tilde{T} (t) = \frac{EA}{L} p(t) \tag{2}$$

This implies large forces on the vessel, usually not quite as
high as the first order wave forces, but their effect might be
felt on the vessel. Also, this dynamic tension build-up is
caused by fluid drag, which is in phase with the velocity ; i.
e., it tends to damp out the imposed motion. As a result, the
dynamic tension supplies damping to the vessel equations. This
may not be significant for heave or pitch, but it certainly
contributes to roll, which is very lightly damped.
A matter of particular concern is the proper modeling of the
mooring lines. As expression (2) shows, the dynamic tension is
inversely proportional to the length. By L we denote the total
length of the line which participates in stretching. As a result, the portion of the line lying on the floor may cause a
reduction in the overall dynamic tension, since it reduce the
equivalent elastic stiffness of the line. At the same time
only the suspended part of the cable (i.e., excluding the portion lying on the floor) is subject to drag forces, and hence
provides damping. This distinction is essential for proper
modeling of mooring lines.

E. Bratteland (ed.), Advances in Berthing and Mooring of Ships and Offshore Structures, 401–403.
© 1988 by Kluwer Academic Publishers.

Finally a word of caution for model testing. When the elastic
stiffness is of importance, as for example in the case of deep
water moorings, proper scaling of the elastic stiffness in model tests is of great importance.
Example: We will supply here some representative values for a
moored vessel in 500 m of water. The vessel has the following
characteristics: length 161 m, beam 16.8 m, draft 5.5 m, displacement 6,860 tons.
It is assumed that the mooring system consists of 6 lines capable of carrying 200 tons of steady force plus the dynamic
tnsion.
Each line, for simplicity, consists of wire 7 cm in diameter,
Young's modulus 7×10^{10} N/M^2 and density 7,800 kg/m^3. If the
subscript 2 denotes sway, 3 heave and 4 roll, the added mass
coefficients of the vessel are at 0.5 rad/sec.

$$A_{22} = 7.4 \cdot 10^6 \text{kg}$$
$$A_{33} = 11.6 \cdot 10^6 \text{kg}$$
$$A_{44} = 69.7 \cdot 10^6 \text{kg M}^2$$

The damping coefficients are

$$B_{22} = 0.61 \cdot 10^6 \text{kg/sec}$$
$$B_{33} = 8.1 \cdot 10^6 \text{kg/sec}$$
$$B_{44} = 4.66 \cdot 10^6 \text{kg} \cdot \text{M}^2 / \text{sec}$$

The restoring coefficients are

$$C_{33} = 18.9 \cdot 10^6 \text{kg/sec}^2$$
$$C_{44} = 86.06 \cdot 10^6 \text{kg} \cdot \text{M}^2 / \text{sec}$$

Finally, the mass of the vessel is $6.92 \cdot 10^6$ kg and the moment
of inertia in roll is $311 \cdot 10^6$ kg M^2. For the same frequency,
an estimate of the forces supplied by the mooring lines is as
follows:

	Stiffness	Damping
Sway	$0.63 \cdot 10^6$ kg/sec^2	$0.8 \cdot 10^6$ kg/sec
Heave	$0.30 \cdot 10^6$ kg/sec^2	$0.46 \cdot 10^6$ kg/sec
Roll	$20.40 \cdot 10^6$ kg·M^2/sec^2	$32.5 \cdot 10^6$ kg·M^2/sec

The force supplied by the mooring lines is subdivided into a
component which is proportional to the motion ("stiffness")
and a component proportional to the velocity ("damping"). From
these calculations it is obvious that the effect of the mooring
lines on heave is negligible. The effect on sway is to add stiffness and significant damping. Finally, roll is damped very
significantly by the mooring lines.
The response of the cables is 77 % elastic.
Results reported in (2) show a large effect of the mooring
lines on the response of vessels in waves. Those results were
derived in model scale (numerical and experimental), where
modeling of the elasticity of the line is impossible; hence,
these results are not applicable for full scale applications.
In (3) an outline of the proper procedure for modeling mooring
lines at reduced scale is presented.

In conclusion, the dynamic tension in mooring lines reaches very large values, and the equivalent spring constant is close to the elastic stiffness of the line. Under normal conditions, the tension affects primarily roll. If the elasticity of the line is not properly modelled, the other motions will be shown to be affected also.

REFERENCES
(1) Triantafyllou, M.S., Bliek, A. and Shin H., "Dynamic Analysis as a Tool for Open Sea Mooring System Design", Trans. SNAME, Vol. 93, 1985, pp. 303-324.

(2) Nakajima, T., Motora, S. and Fujino, M., "On the Dynamic Response of the Moored Object and the Mooring Lines in Regular Waves", Journal of the Society of Naval Architects of Japan, Vol. 150, 1981.

(3) Triantafyllou, M.S., Papazoglou, V. and Mavrakos, S., "On the Proper Dynamic Scaling of Cable Lines", 1987 (submitted).

SHIP DECK MACHINERY
Design forces, arrangement and equipment.

Ivar Krogstad,
A/S Pusnes Marine and Offshore Services,
P.O.Box 111, N-4818 Faervik, Norway

The following is an extract from the report "Mooring" which is
a sub-project in the NTNF (The Royal Norwegian Council for
Scientific and Industrial Research) - supported R & D program:
"Ship operation of the future".

The report was published in 1984. The extract is worked out
by Ivar Krogstad, member of the project team and author of
chapter 5 and 6.

E. Bratteland (ed.), Advances in Berthing and Mooring of Ships and Offshore Structures, 404–440.

2. GLOSSARY, LITTERATURE

Automatic mooring winch = Selftensioning mooring winch

Similar to mooring winch, see below, but incorporating an automatic facility for hauling and rendering mooring lines at certain pre-set loads. A "Constant tension winch" is an automatic mooring winch where the hauling tension is approximately equal to the rendering tension.

Bending ratio

Ratio between bending radius of a rope and rope radius.

Breast

Term for a mooring line atwarth ship, or nearly so, from vessel's side.

Bollard

A static structure comprising one or two bitts (i.e. posts); firmly secured to the deck and used for belaying, fastening and working ropes.

Cable lifter

A deeply grooved drum, shaped to engage the links of a chain cable.

Capstan, Warping capstan, Mooring capstan

A machine having a vertically mounted warping end on which a rope may be wound under power, but not stored.

Centralized control

A control stand common for all mooring winches and/or windlasses. Can be positioned on the bridge.

Chain pipe

Pipe between anchor windlass and chain locker.

Conventional Buoy Mooring (C.B.M.)

The vessel is held by her anchors forward and by mooring lines to one or more buoys aft and at beam (brest-moorings).

Dog clutch

A clutch which can slide axially on a shaft for engagement of drums etc. by means of two or more dogs (claws). The dog clutch is prevented from rotation relative to the shaft by means of keys, splines, squares or similar.

Drum, Barrel, Coiling drum, Rope drum, Rope barrel

A cylinder flanged at both ends. When used the rope is fixed and stored on it.

Drum load or hauling load

The maximum rope tension, measured at the drum exit, when the winch is hoisting or hauling in, with the rope in a single layer on the drum.

Fairlead

A guide to change the direction of a rope without causing damage to it.

Fibre rope handling gear, first line ashore equipment

A power-operated device with one or two drums to ensure that the working part of the fibre rope is reeled in no more than one layer. It may be used in conjunction with a rope storage reel.

Fleet angle

The maximum angle subtended by a rope to a line drawn at right angles to the winch drum or warping end, through the point at which the rope leaves the drum or warping end.

Hawse pipe

Thick walled pipe heeling from deck to ship side into which the anchor stock is hauled.

Head lines

Mooring lines from forecastle, normally 20-60 degrees to ship's centre line.

Holding load or brake load

The maximum tension that the winch brake is capable to hold with rope on the first layer.

IACS

International Association of Classification Societies.

Light-line or Slack-rope speed

The maximum speed the winch can maintain with no tension in rope, measured with the rope in a single layer.

Local control

Winch (windlass) control fitted on or near the actual winch.

Mooring machinery

All machinery used for hauling, rendering, recovery, and holding mooring lines.

Mooring winch

A winch capable of line storage on drums, and ability to tension and hold the tensioned moorings by means of suitable brakes.

Nominal speed

The maximum speed the winch can maintain when applying the drum load.

OCIMF

Oil Companies International Marine Forum.

Panama chock or - fairlead

Closed chock with minimum dimensions according to "Panama Canal Regulations". "Enlarged Panama Fairleads" have larger bending ratio recommended for mooring lines.

Prime mover

An electric or hydraulic motor, steam engine or similar drive, acting directly on the deck machinery.

Recovery load on automatic mooring winches

The rope tension, measured at the drum exit, when the drum starts to rotate in the direction of haul, with the rope in a single layer.

Remote control

Winch (windlass) control from a remote stand, normally from ship's side, or from a platform with view over the controlled winches and lines.

Rendering load on automatic mooring winches

The rope tension at the drum exit, when the drum just starts to rotate in the opposite direction of the applied driving torque, with the rope in a single layer.

Self-locking rotating bollards, Rotobollards

A bollard allowing the bitts (bollard heads) to rotate for tightening the rope. The bitts are locked against rotation in the opposite direction up to a certain tension limit. A separate capstan or warping head must be used for tightening the rope.

Single Point Mooring (S.P.M.)

The vessel is secured to either a single buoy or to a fixed tower structure forward. No stern moorings are used, and the vessel is free to swing.

Ship motions

The six forms of ship movements: Heave, yaw, pitch, sway, roll, and surge are illustrated in Fig. 2.0.

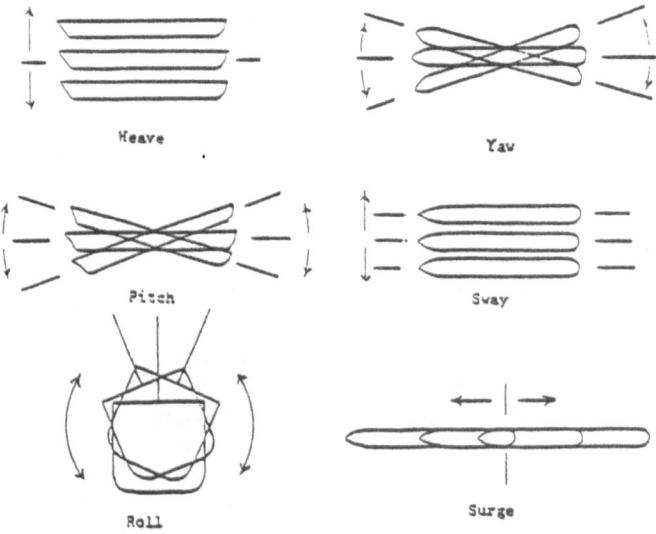

FIG. 2 Ship Motions

Slipping tension
The rope load at which a brake is slipping when it is set to a load less than the holding load.

Split drum
A cylinder arranged with an additional flange at some distance along its length. The additional flange may or may not have a slot.

For mooring one part of the drum is used for storing and the other part for tensioning in order to coil the rope in one layer only.

Spring
Term for a mooring line nearly parallel to vessel's side.

Stalling load
The maximum rope tension, measured at the drum exit, when the drum ceases to rotate in the direction of haul with the rope in a single layer.

Stern lines
Mooring lines from stern, normally 20-60 degrees to ship's centre line.

Twin drum traction winch
A winch with two parallel grooved drums. Frequently used as "Fibre rope handling gear".

Universal fairlead
A roller or sheave fairlead allowing the rope to be led in any direction over rollers or sheaves.

Warping end, Warping head, Warping drum, Drum head
A cylinder having a longitudinally concave surface.
When used the rope is wound around but not stored on it.

Warping winch
Winch fitted with one or two horizontally arranged warping heads.

Literature

The British Ship Research Association:

NS 179	(1967):	Basic consideration and existing drycargo tonnage	I
NS 256	(1969):	Research Investigation for the Improvement of Ship Mooring Methods. Tankers and bulk carriers, existing tonnage and new construction.	II
NS 306	(1971):	Research Investigation for the Improvement of Ship Mooring Methods. Dry cargo vessels, new construction.	III
NS 386	(1973):	Research Investigation for the Improvement of Ship Mooring Methods. Wind and current data for various classes of ship.	IV

Oil companies International Marine Forum:

Standards for equipment employed in the mooring of ships at
single point moorings.
Guidelines and recommendations for the safe mooring of large
ships at piers and sea islands.

3. DECK MACHINERY SPECIFICATION MEMORANDUM LIST

When preparing the specification for a new ship, the following items (at
least) should be included:

General
- List of accepted manufacturers
- Noise limits, vibration limits and other design chriterions
- Materials to be used in machinery and assessories
- Regulations (Suez, Panama, St. Lawrence etc.) to be fulfilled
- Painting and corrosion protection
- Drawings, instruction material, name plates, spare parts, etc. to be delivered
- Testing procedure and documentation

Special
- Estimated number of crew members to be engaged in the mooring operation
- Number of mooring stations (Forecastle, poop etc.)
- Type, length and dimension of ropes to be used (even if this will usually be owners delivery)
- Single point mooring arrangement, if relevant
- Number of mooring drums
- Capacity of drum brakes
- Slack rope speed
- Number of prime moovers (Specify if sheared between windlass and mooring winch)
- Energy carrier (El. hydr., electric, steam, etc.)
- Speed and force requirements. If possible also characteristic speed/force diagram
- Power distribution system
- Degree of automation, monitoring, remote control and instrumentation
- Number of fairleads, chocks, bits etc.
- Description of windlass(es) incl. speed, capacities, dynamic brakes etc.
- Type and length of anchor chain (wire)
- Number, type and weight of anchors
- Anchor stowing position (conventional or on-deck)
- Position of spare anchor
- Type of chain stoppers
- Description of chain lockers, if any

4. DESIGN FORCES AND STANDARD MOORING PATTERN

In order to determine the capacity of the equipment required for safe mooring, it is necessary to calculate the maximum environmental forces expected to act on the ship in a berthed position over its lifetime.

The British Ship Research Association (BSRA) which has studied this extensive and complicated subject, has published the results of the study in Reports Nos. 179, 256, 304 and 386.

In these reports a method is described by which it is possible to calculate the effect of wind and current on a given ship in defined conditions. The resulting forces acting on each half length of the ship in the longitudinal and transverse directions can thereby be estimated.

The calculation is based upon the ship's dimensions above and below the waterline. Use is made of well established hydrodynamic and aerodynamic formulae, taking into account the shallow waters normally found in harbours. The BSRA reports recommend values for wind and current to be used in the calculations.

The following is a simplified summary of numerous calculations carried out as described in the BSRA reports. It is intended for the preliminary estimation of the holding load of mooring winches.

4.1 Environmental actions

All ships may be affected by wind, many by currents, and some by wave motions in berthed condition.

In the method described herein, wind and current are considered to set up static forces when acting on the ship. Gusts and sudden changes in the current are ignored in the first instance.

Waves and other dynamic actions are ignored partly because wave data for ports are not available and partly because most ports do not suffer from swell even in bad weather.

However, if either exposed harbours or sheltered harbours with swell are expected to be served by the ship, the dynamic forces should preferably be taken into account and estimated in order to determine the capacity of the mooring equipment.

4.2. Design conditions

Two conditions are considered. Each condition deals with wind and current of different velocities in order to determine the most severe load acting on the ship in the various directions.

Condition 1:

- a) Wind speed of 60 knots acting in any direction
- b) A current of 5 knots acting longitudinally
- c) A current of 2.5 knots acting at an angle of 10° to the longitudinal axis of the ship. Only the transverse component is considered.

When estimating the longitudinal forces a) and b) are combined
"   "   "   transverse   " a) and c) "   "

Condition 2:

- a) Wind speed of 33 knots in any direction
- b) As in condition 1
- c) As in condition 1

It may be of interest to compare the above conditions with the one adapted by IACS when establishing the unified classification rules for anchoring equipment. In this connection a wind speed of 25 m/sec. and a current of 2.5 m/sec., both acting in the same direction (longitudinally), were chosen as basis.

4.3 Mooring forces

Two groups of ships are considered. One characterized by large super-structure, deck houses or deck cargo (special ships) and the other group comprising those ships where a full underwater hull is the typical feature (conventional ships).

In the course of carrying out numerous calculations to determine mooring forces by the BSRA methods, it was observed that irrespective of ship type, the force in each longitudinal direction is approximately half the value of the total force acting transversely.

Based on this observation, it was possible to predict the approximate forces for different type of ships and varying wind speed as a function of ship size, see fig. 4.3 on next page.

4.4 Mooring pattern

The overall mooring pattern naturally effects the load distribution to individual mooring lines.

In general, the mooring pattern should be symmetrical. The mooring lines must have the ability to absorb the maximum forces occuring in the transverse and longitudinal directions respectively.

The ideal number mooring lines is four. See 5.1.1.

However, for practical purposes, a 6 point mooring as shown in fig. 4.4 is chosen as standard.

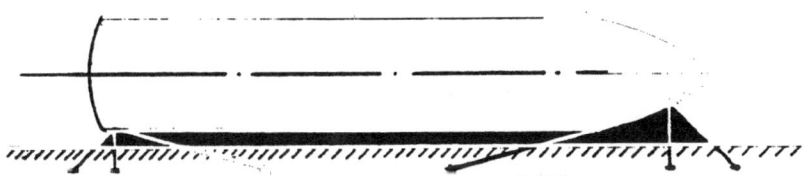

Fig. 4.4, Standard mooring pattern

414

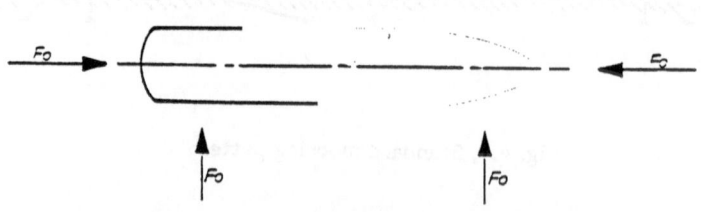

MOORING FORCES

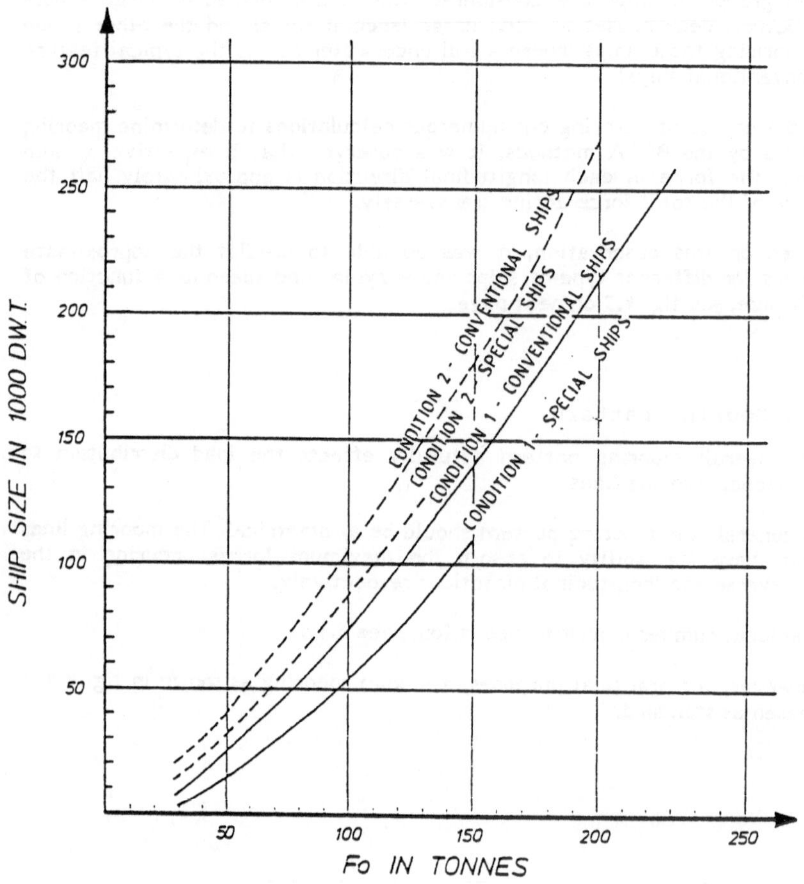

SHIP SIZE IN 1000 DWT.

300

250

200

150

100

50

50   100   150   200   250

Fo IN TONNES

CONDITION 2 - CONVENTIONAL SHIPS
CONDITION 2 - SPECIAL SHIPS
CONDITION 1 - CONVENTIONAL SHIPS
CONDITION 1 - SPECIAL SHIPS

Fo

Fo

Fo

Fo

Fig. 4.3

For preliminary estimation purposes, head and stern lines are assumed to have an angle of 45° to the ship's centre line. Thereby the transverse and longitudinal force components will be equal in magnitude.

Transversely: 2 breast lines
                                           2 lines at 45°

In each longitudinal direction: 1 spring line
                                           1 line at 45°

Additional mooring lines, if required, could be added to this basic layout.

These additional lines should be arranged as any number of head and stern lines, 2, 4, 6 or higher number, or as additional spring and breast lines, in all 4.

For ships with a large freeboard, an average angle to the horizontal of α = 35° is normally chosen for estimation purposes.

For tankers and other ships with smaller freeboard, α = 25° may be chosen.

4.5 Dynamic forces

Since dynamic forces are not included in the curves of fig. 4.3 a utilization factor u = 0.8 or less should be included in the calculations of the holding power.

It is difficult to predict with accuracy the forces resulting from waves and other dynamic actions. The magnitude of these forces on the mooring lines depend on the amplitudes of motions. These in turn depend on those parameters normally associated with forced damped vibration; i.e. mass spring constant, damping and exciting forces.

A ship may move in any combination of the six degrees of freedom. If a mathematical solution is sought it is therefore necessary to establish the equations of motion for each of these movements (heave, pitch, etc.).

Each of the six equations will contain the abovementioned parameters.

- The mass (or moment of inertia) including entrained water, characteristic for the particular motion.

- The spring constant (it may be variable) represented by the restraining forces from pretension of the mooring lines.

- The damping which may partly be a result of the friction between the ship and the water, and partly of the friction between the ship and the fenders.

- The exciting force, its magnitude and frequency or other characteristics.

The various mooring lines (spring, breast, head- and stern lines) will restrain motions in the 6 degrees of freedom with different efficiency. Consequently the dynamic motions in the various degrees of freedom will effect individual mooring line differently.

Surge will primarily effect the spring lines and to a lesser extent the head-and stern lines.

Heave, roll, sway, pitch and yaw will primarily effect breast lines and to a lesser extent head- and stern lines.

Any attempt to simplify and present ready results which have general validity is likely to fail. Programmes are available for theoretical calculations of maximum forces, and the results of such calculations are encouraging when comparing with model- and full scale tests.

4.6 Calculating the holding power

Enter the curve for the relevant ship type at 60 kn, 33 kn, or any intermediate wind speed in fig. 4.3 and read the force F_0.

Design force: $F_1 = F_{0/u}$

For convencience, a tabular calculation of the effectivness of all the mooring lines may be used.

Table 4.6   Effectiveness of Mooring Lines

Transverse		Longitudinal	
Fwd.	Aft	Fwd.	Aft
		Number of springlines from fwd x cos α	Number of spring-lines from aft x cos α
Number of breast lines fwd x cosα	Number of breast lines aft x cos α		
Number of head lines x cos 45° x cos α		Number of head lines x cos 45° x cos α	
	Number of stern lines x cos 45° x cos α		Number of stern lines x cos 45° x cos α

The sum of each vertical column is equal to number of lines to take the load in the main directions.

Then calculate the winch holding load

$$\text{Holding load} = \frac{F_1}{\text{Sum of one column}} \text{ tonnes}$$

Having obtained a preliminary estimate, it will thereafter be natural to choose a standard winch with the nearest possible holding load.

Example
Conventional ship   60,000 TDW
Choosing a wind speed of   33 Knots
Total force from fig. 4.3   65 Tonnes

Choosing 8 mooring lines, i.e. the standard 6 point mooring plus one head line and one stern line, all lines at an angle α = 25° to the horizontal

$(\cos \alpha = 0.91)$ Head- and stern lines at an angle 45° (cos 45° = 0.71)

$(\cos 25° \cdot \cos 45°) = 0.64$

	Transverse		Longitudinal	
	Fwd.	Aft	Fwd.	Aft
2 spring lines			1 x 0.91	1 x 0.91
2 breast lines	1 x 0.91	1 x 0.91		
1 head line	1 x 0.64		1 x 0.64	
1 stern line		1 x 0.64		1 x 0.64
2 additional 45° lines, head and stern	1 x 0.64	1 x 0.64	1 x 0.64	1 x 0.64
Sum of columns = number of lines to take load in the main directions	2.19	2.19	2.19	2.19

$$\text{Holding load} \quad \frac{65}{0.8 \cdot 2.19} = 37.1 \text{ Tonnes}$$

Adapting standard winches   16 Tonnes

with holding load of   470 K

5. DECK MACHINERY ARRANGEMENT
5.1. Number, size and arrangement of winches

The mooring forces according to the guidelines in chapter 4 are the base for selection of number and size of the mooring winches.

From a standardization point of view, all winches should be of the same size, i.e. the same size of prime mover, main shaft etc. This will simplify the maintenance and the number of spare parts. Such standardization is normally done for the same category of winches.

From an operational point of view, the best arrangement is obtained if each chain cable lifter, warping unit and mooring drum has its own driving motor. Then the lead for mooring lines, warping lines and chain cable can be selected independently of the other units. Remote control is simplified, since clutching between the various units is eliminated.

5.1.1 Mooring winches

The ideal number of mooring winches to maintain the ship in position alongside a quay is four, one spring winch and one breast winch foreward and the same aft. However, this arrangement on the ship requires large shore bollards positioned to give perfect lead for the mooring lines. Especially for the breast lines this will create a problem in many harbours. In order to allow a certain flexibility in the mooring line pattern, and to give redundancy in case one winch is damaged, six winches are recommended to maintain the position alongside, that is six point mooring. For smaller ships, say 5-10.000 tdw, four point mooring should be sufficient.

Steel wire rope of 40-44 mm are considered to be the largest size which can be led from the winch over deck and through the ship side fairlead by manpower, say 2-3 men in addition to the winch operator.

A limitation to six winches will therefore require ancillary equipment to handle the heavy mooring lines for large ships, but this is a relatively small investment. Such equipment can be a separate small winch positioned at the fairlead or a reeving line from a warping head.

If a ship is frequently visiting harbours with bollards which are too small for six point mooring or if the harbour authorities so demand, a larger number of mooring lines must be considered. The total holding load for all winches must then be at least the same as for a six point mooring system.

The size of the mooring winches shall preferably be according to ISO 3730 ("Mooring winches") covering nominal size 5-40 (tonnes). The sizes above 40 tonnes can be selected as follows: 50-64-80-100 tonnes.

The definitions of the various loads relative to the ISO design rope breaking load is seen in fig. 5.1.1.A. This fig. also gives the ISO drum dia. and storage capacity. The standard is based on steel cord wire rope with breaking strength 1570 N per square mm.

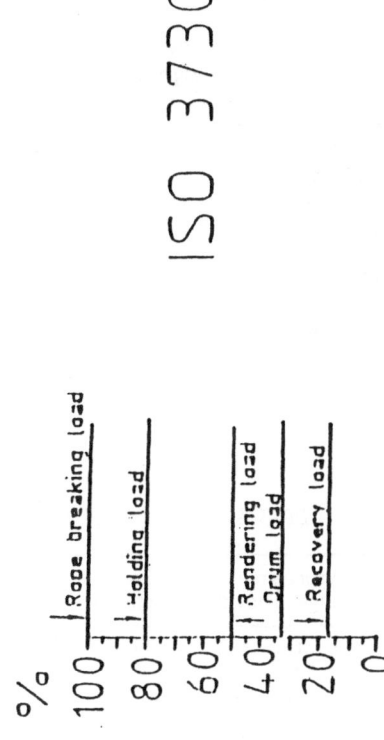

ISO 3730

Nominal size	Drum load	Nominal speed min.	Light-line speed min.	Creep speed max.	Recovery load min.	Rendering load max.	Theoretical diameters of drum and warping-ends	Drum capacity	
								normal	high
	kN	m/s	m/s	m/s	kN	kN	mm	m	m
5	50	0.25	0.5	0.125	25	90	288	180	350
8	80	0.25	0.5	0.125	40	135	352	200	400
12	125	0.20	0.5	0.100	60	189	418	200	400
16	160	0.20	0.5	0.100	80	286	512	250	500
20	200	0.16	0.5	0.080	100	362	576	250	500
25	250	0.16	0.5	0.080	125	447	640	250	500
32	315	0.13	0.5	0.065	155	540	704	250	500
40	400	0.13	0.5	0.065	200	645	768	250	500

For design rope diameter, minimum rope breaking load and holding load, see table 5.1.1.B.

Fig. 5.1.1.A

Wire rope of different qualities can be used, but not wire rope with larger diameter than the design rope. If the steel wire rope used has smaller breaking load than the design rope, the allowed mooring forces must be reduced accordingly.

For fibre rope the ISO 3730 minimum drum diameters are 6 times the rope diameter for polyamide (nylon) and polyester (dacron, terrylene) and 4 times for polypropylene. ISO 3730 says: "It is recommended that synthetic ropes under tension should not be wound on a drum in more than one layer or short life will result".

Split drums enable the possibility of always staying moored on the first rope layer, but the transfer of the mooring rope through the flange slot may require additional manning during the mooring operation. ISO 3730 requires min. 250 meter mooring rope which is much more than that used during normal mooring operations. If the length in use for normal mooring operation is kept on the tension part of the split drum and the rest on the storage part, the advantage of staying moored on the lower rope layers can be maintained without the need for additional manning.

The maximum tension for the highest loaded line during the mooring operation shall be less than the holding load for the winch. The approximate mooring winch size for the various ship sizes is seen from the table 5.1.1.B.

The ship size range is based on:
    Wing - 60 knots - any direction
    Current - 5 knots ahead or astern, 2.5 knots
    0 - 10° or 170 - 180°

Table 5.1.1.B   ISO-Mooring winch size and approximate ship size

Number of winches	Nominal size (tonnes)	Drum load kN	Holding load kN	Design rope dia. mm	Min. breaking strength of rope kN	Approximate ship size range. Tons deadweight	
						Conventional ships. Tankers, bulk-carriers etc.	Special ships w. large wind area. Container-, Ro-Ro-, Passenger etc.
4	12	120	310	26	378	8000	5000
4	16	160	470	32	573	15000	8000
6	16	160	470	32	572	25000	12000
6	20	200	590	36	725	35000	20000
6	25	250	730	40	895	50000	30000
6	32	315	880	44	1080	65000	45000
6	40	400	1050	48	1290	80000	60000
6	50	500	1280	51	1590	110000	85000
6	64	640	1560	57	1980	150000	120000
6	80	800	1940	64	2420	210000	
6	100	1000	2430	77	3400	300000	

The lead from the drum to the overboard fairlead shall preferably be direct. In order to secure efficiency, the mooring lines should be led as horizontal as possible from the ship to the shore bollards. Additional rollers give additional friction with chafing and a larger risk for rope breakage. If a rope breaks in a direct lead system, the danger area is only between the drum and the overboard fairlead, whilst the danger area is considerably increased and is also difficult to foresee, if a pedestal roller is introduced.

The mooring winches should be placed in groups to give a good arrangement. For a six point mooring system it is simple to position the winches in two groups, one forward and one aft.

In a six point mooring system the three winches at each end of the ship should be placed as close to the shipside which is most frequently alongside as the drum fleet angle permits. This gives a minimum distance of 5-6 times the drum length (for split drum winches the tension part length).

Ro-Ro ships, container ships and bulk carriers moor most frequently with the starboard side alongside the quay.

There are several reasons for arranging the mooring winches in groups:

a) When more than one line is used for the same type of mooring (f.inst. "breast") and the lines are running ashore from the same area on the ship, it is simpler to share the load evenly between the winches, especially if the lines are of the same type and have the same length.

b) The danger zone is reduced.

c) A safe manoeuvring stand with view over all the winches in the group and the mooring lines can easier be arranged.

For the spring lines it is an advantage if they can be arranged in way of the parallel midship. They must not, however, interfere with the gangway.

5.1.2 Warping winches - "First line ashore". Side thrusters
The demand for warping equipment is dependent on to what extent the ship is equipped with side thrusters, etc. Vessels without sufficient warping equipment will to a large extent be dependent on tugs for berthing.

Polypropylene rope, which floats and is relatively cheap, is normally used for the warping operation.

Warping by means of warping heads and manual operation of the slack rope part cannot be done safely with fibre rope at tensions above 10 tonnes. If the rope is transferred to a bollard, most of the tension will be lost during the transfer. Used in connection with self-locking rotating bollards, the safety limit will be increased to 15-20 tonnes and this tension will remain on the bollard when slacking off the warping head.

A self-locking rotating bollard can be placed very close to warping head or capstan and fairlead. About 1 m free space is sufficient.

On most ships there will be a demand for some warping heads or capstans to handle tug hawsers.

Single head bollards built into and flush with the ship's side arranged at one or two levels make it easy for the tug crew to handle tug rope.

Limited lengths of fibre rope can conveniently be handled on a rope drum. For "first line ashore" larger lengths are normally required. Ropes can be damaged due to jamming into underlying layers and many rope layers will give reduced tension. Single drums are used for "first line ashore", but other arrangements are preferable for large ships.

Fibre rope handling gear (twin drum traction winch with separate storage reel or bin) has advantages as well when tensioning up a rope as for pulling in rope under tension and it has no relevant limitation where rope length is concerned.

A proposal for approximate ship size range and warping equipment is given in table 5.1.2.

Table 5.1.2 Warping winches

Number of winches	Type of warping equipment WH=Warping head or capstan RB= Roto bollard TW=Fibre rope handling gear	Size Tonnes	Conventional ships Tankers, bulk carriers etc. 1000 TDW	Special ships with large areas (container, Ro-Ro) 1000 TDW
2	WH	10	5	5
			10	10
2 2	WH + RB or Drums 1)	15 20		
			25	25
4 4 2	WH + RB or Drums 1) or TW	15 20 25		
			50	40
2 4	TW or TW	32 20		
			100	
2 4	TW or TW	50 32		
			200	
2 4	TW or TW	64 40		
			400	

The "first line ashore" winches are preferably positioned as near the ships ends and as close to the ship side as practicable.

Twin drum traction winches require very short distance to the fairlead, but space for handling for instance tug hawsers must be provided between the traction winch and the overboard fairlead.

To a certain extent thrusters can be considered as an alternative to warping winches. For comparison a force of 10-15 tonnes per 1000 HK thruster capacity can be used, but this force is reduced to about 50 % at ship speed of 5 knots.

Side thrusters must preferably be positioned below the ballast draft to avoid first lines to enter the thruster inlet.

5.1.3 Compromizes

Economic reasons often result in a deck arrangement which is a compromize between the above mentioned principles.

Frequently a mooring winch has two de-clutchable drums and a warping head in addition. A cable lifter unit and a mooring drum driven from the same motor has been much used.

5.1.4 Total safety, margins

The breast line force against the quay gives a friction force which eases the strain in the spring lines. The deviation from athwartship for the breast lines will also reduce the spring line strain. The size of the winches in the table 5.1.1.B are large enough for "four-point mooring", provided all the mooring lines can be led horizontally, breast lines perpendicular to the ship's side and spring lines parallel with the ship's side. For the specified weather conditions the proposal therefore gives a 50 % margin. This shall also cover dynamic mooring forces and reduction in effective mooring forces due to partly vertical line lead. A vertical angle has to be accepted, and this angle is normally larger for the breast lines than for the spring lines.

Spring lines can in most cases be arranged nearly parallel with the ship side, while the breast lines frequently deviate from the athwartship direction. By introducing 6 point mooring the two additional lines therefore primarily are backup for the breast lines.

5.1.5 Typical deck machinery arrangements (Examples)

Three different deck machinery arrangements are shown. All drums are split drums. If these are replaced by plain drums, the distance to the closest fairlead must be increased in some cases. All mooring lines are led directly from drums to fairleads. The mooring lines in use for the actual mooring operatioe are drawn in thick lines, while alternative rope leads for mooring with the opposite side alongside are shown in thinner lines. The mooring rope direction outside the fairlead is indicated by arrows, spring lines nearly parallel to the ship side, breast lines atwarth-ship and head and stern lines at about 45° angle. First lines ashore (polypropylene or similar) are shown with dotted arrows. If the mooring lines are of fibre rope, the first lines ashore can take a part of the mooring load, but with mooring lines of steel wire rope the contribution from the first lines ashore must be neglected. The remaining symbols used are shown on fig. 5.1.5.1. All mooring line fairleads shown are enlarged Panama fairleads. If fibre rope mooring lines are to be used, the fairleads should be changed to universal roller fairleads.

All arrangements have only two mooring areas, fore ship and aft ship. In some cases this can give unreasonably long spring lines which are difficult to control visually. In such cases it may be necessary to fit additional mooring winches on the forward and aft part of the midship area. The aft deck winch(es) on ships with the bridge aft may preferably be controlled from the bridge. A suitable control position for the forward winch(es) can be at the aft end of the forecastle.

Fig. 5.1.5.1 Arrangement No. 1

This is a 6 point mooring system with 6 gear boxes or driving units. On each drum shaft is fitted a warping head and in addition four of the drum shafts are also driving cable lifter unit or fibre rope handling gear. The warping heads are keyed to the shaft, but the drums, cable lifter units and fibre rope handling gear are declutchable. These clutches are in this case not remotely controlled.

The arrangement is symmetrical except for the fibre rope handling gear winches arranged to starboard of the centreline. The rope lead is arranged for starboard side to the quay. The arrangement is convenient for automatic mooring. Depending on ship size and ancillary equipment 2-3 men in each end of the ship will be required during a mooring operation.

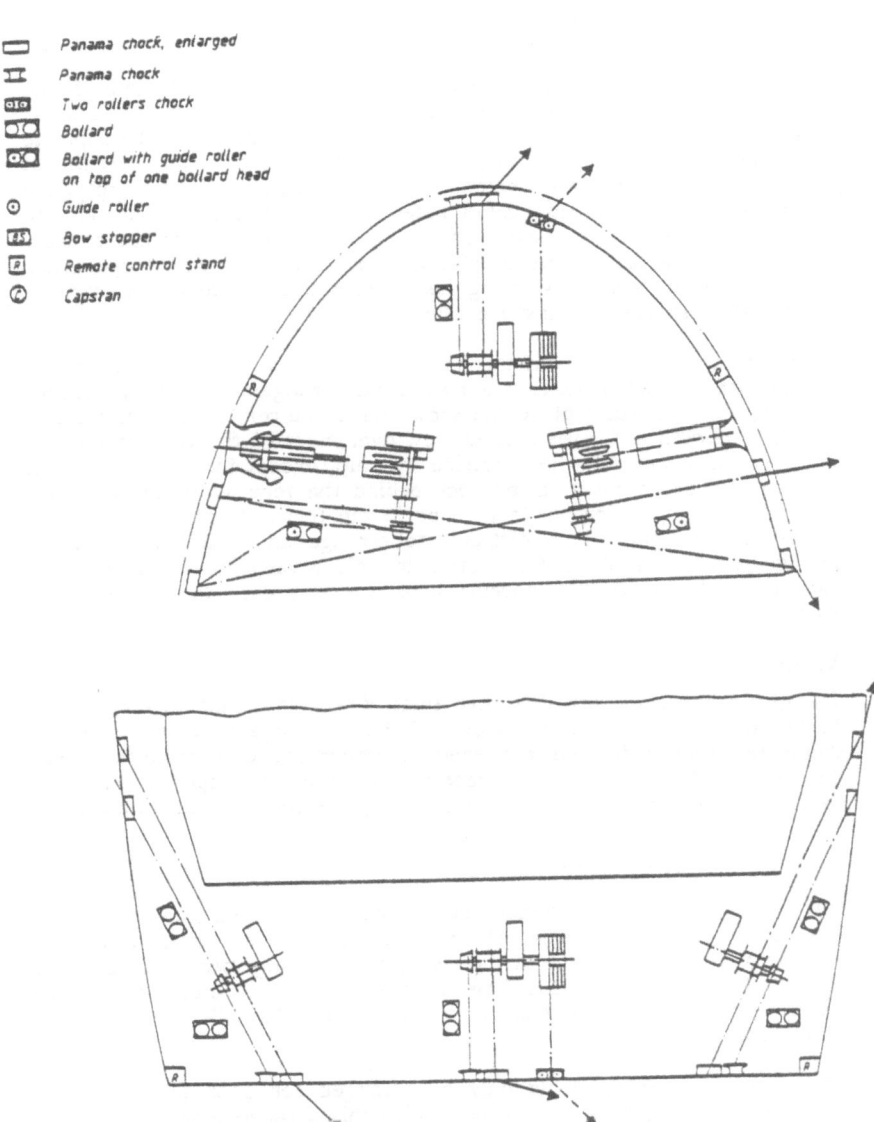

Panama chock, enlarged

Panama chock

Two rollers chock

Bollard

Bollard with guide roller
on top of one bollard head

Guide roller

Bow stopper

Remote control stand

Capstan

Fig. 5.1.5.1. Arrangement no. 1

Forecastle

This "X"-arrangement of the cable lifter unit - mooring drums can be fitted on a relatively small forecastle deck. The normal problem on a short forecastle is to avoid contact between the mooring line led to the opposite ship's side and the warping head on the mooring side winch. Fitting of guide roller(s) to solve this problem is not recommended. Positioning of the driving unit forward of the cable lifter unit contributes to shorten the distance between chain locker and the aft end of the forecast deck. However, a cable lifter unit damage may put the drum out of operation. Replacement of the aft warping heads with a separate driven capstan would make a shorter forecastle possible.

Anchor on deck is shown, but this has no influence on the positioning of the mooring/warping equipment. In a traditional arrangement a chain stopper and a hawse pipe will replace the shown unit. The remote control position is well protected and gives a good view over most of the deck machinery and the mooring/warping lines outside the ship, except for the spring lines. A ro-ro ship often has a raised deck behind the forecastle, and a better remote control position would then be in the forward port and starboard corners of this deck. It is difficult to arrange an ideal Panama Canal arrangement on a short forecastle. In this arrangement the enlarged Panama fairleads for the mooring lines are also used for some of the warping lines.

Aft ship

The spring line and stern line have small directional changes through the fairleads. The breast line is deviating 90° and change to a roller fairlead should be considered, since it is especially important to maintain the breast line tension. Positioning of a storage reel for the fibre rope handling gear below deck can give problems on ships having the steering gear in this area.

5.1.5.2. Arrangement No. 2

This is a 13 point mooring system, but the number of driving units is only 6. The additional mooring drum aft can be justified by higher wind area because of the bridge. Some of the remarks for arrangement No. 1 are also relevant for this arrangement. This arrangement is well suited for Panama Canal passage. Bow stoppers are arranged to fulfil OCIMF's rules for single point mooring.

As automatic mooring could only be arranged for 6 of the 13 drums, automatic mooring could not be used with this arrangement. If a part of the mooring lines is put on selftensioning, this should be all lines operating in the same direction. Because of clutching betwen the various units, this arrangement will require one man fore and aft in addition to the demand for arrangement No. 1, i.e. 3-4 men at each end of the ship.

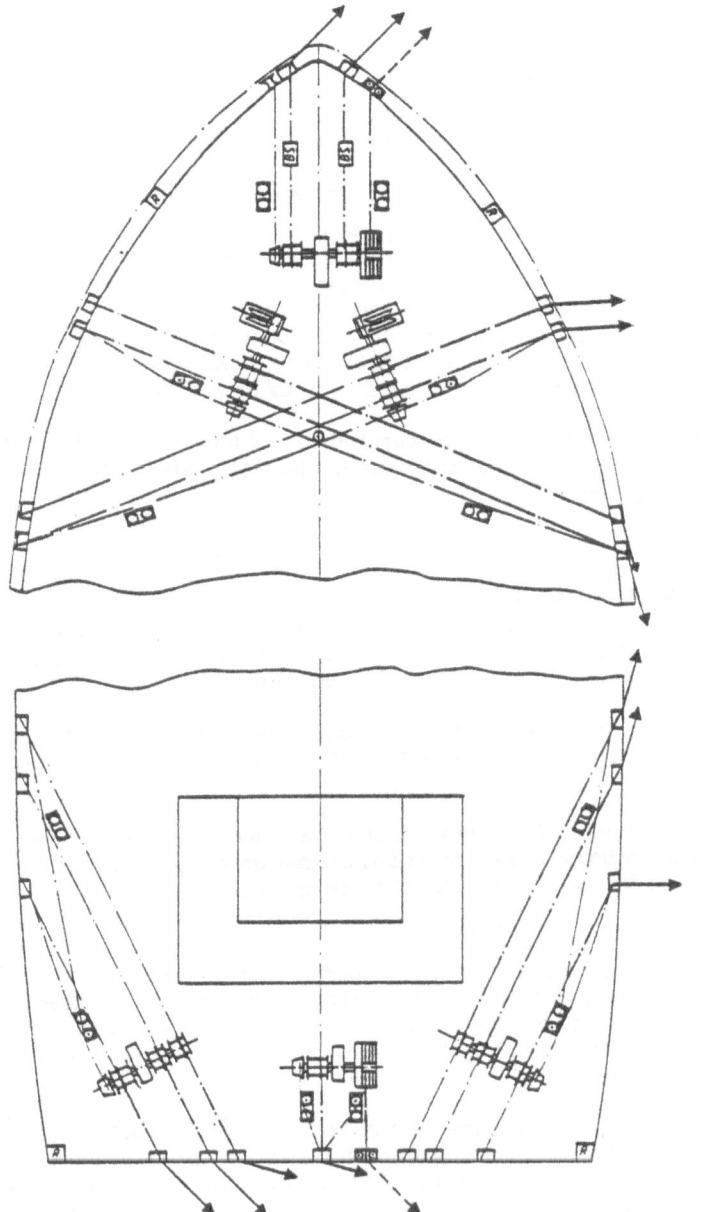

Fig. 5.1.5.2.   Arrangement No. 2.

5.1.5.3 Arrangement No. 3

This is a 6-point mooring system where all mooring machinery has its own motor. The arrangement is based on starboard side to the quay, but the mooring lines can also be led to port side without the use of guide rollers. The cable lifter units are integrated in the anchor on deck arrangement.

Small capstans are arranged for tug hawsers but these capstans can also be used for pulling the mooring ropes from the adjacent drums and through the ship's side fairlead. The tug hawser eye is preferably placed on a quick release hook near the Panama fairlead used for the towing operation. Four such Panama fairleads are shown, but additional units may be necessary. A fairlead with built in quick release hook and a small hawser handling winch is in the market and can replace the shown tug hawser handling system.

This mooring equipment can be operated by 1-2 men at each ship end. It is also well suitable for bridge control after the mooring/warping lines are led through the respective fairleads.

5.2. Mooring crew, Local, remote and automatic control
5.2.1 Mooring crew size

The deck machinery arrangement can be designed in such a way that the number of crew members required for safe and efficient mooring is not increased beyond the number otherwise required.

Optimal safety conditions during mooring is not dependent on on the crew size. It may be obtained with a small mooring crew as long as the deck arrangement is laid out accordingly.

The mooring operation requires more manpower than the warping and anchoring operations. These operations are not done simultaneously and the crew suitable for mooring can do all three operations.

5.2.2 Control positions

The number of control positions is important for determining the mooring crew size. Complete remote control from the bridge (centralized control) is possible, but requires very advanced monitoring equipment. Two positions, one forward and one aft, seems to be the best based on todays technology.

Local control of all functions on the unit is normally required for deck machinery. All controls shall preferably be within reach from the speed control position, and the controls must be outside of the danger zone if a rope should break. If the normal control is done locally, the control stand must be built in as necessary with guards to protect the operator.

Remote control for a group of winches can be situated where wanted.

A remote control stand for a group of winches is normally best positioned at the ship's sides in such a way that the operator can have a good view of all winches, fairleads and the ropes inboard and outboard.

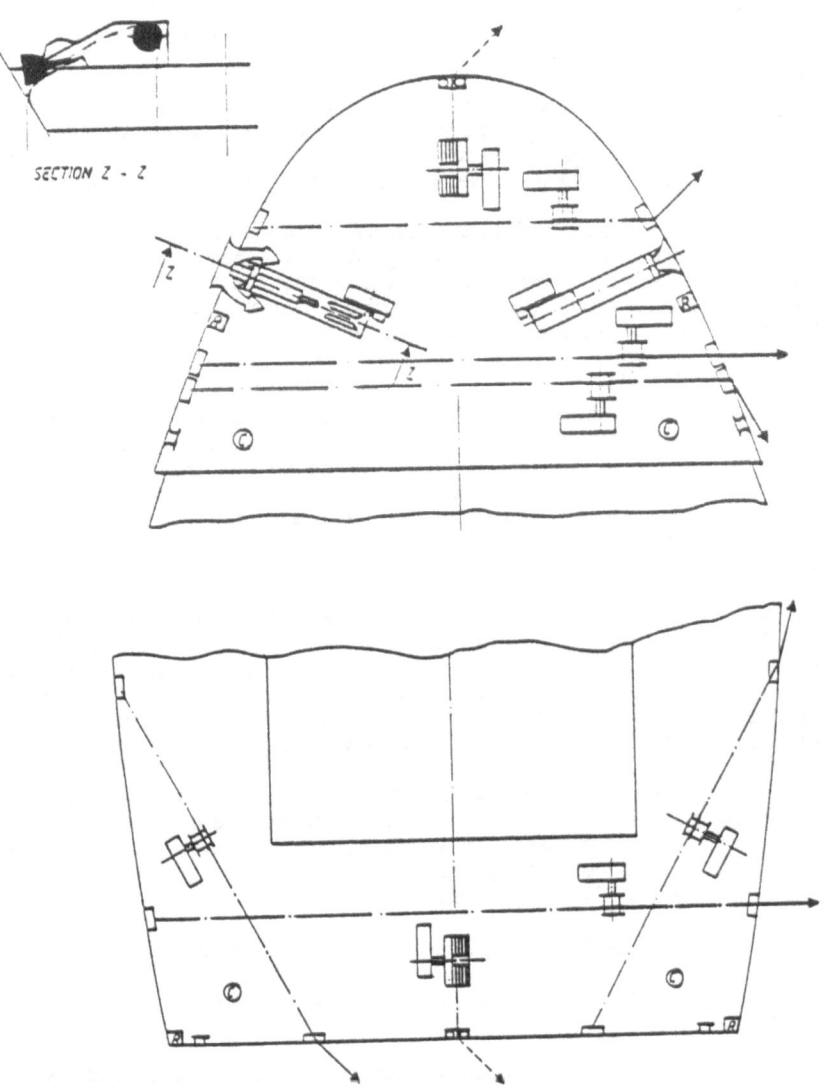

SECTION Z - Z

Fig. 5.1.5.3. Arrangement No 3.

The control stand should be located outside the danger area in case a rope should break. But the control stand should still be protected if it is difficult to foresee the danger area, see 5.1.1.

Bridge control. Centralized control from the bridge is also possible, but this requires TV-monitoring of each drum and possibly also of fairleads. In addition, monitoring equipment for rope tension and rope length as well as alarms for failure in brakes and selftensioning devices would be required.

Bridge remote operated windlasses require not only very good control of the brakes during dropping, but also good motor speed control in both directions. It is most important to monitor the chain length, but chain speed and tension should also be monitored. Clutch and/or gear change should also be arranged and TV-monitoring is preferable.

5.2.3 Mooring operation steps and sequence

If pulling tugs are assisting during the berthing operation, they can do the complete handling of the tug hawsers, provided the ship is equipped with ship's side bollards (see 5.1.2). Normally the tug hawsers are hauled onboard by means of the mooring winches and the mooring crew positions the hawser eye on a bollard. A special winch with a guiding device allows the tug hawser rope handling to be done by one man. The ships own mooring and warping lines are handled as follows:

If an ancillary winch or a reeving line arrangement is used to pull the rope off the drum and through the fairlead, one man can do the whole preparation of a line. The wire rope size can then be increased above 44 mm dia., which is considered maximum for man handling. One line can be prepared at the time if sufficient time is available. When all the lines are prepared for mooring or warping, the remaining steps onboard are:

1. Connect the end of the line from the shore terminal or mooring boat to the mooring or warping line. If the mooring rope end has been paid out through the fairlead and is hanging near the surface, the connection is done by the mooring boat crew. If the ship's own heaving line is used, step No. 1 is to heave this line.

2. Pay out mooring or warping line until the line is fastened ashore.

3. Tighten up and for warping operations pull in warping line as necessary.

4. Set the winch brake and declutch the drum or set the automatic device for automatic operations.

Step No. 1 requires one man for each mooring and warping line which are to be handled simultaneously.

The number of crew members to take care of steps 2-4 is depending on the winch control. If only local control is provided, one operator is required for each winch in addition to one officer for each group of winches. The officer should keep contact with shore, and watch the ropes outboard and inboard.

As the use of rope on bollards requires up to 3 persons and is a very unsafe operation this practice should be avoided as far as possible and be replaced by better solutions.

Self-locking rotating bollards can safely be operated by one man.

Each anchor chain requires one operator for letting go or hoisting. Remote control from the ship's side is preferable in order also to keep an eye on the chain outside the ship.

A mooring crew of 2 men (one officer and one crew member) on foredeck and 2 men on aft deck is normally suitable with the following equipment and arrangement:

-   mooring winches arranged in two groups, one on foreship, one on aftship

-   winch remote control positions on foreship, starboard and port at ship's sides, and similar winch remote control positions aft. Winch speed and drum brakes (on/off) to be controlled.

-   mooring lines on drums

-   an ancillary winch or a reeving line arrangement to pull the rope off the drum and through the fairlead for wire rope sizes of 36 mm and more

-   a special winch for tug hawsers with guiding device allowing the operator to position the hawser eye on the actual bollard.

5.2.4 Automatic control

According to ISO 3730-1976 the mooring winches can be classified in two main groups:

1.   Non-automatic mooring winches
2.   Automatic mooring winches

With automatic mooring winches the operator has the choice of either to leave the tightened mooring line on the drum with the drum brake set (non-automatic) or to switch to automatic mooring. An automatic mooring winch will remain powered with a preset adjustable pull, and the mooring winch will automatically recover or render the mooring rope when the tension varies outside preset limits. The drum brake shall be slipping to avoid rope breaking, if the rope tension exceeds the preset limit.

Automatic mooring winches include previous terms as "selftension", "autotension" and "constant tension". ISO 3730 specifies a maximum rendering load, 50 % of rope breaking strength, and a minimum recovery load 50 % of drum load (about 16 2/3 % of rope breaking strength.)

For a breast line, the rendering-recovering limits could be narrow. It is important that the recovery tension is sufficient to restore the vessel after a rendering operation. The risk that the breast lines in one end of the ship are recovering while the breast winches in the opposite end are rendering and thereby causing the ship to turn away from the quay can be solved by a high rendering-recovery ratio, see 6.3.1.

For spring lines, where one line leading aft is counteracting a line leading forward, a large difference between the rendering and the recovery tension is necessary. The spring lines in one or both directions can also be held on the drum brakes while the breast mooring lines are on "automatic".

Constant tension winches with rendering load equal to recovery load is not recommended for automatic mooring and should never be used for spring lines.

Most tanker terminals do not allow automatic mooring. This restriction is based upon bad experiences with inadequate automatic mooring systems.

A mooring system with few and strong lines is generally more safe than a system with a larger number of lines of the same total strength. This is also the case when automatic control is used.

Today's instrumentation and control equipment allow for all mooring winches to be integrated in one system which can give control signals to keep the ship in the desired position alongside the quay.

6. DECK MACHINERY AND EQUIPMENT
6.1 Types of prime movers

The selection of prime mover type is depending on many factors:

- Rotating speed, torque and control possibilities.

    High speed motors require high gear ratios.
    Limited speed range motors normally require gears with possibility of speed change.
    Requirements for windlass motor are high stalling torque for breaking loose anchors and reasonable speed when lifting chain and anchor.
    A mooring winch motor requires high tension at low speed and high slack rope speed.

    The characteristic RPM/torque curves, which also indicates the control possibilities, are seen in table 6.1 on next page.

- Starting current could be a limiting factor in some cases.

- Efficiency might be of importance.

- Automatic operation possibilities.

- Integrated operation possibilities, if driven with other deck equipment from steam, air, electric or hydraulic systems.

- Noise level

- Maintenance

6.2 Components

The dimensions, capacities and stress levels of main winch and windlass components are to a large extent specified in ISO STANDARDS.

6.2.1 Drums, warping leads, cable lifters

Mooring lines are normally tensioned only during the winding up of the· last turns, and the drum needs not be dimensioned for several layers under tension. A mooring drum is therefore of a relatively light design with moderate barrel thickness and seldom with stiffened flanges. The number of rope layers shall not be more than five on normal drums. On the storage part of a split drum the number of rope layers is not restricted.

Warping head minimum diameters are specified in ISO 3730 and further information is given in ISO 6482 "Warping end profiles".

The cable lifter has to fit the actual chain cable, but the smaller the tolerance, the more sensitive the cable lifter is to deviation from the nominal chain dimension. The pitch circle diameter of the cable lifter has to be smaller than the smallest chain dimension the cable lifter is designed for. Thus the chain will enter higher up on the snugs than it will leave the cable lifter, and some "working" (sliding) has to be accepted. This working will give wear. In order to be able to build up a worn cable lifter by welding, especially larger sizes should be manufactured in steel.

A cable lifter shall have at least 5 snugs according to ISO 4568 "Marine Windlasses and Anchor Capstans".

6.2.2 Brakes

The brake is an important part of any mooring and anchoring system. Each drum and cable lifter should be equipped with a brake, and often the prime mover has a brake. Separate dynamic brakes should be fitted to larger anchor windlasses.

The holding load of a mooring winch drum brake is given by ISO (min. 80% of the breaking load) and for a windlass brake by the classification societies (min. 45% of the chain breaking load).

Band brakes are suitable for winch drums and cable lifters. Band brakes are simple constructions of light weight, and due to the increasing friction around the brake drum (fig. 6.2.2) the setting force is relatively small. To get full benefit from this friction when setting a band brake, the brake drum has to rotate in the braking direction when the brake is set, unless a spring mechanism is inserted in the setting device.

Prime mover type	R.P.M./torque characteristics	Nominal R.P.M. H=high L=low	Remote control S=simple M=medium C=compli- cated	Automatic operation S=simple M=medium C=compli- cated
Electric — Pole changing AC-motors		H	S	C
AC slip ring motors		H	S	C/M
Straight short-circuit AC-motors		H	S	C
DC-motors		H	S	M
Hydraulic — High pressure — Fixed displacement pump and motor (6.1.2.A)		H or L	M	M
Displacement: Pump+Variable Motor+One or two fixed (6.1.2.B)		H or L	M	S
Displacement: Pump-variable Motor-variable (6.1.2.C)		H	M	S
Low pressure. Displacement: Pump-fixed Motor-2or3 fixed		L	M	M
Steam/air		H	C	S

EXPLANATION OF CURVES:

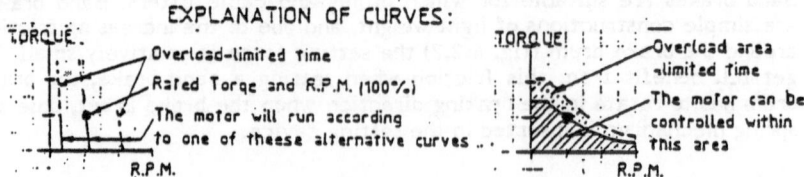

TORQUE
- Overload-limited time
- Rated Torque and R.P.M. (100%)
- The motor will run according to one of theese alternative curves
R.P.M.

TORQUE
- Overload area limited time
- The motor can be controlled within this area
R.P.M.

Table 6.1

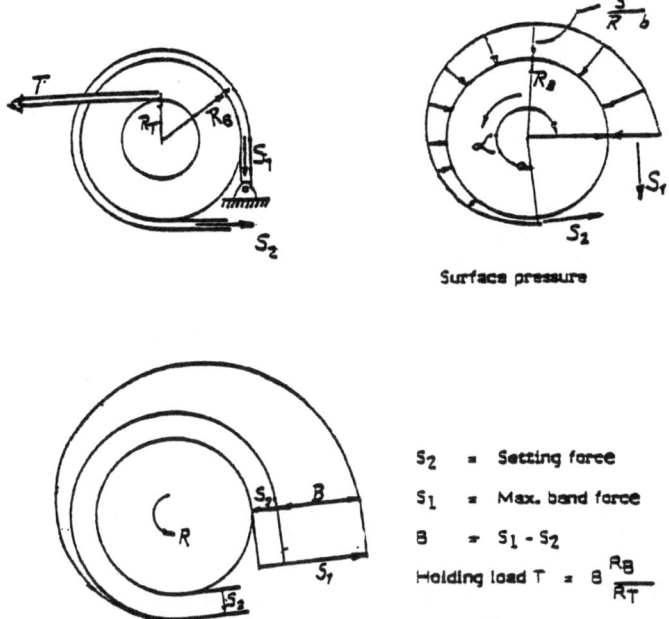

Surface pressure

S_2 = Setting force

S_1 = Max. band force

B = $S_1 - S_2$

Holding load T = $B \dfrac{R_B}{R_T}$

Fig. 6.2.2   Band brake   principles

It is therefore important that the rope is correctly wound on a drum with band brake. If not, the holding load will be only a fraction of the rated value, say 20-40%.

Large windlasses, say for over 90 mm chain, should preferably be fitted with additional brakes for dynamic braking. The dynamic brake is resisting the wear better than the band brake, and is also absorbing kinetic energy much better. The band brake will for large windlasses be used for static purposes only, and be a backup for the dynamic brake.

A dynamic brake can be a vane type water brake, or a disc brake without or with water cooling. A noncooled disc brake gives the simplest installation and will have sufficient capacity at anchoring depths up to about 100 meters also for the largest chain sizes. Replacing the caliper linings of a disc brake is simple.

The windlass prime mover has also been used for dynamic braking. The lowering speed is then limited to the max. R.P.M of the prime mover, and the brake effect at any windlass will be depending on the capacity of the prime movers of the power transmission system.

A mooring drum brake and a windlass brake shall preferably be adjustable up to its rated value +/- 10%. This tolerance covers dynamic and static conditions. The adjustment may be done locally, but remote on/off control may be considered. If the prime mover power fails, the brakes should be automatically set. Brake screws, nuts and bushing, etc. in contact with sea atmosphere should be of stainless steel.

6.3 Automation and Instrumentation

The quality of sensors in marine environments is of vital importance. For electronic sensors stability, shock resistance and protection against humidity and salt is more important than accuracy below 1 %.

6.3.1 Automatic mooring winches

The drum brake is a vital part of a mooring drum. Possibility for reading the slipping tension is important, see 6.2.2.

Live motor principle

According to ISO 3730 the rendering tension can be up to 3 times the recovery tension. A mooring winch with a prime mover which can stall and be overhauled in reverse direction without overheating will normally meet these requirements, and the ratio will be:

$$\frac{\text{Rendering tension}}{\text{Recovery tension}} = \frac{\text{motor overhauling torque}}{\text{motor stalling torque x mech. efficiency}^2}$$

The torque may vary during one revolution of a motor. The overhauling torque will be the max. torque for a motor overhauled slowly in direction out, while it is set for direction "in". Correspondingly the stalling torque will be the minimum torque when the motor is set for and rotating slowly in direction "in". It may be necessary to reduce the prime motor power for such operation (electric current, hydraulic or steam pressure).

If the rendering-recovery ratio is too small, the ratio can be increased by inserting a brake in the system. In some high pressure hydraulic winches separate valves are controlling rendering and recovery tension, but the ratio cannot be smaller than given by the formula.

If the rendering-recovery ratio is too large, it is necessary to incorporate a load cell system which is controlling the prime mover.

Dead motor principle

For winches operating with "dead motor" the rope tension is held by the motor brake in static mode. If the tension is varying outside the set range, a load cell system will give signals to start the motor for rendering or recovery until the tension is within the set range. If this range is too narrow, frequent starts and stops of the motor will be the result. Such operation could cause resonance vibrations in the mooring system.

Motor	"Live"	"Dead"
Pole changing AC short circuit motors		x
AC slip ring motors	x	x
DC motors	x	
High pressure hydraulic motors	x	
Low pressure hydrulic motors	x	
Steam and air motors	x	

Table 6.3.1   Principles for motor operation of automatic mooring winches

Pole changing AC short circuit motors. In addition to the creep speed step, the next speed step is coupled in if the tension is varying outside a second and larger tension range. If the tension exceeds a limit larger than the motor max. rendering limit, the motor brake will lift. The brake will also lift if the motor current for "automatic mooring" is switched off.

AC slip ring motors. Live motor drive requires a special motor and control system.

High pressure hydraulic motors are normally arranged in an integrated system according to fig. 6.1.2.B.1. The control valve is set for "heave", and normally separate "mooring valve(s)" are inserted between the control valve and the hydraulic motor.

Low pressure hydraulic motors. Since these motors are working according to the "constant flow" principle (Fig. 6.1.2.A), a separate small "mooring pump" is often inserted in the system to save energy.

It is possible to arrange a computerized mooring control system either based on the angles of the mooring lines or on the relative movement of the ship to the quay. Monitoring sensors for such systems are available. Load sensors must also be incorporated.

6.3.2 Load sensors
The main principles for load sensors are based on mechanical springs, hydraulic pressure or electric bridge systems with strain gauges. Mechanical springs and hydraulic pressure systems gives substantial deflection and are normally relatively complicated to build into deck machinery. Strain gauge based sensors have deflection similar to the remaining parts in a mechanical transmissing system, and have also the advantage of an electric output signal which is easily transferred over a long distance to several stations.

For build in purposes bolt type load cells are simple. Such load sensors are normally indicating shear forces, and the requirements for accurate build in are moderate. It is important that these load cells are turned to the correct indication direction and locked against rotation and further that the bushings being in contact with the load cell have sufficient hardness.

The tolerance of the load cell is normally about +/-0.5%, but the mechanical errors are normally larger. A total tolerance of +/-5% will be required.

Load reading based on electric motor current or hydraulic motor pressure is less accurate but can be used.

6.3.3 Speed and length sensors

Speed and length sensors are normally electric of inductive type, but mechanical types can also be used, especially for back up purposes.

The inductive type requires steel vanes to pass by in a distance of up to about 15 mm and two sensors are necessary to record the direction.

The counting of vanes can be accurate. For instance a five snug cable lifter will lift 10 chain links for each revolution and the accuracy will then be that of the chain (-0/+2.5%).

For rope passing over a sheave or reeled on first drum layer the length reading can be within +/-1%, but correction should be adapted for upper layers. A tolerance of +/-2.5% should be specified for steel wire rope and +/-5% for fibre rope.

6.4 Fairleads (checks), bollards (bitts)

Fairleads (chocks) used for leading the mooring ropes should have a smooth bearing surface without any obstruction or roughness that may cause damage to the ropes.
"Wear and tear" is the most common reason for worn out fibre ropes, not the bending stresses.
When used in connection with wire ropes a bending ratio of 12 is satisfactory (BSRA report 256).

6.4.1 Fairleads (chocks)

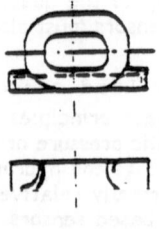

Panama type closed fairleads would normally be fitted according to Panama Canal Regulations. fig. 6.4.1.A. They should be installed with an accompanying bollard, and arranged for power-operating the hawsers by capstans, warping winches or warping heads on windlasses or winches.

Fig. 6.4.1.A

This type of "fairleads" should also be fitted for towing bollards, and warping purposes along vessel's side.

Enlarged Mooring Pipes or "Shell" type enlarged fairleads are big, closed fairleads with a wide bearing surface, which must be kept very smooth. Fig. 6.4.1.B.

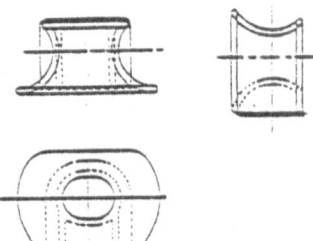

Fig. 6.4.1.B

Universal Roller Fairleads, Fig. 6.4.1.C designed for St. Lawrence Seaway, are the most commonly used fairleads for fibre ropes in connection with mooring winches.
They may be designed as single or double type closed fairleads, having vertical and horizontal rollers, where the hawser may enter from any inlet angle, and pass through.

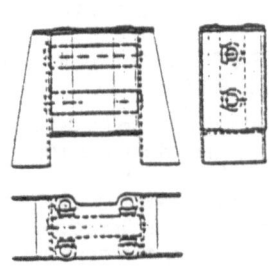

Fig. 6.4.1.C

Open type roller fairleads, Fig. 6.4.1.D located in bulwark openings for warping purposes, are a more simple type, and are not able to take hawser inlet above the horizontal plane.

Vertical Lead Roller Fairleads, Fig. 6.4.1.E (Pedestal roller) are installed as inboard fairleads.
The fairlead, operating approximately in the horizontal plane, are used to lead the hawser and change the direction of haul. Roller fairleads with nominell dia. 150-450 mm are standardized in NS 2585.

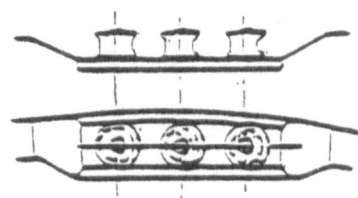

Fig. 6.4.1.D

6.4.2   Bollards

Bollards, Fig. 6.4.2, will generally be installed in connection
with a Panama fairlead, a warping head, or a capstan for
hauling-in purposes. The bollard
should be equipped with a
securing ring on each side for
holding the rope when tightening
up on the bollard.

Distance between bollard and
accompanying chock should be at
least 1.5 -2.0 m, depending on
the size of equipment.

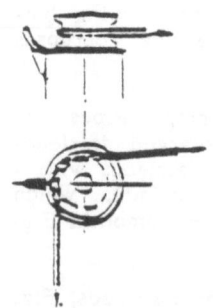

Fig. 6.4.1.E

SELF-LOCKING   ROTATING BOLLARDS

A bollard with bollard-heads that
can rotate when the rope is being pulled in over the warping
head, but locked against rotation
in the opposite direction.
If the pull in the opposite direction is increased above a certain
limit, a frictional brake will slip
to prevent rope breakage.
If it is desired to render rope the
brake can be hydraulically released.

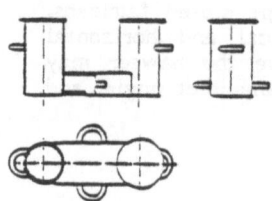

Fig. 6.4.2

MOORING SYSTEMS FOR SHIPS HARBOURING IN PORT'

S.Ueda
Chief of the Offshore Structures Laboratory, Structures
Division, Port and Harbour Research Institute, Ministry
of Transport, Japan

1. INTRODUCTION

Currently in Japan, it is recommended for ships larger than
1,000GRT to go out of the port and anchor in a bay when a
typhoon is approaching to the district where the port is
located. This recommendation has been enforced since the
Kitty Typhoon (1950) and the Jane Typhoon(1951) attacked the
Port of Yokohama and the Port of Kobe, because a great deal
of damage was done to ships moored to quay walls or anchored
inside or outside the port. For more than thirty years, these
large sized ships thus have to undock for surviving a storm.
It is true that ship damage has been considerably reduced following enforcement of this recommendation. Recently, however,
it has become difficult for those ships to find enough space
for anchoring in a bay, not only due to increase of ship size
and the number of arriving ships but also due to complex utilization of the marine space. It is estimated that anchoring
space for about 240 vessels is needed in Tokyo Bay and in
Osaka Bay, and 220 vessels in Ise Bay. But investigation carried out by the Second District Port Construction Bureau.
Ministry of Transport, concluded that anchoring space was available only for 90 to 120 vessels in Tokyo Bay. In fact, in
disregard of the recommendation, a fair number of ships had
stayed inside the port during typhoons passing over or near
the district where the port is located. Among these, a number
of ships larger than 10,000DWT were included, especially in
such ports as the Port of Kobe, the Port of Yokohama and the
Port of Tokyo. Fortunately, no serious damage has occured.
It is said that there are several quay walls which could serve
for harbouring in such ports as the Port of Tokyo and the Port
of Yokohama if the typhoon is not too strong. And captains
would require to stay inside a port if the port facilities are
improved so that ships can moor safely during a typhoon approaching and passing over or near the district where the port is
located. On the other hand, Tokyo Bay, Osaka Bay and Ise Bay
are nowadays subject to complex utilization. In Osaka Bay,
the Kansai International Air Port is now under construction.
Port Islands such as Kobe, Rokko and Osaka Nannko are already
constructed and operating new city functions such as transportation of goods, traffics, business centers, housings, parks,
meeting halls, hospitals,community centers and so on. These
functions are established by constructing man made islands in
the bay. Anchoring space for ships become reduced as a necessary consequence.

E. Bratteland (ed.), Advances in Berthing and Mooring of Ships and Offshore Structures, 441–448.
© 1988 by Kluwer Academic Publishers.

Therefore, some counter-measures are needed for harbouring
ships inside port. We must obtain solutions how to improve
the port facilities and how to evaluate quay walls whether
they can serve or not for harbouring ships. The key to the
solutions is to know motions and mooring forces of the moo-
ed ships subjected to strong gusty wind and to irregular waves
inside the port. The nummerical simulation mehod are most useful for that purpose.

2. CONDITIONS OF CASE STUDIES

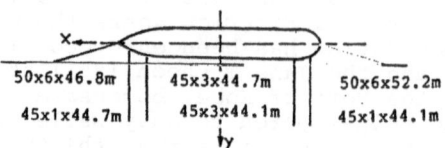

Figure 1 Arrangement for Mooring Ropes

Case studies were carried out by means of numerical simulation
method for 10,000DWT, 5,000DWT and 3,000DWT cargo ships. The
major discussion point of the case study is the improvement of
mooring system which consists of fenders, mooring ropes and
bollards. First of all, mooring ropes size and number were
determined on the assumption that all of the capacity of mooring equipments installed on the ship shall be mobilized during
a typhoon as shown in Figure 1. This simplified arrangement of
mooring ropes consists of bow lines, stern lines and breast
lines for the purpose of saving computation time during the
numerical simulations. Mooring conditions are examplified in
the figure for 10,000DWT cargo ship.Here, spring lines are not
included   because there is a risk for these to break and
spring lines are considered to be ineffective if a ship has
large motions. All mooring lines are nylon ropes with diameter
in the range from 40 to 50 mm.
Next, fender sizes and arrangements are discussed. As the wind
blows from all directions around a ship whenever any wind protection boards or warehouses are constructed on the wharf.Re-
garding fenders, the on-shore wind blowing from the sea toward
the wharf induces the maximum reaction force. Considering type
No.1 fenders (buckling type), fender size is determined so that
the deflection of the fender shall be less than 10% for the
wind load resulting from the mean wind speed. And the total
deflection, adding the deflection caused by ship motions, shall
be less than 35% of its height. Regarding type No.3 (pneumatic
type) fenders, the fender size is determined so that the reaction force against the deflection of 50% of its height shall be
less than half of the reaction force caused by the wind load
for the mean wind speed. And the total deflection, adding the
deflection caused by ship motions according to the wave forces,
shall be less than 50% of its height. The mean wind speed is
35m/s. Table 1 lists fender size and the number of fenders to
be used for both types of fenders. Case study was carried out

under the conditions that the wave heights were $H_{1/3} = 0.3, 0.5,$ 0.7, 1.0m, the wave periods being $T_{1/3}$ = 8, 10, 12s, and the wave directions are 30, 45, 60, 90 degrees. The wind speeds were 30, 35m/s, and the wind directions 90, 120, 150, 180, 210, 240, 270 deg ees. It must be emphasized that large sized fenders be installed at quay walls which should serve for harbouring ships. Usually, fender size is about 300 to 500 mm for buckling type fenders and 600 to 1,000mm for pneumatic type fenders for ships ranging from 3,000 to 10,000DWT. According to the results, fender size should be 1,150 to 1,600 mm for buckling type fenders and 1,500 to 2,000mm pneumatic type fenders.

Table 1 Fender size, Allocation and Allowable
Deflections and Reactions

	Buckling Type		Pneumatic Type	
	Size and Number	Allowable Deflection Reaction	Size and Number	Allowable Deflection Reaction
10,000dwt	1,600mm 2	560mm 147.7tf	2,000mm 2	1,000mm 191.5tf
5,000dwt	1,250mm 2	437.5mm 94.2tf	1,800mm 2	900mm 156.4tf
3,000dwt	1,150mm 2	402.5mm 78.2tf	1,500mm 2	750mm 109.4tf

3. RESULTS OF CASE STUDIES
3.1. For 10,000dwt Cargo Ships
Harbouring ships moored to quay walls may be feasible under the conditions that the wave direction is in the range from 30 to 45 degrees, the wave height $H_{1/3}$ is less than 0.5m, the wave period $T_{1/3}$ = 10s, the wind direction is in the range from 0 to 180 degrees, which means that the wind blows from the sea towards the wharf and pushes the ship toward the quay wall, and the wind speed is less than 30m/s. For the conditions mentioned above, the fender reaction forces, fender deflections and the tensions of mooring ropes do not exceed the allowable values. If the wave direction is about 30 degrees, the fender deflection is less than the allowable value for wind with mean wind speed 35m/s. Surge motion, however, becomes 5 to 12m for wind direction from 120 to 180 degrees. If the wind directions are in the range from 180 to 360 degrees, meaning that the wind blows from the wharf to the sea, a moored ship will drift about 15 to 17m away from the quay line and the tension exceeding the allowable value will be induced in some mooring ropes. This means that the capacity of the mooring equipments installed on the ship is less than the mooring forces induced by the wind with mean speed 30m/s. Therefore, if the rules for steel ships

are not revised, some counter-measures to reduce the wind forces should be taken. There is no difference of ship motions moored to quay walls with different types of fenders under the conditions that the wave direction is 30 degrees, the wave height $H_{1/3}$ is less than 0.5m, the wave period $T_{1/3} = 10s$, and the wind speed is 30m/s.

3.2. For 5,000dwt Cargo Ships
Harbouring ships moored to quay walls may be feasible under the conditions that the wave direction is 30 degrees, the wave height $H_{1/3}$ is less than 0.5m, the wave period $T_{1/3} = 10s$, the wind direction is in the range from 0 to 180 degrees, and the wind speed is less than 35m/s. But, surge motion becomes 5 to 12 m for wind directions ranging from 120 to 180 degrees. If the wind directions are in the range from 180 to 360 degrees, a moored ship will drift about 15 to 17m away from the quay line and tension exceeding the allowable value will be induced in some mooring ropes. Regarding fender characteristics, almost the same results as for the 10,000DWT cargo ships were obtained.

3.3. For 3,000dwt Cargo Ships
Harbouring ships moored to quay walls may be feasible under the conditions that the wave direction is in the range from 30 to 45 degrees, the wave heigth $H_{1/3}$ is less than 0.5m, the wave period $T_{1/3} = 10s$, the wind direction is in the range from 0 to 180 degrees, and the wind speed is less than 35m/s. But, surge motion becomes 2 to 8m for wind direction in the range from 120 to 180 degrees. If the wind direction is in the range from 180 to 360 degrees, a moored ship will drift about 8 to 11m away from the quay wall and tension exceeding allowable value will be induced in some mooring ropes. Regarding characteristics almost the same results as for the 10,000 DWT cargo ships were obtained.

Some counter-measures are needed for the wind blowing from the directions in the range from 180 to 360 degrees. Some experiments were carried out with a wind protection screen set at location 1.2m and 2.0m from the quay line. The heights of the wind protection screens were 0.2, 0.4, and 0.6m. Here the results for case denoted FBQHVIR30 high is presented. The wave height $H_{1/3} = 1.67cm$, the wave period $T_{1/3} = 1.63s$, the wind speed is 5.48m/s, the wave direction is 90 degrees and the wind direction is 270 degrees. Figure 2 shows the distribution of wind speed behind the wind protection screen. Wind speed decreased to about 30% of incident wind speed, and it is supposed that the wind direction becomes adverse according to comparison of ship motion for both hydraulic model tests and numerical simulations as shown in Figures 3 and 4. Here, the computations were made under the conditions that the wind speeds were 2.19 and 0m/s, and the wind directions were 90 and 270 degrees. It seems that the computation results under the condition when the wind speed was 2.19m/s and the wind direction was 90 degrees give the best fit to the results of the

4. PROTECTION AGAINST WIND

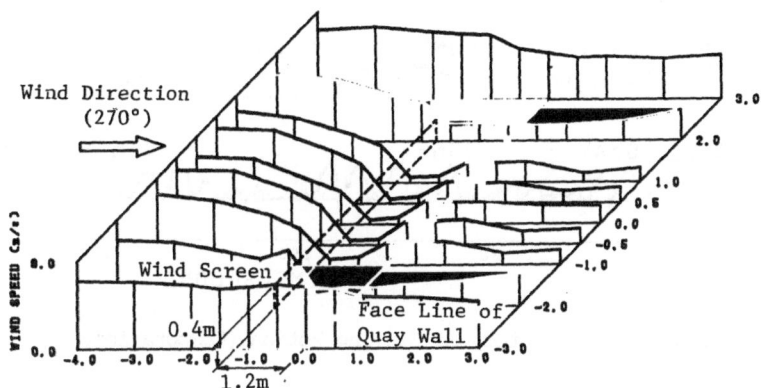

Figure 2 Distribution of Wind Speed behind Wind Screen

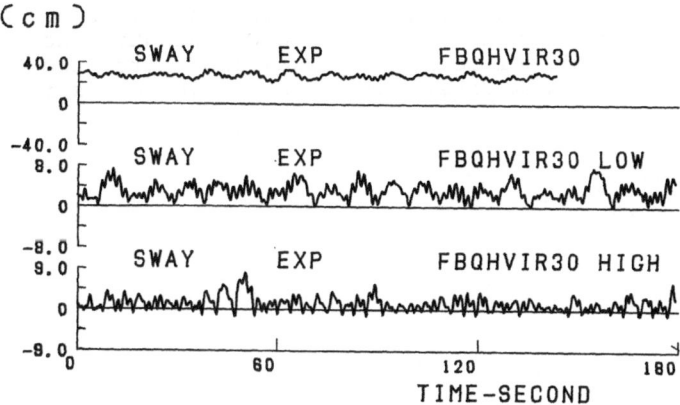

Figure 3 Time History of Swaying (Model Test)

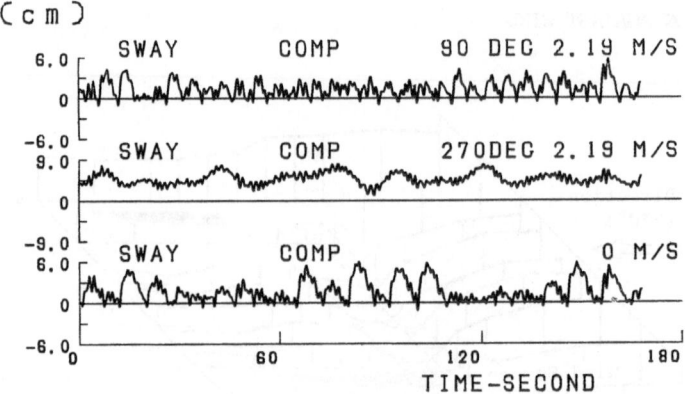

Figure 4 Time History of Swaying Motion (Computation)

hydraulic model tests. Accordingly, it may be said that the wind direction is adverse behind the wind protection screen as long as the ratio of distance to height of the wind protection screen is about four or five. In practice, warehouses may perform the role of wind protection screen.

5. SELECTION OF QUAY WALLS FOR HARBOURING SHIPS
As an example, some quay walls were selected for the quays which may serve for harbouring ships. With the aid of computer-program, calculating the wave height inside the port, quays were selected where the wave height was less than 50cm. Figure 5 shows the distribution of wave heights inside the breakwaters at the Port of Yokohama for the layout existing in 1980. It may be said that there are such quays at the Honmoku Wharf, the Yamashita Wharf, the Ohsanbashi, the Shinkou Wharf, the Mitsubishi Dock, the Takashima Wharf, the Pier of Asia Sekiyu, and the Takaramachi Wharf which may serve for harbouring ships.

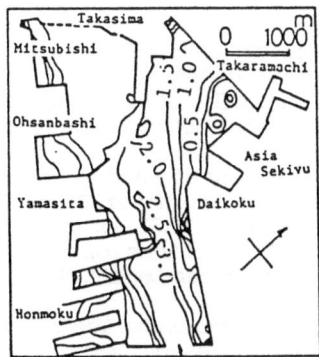

Figure 5 Distribution of Wave Height in Port (Yokohama)

6. CONCLUSION

Harbouring ships is an important sucject in Japan especially in Tokyo Bay, Osaka Bay and Ise Bay according to increase in both ship size and the number of arriving ships, and also related to complex utiliazation of the marine space. We have to make effective counter-measures for that purpose. This nurerical simulation method could effectively be used in this respect.

448

REFERENCES

1.Brunn,P.[1981]:"Breakwater or mooring system ?", The Dock & Harbour Authority, September. 1981, PP.126-129.

2.Hsu,F.A.[1970]:"Analysis of Peak Mooring Forces by Slow Vessel Drift Oscillations in Random Seas", 2nd Offshore Technology Conference #1159, pp.1=135-1=146.

3.Ijima,T. et al[1972]:"Scattering of Surface Waves and the Motions of a Rectangular Body by Waves in Finite Water Depth", Transaction of Japan Society of Civil Engineers, No.202, PP.33-48.

4.Ijima,T. et al:[1975]:"On the Motions of Elliptical or Rectangular Floating Body in Waves of Finite Water Depth", Transactsion of Japan Society of Civil Enginees, No.244,PP.91-105.

5.Ito,Y. and Chiba,S.[1972]:"An Approximate Theory of Floating Breakwater", Report of the Port and Harbour Research Institute, Vol.11, No.2, PP.141-166.

6.Kubo,M:"Fundamental Study on the Sheltering Dgree of the Port Regarding to the Critical Conditions of Loading", 241p.

7.Lean,G.H.[1971]:"Subharmonic Motions of Moored Ships Subjected to Wave Action", Transaction of Royal Institute of Naval Architects, London, No.113, PP.387-399.

8.Oortmerssen,G.Von[1976]:"The Motions of Moored Ship in Waves", Publication No.510, Netherlands Ship Model Basin, 138p.

9.Pinkster,J.A.[1980]:"Low Frequency Second Order Wave Excitating Forces on Floating Structures", Publication No.650, Netherland Ship Model Basin, 204p.

10.Russel,R.C.H.[1958]:"A Study of the Movement of Moored Ships Subjected to Wave Action", Proceedings of Institution of Civil Engineering. Vol.12, pp.379-398.

11.Sawaragi,T. et al[1983]:"New Mooring Systems to Reduce Ship Motion and Berthing Energy",Proceedings of 30th Japanese Conference on Coastal Engineering, PP.460-464.

12.Slinn,P.J.B:"Effect of Ship Movement on Container Handling Rates", The Dock & Harbour Authority, August 1979, pp.117-120.

13.Takaishi,Y. and Kuroi,M.[1977]:"Practical Calculation Method of Ship Motions in Waves", 2nd Symposium on Sea Keeping, pp.109-13.

Tsuji,T. et al[1972]:"Model Test about Wind Forces Acting on the Ships",Report of the Ship Research Institute, Vol.7, No.5, pp.13-37.

14.Ueda,S. and S.Shiraishi[1983]:"Method and its Evaluation for Computation of Moored Ship's Motions", Report of the Port and Harbour Research Institute, Vol.22, No.4, pp.188-218.

15.Ueda,S.[1984]:"Analytical Method of Ship Motions Moored to Quay Walls and the Applications", Note of the Port and Harbour Research Institute, No.504, 371p.

16.Ueda,S.[1986]:"Analytical Treatment of Ship's Mooring Problems in Rough Weather", Proc. of 5th OMAE, Tokyo, 1986, pp.522-528.

17.Ueda,S.[1987]:"Analytical Treatment of Ship Motions Moored to Quay Walls and Application", Proc. of 5th Coastal Zone 87, Seattle, 1987,pp.1545-1559.

FENDERS INSTALLED IN JAPAN

S. Ueda
Chief of the Offshore Structures Laboratory, Structures
Division, Port and Harbour Research Institute, Ministry
of Transport, Japan.

1. INTRODUCTION

This paper describes fenders installed on wharves available
for conventional cargo ships and container ships in Japan.
Data were obtained from design document of port facilities
in Japan and also from port authorities of both Yokohama
and Kobe. Some statistical analysis has been done and relations between such factors as fender type, fender size,
design berthing speed, fender setting interval, fender setting
height, and ship size and so on are obtained.

2. INVESTIGATION

2.1 Data Obtained from Design Document of Port Facilities.
Data were obtained from the document of design of port
facilities in Japan. Here, only rubber fenders are investigated. Items concerned with fendering systems are fender
type, fender name, shape, dimension, setting height and setting interval. The energy absorption and the reaction of
fenders are calculated according to catalogues issued by manufacturing companies.

2.2 Investigation at Port of Yokohama and Port of Kobe.
Data were obtained from port authorities at both Yokohama and
Kobe with the assistance of District Port Construction Bureau,
Ministry of Transport. Water depths for the facilities con-
sidered are more than 4.0m. However, berths belonging to
private sectors are not necessarily included.

2.3 Number of Data
Number of data are as follows.
1) Data from the document of design   831
2) Data obtained at Port of Yokohama   96
3) Data obtained at Port of Kobe   100
                              Total 1027

3. ANALYSIS OF DATA

According to data obtained, relations between items given below are concerned with size and allocation of rubber fenders.
Items are classified in two groupes.

E. Bratteland (ed.), Advances in Berthing and Mooring of Ships and Offshore Structures, 449–453.
© 1988 by Kluwer Academic Publishers.

A group	ship size	dwt
	ship type	general cargo ship
		bulk cargo ship
		container ship
		oil carriers
		car carriers
		ferry boat
		others
	structures	caisson
		sheet pile
		cellular
		pile
		dolphin
	wave	
	tide	
B group	fender type	
	fender name	
	fender size	
	setting height	
	setting interval	
	berthing speed	
	reaction force	
	energy absorption	

4. RESULTS OF ANALYSIS

Figure 1 to 6 are relations between dwt and items of B group. It seems that the size, reaction force and energy absorption per set of rubber fender increases proportional to dwt. As for the other items as berthing speed, setting interval and setting height, data dispersion is large and only the range of values are given.

REFERENCE
1. Ueda, S. and E. Ooi: On the Design of Fendering Systems for Mooring Facilities in Port, Note of the Port and Harbour Research Institute, No. 596, 1987.

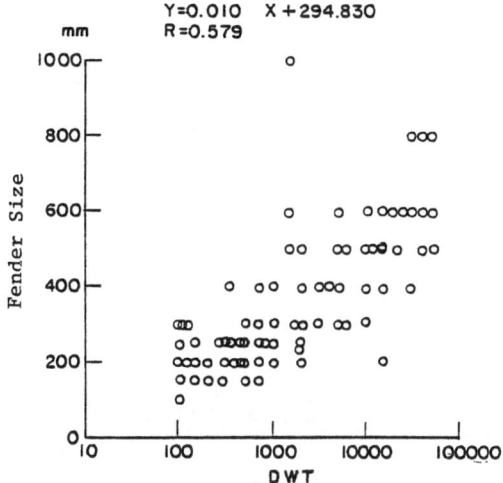

Figure 1   Relation between Fender size and dwt
(General Cargo Ship, V type Fender)

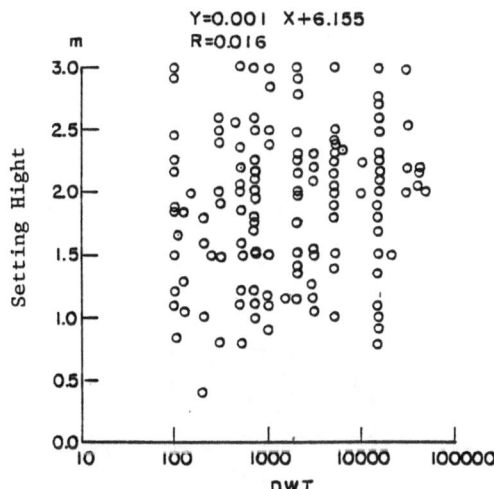

Figure 2   Relation between Setting Height and dwt
(General Cargo Ship, V type Fender)

452

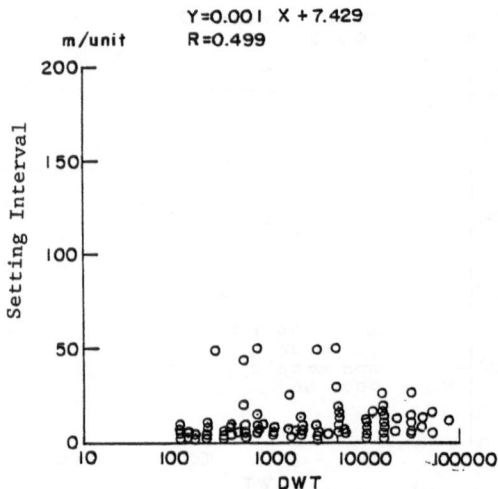

Figure 3   Relation between Setting Interval and dwt
(General Cargo Ship, V type Fender)

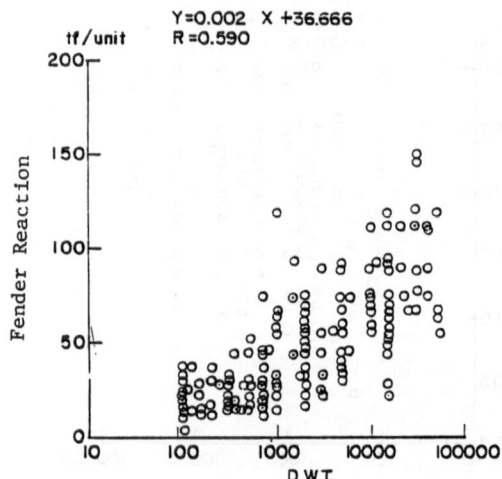

Figure 4   Relation between Reaction and dwt
(General Cargo Ship, V type Fender)

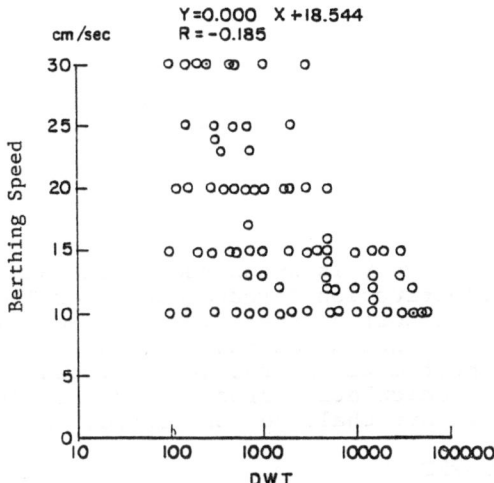

Figure 5   Relation between Berthing Speed and dwt
(General Cargo Ship, V type Fender)

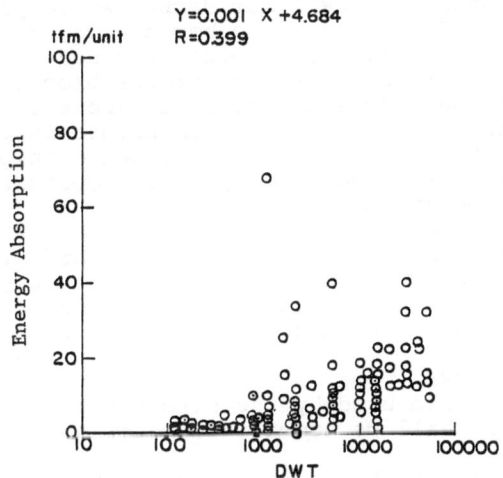

Figure 6   Relation between Energy Absorption and dwt
(General Cargo Ship, V type Fender)

FENDERS USED IN TURKEY

ALI RIZA GÜNBAK

Assoc. Prof. Dr., Coastal and Harbor Eng. lab.,
Middle East Technical University, Ankara, Turkey.

1. INTRODUCTION
Port construction in Turkey is sponsored by the Railways-
Ports and Airports Construction Directorate under The Ministry
of Transportation. All port construction projects whether
governmental or private should get the approval of this Directorate. In this short article, the fender types used in
Turkish Ports and the calculation procedure of Fender Forces
as used by the Directorate shall be summarized.

2. FENDERS USED IN TURKEY
Excluding some factory and refinery port owned by the private
sector, all ports and berthing facilities in Turkey are owned
by the government. Until now almost all of these berthing
places were designed for ships of up to 20.000 DWT. Some of
the berthing places belonging to refineries access tankers of
the order of 250.000 DWT and steel factories access bulk
carriers of the order of 120.000 DWT.
Except the new extension of the Izmir Harbor where patented
fenders shall be used directly on the berth face, truck tires
were used as fender units at all government owned harbors.
They were sometimes used together with timber blocks both
located directly on the concrete face of the berth. The tires
were sometimes filled in by rubber pieces before installed.
At the two steel port harbors where they receive ships up to
120.000 DWT bulk carriers, they used patented cylindrical
shaped fenders located on the berth face.
The three refinery harbors which may receive tankers ut to
250.000 DWT is mainly installed by steel piled breasting dolphins. One of these ports is installed by gravity type fender.
At the rest of the private harbors, the most usual type of
fender used were tires and patented cylindrical fenders.

3. CALCULATION PROCEDURE
The calculation procedure used by the Directorate for calcu-
lating fender forces shall be summarized below as taken from
Aktürel (1978).

  - Kinetic Energy of the berthing ship during impact, E_s

$$E_s = \frac{W\ V^2}{2g} \tag{1}$$

where

$W \quad = W_1 + W_2$

$W_1 \quad =$ Dead Weight Tonnage

E. Bratteland (ed.), Advances in Berthing and Mooring of Ships and Offshore Structures, 454–456.
© *1988 by Kluwer Academic Publishers.*

W_2 = Hydro Dynamic Weight = $\frac{\pi}{4} D^2$ L

D = Draft of ships

L = Length of ship

g = acceleration of gravity

= density of water

V = approach velocity of the ship perpendicular to the quay as given by Table (1)

Table 1. Recommended Approach Velocities of Ships Perpendicular to the Berth to be used in Eqn (1).

Condition	Approach	Berthing Velocity Transverse to Berth (m/sec)		
		Up to 1500DWT	1500DWT to 7500DWT	More than 7500DWT
Strong wind heavy sea	Difficult	0.75	0.55	0.40
Strong wind- heavy sea	Favourable	0.60	0.45	0.30
Moderate wind- heavy sea	Moderate	0.45	0.35	0.20
Protected	Difficult	0.25	0.20	0.15
Protected	Favourable	0.20	0.15	0.10

- Energy lost during rotation of the ship, E_r is approximately half of the ships kinetic energy

$$E_r = \frac{1}{2} E_s \tag{2}$$

- Effective berthing energy, E, is therefore;

$$E = E_s - E_r = \frac{1}{2} \frac{Wv^2}{2g} \tag{3}$$

- Energy absorbed by the berthing structure is assumed to be 20% of the effective berthing energy and the remaining energy is assumed to be absorbed by the fender.

- It is recommended to use the characteristic force-displacement curves of patented fenders of different shapes with the following maximum displacement values.

Fender type	Displacement,
Cylindrical	0.5 H
V shaped	0.45 H
square cross-section	0.40 H

H = thickness of the fender in the direction perpendicular to the berth face.

- for the truck tires directly installed on the berth face; the work done by the fender, E_p is given as;

$$E_p = \frac{1}{2} R \delta \qquad (4)$$

where

R = Impact force on the fender

δ = Total displacement for the fender = $\frac{R}{A} \frac{t_1}{E_1}$ (5)

A = Compression surface of the tire

t_1 = Total wall thickness of the tire in the direction of compression

E_1 = Modulus of elasticity of the tire

Replacing Eqn (5) in Eqn (4) and equating to 0.8 E the force on the fender may be calculated as;

$$R = \sqrt{\frac{1.6 \, A \, E_1}{t_1}} \, \sqrt{E} \qquad (6)$$

- For truck tires used on top of timber blocks, the total displacement of the fender is given as;

$$\delta = \delta_1 + \delta_2$$

where

δ_1 = displacement of tire

δ_2 = displacement of timber = $\frac{R}{A} \frac{t_2}{E_2}$ (7)

t_2 = timber block thickness

E_2 = modulus of elasticity of the timber.

4. KNOWN BERTHING - MOORING PROBLEMS

- A 250.000 DWT tanker over-turned a breasting dolphin during berthing operation at a refinery port.

- At a steel port harbor, installed with cylindrical fenders, very serious surge motions are recorded at the moored ship with bow or stern lines, under North Westerly storms. They caused several damages on the ships of sizes 40.000 - 50.000DWT and on the structure

5. DISCUSSION

A rather simplified energy method is used to predict the forces exerted on the structure during a berthing operation. Eqns.(4), (5) and (6) should be used with caution without checking the force - deflection character of the fender which may also be a function of the aging of the tire. It is believed that, if there is not many damages recorded at the Turkish Ports installed with truck tires, it is due to their well protected character against waves where the ships berthed to the structures at velocities far lower than given in Table (1).

6. REFERENCE
Aktürel Tuncay(1978),Calculation Procedure for Berthing Struct tures, Internal Report of the Directorate of Port Ankara, Turkey.

EXPERIMENTAL AND THEORETICAL INVESTIGATION OF AN ADVANCED FENDER/ MOORING SYSTEM BASED ON ENERGY ABSORBING PRINCIPLES

PER TRYDE
INSTITUTE OF HYDRODYNAMICS AND HYDRAULIC ENGINEERING THE TECHNICAL UNIVERSITY OF DENMARK

Contribution to discussion.

The aim of the project was to investigate and develop a fender/mooring system, which would be able to limit the movements of a moored ship subject to incoming waves, by use of energy absorbing units.

Model tests were performed with irregular waves with a ship moored against a energy absorbing fender system simulating a hydraulic computer guided prototype developed and patented by Irving Brummenæs (Norway). It was shown in the tests that it is possible to limit the movements of the ship by applying the energy absorbing principle. In order to increase the effect a prestressing of the mooring lines was tested and it proved that the damping was inreased.
As an example the results of the sway reduction are shown in Fig. 4.30 (see next page).
The tests were made by Lars Jensen and Torben Michael Jensen as part of a master's thesis, June 1984. It was written in Danish. The full report is available at the institute.
P. Tryde was supervisor at the project.

E. Bratteland (ed.), Advances in Berthing and Mooring of Ships and Offshore Structures, 457–458.

458

Institute of Hydrodynamics and Hydraulic Eng.
Technical University of Denmark

Prototype: Ship 150,000 DWT

Prototype data: Wave height Hs = 1.3 m
                   Sway 1 : prestressing 2 x 716 kN
                   Sway 2 : prestressing 2 x 1074 kN
                   Sway 3 : prestressing 2 x 1432 kN

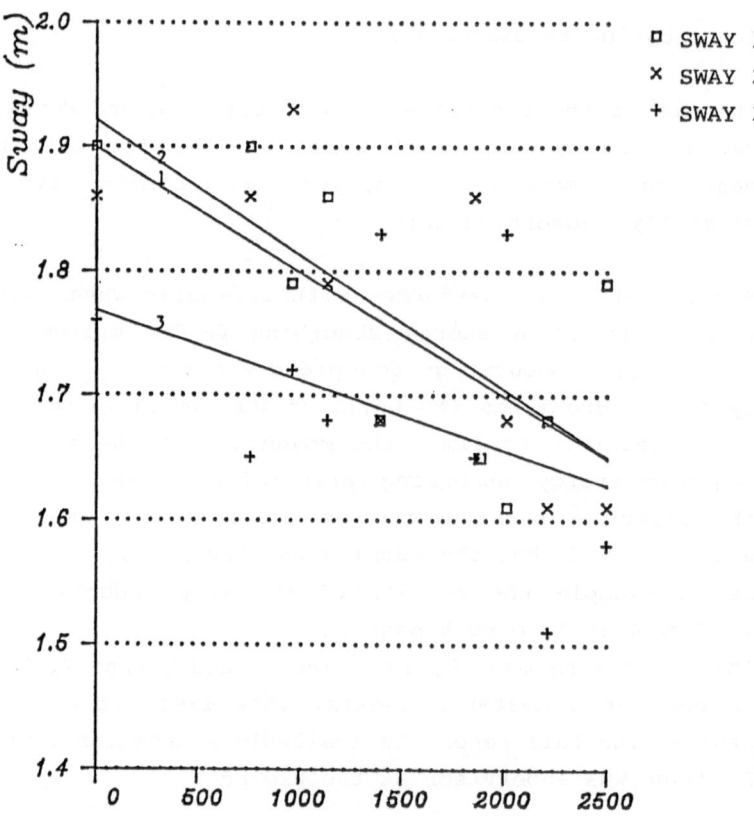

Force increment (kN) Model scale 1:90

Fig. 4.30. SWAY REDUCTION

TO WHOM IS THE RESPONSIBILITY FOR BERTHING AND MOORING
OPERATIONS TO BE ENTRUSTED?

F. Vasco Costa

The full responsibility for the operations of berthing and of
mooring a ship has traditionally been entrusted to the crew
on board assisted by a local pilot.
But who is in a better position to know how to bring a ship
safely to each particular berth of a harbour and how to keep
her quiet at berth during loading and unloading operations?
Those who are moving continuously from one harbour to another
or those who remain year after year attending to berthing and
mooring operations at a few berths within the same harbour?
Who is in a better position to know about currents and wind
gusts and the locations, and times of the year, when they are
most likely to occur? Who knows better about the characteristics of the fenders installed on the different berths and the
forces they exert on a ship's hull not only while they deflect
under the impact of a ship but, in particular, while they
recover their original form? Who knows better, or who can
know better, about the specific modes of oscillation of each
basin and the correspondent periods? Who can know better about the particular sea conditions that give occasion to the
drift forces Dr.Oortmerssen told us about? Who can know more
about the influence of oblique currents on a ship about to
berth, as described by Prof.Tryde? Those on board or those on
the ship?
To the authors of Report NS179 of the British Ship Research
Association (pgs. 14 and 29), those on board belong to the
"deck crew" or the "complement", while those on land are just
the "shore gang".
In the old days there is no doubt that the deck crew were in a
much better position than the shore gang to bring a sailing
ship to berth. Not only were they much smarter but they knew
much better how to manoeuver the ship which they sailed from
harbour to harbour. But conditions have changed. Nowadays
there is no reason to assume that those on shore cannot be as
smart, as careful and as reliable as those on board.
There is no reason now to assume that a "shore gang" will not
be able to attain the expertise of a "deck crew" in bringing
any type of ship to berth under the action of the harbour's
own tug-boats and of mooring ropes passed to shore and manned
by the shore gang.
Who can better select, from among the great variety of available monitoring equipment for berthing operations, the most
adequate for the particular local conditions? Who is in a better
position to keep the equipment in good operational conditions?

E. Bratteland (ed.), Advances in Berthing and Mooring of Ships and Offshore Structures, 459–460.
© 1988 by Kluwer Academic Publishers.

460

Who knows better not only the language but also the particular
skills of the crews of local tug-boats and line-boats? Who is
in a better position to know when to use long or short, smooth
or stiff ropes, taking into due account local weather and sea
conditions and the characteristics of the fenders installed at
each particular berth?
Are harbour authorities so conservative they cannot realize
that conditions have changed since the time of small ships manoeuvering under their own means? Are they to continue to prohibit the use of self-tension winches by the ship's crews, without realizing that such winches can contribute to keeping
the ships quieter and to reducing the frequency of accidents?
If installed on shore it will be easy to keep such winches in
good operational conditions. Self-tension winches, if properly operated, can contribute to reduce "downtime" at a harbour,
and therefore contribute to attract traffic.
If the responsibility for all the berthing and mooring operations
is passed from those on ship to those on shore, it can be taken
for granted that not only the time required for such operations
but, also the frequency of accidents will be reduced. Both
shipowners and harbour authorities will greatly benefit from
the transfer of the responsibility for berthing and mooring of
ships from the "deck crew" to the "shore gang".

MOORING SYSTEMS OF THE WORLD LARGEST FLOATING OIL STORAGE
BASE

S.Ueda
Chief of the Offshore Structures Laboratory, Structures
Division Port and Harbour Research Institute, Ministry of
Transport, Japan.

1. INTRODUCTION
Stockpiling of oil in Japan has been executed by private
enterprises to maintain 90 days amount of consumption, however,
after the Oil Crisis in 1974, Japanese government decided to
add thirty million kl as national stockpiling. Then the scheme of national stockpiling of oil was established and the construction of oil storages bases was started. Now oil stockpiling has been begun in such two bases as Tomakomai and
Mutsuogawara, construction has been executed in such bases as
Akita, Fukui, Shibushi, Kamigoto Shirashima, Kuji, Kikuma, and
Kushikino.
Among those bases mentioned above, both Kamigoto and Shirashima bases are offshore oil storage bases. Formerly, several
offshore oil storage bases have been constructed, however, almost all of them are used for temporary oil storage until a
shuttle tanker arrives. Therefore, the amount of each oil
storage bases is at most two hundred thousand
The oil storage bases on the subject are offshore floating oil
storage bases consisting of several large-sized steel-made
floating tanks which can store about eight hundred thousand kl
each, and total amounts of oil storage of each base is about
four or five million kl, respectively. These floating tanks
are moored to concrete dolphins by use of large sized rubber
fenders. In this paper, the outline of design of mooring
systems regarding the Kamigoto Oil Storage Base are described.

2. OUTLINE OF OFFSHORE FLOATING OIL STORAGE BASES
Two offshore floating oil storage bases are now under construction namely Kamigoto Base and Shirashima Base. Locations of
these bases are shown in Figure 1. Figures 2-a, show plan
views of the bases. Each of them   consists of five or eight
large sized floating steel tanks, sea-berth available for up to
three hundred thousand dwt tankers, pipelines, breakwaters, oil
spill preventing fences and other facilities concerned. As for
the Kamigoto Base, the floating steel tank is 390m in length,
97m in width, 27.6m in height and 24.5m in draft fully laden
and the maximum capacity of oil storage is about eight hundred
and eighty thousand kl. At the beginning of planning, the total amount of oil storage of this base was nearly six million
kl by use of seven floating tanks. The plan was, however,
revised to nearly four million kl in total amount by use of
five floating tanks at first stage of construction. Remaining
two tanks will be constructed in the second future stage.

461

E. Bratteland (ed.), Advances in Berthing and Mooring of Ships and Offshore Structures, 461–473.
© *1988 by Kluwer Academic Publishers.*

And as for the Shirashima Base, the floating steel tank is
397m in length, 82m in width, 25.4m in height and 22.7m in
draft fully laden and the maximum capacity of oil storage is
about seven hundred thousand kl. The total amount of oil
storage in this base is planned to nearly five million kl by
use of eight floating tanks. The construction of the Kamigoto
Base is   ahead of the Shirashima Base. Three floating tanks
have been moored at the Kamigoto site, and the prototype field
measurement of motions of floating tanks and fender deflections is now in progress. After the qualification of data
obtained through the field observation and confirmation of
safety of mooring systems of those floating tanks, oil storage
will starts. This is scheduled in autumn 1988 at the moment.
Therefore, description hereafter will be made on the mooring
systems of the Kamigoto Base.

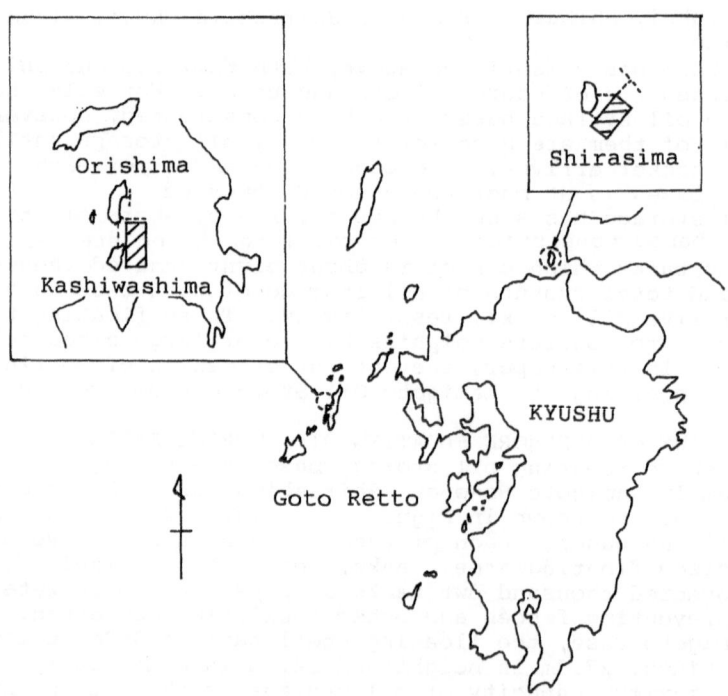

Figure 1. Location of Oil Storage Bases

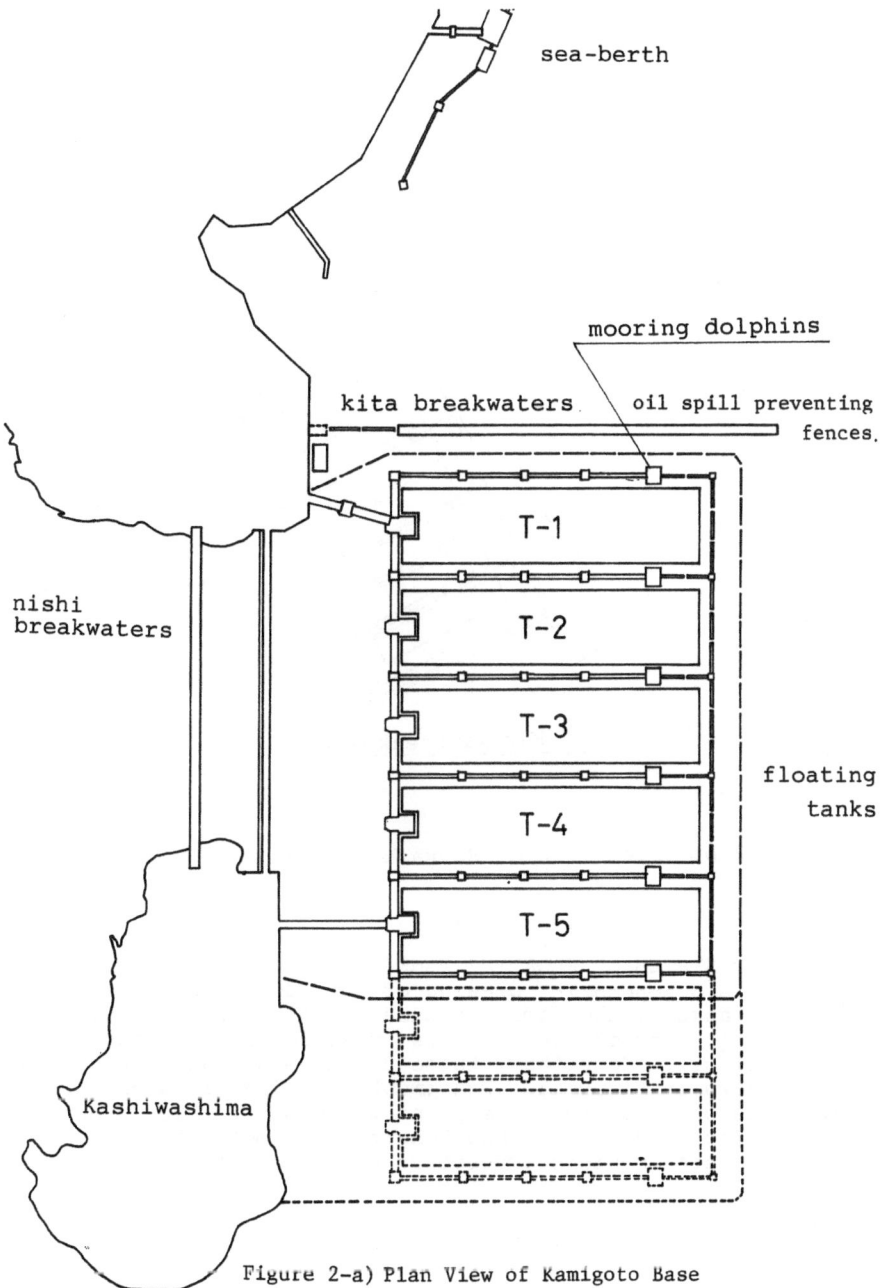

Figure 2-a) Plan View of Kamigoto Base

464

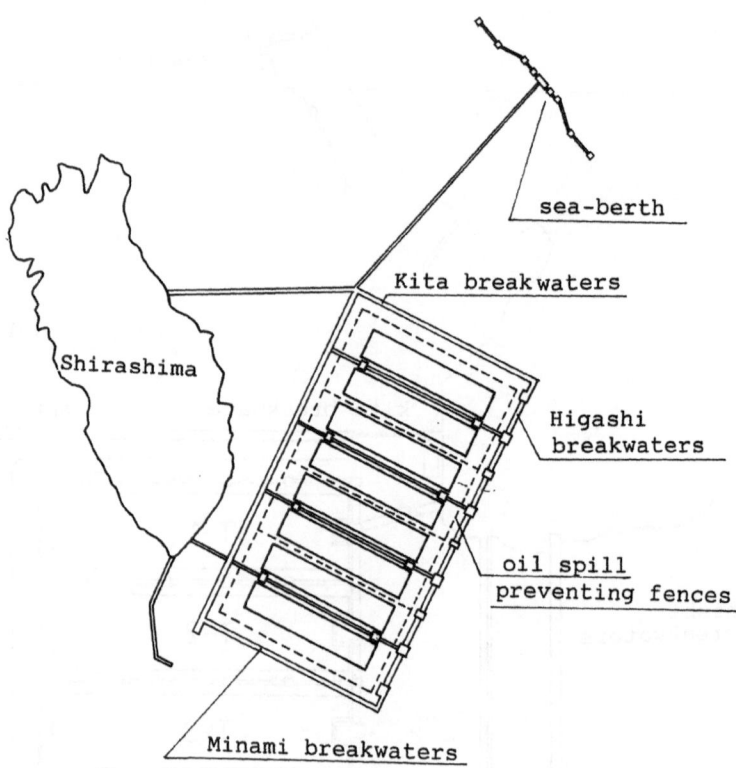

sea-berth

Kita breakwaters

Shirashima

Higashi
breakwaters

oil spill
preventing fences

Minami breakwaters

Figure 2-b)   Plan View of Shirashima Base

3. MOORING SYSTEM OF FLOATING TANKS
Figure 3 is a schematic drawing of the mooring systems of
floating tanks at the Kamigoto Base. There are three dolphins
for mooring. One of these is also used for loading and unloading of oil. Hereafter, these dolphins are named L-5, B-5,
B-6 as shown in Figure 3. Equipments are installed on the L-5
dolphin as on a loading tower, and fenders for absorption
included for both lateral and longitudinal motions. Fenders
against lateral motions are set on both sides of the dolphin.

465

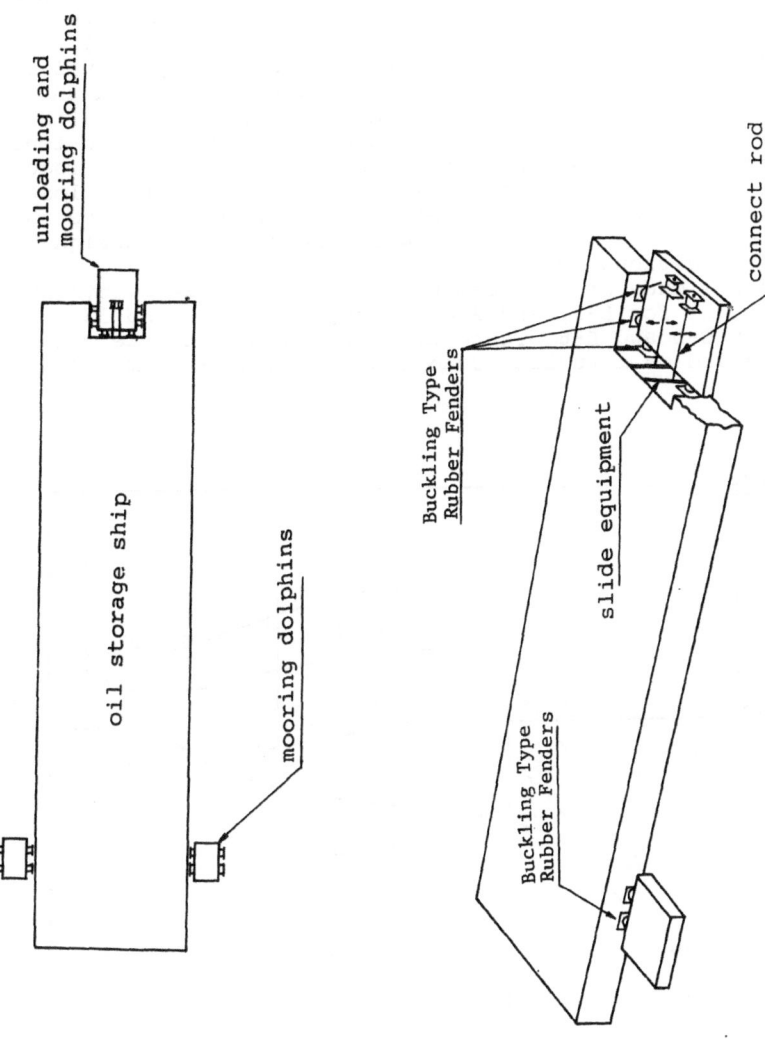

unloading and
mooring dolphins

oil storage ship

mooring dolphins

Buckling Type
Rubber Fenders

Buckling Type
Rubber Fenders

slide equipment

connect rod

Figure 3 Schematic drawing of mooring System

466

And fenders against longitudinal motions are set on the side
and the top of the dolphin. Fenders installed on the top of
the dolphin are connected to the floating tank by steel rods
with 195 mm in diameter. These steel rods are kept level by
means of devices to make them slide following tide and the
draft of the tank. As described later, size of fenders is
determined according to the results of numerical simulations
which compute motions and mooring forces considering irregularity of waves, fluctuation of wind and the non-linear load
fluctuation deflection characteristics of fenders. Size and
numbers of fenders vary according to the different design wave
and wind conditions for each tank.
Dolphins are made of concrete with 27m length, 24m width and
33.5 m height for L-5, and 23m length, 19m width and 33.5 m
height for B-5 and B-6, respectively. These concrete caissons
are installed on a rubble mound base made at -28.5m depth.

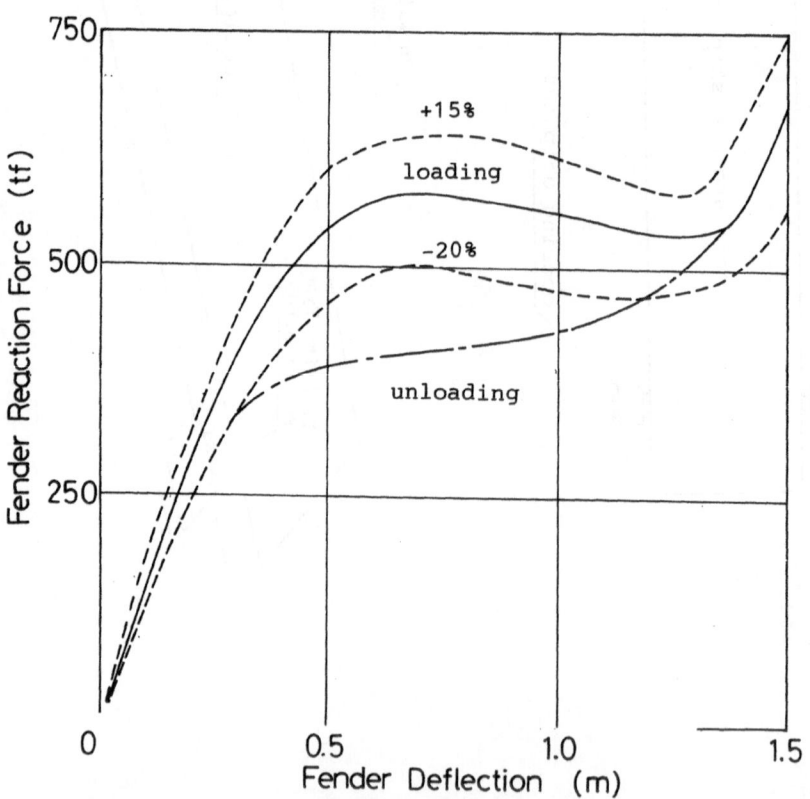

Figure 4 Load-Deflection Characteristic of Rubber Fender

Fenders used for this mooring system are so-called buckling type rubber fenders. Figure 4 shows the load-deflection characteristics of these fenders. When a fender is loaded, the reaction force increase at a high rate and it reaches nearly the maximum reaction force against the deflection for about 20% of the maximum deflection. Then, the reaction force is kept almost constant for any deflection between 20 to 45 or 50%. However, the reaction force increases at a higher rate than previous loading for deflection exceeding 45 to 50%, as shown by full line in Figure 4. This type of fender also exhibits large hysteresis in unloading as shown in Figure 4 with dotted and dashed line. Dotted lines drawn on both upper and lower side of full line denoted as + 15% or - 20% are load-deflection characteristics used in design of mooring systems described later. Size of rubber fenders used for this system are in the range from 2250mm to 3000 in height. The maximum reactions are in the range from 385tf to 675tf each.

4. SELECTION OF FENDERS FOR MOORING SYSTEMS

Mooring systems was designed by use of numerical simulation method because a floating tank is to be subjected to irregular waves and gusty wind and also the load-deflection characteristics of mooring systems are non-linear. Therefore, motions of floating tank and deflections of fenders can not be obtained by superpositioning of those values caused by each external forces. This numerical simulation method is qualified through model experiments.

Table 1 lists major design conditions for mooring systems. Wind and wave conditions are determined by statistical methods, and such that two levels of conditions were used for ordinary and extraordinary conditions. The ordinary condition means that a floating tank stores oil and is fully or half laden. Wind and wave conditions for the ordinary condition are equivalent to those of one hundred years recurrence. The extraordinary condition means that a tank for some reason is in ballasted condition including the emergency discharge. Wind and wave conditions for the extraordinary condition are equivalent to those of twenty years recurrence. The maximum wind speed for ordinary and extraordinary conditions are 51m/s and 44m/s, respectively, in mean wind speed during ten minutes. Waves at the mooring site of floating tanks are calculated as listed in Table 1 by means of wave diffraction computation program. Wave heights at the mooring site are less than 1.0m with periods of 13s and 10s.

Formerly, there were drawings of two dotted lines in Figur 4. These deviations of load-deflection characteristics were disccreetly determined according to the results of investigations and tests carried out by manufacturing company of these fenders. Such factors are considered as individual differences, deterioration, dynamic response, creep against stationary loading, bi-axial loading and temperature. Table 2 lists factors and deviations to normal load-deflection characteristics.

Table 1 Major Design Conditions

		100 years	20 years
Wind	Max. Mean Wind Speed	51m/s	44m/s
	Max. Instantaneous Wind Speed	70m/s	60m/s

		West Ward $H_{1/3}$	West Ward $T_{1/3}$	North ward $H_{1/3}$	North ward $T_{1/3}$
Wave	Entrance of bay	7.6m	13s	5.5m	10s
	Mooring Basin East	0.4m	13s	1.0m	10s
	Mooring Basin South	0.6m	13s	1.0m	10s
	Mooring Basin Inside	0.4 0.5m	13s	0.4 1.0m	10s

Tide	HHWL	DL +3.8m
	HWL	DL +2.83m
	LWL	DL +0.0m

Current	0.5knot

Temperature of Sea Water	Highest	30°C
	Lowest	10°C

Table 2 Deviation of Fender Characteristics
Deviation of Fender Characteristics (Deviation of Reaction Force for the Same Deflection)

Individual difference	0.9	1.10
Deterioration	1.0	1.05
Dynamic Response	1.0	1.10
Creep	Load caused by stationary loads shall be less than the fender reaction force for 10% deflection.	
Repeat Loading	0.8-0.9 (loading with load equivalent to reaction for 40% deflection with ten times of repetition)	
Bi-axial Loading Temperature	Compression test with shear 10% of the compression 0.95-1.25 (50 to 0C, is assumed standard)	

In the design procedure of mooring systems of floating tanks, first of all, fender size is to be tentatively determined. Considering the distinctive mark of load-deflection characteristics of buckling type rubber fenders, fenders will be selected so that loads caused by stationary wind and current loads at each mooring point shall be less than the fender reaction force for the 10% deflection. The deflection of 10% is determined according to the results of creep tests of fenders. The test were carried out against several stationary loads working for several hours. When the load was equivalent to or less than the reaction force for 10% deflection, the fender deflection did not increase with time. But, when the load was larger than the reaction force for 10% deflection, the fender deflection increased with time.

As previously mentioned two loading conditions of floating tanks were chosen as design conditions for mooring systems. Those are ordinary and extraordinary conditions.

Properties of the floating tank under those conditions are listed in Table 3. According to the results of calculations, the latter condition is rather significant for determination of fenders, because wind force is dominant among the stationary loads. Mooring points are denoted as in Figure 5. Table 4 lists loads from wind and current subjected to each mooring point with T-5 as an ecample. T-5 subject to the most severe wave and wind loads and consequently fenders installed to these dolphins are of the maximum size among all of the dolphins used for mooring floating tanks. The wind attack angle is taken clockwise. As this load is static, the deviation is set as ± 10% to normal load-deflection characteristics. Here, the maximum loads to mooring points 1 and 2 are 589tf and 443tf, respectively. By using C3000H (R_H) or equivalent ones, total load from wind and current subjected to the mooring point are less than 90% of the fender reaction force of 780tf for 10 % deflection. As for the mooring point 3, the maximum load from wind and current is 283tf. By using C2250H(R_H) or equivalent ones, load from wind and current subjected to each mooring point is less than 90% of the fender reaction force of 448tf for 10% deflection.

Table 3   Properties of Floating Tank

L x B x D	390m x 97m x 27.6m	
Operating Condition	Ordinary	Extraordinary
Loading Condition	Half laden	Ballasted
Displacement	560,000tf	180,000tf
Free board	14.0m	23.9m
Draft	14.6m	4.7m
KG	8.5m	9.5m

Table 4 Load from Wind and Waves Subject to Mooring Point

Wind Direction	-30	-20	-10	0	10	20	30	90
C_x	-0.13	-0.09	-0.04	0.0	0.04	0.09	0.13	0.25
C_y	0.71	0.77	0.81	0.82	0.81	0.77	0.71	0.0
C_m	-0..071	-0.051	-0.034	0.0	0.034	0.051	0.071	0.0
MP1	371tf	424tf	487tf	540tf	579tf	589tf	563tf	0tf
MP2	428tf	443tf	425tf	383tf	332tf	277tf	235tf	0tf
MP3	-146tf	-101tf	-45tf	0tf	45tf	101tf	146tf	280tf

Load from Current MP1   7tf,   Mp2   5tf,   MP3   3tf

Table 5   Wave and Wave Conditions for design of mooring system

Case	Loading Condition	$H_{1/3}$	$T_{1/3}$	U_{10}	Current Speed
S-1	Half	0.83m	10S	51m/s	0.5knot
S-2	Half	0.45m	13s	51m/s	0.5knot
S-3	Ballasted	0.68m	10s	44m/s	0.5knot
S-4	Ballasted	0.38m	13s	44m/s	0.5knot

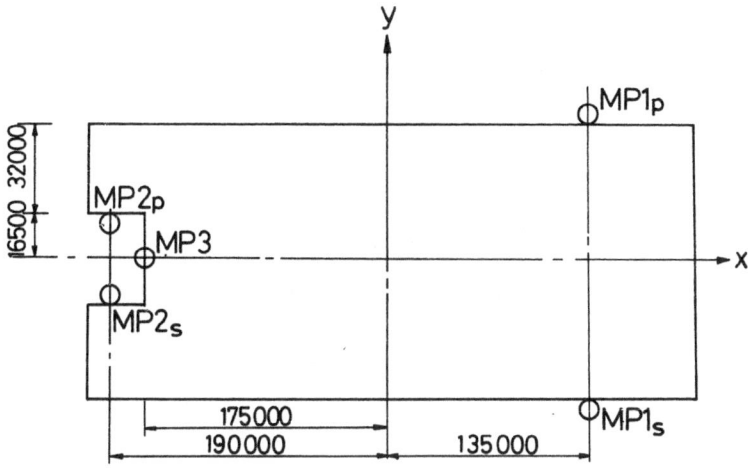

Figure 5 Location of Mooring Point

5. NUMERICAL SIMULATIONS FOR DESIGN AND QUALIFICATION OF MOO-RING SYSTEMS

Numerical simulations were carried out to compute both motions of floating tanks and mooring forces. Table 5 lists wind and wave conditions for each case. S-1 and S-2 are cases for ordinary conditions, and S-3 and S-4 are cases for extraordinary conditions. As the numerical simulation program yields solutions in time domain, some statistical treatment are needed to estimate the maximum motions and mooring forces. Figure 6 schematically shows the time history of resultant motion at the mooring point. The base line means the initial position of the floating tank. In this figure upper side of the base line means the motion toward fender compression. Peak values are defined as the difference between the peak of the resultant motions and base line as shown in Figure 6. Analyzing these peak values, one can obtain the mean and the standard deviation. It is confirmed that the distribution of peak values fits well to Reyleigh Distribution. The maximum resultant motion is defined as either the maximum of peak values, or the equivalent to one thousand occurence which is calculated as the addition of mean and 2.738 times the standard deviation of peak values. Table 6 lists the results of computations and the estimated maximum fender delections for each mooring point. Here, the maximum fender deflection in case S-3 is 34% of its height and the maximum reaction force is S-3 is 1323tf for mooring point 1, as an example. The maximum fender deflection is less than the allowable deflection which is 35% of the fender height.

472

According to the results of the numerical simulations, the
mooring systems were qualified. The design load of dolphins
are determined as 1.5 times the maximum mooring forces computed.

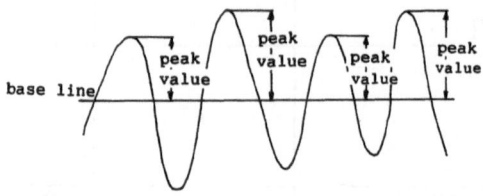

Figure 6. Definition of Peak Value

6. CONCLUSION
The world largest offshore floating oil storage bases are now
under steady construction. At the Kamigoto base, the prototype field observation of motions of floating tank and
fender deflections is carried on. These results may be used
not only to qualify the design method of this system but also
 effectively used for future development of same kind of offshore structures.
At the end of this paper, the author expresses his appreciation to the Kamigoto Oil Storage Company for substituting
valuable data and allowing him the presentation, and also to
Dr. Shuku, Mitubishi Heavy Industry Company for his kind
advices.

Table 6 Results of Computation

Mooring Point	Case No.	Wind Direction	Deviation	Max.Fender Deflection	Max.Reaction Force
MP-1	S-1	20	-20%	20%	915tf
			+15%	18%	1270tf
		-20	-20%	18%	883tf
			+15%	16%	1214tf
	S-2	20	-20%	16%	1230tf
			+15%	4%	272tf
	S-3	20	-20%	34%	928tf
			+15%	20%	1323tf
		-20	-20%	24%	928tf
			+15%	15%	1196tf
	S-64	20	-20%	31%	928tf
			+15%	19%	1295tf
MP-2	S-1	20	-20%	17%	872tf
			+15%	17%	1237tf
		-20	-20%	21%	928tf
			+15%	19%	1288tf
	S-2	20	-20%	21%	928tf
			+15%	20%	1316tf
	S-3	20	-20%	29%	928tf
			+15%	15%	1196tf
		-20	-20%	20%	1327tf
			+15%	24%	928tf
	S-4	20	-20%	24%	928tf
			+15%	12%	1058tf
MP-3	S-1	-90	-20%	13%	448tf
			+15%	13%	663tf
	S-2	-90	-20%	14%	456tf
			+15%	14%	656tf
	S-3	-90	-20%	21%	509tf
			+15%	16%	685tf
		-20	-20%	11%	406tf
			+15%	9%	518tf
	S-4	-90	-20%	25%	520tf
			+15%	15%	679tf

REFERENCES

1.Shuku,M. et al: The Motions of Moored Floating Storage Barge in Shallow Water, Journal of the Society of Naval Architects of Japan, Vol.146, 1979, pp.245-254.
2.Toyoda,S. et al: On the Initial Planning of Floating Oil Storage System, Journal of the Society of Naval Architects of Japan, Vol.146, 1979, pp.365-374.

The Use of Marine Fenders as Energy Absorbing Damper Units in Mooring Systems.

H. Kawakami, Manager, Marine Fender Engineering and Development

BRIDGESTONE CORPORATION, Tokyo, Japan.

1. INTRODUCTION

Engineers and scientists in many countries have pointed out that the mooring conditions of a ship as well as berthing is affected by fender characteristics.

For the purpose of clarifying how fender characteristics affect mooring condition, Bridgestone Corporation has developed a new simulation method combining computer analysis with a triaxial testing machine.

In this paper, we as a fender manufacturer, first discuss factors affecting fender characteristics, then, the concept of new simulation methods and the actual simulation results are explained.

2. THE CHARACTERISTICS OF RUBBER FENDERS

2.1 Factors Affecting Fender Characteristics:

It is usually assumed that the reaction force of a fender may only be found from the load - deflection curve for a given deflection. Several external factors, however, are considered to influence the performance of rubber fenders. These factors are simply classified into four main categories as follows:

2.1.1 Mechanical Factors:

The compression angles and directions can usually reduce the reaction force of a fender by bending and shearing the fender unit. A series of tests and an extensive research program has been done to establish a set of coefficients which can be used to determine these performance characteristics. Friction, torsion and other mechanical factors often combine with them, which will make the system more complicated. Until the recent study was started by Bridgestone it has been difficult to accurately simulate the actual behaviour of fender units under such conditions.

2.1.2 Dynamic Factors:

The deflection speed (frequency), the total number of compressions (fatigue), the creep and the stress relaxation which are called dynamic factors also influence the fender performance.

These dynamic factors are considered to be caused by the visco-elastic properties of rubber. The hysteresis loss of a fender is also considered as one of the dynamic characters when it is regarded as the speed dependency at the negative velocity in the recoil cycle. Each of these factors can be independently studied by the relevant model or actual size test. And the combination of them can be simulated by the 3-axis simulator introduced in Section 3.

FIGURE 1 gives an example of results from the Berthing Simulation studies which gives an indication of the velocity dependency of the reaction force of a cell type fender.

E. Bratteland (ed.), Advances in Berthing and Mooring of Ships and Offshore Structures, 474–490.
© 1988 by Kluwer Academic Publishers.

FIGURE 1
Velocity Dependency of Cell Type Fender. (C 2000H) [4]

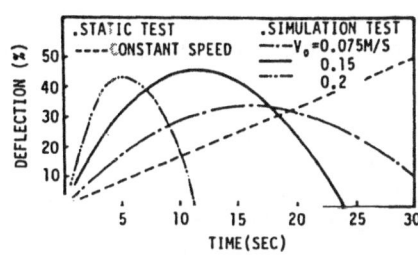
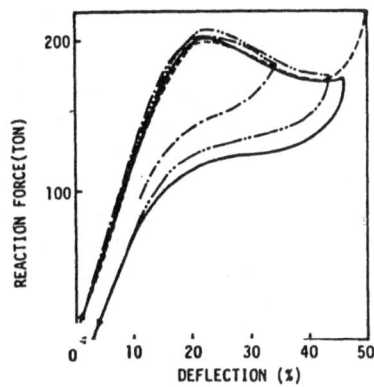

2.1.3 Chemical Factors:

The environmental factors such as: temperature, ozone, ultraviolet rays etc., influence the physical properties of rubber and, consequently, change the performance of fenders. Aging can be considered as one of the chemical factors although it is also affected by other factors.

2.1.4 Statistical Factors:

The performance deviation resulting from production tolerances should be considered statistically. Usually the range ±10% is adopted according to the record of the pre-delivery tests conducted in the plant.

2.1.5 The Range of Performance Deviations:

When a rubber fender is used as a mooring damper, the fact that some factors are independent and some are not makes it hard to grasp the magnitude of the total deviation.

TABLE 2.1 shows the range of deviations which is taken as a guideline in Japan.

TABLE 2.1 The range of performance deviations

Factor	Range	Remark
Deviation of Production	0.90 ~ 1.10	
Aging	1.00 ~ 1.05	
Speed	1.00 ~ 1.10	
Fatigue	0.80 ~	10th comp. at 40% amp.
Angle	0.90 ~	appox. 6°
Temp.	0.95 ~ 1.25	50°C ~ 0°C

476

2.2   Material Development for the Specific Properties:

For specific requirements, the characteristics of rubber should sometimes be changed. In other words, the factors which were mentioned previously
are sometimes regarded as the target of material developments. The following are
examples of specific cases; one is from the chemical factors (temperature), and the
other is from the dynamic factors.

2.2.1   Low Temperature Rubber:

There are many cases in which rubber fenders are used in very cold
environments such as the Arctic or Polar regions. But as shown in FIGURE 2,
rubber is a typical temperature dependent material. Below the brittle point -
usually around -40°C - rubber loses its elasticity and the fender becomes a fragile
structure. So the target of material development is to keep the magnitude of
temperature dependency as small as possible, maintaining the elasticity of rubber.
FIGURE 2 shows the recently developed material which maintains its elasticity
even at -65°C. The load-deflection curve of the fender made of this rubber is
shown in FIGURE 3.

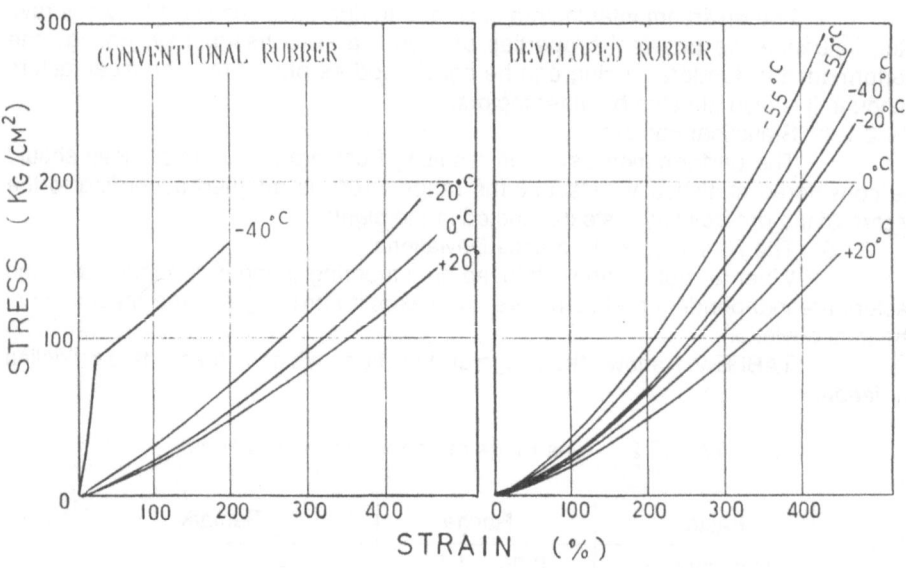

FIGURE 2   The Properties of Rubber in Low Temperature

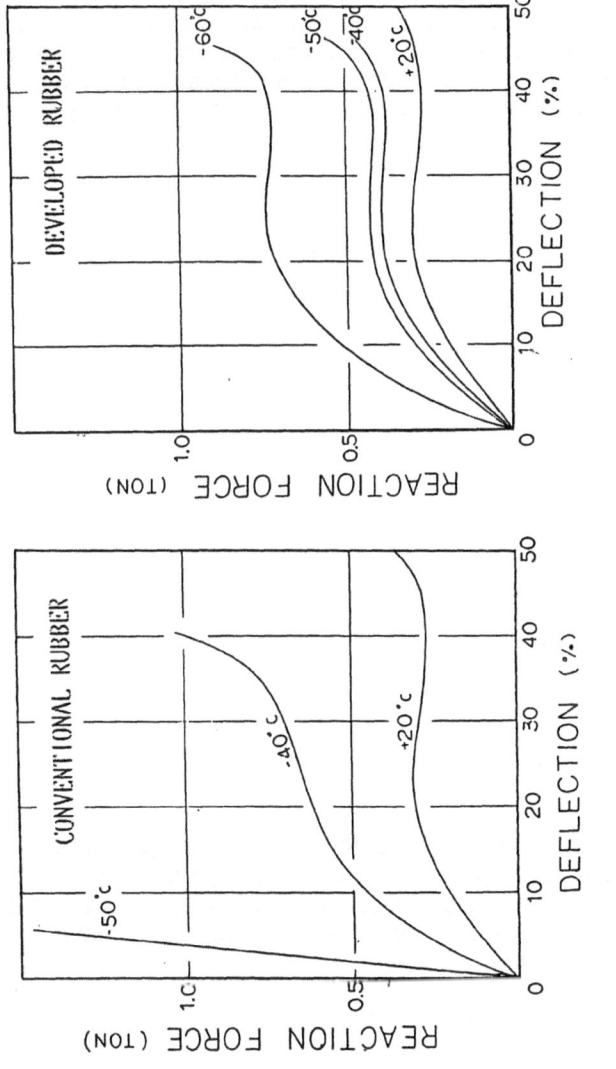

FIGURE. 3   The Fender Performance in Low Temperature

2.2.2 High Viscosity Rubber:

When the viscosity of rubber becomes dominant the performance of the fender changes giving a higher hysteresis loss and larger velocity dependency of deflection. This performance is sometimes referred to as "non" or "low recoiling". One of the main difficulties to develop this kind of material is to maintain the minimum elasticity for the recovery of its shape after the load is released. FIGURE 4 shows the successful example of 100mm height fender model made of newly developed high viscosity material. About 80% of the absorbed energy is lost in the recoil cycle.

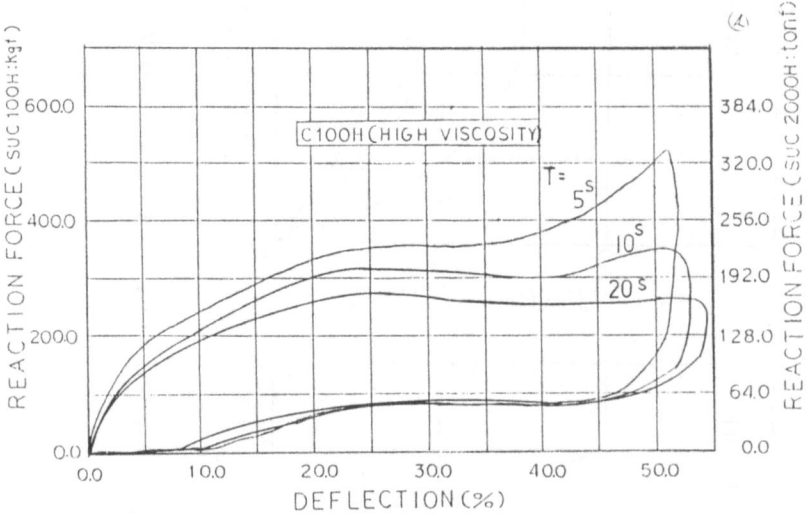

FIGURE 4   The High Viscosity Fender (Hysteresis)

It must be noted, however, that the improvement gains of one factor of the rubber materials are usually offset by other physical property losses. This high viscosity rubber is more temperature dependent than tconventional material. In reality, a little more improvement may be necessary to achieve such high hysteresis losses in a marine environment.

3.   TECHNOLOGY FOR FENDER EVALUATION

3.1   The Concept of the 3-Axis Simulator: [2], [3]

As mentioned above in Section 2, it is difficult to accurately simulate the load - deflection performance of a fender. Usually in a numerical analysis, fenders are treated as simple elements whose loads are only the function of deflection. On the other hand, hydraulic experiments with scale models have other limitations. For the sake of hydraulic similarity, the scale down test requires the fender model to have very much smaller elasticity than can actually be produced with rubber.

Whichever the simulation would be - numerical or experimental - the substitute spring elements make it hard to get enough information for the improvement or new development of rubber fenders. The 3-axis simulator was developed in this background, as a hybrid method of a numerical simulation and a model experiment.

FIGURE 5 shows the concept of the 3-axis simulator. The four hydraulic exciters can simulate motion in 3 degrees of freedom - X, Z and θ in the Sway, Heave, and Roll directions. So the fender model can be compressed, sheared and bent. These exciters send the data about the loads from the fender and the positions of each exciter to the computer. The computer which is programmed with the integration alogarithms of the equations of motion, calculates the excursion of the ship substituting the fender loads into the equation. Then the positions of each exciter in the next time step are quickly calculated and fed to each exciter in the form of appropriate analogue signals. Thus the 3-axis exciters move as the actual ship hull would move with the effect of the fender model.

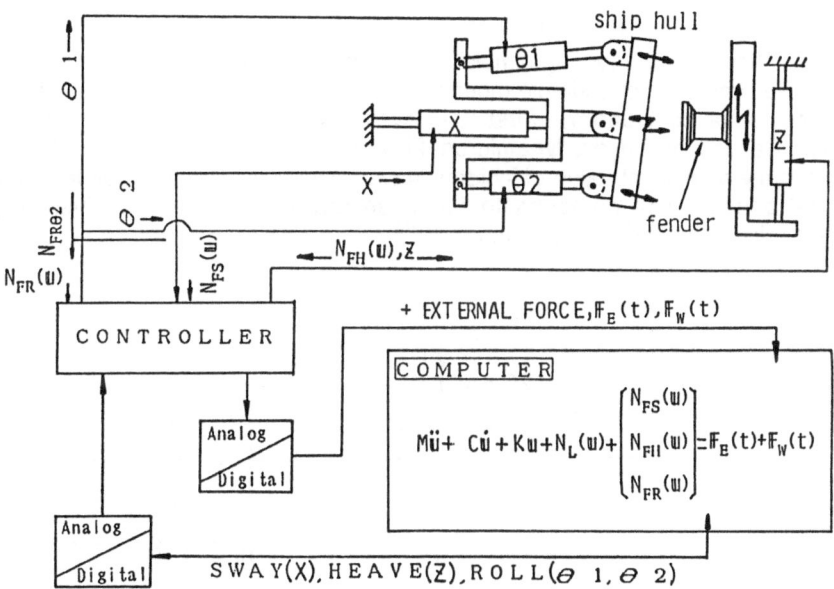

FIGURE 5   The Concept of 3-axis Simulator

This system gives a two-dimensional view of a mooring system (See FIGURE 7). But it can also be extended to the berthing simulation when the equations of motion in the computer are properly assumed for the two dimensional berthing phenomena (FIGURE 6). [4]

480

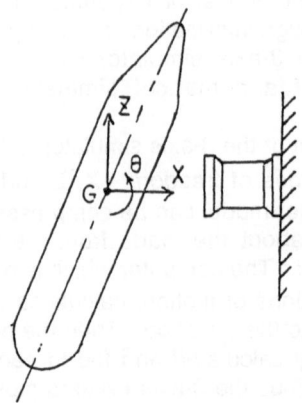

FIGURE 6   The Berthing Simulation

3.2   The Mooring Simulation:

Although the mooring simulation by means of the 3-axis exciter enables the accurate simulation of a simple mooring damping system, this system still has a few things to be considered. First, this system is limited to two dimensional analysis. Second, the hydrodynamic coefficients of the equations are assumed constant at the significant value of incident wave frequency. The three dimensional numerical analysis can compensate for these drawbacks if it is sophisticated enough to take account of these factors. [5], [6]

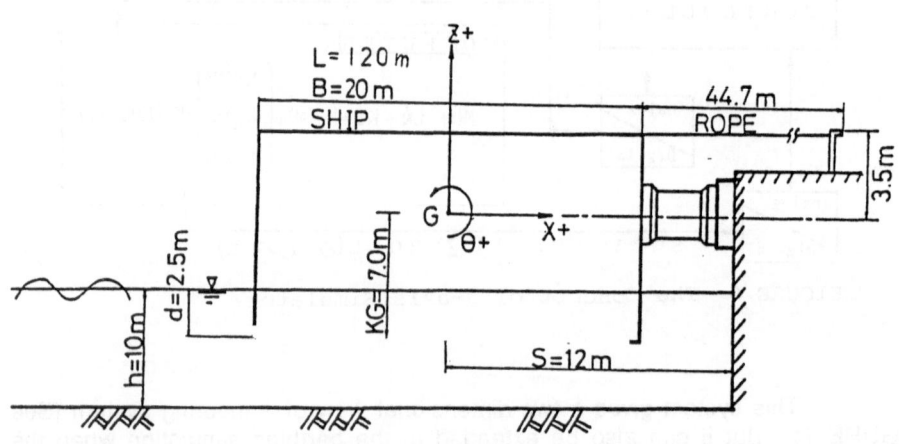

FIGURE   7   The Mooring Conditions

FIGURE 7 shows the moored ship at a quay wall. In this paper the following conditions were used as an example.

> ship   ;   rectangular box equivalent to 10,000 GT
> wave   ;   0.5m (1/3 significant)
> wind   ;   30m/s or 0m/s (average)

Added-mass, damping coefficients and wave exciting forces are determined as the solution of boundary-value problem based on the Eigen function expansion méthod [7]. Random wave forces are calculated as a linear superposition of sinusoidal wave force components with random phase differences. The Bretschneider - Mitsuyasu spectrum is applied for the frequency spectrum of incident waves [8]. The Davenport wind spectrum is used for the wind forces in the same manner [9]. All of these data were calculated and stored in a floppy disk in the time domain before the simulation.

Figure 8 shows the comparison of sway motions obtained from the 3-axis mooring simulator and from the numerical simulation under the same conditions; Fender : C 2000H, Wave Period : 9s, Wind : 30m/sec. These two results look similar in the time domain. But the power spectrums showi slight differences. These differences are caused by the difference in the fender reaction forces which are shown in FIGURE 9. In FIGURE 9, the fender used for the 3-axis simulator has a lower reaction force than the estimated reaction force of the numerical simulation which are originally from the same fender model. The mechanical factors - Shearing, Bending, Friction, and the dynamic factors - Fatigue, Velocity, Stress Relaxation - dissipate the compression force into other kind of energy.

FIGURE 10 shows the fender forces in three directions. In FIGURE 10, the shearing force is approximately 15% of the compressive force which suggests the practical friction coefficient between the ship's hull and the protector panel of the fender should be considered around this value.

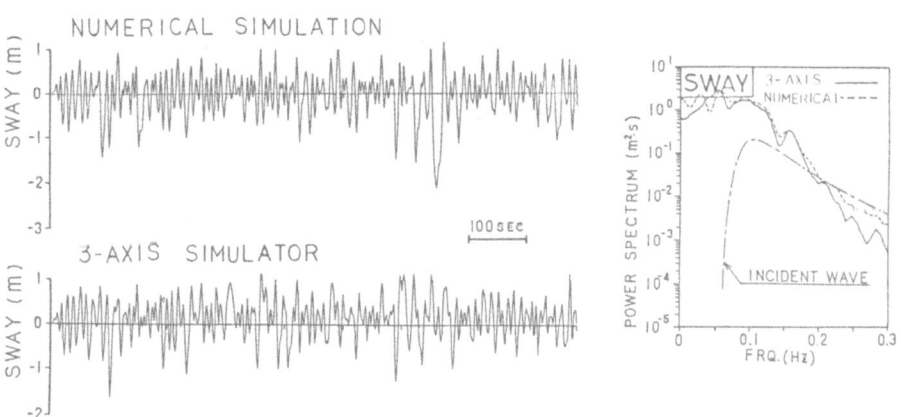

FIGURE 8 The Sway Motions of Two Methods

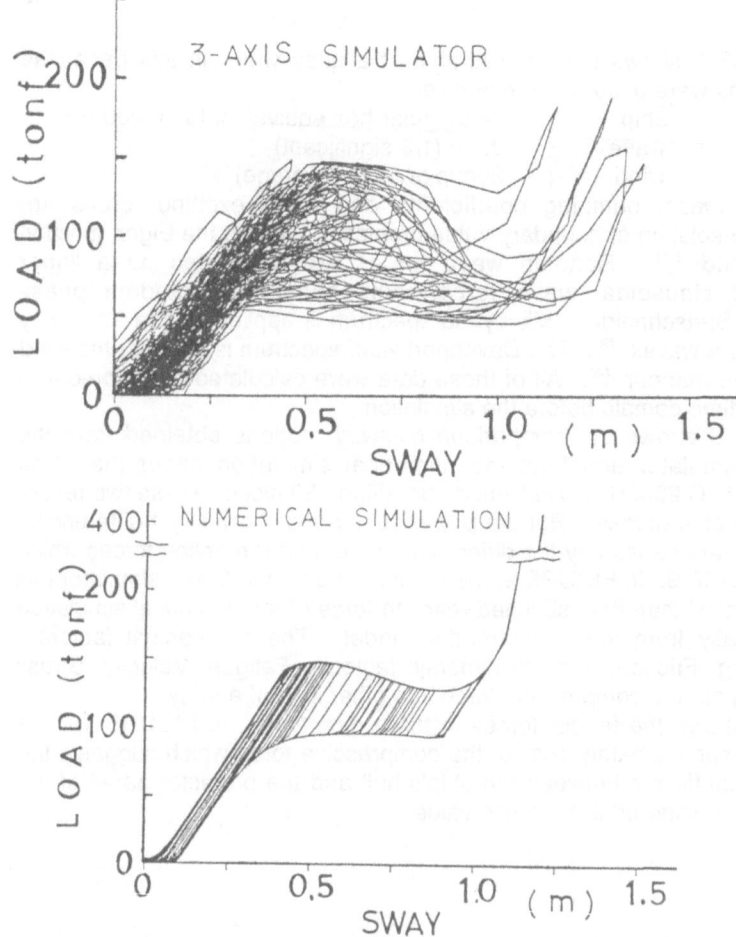

FIGURE 9   The Fender Reaction Forces of C2000 H

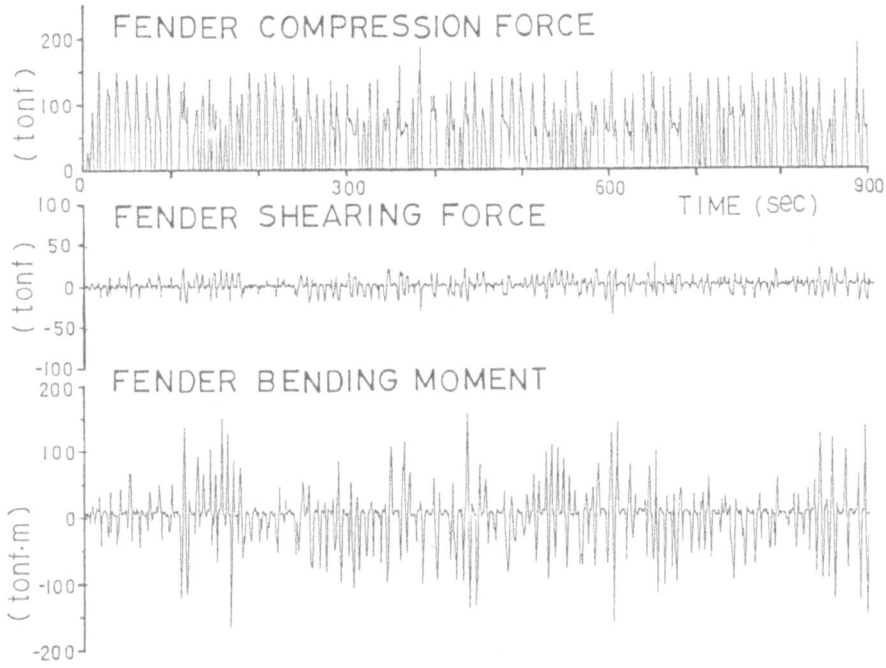

FIGURE 10   The Fender Forces

3.3   Case Studies of Mooring Systems:
       In this section, a few examples of case studies on new ideas for mooring systems are introduced.
       3.3.1   Improvement on Mooring Ropes:
       The simulation was conducted as shown in FIGURE 7, changing the performance of the ropes.
       As an example of a conventional mooring rope, 80 mm nylon rope of 44.7 m length was used. The basic problem with this type is that the spring rate of the rope is very much lower than that of the fender. It increases the non-linearity of the whole mooring system and, consequently, causes the long period excursion of the ship which is known as subharmonic motion. In order to improve on this point, the following two ideas were tested.
       Pre-tensioning: At its initial position, the ship is already pulled towards the quay with a 20 tonnes winch load. This means that the initial length of the same nylon ropes are shorter. The tension - deflection performances of these are shown in FIGURE 11.

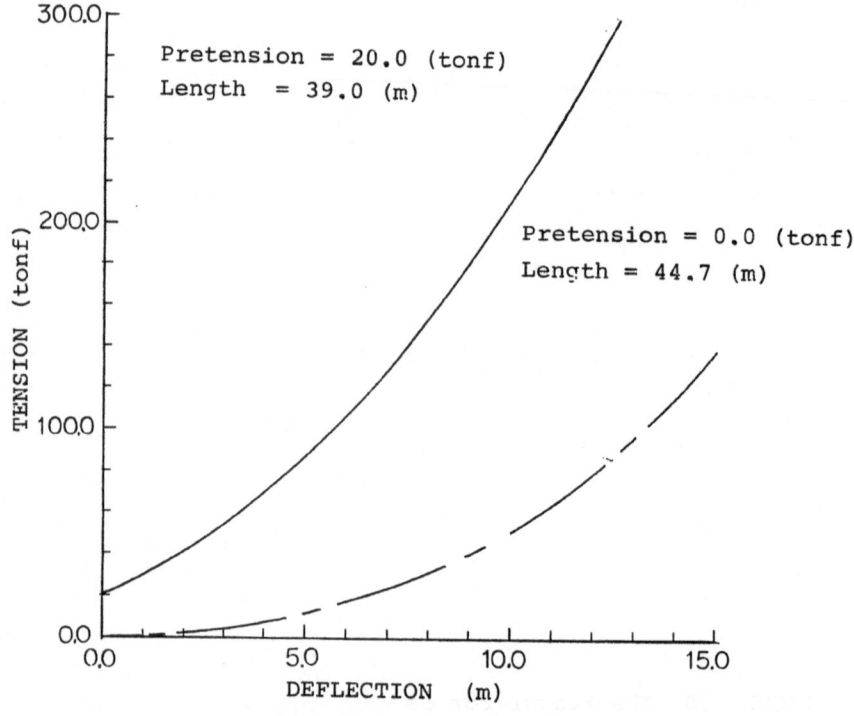

FIGURE 11 The Performance of Mooring Rope

Rubber Tail: A cell fender of 1.0 m height (100 mm at model) is used as a rubber tail extending its stroke 4 times of fender deflection using 4 pulleys. Steel wire is used as the mooring rope. The profile and the performance are shown in FIGURE 12 and 13.

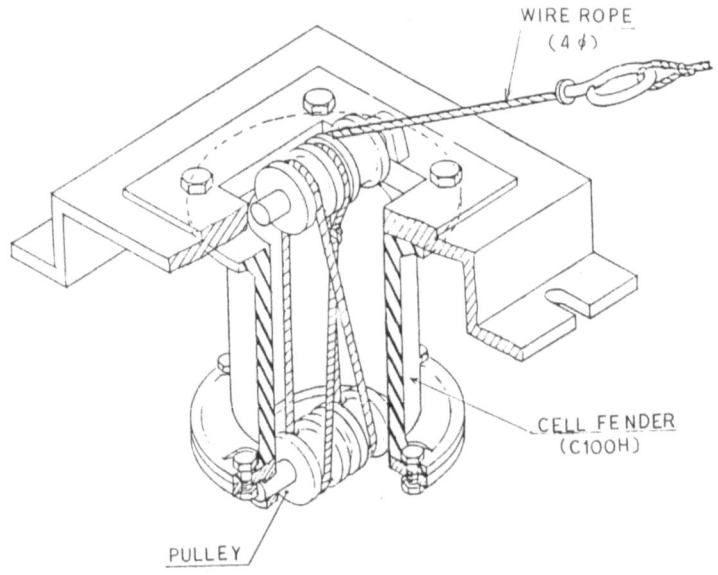

WIRE ROPE
(4 φ)

CELL FENDER
(C100H)

PULLEY

FIGURE   12 The Rubber Tail with Cell Fender

The periods of the incident waves are 7s and 10s. The moored ship does not make subharmonic motion in 7s but does in 10s. The result of sway, maximum reaction forces of the fender, and maximum rope tensions are shown in Table 3.1. It is difficult to clearly define the results or advantages of one system over another from this limited number of tests. However, it can be said that the pretensioning of the rope reduces both sway and fender reaction force if the wind velocity is 0 m/s and the wave period is 10s which are the conditions under which subharmonic motion occurs.

486

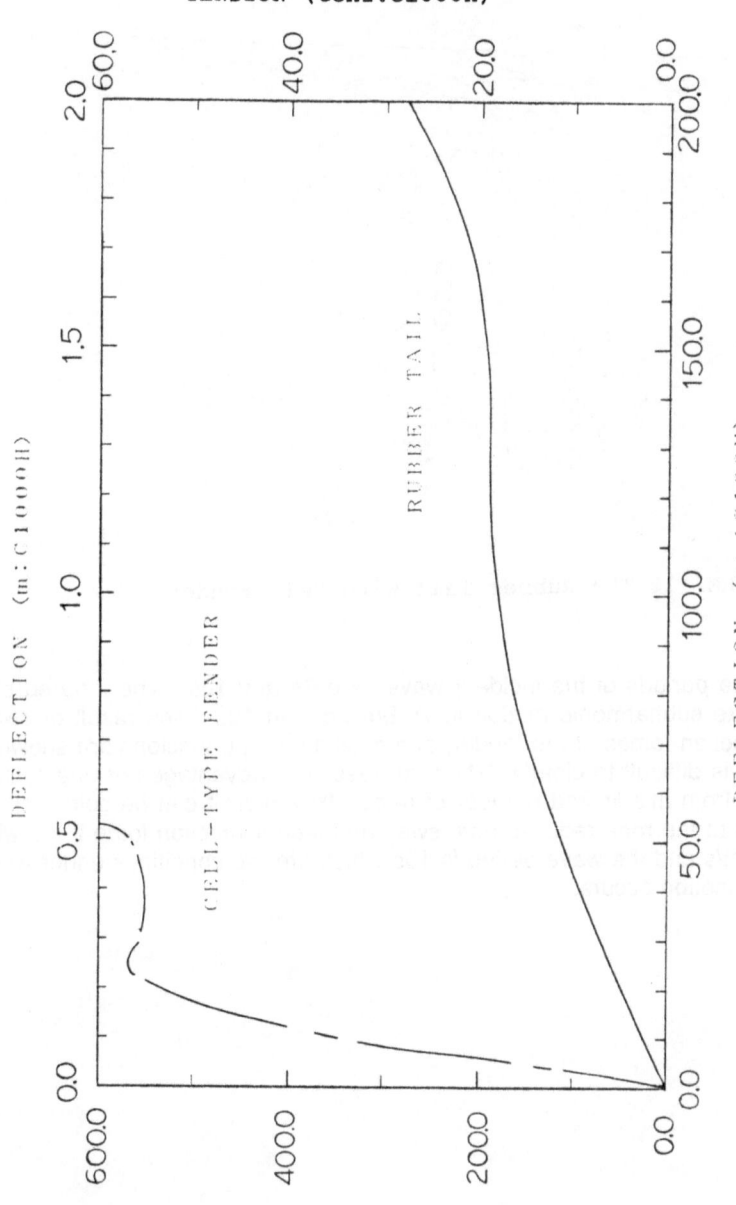

FIGURE 13 The Performance of Rubber Tail

Table 3.1 Results of Rope Improvement

Case No.	Wave Period (sec)	Rope	Wind Velocity =0.0(m/s)			Wind Velocity =30.0(m/s)		
			Sway 1/3 (m)	RF (tonf)	TR (tonf)	Sway 1/3 (m)	RF (tonf)	TR (tonf)
1	7.0	NYLON	0.62	110.0 (0.21)	<1.0 (1.00)	1.15	180.0 (0.64)	<1.0 (1.32)
2	7.0	NYLON PRET.20tonf	0.85	140.0 (0.38)	27.0 (1.00)	0.95	150.0 (1.66)	27.0 (1.00)
3	7.0	RUBBER TAIL	0.93	130.0 (0.55)	47.0 (0.92)	1.11	180.0 (1.25)	49.0 (1.28)
4	10.0	NYLON	3.89	180.0 (0.40)	25.0 (7.40)	1.89	220.0 (1.25)	1.4 (1.93)
5	10.0	NYLON PRET.20tonf	2.49	140.0 (0.50)	49.0 (2.90)	1.65	330.0 (1.36)	30.0 (1.20)
6	10.0	RUBBER TAIL	3.69	180.0 (0.79)	60.0 (4.00)	1.75	265.0 (1.25)	25.0 (1.60)

Sway 1/3:1/3Significant Value of Sway Amplitude
RF   :Maximum Reaction Force of Fender (Maximum Deflection[m])/1Unit
TR   :Maximum Tension of Rope (Maximum Deflection,[m]) /1Unit
PRET.20 :Pretension 20 (tonf)

3.3.2   Improvements on Fenders:

In order to evaluate improvement ideas compared with conventional fenders, simulation tests were done. The simulation conditions are shown in FIGURE 7. The focus of the improvement ideas centre on the increase of the viscosity or, in other words, optimising on the hysteresis effect.

• High Viscosity Rubber Cell Fender:

The high viscosity rubber which was introduced in Section 2.2 is used. The size of the fender is Cell C 2000H and the performance of this is shown in FIGURE 4.

• Water Filled Cell Fender:

The mechanical way to increase the viscosity of the cell fender was considered. It was an attempt at taking advantage of the hydraulic response effects of water. The basic concept is similar to the hydraulic fenders which have been in use for a number of years on passenger ferry berths in Japan. A coventional cell fender is filled with water from a reservoir through an orifice. This method can be taken to improve the capacity of a conventional fender as if it has a closed chamber inside. The concept is shown in FIGURE 14. The performance of conventional C 2000H and the C 2000H with water is shown in FIGURE 15. In FIGURE 15 the hysteresis of the fender was not increased by water as much as expected. The reason is that the inside volume of the cell fender does not change proportionally to the deflection of the fender.

488

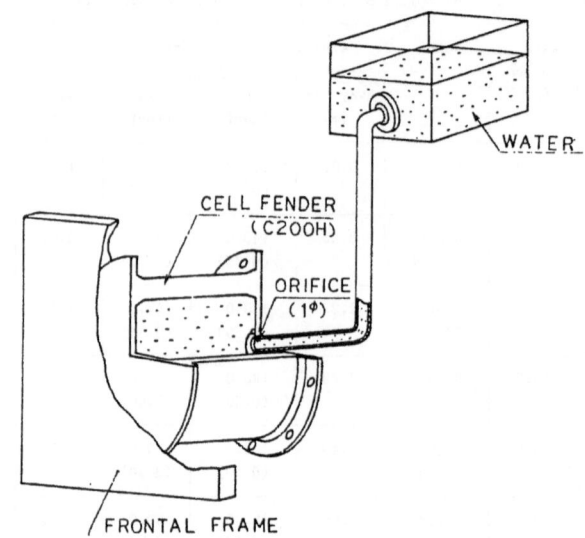

FIGURE   14   The Water Filled Cell Fender

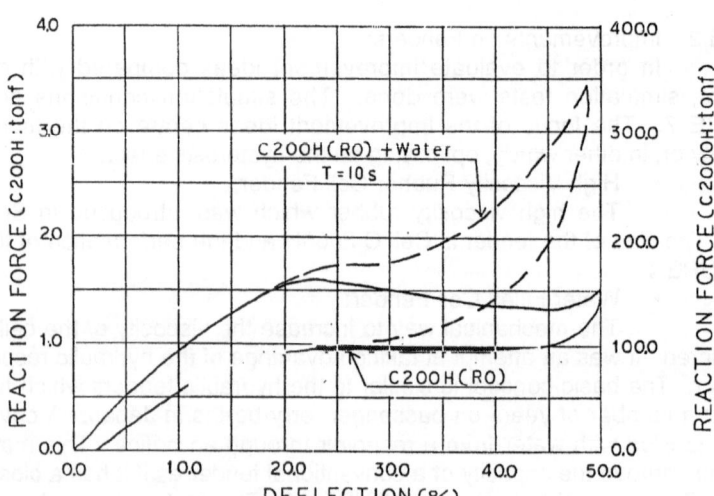

FIGURE 15   The Performance of Cell Fender
(C200H, C200H+WATER)

The results are shown in Table 3.2

Table 3.2   Results of Fender Improvement

Case No.	Wave Period (sec)	Fender	Wind Velocity =0.0(m/s) ROPE PRET.=20.0(tonf)			Wind Velocity =30.0(m/s) ROPE PRET.=0.0(tonf)		
			Sway ¹/₃ (m)	RF (tonf)	TR (tonf)	Sway ¹/₃ (m)	RF (tonf)	TR (tonf)
7	7.0-	C2000H×2	0.85	140.0 (0.38)	27.0 (1.00)	1.15	180.0 (0.64)	<1.0 (1.34)
8	7.0	C2000H +WATER×2	0.87	130.0 (0.34)	28.0 (1.10)	0.96	170.0 (0.61)	<1.0 (1.18)
9	7.0	C2000H(H.V) ×2	0.23	122.0 (0.31)	26.1 (0.91)	0.76	144.4 (0.71)	<1.0 (0.71)
10	10.0	C2000H×2	2.49	140.0 (0.50)	49.0 (2.90)	1.89	220.0 (1.25)	1.40 (1.93)
11	10.0	C2000H +WATER×2	3.26	260.0 (1.00)	60.0 (3.50)	2.87	480.0 (1.23)	3.0 (2.79)
12	10.0	C2000H(H.V) ×2	1.60	144.0 (0.46)	34.4 (1.90)	1.34	298.0 (1.38)	<1.0 (1.21)

Sway ¹/₃:¹/₃Significant Value of Sway Amplitude
RF   :Maximum Reaction Force of Fender (Maximum Deflection[m])/1Unit
TR   :Maximum Tension of Rope (Maximum Deflection,[m]) /1Unit
H.V   :Rubber for High Viscosity
PRET.20 :Pretension 20 (tonf)

In order to reduce the magnitude of subharmonic motion, the mooring rope was pre-tensioned in case of wind velocity is 0 m/s.

In the results of the high viscosity rubber fender, the sway motion is decreased and the reaction forces are generally lower. However, the fender does not always fully recover before the next compression so that the ship is allowed to drift closer to the quay wall and this causes larger reaction forces at higher deflections. This is the problem to be considered for this type of fender.

On the other hand, the water filled fender allows larger sway motions and shows bigger reaction forces. This is contrary to the desired targets and means that further improvement work is required.

These results are not sufficient to tell conclusively which idea is successful and which is not. Many other results of studies over a number of years, which are not given in this paper, are still far from matching theoretical ideals with practical solutions.

Further studies are required to confirm the feasibility and the suitability of mooring systems in variable mooring conditions.

4. CONCLUSIONS AND ACKNOWLEDGEMENTS

Recognising the basic fact that both berthing and mooring of a ship are affected by fender characteristics was the starting point of our study. To confirm this, a new simulation method was developed.

Simulation results are still quite limited, and also several problems such as the determination of hydrodynamic coefficients exist at the present time.

However, we believe that this method is quite effective in developing new fender systems, which will be necessary for designing improved marine facilities.

Finally, the authors would like to extend their great appreciation to Dr. Ueda of the Ministry of Transport and Dr. Oda of Osaka City University for their kind cooperation and valuable guidance in the execution of these studies.

REFERENCES.

1). "Technical Standards on Port and Harbour Facilities - II Floating Oil Storage Facilities" The Japan Port and Harbour Association, September, 1980, pp. 127 (Japanese.)

2). Oda, K., Ishigami, M., Akiyama, H., "Ship behaviour analysis by means of a hybrid type numerical simulation," Proceedings of 32nd., JCCE, 1985, pp. 662 ~ 666. (Japanese).

3). Oda, K., Ishigami, M., Akiyama, H., "Analysis of ship motion under irregular winds and waves by means of a hybrid type numerical simulation," Proceedings of 33nd., JCCE, 1986, pp. 541 ~ 545. (Japanese).

4). Tagomori, S., Kuroda, K., "Berthing Simulation Methods for Fender Systems," Proceedings of "PORTS 86", ASCE, May 1986, pp. 735 ~ 748.

5). Oortmerssen, G. van., "The motions of a Moored Ship in Waves," NSMB Publication No: 510, Netherlands, 1976.

6). Ueda, S., Shiraishi, S., "Analytical Treatment of Ships' Mooring Problems in Rough Weather," Proceedings of "OMAE '86", Vol. III, 1986 pp. 522 ~ 528.

7). Iijima, T., Tabuchi, T., Yumura, Y., "Scattering of Surface Waves and the Motion of a Rectangular Body by Waves in Finite Water Depth," Proceedings of JSCE, No: 202, June 1972, pp. 33 ~ 48. (Japanese).

8). Bretschneider, C. L., "Significant Waves and Wave Spectrrum, Fundamentals of Ocean Engineering - Part 7," Ocean Industry, Feb., 1968, pp. 40 ~ 46.

9). Davenport, A. G., "Gust Loading Factors," Journal of the Proceedings of ASCE, ST3, June, 1967, pp. 11 ~ 34.

POSSIBLE FUTURE DEVELOPMENTS

F. Vasco Costa

The careful study of local conditions related to wave trains and the drift forces they can exert on ships, where and when sudden changes in direction and magnitude of currents are likely to occur, and where and when wind gusts can affect manoeuvring are suggested as ways to improve berthing operations.
The long heavy mooring ropes in use require two men for their handling. The use of remotely controlled launches, instead of manned mooring launches, is therefore suggested for the passing of mooring lines from shore to ship, instead of from ship to shore.
Considering ways to keep moored ships quiet at berths, the following are suggested:

- The careful study of the positions of the nodes and anti-nodes for the distinct possible modes of oscillations of the basins when the location of berths for different types of ships are to be selected.

- The use of damping devices intended to reduce the amount of energy given back to the ship while fenders and mooring ropes recover their original form after being stressed.

- The use of monitoring equipment permitting knowledge of when to relax or to stress mooring lines, as suggested in the lecture by Dr.van Oortmerssen.

E. Bratteland (ed.), Advances in Berthing and Mooring of Ships and Offshore Structures, 491.
© 1988 by Kluwer Academic Publishers.

CONCLUSIONS AND RECOMMENDATIONS

Based on evaluation and discussions at the end of the ASI

Previous ASI on berthing and mooring of ships (Lisbon 1965, Wallingford 1973), together with other recent work as Guidelines and Recommendations for the Safe Mooring of Large Ships at Piers and Sea Islands by Oil Companies International Marine Forum (1978), and PIANC report on Improving the Design of Fender Systems (1984) have significantly contributed to the understanding and development of mooring and fenders.
The scope of this ASI has been to introduce recent advances, models and tools in the system approach - emphasizing safety when berthing, moored and deberthed, with a view towards reducing frequencies and concequences of accidents. Further developments on integrated design, probabilistic approach and criteria to be applied are encouraged. All work must consider and include local conditions in the respective ports.

BERTHING / DEBERTHING
As tug boat assistance is expensive and is often difficult to obtain, the introduction of new and more effective systems of ship handling including the use of quay based winches and advanced fender systems seem appropriate. Berthing and deberthing depend on the environmental conditions, on the manoeuverability of ships, including effect of bow and/or stern thrusters, and on possible assistant measures. Better manoeuverability results in more economical approach channels, lesser downtime for berthing due to bad weather conditions, and more reliability when ships are berthed.

MOORING
Available experiences on mooring confirm the usefulness of tension moorings, rightly applied, to obtain higher operational efficiency and safety at berth. Reliable mooring drum brakes with slipping tension readout is also important. Instrumentation, monitoring and adjustment of mooring line tension are particularly important when multiple, parallel mooring lines are used.

FENDERING AND BERTH STRUCTURES
Particularly for exposed areas, fenders of lower recoilability and tension winches seem to be useful in reducing ship motions, mooring forces, and forces on vessels and berth structures. Fender systems applicable for a wide range of ship sizes could be developed by the introduction of adjustable stiffness or a step fender system. Further research and development work in these areas are encouraged. There is also a need for further development of more simplified berth structures.

SAFETY
Future developments should emphasize on increase of safety at berth and efficiencies of operation. Berthing aid systems and

monitoring systems of combined mooring and fendering actions
are in this respect important.

EQUIPMENT
Further development of equipment to secure higher safety and
efficiencies of operation is encouraged.

RESEARCH AND DEVELOPMENTS
Many problems remain to be solved for proper modelling. The
mathematical models are becoming an important tool, but they
require reliable hydrodynamic and mechanical inputs and criteria. Modelling of initial conditions of berthing and of all
external forces is important. Environmental forces must be
known in great detail. So must mooring and fendering as well
as structural characteristics.
An integrated approach to the design of all berthing, mooring,
fendering facilities and berth structures are strongly recommended. This approach should include inputs throughout the
design from all parties involved and a real time feedback on
motions and loads. Simplicity of design and ease of maintenance and repair should be stressed. Field measurements should
be undertaken to confirm and improve fender/mooring systems
and to collect reliable hydrodynamic and mechanical inputs and
criteria for developing the mathematical and physical models.
Future advances are possible by coupling advances in mathematical modelling with real time measurements and micro-computer
technology.
Additional development and testing of more advanced fenders including low-and non-recoiling fenders is required to fully realize the potential of these devices.
It is to be recognized that considerable advances in the understanding of the forces acting on, and the response from, berthing vessels and moored vessels or structures has been made.
However, a wider distribution of the information and techniques
now available is to be encouraged. The information should be
made available to every day users and designers of port and
harbour facilities in a form that can be easily understood and
readily applied to practical situations.
Existing international organizations should encourage coopera-
tive research on mathematical and physical models, full scale
measurements and development of new concepts.

SUBJECT INDEX

496

LIST OF PARTICIPANTS

Organizers:

1.   E. Bratteland, professor
          Division of Port and Ocean Engineering
          Norwegian Institute of Technology
          Norway

2.   A. Tørum, professor
          Norwegian Hydrotechnical Laboratory
          Norway

Lecturers:

3.   T. Andersen, principal engineer
          Veritas Offshore Technology
          and Services A/S
          Norway

4.   P. Bruun, professor emeritus
          Norwegian Institute of Technology
          Norway

5.   H.L. Fontijn, Dr
          Delft University of Technology
          The Netherlands

6.   I. Fylling, chief research engineer
          The Norwegian Marine Technology
          Research Institute
          Norway

7.   D.H. Lenschow, senior scientist
          National Center for Atmospheric
          Research
          USA

8.   M. Mathiesen, senior research engineer
          Norwegian Hydrotechnical Laboratory
          Norway

9.   G.van Oortmerssen, Dr
          Netherlands Ship Model Basin
          The Netherlands

10.   P. Tryde, professor
          Technical University of Denmark
          Denmark

11.   F. Vasco Costa, professor emeritus
          Instituto Superior Técnico
          Portugal

12.   G. Viggosson, senior engineer
          Icelandic Harbour Authority
          Iceland

Participants:

13.   V. Aanesland, Dr
          Marintek A/S
          Norway

14.   C. Aquilina, eng. associate
          Esso Engineering (Europe) Ltd
          United Kingdom

15.   J.B. Christoffersen, research engineer
          Danish Maritime Institute
          Denmark

16.   I. Datta, Dr
          National Research Council of Canada,
          Canada

17.   D. Davis, project leader
          U.S. Naval Civil Engineering Laboratory,
          USA

18.   U. Demirørs, civil engineer
          Eregli Iron and Steel Works
          Turkey

19.   A.R. Günbak, Dr
          Middle East Technical University
          Turkey

20.   A.J. Guthrie, Ass. manager
          Bridgestone Corporation
          United Kingdom

21.   E. Hess Thaysen, engineer
          Danish State Railways
          Denmark

22.   W.J. Huffman, engineer
          Thomas & Hutton Eng. Co.
          USA

23.   H. Ivarson, chief design engineer
          Port of Gothenburg
          Sweden

24.   R. Julian, project engineer
          US Naval Civil Engineering Laboratory,
          USA

25.   J.B. Jørgensen, engineer
          Danish State Railways
          Denmark

26.   H. Kawakami, manager marine fender
          Bridgestone Corporation
          Japan

27.   J. Korsvold, senior engineer
          Norsk Hydro A/S
          Norway

28.   I. Krogstad, marine engineer
          A.S. Pusnes Marine and
          Offshore Services
          Norway

29.   A.C. Lara, professor
          Centro de Ingeniena Oceanica
          Argentina

30.   T. Lundestad, graduate student
          Norwegian Institute of Technology
          Norway

31.   M. Marcos Rita, engineer
          Laboratorio Nacional de Engenharia Civil
          Portugal

32.   R. Medina, engineer
          Escuela de Caminos Canales
          Y Porto
          Spain

33.   S. Munthe, engineer
          Scandiaconsult
          Sweden

34.   H.M. Noble, president
          Noble Consultants Inc.
          USA

35.   B.C. Persson, president
          Trelleborg AB
          Denmark

36.   T. Poulsen, design engineer
          Port of Esbjerg Authority
          Denmark

37.   C. Pupo Pesce, M. Sc
          IPT - DINAV
          Brazil

38.   J. Ralstone, chief sales engineer
           The Yokohama Rubber Co
           United Kingdom

39.   W. Rankka, M.Sc
           Chalmers University of Technology
           Sweden

40.   L. Rydberg, chief designer
           Trelleborg AB
           Sweden

41.   S. Sigurdarson, M.Sc
           Icelandic Harbour Authority
           Iceland

42.   J. Skare, civil engineer
           Siviling. Elliott Strømme A/S
           Norway

43.   M. Triantafyllou, professor
           Massachusetts Institute of Technology
           USA

44.   S. Ueda, Dr
           Port and Harbour Research Institute
           Japan

45.   A. Vagle, graduate student
           Norwegian Institute of Technology
           Norway

46.   A. Van der Eijk, engineer
           Delta Marine Consultants b.v.
           The Netherlands

47.   F.F.M. Veloso Gomes, associate professor
           Universidada do Porto
           Portugal

48.   P.C. Viana, graduate student
           Instituto Superior Técnico
           Portugal